Mathematical Modeling and Soft Computing in Epidemiology

Information Technology, Management, and Operations Research Practices

Vijender Kumar Solanki, Sandhya Makkar, and Shivani Agarwal

This new book series will encompass theoretical and applied books, and will be aimed at researchers, doctoral students, and industry practitioners to help in solving real-world problems. This bookseries will help in the various paradigms of management and operations. This bookseries will discuss the concepts and emerging trends on society and businesses. The focus is to collate the recent advances in the field and take the readers on a journey that begins with understanding the buzz words such as employee engagement, employer branding, mathematics, operations, technology and how they can be applied in various aspects. It walks readers through engaging with policy formulation, business management, and sustainable development through technological advances. It will provide a comprehensive discussion on the challenges, limitations, and solutions of everyday problems like how to use operations, management, and technology to understand the value-based education system, health and global warming, and real-time business challenges. This book series will bring together some of the top experts in the field throughout the world who will contribute their knowledge regarding different formulations and models. The aim is to provide the concepts of related technologies and novel findings to an audience that incorporates specialists, researchers, graduate students, designers, experts, and engineers who are occupied with research in technology, operations, and management-related issues.

Performance Management
Happiness and Keeping Pace with Technology
Edited by Madhu Arora, Poonam Khurana, and Sonam Choiden

Soft Computing Applications and Techniques in Healthcare
Edited by Ashish Mishra, G. Suseendran, Trung-Nghia Phung

Mathematical Modeling and Soft Computing in Epidemiology
Edited by Jyoti Mishra, Ritu Agarwal, Abdon Atangan

For more information about this series, please visit: https://www.crcpress.com/Information-Technology-Management-and-Operations-Research-Practices/book-series/CRCITMORP

Mathematical Modeling and Soft Computing in Epidemiology

Edited by
Jyoti Mishra, Ritu Agarwal, and Abdon Atangana

CRC Press
Taylor & Francis Group
Boca Raton London New York

CRC Press is an imprint of the
Taylor & Francis Group, an **informa** business

First edition published 2021
by CRC Press
6000 Broken Sound Parkway NW, Suite 300, Boca Raton, FL 33487-2742

and by CRC Press
2 Park Square, Milton Park, Abingdon, Oxon, OX14 4RN

CRC Press is an imprint of Taylor & Francis Group, LLC

Library of Congress Cataloging-in-Publication Data
Names: Mishra, Jyoti (Professor of Mathematics), editor. |
Agarwal, Ritu (Professor of Mathematics), editor. | Atangana, Abdon, editor.
Title: Mathematical modeling and soft computing in epidemiology / edited by
Jyoti Mishra, Ritu Agarwal and Abdon Atangana.
Description: First edition. | Boca Raton, FL : CRC Press, 2021. |
Series: Information technology, management and operations research practices |
Includes bibliographical references and index.
Identifiers: LCCN 2020026696 (print) | LCCN 2020026697 (ebook) |
ISBN 9780367903053 (hbk) | ISBN 9781003038399 (ebk)
Subjects: LCSH: Epidemiology—Mathematical models. | Soft computing. |
AMS: General — General and miscellaneous specific topics — Theory of
mathematical modeling. | Biology and other natural sciences —
Mathematical biology in general. | Mathematics education — Mathematical
modeling, applications of mathematics — Biology, chemistry, medicine.
Classification: LCC RA652.2.M3 M39 2021 (print) |
LCC RA652.2.M3 (ebook) | DDC 614.401/5118—dc23
LC record available at https://lccn.loc.gov/2020026696
LC ebook record available at https://lccn.loc.gov/2020026697

ISBN: 978-0-367-90305-3 (hbk)
ISBN: 978-1-003-03839-9 (ebk)

Typeset in Times
by codeMantra

Contents

Preface

The edited book *Mathematical Modelling and Soft Computing in Epidemiology* describes different mathematical modeling and soft computing techniques in epidemiology for experiential research in project how *infectious diseases* progress to show the likely outcome of an *epidemic* and help inform *public health* interventions. This book will highlight that the models use some basic assumptions and mathematics to find *parameters* for various *infectious diseases* and use those parameters to calculate the effects of different interventions, such as mass *vaccination* programmers, cancer, and tuberculosis. This book will walk through the emerging trends in modeling of infectious diseases is a tool which has been used to study the mechanisms by which diseases spread, to predict the future course of an outbreak and to evaluate strategies to control an epidemic. It will help the researchers to appreciate the *Mathematical Modeling and Soft Computing in Epidemiology*. This book will provide a comprehensive discussion of epidemiological modeling that refers to dynamic, deterministic modeling where the population is divided into compartments based on their epidemiological status (e.g., susceptible, infectious, recovered), for which movements between compartments by becoming infected, progressing, recovering, or migrating are specified by differential or difference equations. Overall, this book will develop an understanding of the need of work epidemiologists and public health workers that mathematical modeling can be of use to them.

The objective is to bring the mathematical modeling and soft computing in epidemiology methods in a single volume, which can add to the existing knowledge of undergraduate and postgraduate students, researchers, academicians, and industry people. This book intends to cover the main aspects of mathematical modeling and soft computing in epidemiology, and its goal is to persuade epidemiologists and public health workers that mathematical modeling can be of use to them. The primary users for this book include researchers, academicians, postgraduate students, and specialists. This edited book will have separate chapters to facilitate readers of epidemiology will ensuring its continued popularity in adapting this book.

CHAPTER 1

This chapter proposed an approximation-based evolutionary computing framework for predicting and analyzing the dynamics of virus propagation of dengue disease involving virus incubation period. The proposed framework consolidated distinguishes techniques of Padè approximation, penalty function approach, Nelder–Mead simplex (NMS) algorithm, and differential evolution (DE) for solving the underlying model numerically.

CHAPTER 2

This chapter is an attempt to describe mathematically the functioning of fuzzy recombination of chromosomes. Chromosomal diseases happen when either entire chromosome or enormous fragment of chromosome is duplicated or missing or transformed. In this chapter, the fuzzy topological features of the recombination space are analyzed. Various crossover models are spontaneously produced in the recombination space, and this can be structured using fuzzy pretopology. The outcomes revealed in this chapter of unequal crossover replicate the connectivity of the fuzzy recombination space in genetic epidemiology.

CHAPTER 3

This chapter aims that sufferers of the disease are provided with some initial support and the medical practitioners may get some preliminary ideas about the fatal diseases. Our knowledge in this domain still remains inadequate (as in the case with many other biomedical issues) compared to the information obtained from the study of similar technological structures, and mathematical models are constructed and presented in the investigations related to cause and remedy of the human physiological disorders.

CHAPTER 4

This chapter provides a machine-learned regression assessment of the HIV epidemiology development in Asian region. The presented model with randomperturbations aids to advance the accepting of dynamical behavior, and with time, the growth of virus particles can be gradually decreased through these differential equations and then the infection could tend towards zero level, which will after sometimes be more or less not effective inside the body. Great accomplishments have been made for decreasing morbidity and mortality via first decade of 2000, continuously emphasizing for the access to region-appropriate preventive scales.

CHAPTER 5

This chapter presents general mathematical modeling to find the potential number of ways to distribute certain things to certain places in the medical field. Mathematical model is nothing but finding the double twin domination number of a graph theoretical representation of the real-life situation. In this chapter, we provide a mathematical model to obtain the potential number of ways to perform certain tasks in the medical field. We obtain this number for many special types of graphs.

CHAPTER 6

This chapter focuses on addressing an analysis of an epidemic model in the context of fractional calculus. We consider fractional SIRI model with delay described by Caputo–Liouville-generalized fractional derivative. The existence and uniqueness of the SIRI epidemic model have been investigated. We determine the reproduction number R_0. We establish the disease-free equilibrium point and the endemic equilibrium point. We present the stability analysis of the disease-free equilibrium point and the endemic equilibrium points in terms of the reproduction number R_0.

CHAPTER 7

This chapter aims to develop a Nipah virus transmission model in human population. It is assumed that the Nipah virus is transmitted in human population in two possible ways: first through the contact of susceptible human with infected pig and second through the contact of susceptible human with infected human. Boundedness of all solutions of our proposed system has been investigated from the theoretical and numerical results; it can be concluded that in order to reduce Nipah virus transmission from human population, combined effects of vaccination and treatment controls can be taken into consideration concurrently.

CHAPTER 8

This chapter focuses on the application of eternal domination in epidemiology. This concept plays a vital role in curing diseases in a particular area and controlling them. In this chapter, it is proposed to provide a real-life application in epidemiology.

CHAPTER 9

In this paper, the coupled fractional differential equations that arise in various nonlinear epidemiological models are examined by a hybrid and innovative homotopy perturbation method via the Laplace transform (HPTM).Important applications of these equations are to model the transmission dynamics of epidemic models and to model the phenomena of electrical activity in heart.

CHAPTER 10

This chapter focuses on stabilizing the content of nitrogen mass cycle using its mathematical model. The mathematical model for the nitrogen mass cycle is formulated. Mathematical properties of the model in both deterministic and non-deterministic models are also discussed.

CHAPTER 11

This chapter focuses on the study of nitrogen mass cycle model when there is exponential growth. Framing of the exponential growth mathematical model for nitrogen mass cycle is followed by the analysis for its boundedness, local stability, global stability, and bifurcation. Numerical simulation describing the stability of nitrogen mass cycle with exponential growth is accomplished.

CHAPTER 12

This chapter tells about the application of blockchain technology in Hospital Information System (HIS). This study will help other researchers and practitioners to gain better understanding of blockchain technology and HIS in the healthcare sector.

CHAPTER 13

This chapter focuses on to explore complexity analysis of pathogenesis of coronavirus.

Mathematical model for the spread of coronavirus is developed. With the help of time-series analysis and phase space, the spread of disease is studied. Also, the case study for China country is provided for the spread of the disease.

CHAPTER 14

The aim of this chapter is to present a fractionalized model for HBV infection, and formulate the fractionalized model for the epidemic problem. The suggested technique is obtained by merging the homotopy analysis technique, the Laplace transform, and the homotopy polynomials.

CHAPTER 15

In this chapter, the author analyzed the precautionary measures taken by Government of India to combat coronavirus mathematically, concluded the result, and established the nonlinear dynamics of SARS-CoV2 virus: India and its government policy.

CHAPTER 16

This chapter discusses ethical and professional issues in epidemiology.

The authors term epidemiology as what causes epidemics, and how they spread and how they can be prevented or reduced.

CHAPTER 17

This chapter provides a cloud virtual image security for medical data processing. Security in cloud computing is one of the fundamental issues that prevent the quick reception of the innovation and research.

CHAPTER 18

This chapter analyzes medical data security using blockchain and machine learning techniques in cloud computing. The research is about data privacy, integrity, and gained access control over the shared data with better efficiency for medical data records in cloud computing environment.

CHAPTER 19

This chapter provides a mathematical model to avoid delay wound healing by infinite element method .In this regard, mechanical interactions between the tissues and the fluid have been analyzed.

CHAPTER 20

This chapter analyzes data classification framework for medical data through machine learning techniques in cloud computing. In this study, data classification with machine learning and secure storage of classified data are the main concern area, and different previous researches and security mechanisms are going to review for extracting key idea and knowledge for designing efficient systems for the concerned issues.

We give sincere thanks to Ms. Erin Harris, Senior Editorial Assistant, CRC Taylor & Francis Group, for giving us an opportunity to convene this book in his esteemed publishing house; and Dr. Vijender Kr. Solanki, Sandhya Makkar, and Shivani Agarwal, Series Editors in IT, Management and Operation Research, for their kind cooperation in completion of this book. We thank our esteemed authors for having shown confidence in this book and considering as a platform to showcase and share their original research work. We would also wish to thank the authors whose papers were not published in this book, probably because of the minor shortcomings.

Editors

Dr. Jyoti Mishra is currently an Associate Professor at Department of Mathematics, Gyan Ganga Institute of Technology and Sciences, Jabalpur. She is a qualified individual with around 10 years of expertise in teaching, R&D with specialization in Mathematics. She completed BSc, MSc and BEd. She received her PhD degree in special function from APS University, Rewa, India. Her research interests include special function, fractional calculus, partial differential equations, and mathematical modeling. She has been a part of various seminars, paper presentations, research paper reviews, conferences as co-convener, Member of Organizing Committee, Member of Advisory Committee, Member of Technical Committee, and Member of Organizing in organizing INSPIRE Science Internship Camp. She has published more than 25 research papers in reputed journals and authored three books. She has been actively involved in professional bodies Indian Mathematical Society, IAPS, and FATER. She is a reviewer in many SCI journals, IEEE international conferences, CSNT-2015, CICN-2016, CICN2017, INDIACom-2019, ICICC-CONF 2019, and many more. She has published 08patents in Intellectual Property, India.

Dr. Ritu Agarwal has done PhD, MPhil, MSc, and JRF-NET. She is an Assistant Professor of mathematics in Malaviya National Institute of Technology, Jaipur. She is having 15 years of teaching and research experience. Her areas of research include biocomplex analysis, geometric function theory, mathematical modeling, and fractional calculus. She has published 45 research papers in journals of repute. Under her guidance, two PhD completed. She has awarded a research project by NBHM. She attended 12 international conference. She has organized various events like short-term courses, FDP, and national- and international-level conferences. She has been actively involved with professional bodies Indian Mathematical Society, IAENG, ISTE, ISCA, FIM, IACSIT, RMS, SSFA, VIJNANA, Gwalior Academy of Sciences, TIMC, and IEEE. She has 11 text books for BE/BTech to her credit and has supervised approximately 20 MSc dissertations.

Dr. Abdon Atangana received PhD degree from University of the Free State, South Africa. He is a Professor of Applied Mathematics Institute for Groundwater Studies, University of the Free State. He has awarded the youngest leading international researcher in fractional calculus in 2019, Obada Award 2018, African Academic of Science, Affiliate 2018–2022 most cited mathematics paper in the world 2017, and African Award of applied Mathematics 2017. He is the number 1 top overall contributor to the peer review of the field of mathematics. The sentinel of science awards honor the elite contributors to scholar peer review and editorial pursuits internationally. Recipients have demonstrated an outstanding expert commitment to protecting the integrity and accuracy of published research in their field in 2017. He has received many best researcher awards in different universities. In the field of applied mathematics, he has introduced more than 15 mathematical operators and numerical schemes that are nowadays used in all fields of science, technology, and engineering. He has published approximately 250 SCI papers in different reputed journals, and H-index is 36 and citation is 5148.

Contributors

Javaid Ali
Department of Mathematics
University of Management and Technology
Lahore, Pakistan

Yaser Mohammed Al-Worafi
College of Pharmacy
University of Science and Technology
Taizz, Yemen
and
College of Pharmacy
University of Science and Technology of Fujairah
Fujairah, United Arab Emirates

S. Anuthiya
Department of Mathematics
Gandhigram Rural Institute – Deemed to be University
Gandhigram, India

Selvam Avadayappan
Department of Mathematics
VHNSN College
Virudhunagar, India

Faizah Safina Bakrin
School of Pharmacy
KPJ Healthcare University College
Nilai, Malaysia

Aashima Bangia
University School of Basic & Applied Sciences
GGS Indraprastha University
Delhi, India

Ramakant Bhardwaj
Department of Mathematics
Amity University Kolkata
Kolkata, India

Rashmi Bhardwaj
Non-Linear-Dynamics-Research-Lab
University School of Basic & Applied Sciences
GGS Indraprastha University
Delhi, India

Jignesh Pravin Chauhan
Department of Mathematics
Marwadi University
Rajkot, Gujarat

Manoj Dubey
Department of Mathematics
IES-IPS Academy
Indore, India

Deepa Elangovan
School of Pharmacy
KPJ Healthcare University College
Nilai, Malaysia

Khang Wen Goh
Faculty of Science and Technology
Quest International University Perak
Ipoh, Malaysia

Manish Goyal
Department of Mathematics
Institute of Applied Sciences and Humanities
GLA University
Mathura, India

Shivangi Gupta
Department of Mathematics
Institute of Applied Sciences and Humanities
GLA University
Mathura, India

Gautam Suresh Hathiwala
Department of Mathematics
Marwadi University
Rajkot, Gujarat

Minakshi Biswas Hathiwala
Department of Mathematics
Marwadi University
Rajkot, Gujarat

Zahid Hussain
Faculty of Health
University of Canberra
Canberra, Australia

Manisha Jain
Division of Mathematics School of Advanced Sciences and Languages
Vellore Institute of Technology
Bhopal, India

Ranjan Kumar Jana
Department of Applied Mathematics & Humanities
Sardar Vallabhbhai National Institute of Technology
Surat, India

Kamlesh Jangid
Department of HEAS (Mathematics)
Rajasthan Technical University
Kota, India

Yaman Walid Kassab
Faculty of Pharmacy
University of Cyberjaya
Cyberjaya, Malaysia

Kritika
Department of Mathematics
Malaviya National Institute of Technology
Jaipur, India

Kah Seng Lee
Faculty of Pharmacy
University of Cyberjaya
Cyberjaya, Malaysia

Chiau Soon Long
Faculty of Science and Technology
Quest International University Perak
Ipoh, Malaysia

G. Mahadevan
Department of Mathematics
Gandhigram Rural Institute – Deemed to be University
Gandhigram, India

Biswadip Basu Mallik
Basic Science & Humanities Department
IEM
Kolkata, India

Long Chiau Ming
PAPRSB Institute of Health Sciences
University Brunei Darussalam
Gadong, Brunei

Aditya Mani Mishra
Department of Mathematics
Institute of Physical Sciences, for Study and Research
VBS Purvanchal University
Jaunpur, India
and
Department of HEAS (Mathematics)
Rajasthan Technical University
Kota, India

Ashish Mishra
Computer Science and Engineering
Gyan Ganga Institute of Technology and Sciences
Jabalpur, India

Jyoti Mishra
IPS College of Technology & Management
Gwalior, India

Pankaj Kumar Mishra
Department of Applied Physics
Amity School of Engineering and Technology
Amity University Madhya Pradesh
Gwalior, India

Kala Raja Mohan
Department of Mathematics
Vel Tech Rangarajan Dr. Sagunthala R&D Institute of Science and Technology
Chennai, India

Saktipada Nanda
Department of Electronics & Communication Engineering
IEM
Kolkata, India

Subhrendu Guha Neogi
Computer Science and Engineering
Amity School of Engineering and Technology
Gwalior, India

Prabir Panja
Department of Applied Science
Haldia Institute of Technology
Purba Midnapore, India

T. Ponnuchamy
Department of Mathematics
Gandhigram Rural Institute – Deemed to be University
Gandhigram, India

Amit Prakash
Department of Mathematics
National Institute of Technology
Kurukshetra, India

Sunil Dutt Purohit
Department of HEAS (Mathematics)
Rajasthan Technical University
Kota, India

Deepak Singh Rajput
Computer Science and Engineering
Amity School of Engineering and Technology
Gwalior, India

Suresh Rasappan
Department of Mathematics
Vel Tech Rangarajan Dr. Sagunthala R&D Institute of Science and Technology
Chennai, India

Muhammad Bilal Riaz
Department of Mathematics
University of Management and Technology
Lahore, Pakistan
and
Institute for Groundwater Studies
University of the Free State
Bloemfontein, South Africa

Muhammad Saeed
Department of Mathematics
University of Management and Technology
Lahore, Pakistan

Ndolane Sene
Laboratory Lmdan
Cheikh Anta Diop University
Dakar Fann, Senegal

Harish K. Shakya
Computer Science and Engineering
Amity School of Engineering & Technology
Gwalior, India

Saurabh Sharma
Computer Science and Engineering
Amity School of Engineering and Technology
Gwalior, India

M. Vimala Suganthi
Department of Mathematics
Gandhigram Rural Institute – Deemed to be University
Gandhigram, India

Ching Siang Tan
School of Pharmacy
KPJ Healthcare University College
Nilai, Malaysia

Shiv Kumar Tiwari
Computer Science and Engineering
Amity School of Engineering and Technology
Gwalior, India

A.K. Upadhyay
Computer Science and Engineering
Amity School of Engineering and Technology
Gwalior, India

1 Evolutionary Modeling of Dengue Fever with Incubation Period of Virus

Javaid Ali
University of Management and Technology

Muhammad Bilal Riaz
University of Management and Technology
University of the Free State

Abdon Atangana
University of the Free State

Muhammad Saeed
University of Management and Technology

CONTENTS

1.1 INTRODUCTION

Aedes mosquitoes predominantly cause the transmission of Dengue virus. These mosquitoes generally dwell in the areas with and elevation below 1,000 m (3,300 ft.) and having latitudes within 35° South and 35° North. Early morning and evening hours are two typical timings at which Aedes mosquitoes are more likely to bite. However, they can possibly bite at other day timings, and hence, a spread of infection may occur. The reports of World Health Organization (WHO) (2012) disclose that around 390 million sufferers from dengue disease do exist around the globe and 96 million necessitate medical treatment. About 500,000 victims of the severest category of dengue hemorrhagic fever require hospitalization every year. About 40% of the population of the world live in dengue-prevalent regions.

In line with the WHO assessments, approximately 22,000 deaths take place each year, involving a notable number of children. Dengue disease is an ancient disease. Earliest record of this disease was first found in Chinese encyclopedia of disease symptoms and remedies, first published in 610 A.D and second in 922 A.D. The Dengue virus spread to new geographic areas during the development of port cities in the eighteenth and nineteenth centuries. Following World War II, global pandemic of dengue began. Dengue virus was first quarantined in Japan during 1943. Within a period of one year, several parts of the world became the Dengue virus colonies. From 1953 to 1954, its most severe form, dengue hemorrhagic fever, befell first time in Manila city. The transmission of dengue disease to Asia, especially, Sri Lanka, India, Pakistan, and Maldives, occurred geographically from the other Southeast Asian countries. American region experienced this disease by the 1980s. Very little was recognized about the transmission of dengue disease in Africa before 1980. In 1997, Dengue viruses and *Aedes aegypti* had a global dissemination in tropical as well as subtropical regions.

From 2000 to 2012, various data analyses and literature surveys were carried out to describe insight of the dengue fever disease within Malaysia. From a list of 237 diagnosed relevant cases, the insertion criteria were satisfied by 28 data sources. A survey describing the dynamics of dengue disease in Thailand during 2000 and 2011 was reported. Out of 610 identified pertinent searches, 40 individuals fulfilled the criteria for inclusion. The registered cases of dengue disease during 2001, 2002, 2008, and 2010 were more deadly in youngsters. A survey revealed that 51 individuals out of 714 recognized pertinent searches in Brazil satisfied the inclusion criteria between 2000 and 2010. In fact, the epidemiological design of Dengue disease in Brazil is complicated because of the influence of several social and environmental factors.

This study investigates a dynamical model involving incubation periods of vector populations and population of humans with Dengue virus. The susceptible human bitten by an infectious vector is unable to transmit Dengue virus promptly. Such kind of individual is named as infected human through this period. In humans, the inherent incubation period consumed by Dengue virus is around five days (Gubler, 1998). When a susceptible vector bites the infectious person, it is considered as an infected vector until it turns to be the infectious one. On the other hand, in vector population, the extrinsic gestation period consumed by Dengue disease virus is approximately ten days (Gubler 1998). A person is named as susceptible if he is infection-free but does not possess immunity to fight the Dengue disease virus. A victim of the infection that restores the immunity is labeled as the recovered person.

The emergence of efficient mathematical modeling techniques has made it possible to extract meaningful and comprehensive insight knowledge about epidemics. The development of the epidemiological model and numerical simulations allow comparing and analyzing the sensitivity and conjuncture paradigms. Moreover, it is possible to predict the arbitrator, the host, and the ecological hazards influencing community health, and therefore, health authorities can scientifically recommend and implement appropriate health services (Cooley et al. 2016). Several studies have been conducted on various models of dengue transmission dynamics (Halstead 1998; Naowarat and Rajabhat 2011; Pongsumpun 2008; Rafiq and Ahmad 2015). It is an established fact that nonlinear and implicit initial value problems (IVPs) generally lack analytical solutions. Existing traditional finite difference techniques such as Euler (EU) and Runge–Kutta (RK4) methods may generate deceitful chaos and misleading oscillations for specific choices of discretization parameters (Rafiq et al. 2017; Rafiq and Raza 2017; Zafar et al. 2017). Because of these challenges and various other instabilities of numerical schemes, these methods appear to be less confidential options.

Over the recent years, computational intelligence (CI) has been enriched by the emergence of a number of efficient metaheuristics emulated from intelligent natural phenomena. Some notable adaptations are based on swarm behaviors (Kennedy and Eberhart 1995; Wang et al. 2016), evolution (Goldberg 2005; Storn and Price 1997), sports strategies (Alatas 2017), water dynamics (Ali et al. 2015, Sadollah et al. 2015), food foraging behavior (Karaboga and Basturk 2007; Passino 2002), etc. In spite of all the applicability and effectiveness, these methods inherit drawbacks such as high computational cost and slow convergence. Conventionally such issues are tried to resolve by performing problem-specific tunings or hybridizing the optimizers. Nelder–Mead simplex algorithm (NMSA) (Nelder and Mead 1965) is a popular local search solver, which has been considered for hybridization to design an efficient low-cost hybrid optimizer.

Metaheuristics-based techniques for the treatment of differential equations fall in the category of meshless numerical methods. Over the recent years, several differential equations have been solved numerically by metaheuristic approaches (Babaei 2013; Ara et al. 2018; Panagant and Bureerat 2014; Lee 2006; Karr and Wilson 2003), but they have rarely been applied to epidemic models. The purpose of this work is to design a meshfree numerical scheme for finding the closed-form approximate solution of dengue disease dynamical model involving virus incubation period. The proposed method consists of the following novel components:

- The state variables are approximated by nonlinear Padé rational functions.
- The governing equations are transformed to residual functions.
- Initial conditions and other solution requirements are incorporated by penalty function approach by viewing them as problem constraints.
- An equivalent optimization problem has been formulated whose solution is analogous to that of the considered model.

- The notion of multiple simplex searches under the cooperative evolutionary crossovers and mutations by embedding their best vertices to the set of trial solutions has been introduced.
- A hybrid of differential evolution (DE) and cooperative multi-simplex search has been proposed.
- The proposed method aims to find the closed-form solution of the dynamical model of dengue fever disease that has not been solved analytically so far.

The rest of this chapter is organized as follows. Section 1.2 revisits the basic definition of Padé approximation, NMSA, and DE. Section 1.3 describes and examines the dynamic model of dengue fever disease involving virus incubation period. Section 1.4 presents the complete framework of the proposed differential cooperative multi-simplex Padè approximation (DCMP) scheme. Section 1.5 discusses the simulation results that have been produced. The end of this chapter provides the concluding remarks as well as some possible research directions.

1.2 BASIC NOTIONS

1.2.1 Padé Approximation

In general, Padé approximation with order (n_1, n_2) in normalized form is a rational function of the following shape (Padé 1892; Vijta 2000):

$$P_{n_1,\,n_2}(t) = \frac{\sum_{i=0}^{n_1} a_i t^i}{1 + \sum_{j=1}^{n_2} b_j t^j}, \tag{1.1}$$

where the polynomials $\sum_{i=0}^{n_1} a_i t^i$ and $\sum_{j=0}^{n_2} b_j t^j$ are called Padé approximants. There are $(n_1 + n_2 + 1)$ unknown coefficients that are involved in it and are to be found so that a goal function and the Maclaurin series expansion of $P_{n_1,\,n_2}(t)$ coincide as much as possible (Bojdi et al. 2013).

1.2.2 Non-Stagnated Nelder–Mead Simplex Algorithm (NS-NMSA)

A simplex in $\mathbb{R}^n$ is a convex hull $\{x_1, x_2, x_3, \ldots, x_{n+1}\}$ of $n+1$ points of $\mathbb{R}^n$. NMSA is a local search algorithm which exploits a nondegenerated simplex to perform minimization of a bounded below objective function $f : \mathbb{R}^n \to \mathbb{R}$. It starts by sorting the simplex vertices from the best (the one with the smallest objective function value) to the worst (the one with the largest objective function value) and tries to improve the best vertex by replacing the worst vertex using the following relation:

$$x^{(\text{new})} = x_{n+1} + \psi \cdot (\bar{x} - x_{n+1}), \tag{1.2}$$

where $\bar{x}$ is the aggregate point of all the vertices excluding the worst vertex. Equation (1.2) results in the operation of reflection, expansion, contraction inside, or contraction outside, in which ψ is 2, 3, $-\frac{1}{2}$, or $\frac{3}{2}$, respectively. If none of these steps is successful, a shrink step is used. Theoretical and geometrical details of NMSA can be found in Nelder and Mead (1965). Since some particular functions have been reported on which NMSA converges to a nonstationary point (McKinnon 1998; Ali et al. 2017), a convergent and non-stagnated variant (NS-NMSA) proposed by Ali et al. (2017) has been used for guaranteed convergence of NMSA.

1.2.3 Differential Evolution (DE)

DE is a popular and effective metaheuristic algorithm in global optimization. The main operations of DE are mutation, recombination, and selection. Applying mutation operation on the solution $\boldsymbol{x}$ generates a new trial solution $\boldsymbol{v}$ by utilizing three other mutually distinct members $\boldsymbol{w}$, $\boldsymbol{y}$, and $\boldsymbol{z}$ as given below:

$$v_m = w_m + F.\left(y_m - z_m\right), \text{for each } m = 1, 2, 3, \ldots, n.$$

The parameter F is called a differential constant. Recombination operation probabilistically merges the characters of $\boldsymbol{v}$ and $\boldsymbol{x}$. A predefined positive real number $CR \in (0,1)$, called recombination probability, a random number rand $\in (0,1)$, and a random integer $j_{\text{rand}} \in \{1,2,3,\ldots, n\}$ are used for a recombination step applied on each dimension $m: 1,2,3,\ldots, n$ according to the following relation:

$$u_m = \begin{cases} v_m & \text{if rand} < CR \text{ or } m = j_{\text{rand}} \\ x_m & \text{otherwise} \end{cases}.$$

If $f(\boldsymbol{u}) < f(x)$, the selection stepassigns $\boldsymbol{u}$ to $\boldsymbol{x}$otherwise discards it. The iterations of DE algorithm are stopped on the fulfillment of some prescribed termination criteria.

1.3 MODEL OF DENGUE DISEASE WITH INCUBATION PERIOD OF VIRUS

The modeling procedure involves the following assumptions:

- The size of human population and that of mosquito remain constant.
- The effected individual becomes permanently safe from the specific Dengue virus but remains sensitive for the rest of mosquito population.
- Birth and death rates of both the mosquito and human populations behave in a similar fashion.

Let at any time t, the variables $\bar{S}(t)$, $\bar{X}(t)$, $\bar{I}(t)$, and $\bar{R}(t)$ represent the numbers of susceptible, infected, infectious, and recovered human population, respectively, whereas $\bar{U}(t)$, $\bar{V}(t)$, and $\bar{W}(t)$ represent the numbers of susceptible, infected, and

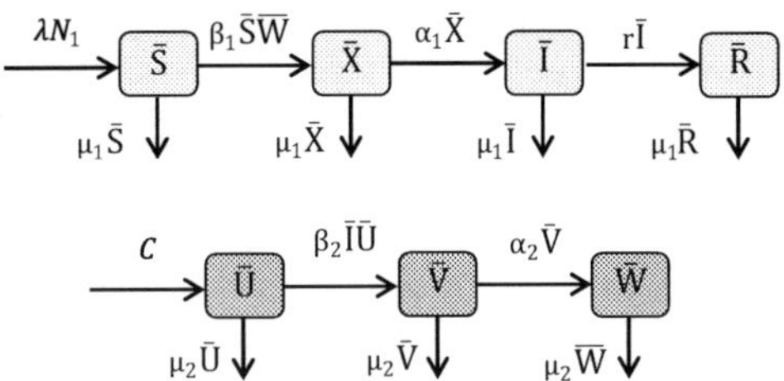

FIGURE 1.1 Block diagram representing dynamics of Dengue virus model.

infectious mosquito, respectively. The transmission dynamics of Dengue virus are shown in Figure 1.1.

The model parameters are N_1 (human population size), N_2 (size of vector population), λ (birth rate of human population), μ_1 (death rate of human population), α_1 (rate of change of infected into infectious human populations), β_1 (rate of infectious Dengue virus from vector to human population), R (human population recovery rate), μ_2 (death rate of vector population), β_2 (rate of infectious Dengue disease virus from human population to vector population), α_2 (rate of change of infected vector to infectious vector population), and C (constant recruitment rate of vector population).

The equations governing the transmission dynamics of Dengue disease are given below.

$$\left.\begin{aligned}
\frac{d\bar{S}}{dt} &= \lambda N_1 - \beta_1 \bar{S}\bar{W} - \mu_1 \bar{S} \\
\frac{d\bar{X}}{dt} &= \beta_1 \bar{S}\bar{W} - (\alpha_1 + \mu_1)\bar{X} \\
\frac{d\bar{I}}{dt} &= \alpha_1 \bar{X} - (r + \mu_1)\bar{I} \\
\frac{d\bar{R}}{dt} &= r\bar{I} - \mu_1 \bar{R} \\
\frac{d\bar{U}}{dt} &= C - \beta_2 \bar{I}\bar{U} - \mu_2 \bar{U} \\
\frac{d\bar{V}}{dt} &= \beta_2 \bar{I}\bar{U} - (\alpha_2 + \mu_2)\bar{V} \\
\frac{d\bar{W}}{dt} &= \alpha_2 \bar{V} - \mu_2 \bar{W}
\end{aligned}\right\}. \tag{1.3}$$

The constant sizes of human and vector populations are, respectively, as follows:

$$N_1 = \bar{S} + \bar{X} + \bar{I} + \bar{R} \tag{1.4}$$

$$N_2 = \bar{U} + \bar{V} + \bar{W}. \tag{1.5}$$

Therefore,

$$\frac{dN_1}{dt} = 0 \Rightarrow \lambda = \mu_1$$

$$\frac{dN_2}{dt} = 0 \Rightarrow N_2 = \frac{C}{\mu_2}.$$

The normalized state variables are defined as follows:

$$S = \frac{\bar{S}}{N_1},\ X = \frac{\bar{X}}{N_1},\ I = \frac{\bar{I}}{N_1},\ R = \frac{\bar{R}}{N_1},\ U = \frac{\bar{U}}{N_2},\ V = \frac{\bar{V}}{N_2},\ W = \frac{\bar{W}}{N_2}.$$

Using (1.4) and (1.5), the dynamical systems of two populations can be expressed as the following reduced form:

$$\frac{dS}{dt} = f_1(t);\ \frac{dX}{dt} = f_2(t);\frac{dI}{dt} = f_3(t);\frac{dV}{dt} = f_4(t);\ \frac{dW}{dt} = f_5(t)\ , \tag{1.6}$$

where

$$f_1(t) = \mu_1 - \beta_1 SW - \mu_1 S;\ f_2(t) = \beta_1 SW\left(\frac{C}{\mu_2}\right) - (\alpha_1 + \mu_1)X;$$

$$f_3(t) = \alpha_1 X - (r + \mu_1)I;\ f_4(t) = \beta_2 IN_1(1 - V - W) - (\alpha_2 + \mu_2)V;$$

$$f_5(t) = \alpha_2 V - \mu_2 W.$$

With the conditions:

$$S(0) = S_0;\ X(0) = X_0;\ I(0) = I_0;\ V(0) = V_0;\ W(0) = W_0$$

$$S + X + I + R = 1 \tag{1.7}$$

$$U + V + W = 1\ . \tag{1.8}$$

1.3.1 Steady States of the Model

Disease-free equilibrium state is:

$$D_0 = (1,0,0,0,0).$$

Endemic equilibrium state is:

$$D_* = \left(S^*, X^*, I^*, V^*, W^*\right),$$

where

$$S^* = \frac{(\alpha_2 + \mu_2)\left(\alpha_1\gamma_2\mu_1 + MN\mu_1^2\mu_2\right)}{\alpha_1\gamma_2\left(\mu_1(\alpha_2 + \mu_2) + \alpha_2\gamma_1\right)}$$

$$X^* = \frac{M\mu_1^2\mu_2(\alpha_2 + \mu_2)(E_0 - 1)}{\alpha_1\gamma_2\left(\mu_1(\alpha_2 + \mu_2) + \alpha_2\gamma_1\right)}$$

$$I^* = \frac{\mu_1\mu_2(\alpha_2 + \mu_2)(E_0 - 1)}{\gamma_2\left(\mu_1(\alpha_2 + \mu_2) + \alpha_2\gamma_1\right)}$$

$$V^* = \frac{MN\mu_1^3\mu_2^2(E_0 - 1)}{\alpha_2\gamma_1\left(\alpha_1\gamma_2\mu_1 + MN\mu_1^2\mu_2\right)}$$

$$W^* = \frac{MN\mu_1^3\mu_2(E_0 - 1)}{\gamma_1\left(\alpha_1\gamma_2\mu_1 + MN\mu_1^2\mu_2\right)},$$

with

$$\gamma_1 = \beta_1\left(\frac{C}{\mu_2}\right),\ \gamma_2 = \beta_2 N_1,\ M = \frac{r + \mu_1}{\mu_1},\ N = \frac{\alpha_1 + \mu_1}{\mu_1},$$

$$E_0 = \frac{\alpha_1\alpha_2\gamma_1\gamma_2}{\mu_2(r + \mu_1)(\alpha_1 + \mu_1)(\alpha_2 + \mu_2)}$$

1.3.2 Sensitivity of Basic Reproductive Number

$$\frac{\partial}{\partial\alpha_1}(E_0) = \frac{\alpha_2\gamma_1\gamma_2\mu_1}{\mu_2(r + \mu_1)(\alpha_1 + \mu_1)^2(\alpha_2 + \mu_2)} > 0$$

$$\frac{\partial}{\partial\alpha_2}(E_0) = \frac{\alpha_2\gamma_1\gamma_2\mu_2}{\mu_2(r + \mu_1)(\alpha_1 + \mu_1)(\alpha_2 + \mu_2)^2} > 0$$

$$\frac{\partial}{\partial\mu_1}(E_0) = -\frac{\alpha_1\alpha_2\gamma_1\gamma_2(2\mu_1 + \alpha_1 + r)}{\mu_2(r + \mu_1)^2(\alpha_1 + \mu_1)^2(\alpha_2 + \mu_2)} < 0$$

$$\frac{\partial}{\partial \mu_2}(E_0) = -\frac{\alpha_1\alpha_2\gamma_1\gamma_2(\mu_2+\alpha_2)}{\mu_2^2(r+\mu_1)(\alpha_1+\mu_1)(\alpha_2+\mu_2)^2} < 0$$

$$\frac{\partial}{\partial r}(E_0) = -\frac{\alpha_1\alpha_2\gamma_1\gamma_2}{\mu_2(r+\mu_1)^2(\alpha_1+\mu_1)(\alpha_2+\mu_2)} < 0$$

$$\frac{\partial}{\partial \gamma_1}(E_0) = \frac{\alpha_1\alpha_2\gamma_2}{\mu_2(r+\mu_1)(\alpha_1+\mu_1)(\alpha_2+\mu_2)} > 0$$

$$\frac{\partial}{\partial \gamma_2}(E_0) = \frac{\alpha_1\alpha_2\gamma_1}{\mu_2(r+\mu_1)(\alpha_1+\mu_1)(\alpha_2+\mu_2)} > 0,$$

which demonstrate that E_0 is increasing with respect to $\alpha_1, \alpha_2, \gamma_1,$ and γ_2, and decreasing with respect to μ_1, μ_2, and r.

1.4 THE PROPOSED DCMP FRAMEWORK

The architecture of the proposed DCMP framework consists of the following main steps.

Step 1: Construction of Padé approximation-based residual functional.

Suppose that $S(t)$, $X(t)$, $I(t)$, $V(t)$, and $W(t)$ are approximated by Padé rational functions according to the following expressions:

$$S(t) \approx \frac{\sum_{i=0}^{n_1} a_{1i}t^i}{1+\sum_{j=1}^{n_2} b_{1j}t^j};\; X(t) \approx \frac{\sum_{i=0}^{n_1} a_{2i}t^i}{1+\sum_{j=1}^{n_2} b_{2j}t^j};\; I(t) \approx \frac{\sum_{i=0}^{n_1} a_{3i}t^i}{1+\sum_{j=1}^{n_2} b_{3j}t^j};$$

$$V(t) \approx \frac{\sum_{i=0}^{n_1} a_{4i}t^i}{1+\sum_{j=1}^{n_2} b_{4j}t^j};\; W(t) \approx \frac{\sum_{i=0}^{n_1} a_{5i}t^i}{1+\sum_{j=1}^{n_2} b_{5j}t^j}.$$

Imposing initial conditions:

$$S(0) = S_0,\; X(0) = X_0,\; I(0) = I_0,\; V(0) = V_0,\; W(0) = W_0,$$

we obtain

$$a_{10} = S_0,\;\; a_{20} = X_0,\;\; a_{30} = I_0,\;\; a_{40} = V_0,\;\; a_{50} = W_0.$$

For each of the discrete time step t_q; $q = 0, 1, 2, 3, \ldots, q_{\max}$, the system (1.6) reduces to the following form:

$$\varepsilon_1(t_q) = 0;\; \varepsilon_2(t_q) = 0;\; \varepsilon_3(t_q) = 0;\; \varepsilon_4(t_q) = 0;\; \varepsilon_5(t_q) = 0, \tag{1.9}$$

where each $\varepsilon_\ell : \ell = 1, 2, \ldots, 5$ is the residual function and is defined by:

$$\varepsilon_\ell(t_q) = \left(1 + \sum_{j=1}^{n_2} b_{\ell j} t_q^{\,j}\right)\left(\sum_{i=0}^{n_1} i a_{\ell i} t_q^{\,i-1}\right) - \left(\sum_{i=0}^{n_1} a_{\ell i} t_q^{\,i}\right) \times \left(\sum_{j=1}^{n_2} j b_{\ell j} t_q^{\,j-1}\right) - f_\ell(t_q)\left(1 + \sum_{j=1}^{n_2} b_{\ell j} t_q^{\,j}\right)^2.$$

System (1.9) is comprised of $5q_{\max}$ nonlinear simultaneous equations in $5(n_1+n_2)$ unknown coefficients of polynomials of Padé approximation. By assuming $x = \left(a_{11}, \ldots, a_{1n_1}, b_{11}, b_{12}, \ldots, b_{1n_2}; \ldots; a_{51}, \ldots, a_{5n_1}, b_{51}, b_{52}, \ldots, b_{5n_2}\right)^T \in \mathbb{R}^{5(n_1+n_2)}$, system (1.9) is transformed to a global optimization problem of the following form:

$$\text{Minimize } \varepsilon(\underline{x}) = \sum_{\ell=1}^{5} \sum_{q=0}^{q_{\max}} \left[\varepsilon_\ell(t_q)\right]^2 . \tag{1.10}$$

Step 2: Problem constraints formulation

The initial states (1.11)–(1.15) are imposed as equality constraints of optimization problem:

$$h_1(t) = S(t) - S_0 = 0 \tag{1.11}$$

$$h_2(t) = X(t) - X_0 = 0 \tag{1.12}$$

$$h_3(t) = I(t) - I_0 = 0 \tag{1.13}$$

$$h_4(t) = V(t) - V_0 = 0 \tag{1.14}$$

$$h_5(t) = W(t) - W_0 = 0 . \tag{1.15}$$

The inequality constraints related to the positivity of the solution are described by (1.15) for $1 \le \ell \le 5$, whereas (1.17) and (1.18) incorporate the boundedness of the numerical solution:

$$g_{\ell q} = \frac{\sum_{i=0}^{N} a_{\ell i} t_q^{\,i}}{1 + \sum_{j=1}^{M} b_{\ell j} t_q^{\,j}} \ge 0 \tag{1.16}$$

$$g_{1q} + g_{2q} + g_{3q} = 1 \tag{1.17}$$

$$g_{4q} + g_{5q} = 1 . \tag{1.18}$$

Step 3: Implementation of penalty function technique

The unconstrained optimization problem is formulated by employing the following type of penalty function:

$$\mathrm{H}(x) = M \times \max_{1 \le q \le q_{\max}}$$
$$\times \left\{0, (h_1)^2, (h_2)^2, (h_3)^2, (h_4)^2, (h_5)^2, -g_{1q}, -g_{2q}, -g_{3q}, -g_{4q}, -g_{5q}\right\}.$$

The scalar $M \in \mathbb{R}$ is a sufficiently large number called the penalty factor. Consequently, the following unconstrained function is obtained:

$$\text{Minimize } \varphi(\boldsymbol{x}) = \varepsilon(\boldsymbol{x}) + \mathcal{H}(\boldsymbol{x}) . \quad (1.19)$$

Step 4: Use of differential cooperative multi-simplex search

The iterative steps of optimization process on the penalized function (1.19) are given as below:

1. Generate randomly an initial population of K trial solutions $\left(x_j \in \mathbb{R}^{5(n_1+n_2)}; 1 \le j \le K\right)$, and for each $1 \le j \le ns$, generate a simplex $\boldsymbol{Y}^{(j)}$ around $\boldsymbol{x}_j$, sort the simplex, and assign the best vertex to $\boldsymbol{x}_j$.
2. For each trial solution, evaluate the fitness $\varphi_j = \varphi(\boldsymbol{x}_j)$. Set $T = 0$.
3. Set $T = T + 1$.
4. For each solution $j \in \{1, 2, 3, \ldots, K\}$, select three different solutions $\boldsymbol{x}_A, \boldsymbol{x}_B$, and $\boldsymbol{x}_C$ from the population excluding $\boldsymbol{x}_j$. Set $\boldsymbol{y} = \boldsymbol{x}_j$ and choose $j_{\text{rand}} \in \{1, 2, 3, \ldots, n\}$ randomly.
   - For each of the dimensions $i = 1, 2, 3, \ldots, 5(n_1 + n_2)$, alter the ith coordinate according to the following expression:
   $$y_i = \begin{cases} x_{Ai} + F \times (x_{Bi} - x_{Ci}), & \text{if rand} < CR \text{ or } i = j_{\text{rand}} \\ x_{ji}, & \text{Otherwise} \end{cases}.$$
   - If $\varphi(\boldsymbol{y}) < \varphi_j$, then $\boldsymbol{x}_j \leftarrow \boldsymbol{y}$; otherwise, discard $\boldsymbol{y}$.
   - If $j \le ns$, update the best vertex of $\boldsymbol{Y}^{(j)}$ by $\boldsymbol{x}_j$, apply an iteration of NMSA on $\boldsymbol{Y}^{(j)}$, and then assign the best vertex to $\boldsymbol{x}_j$ for cooperation with the current population.
5. Update the current best solution.
6. If $T > T_{\max}$, preserve the best solution and terminate; otherwise, go to step 3 for the new iteration.

In step 4, the function rand returns a randomly selected real number from the interval $(0, 1)$, the parameter F is called the differential constant, and CR denotes the crossover fraction. In step 6, $T_{\max}$ is the maximum number of allowed iterations. Figure 1.2 presents the complete schematic flow of the proposed DCMP framework.

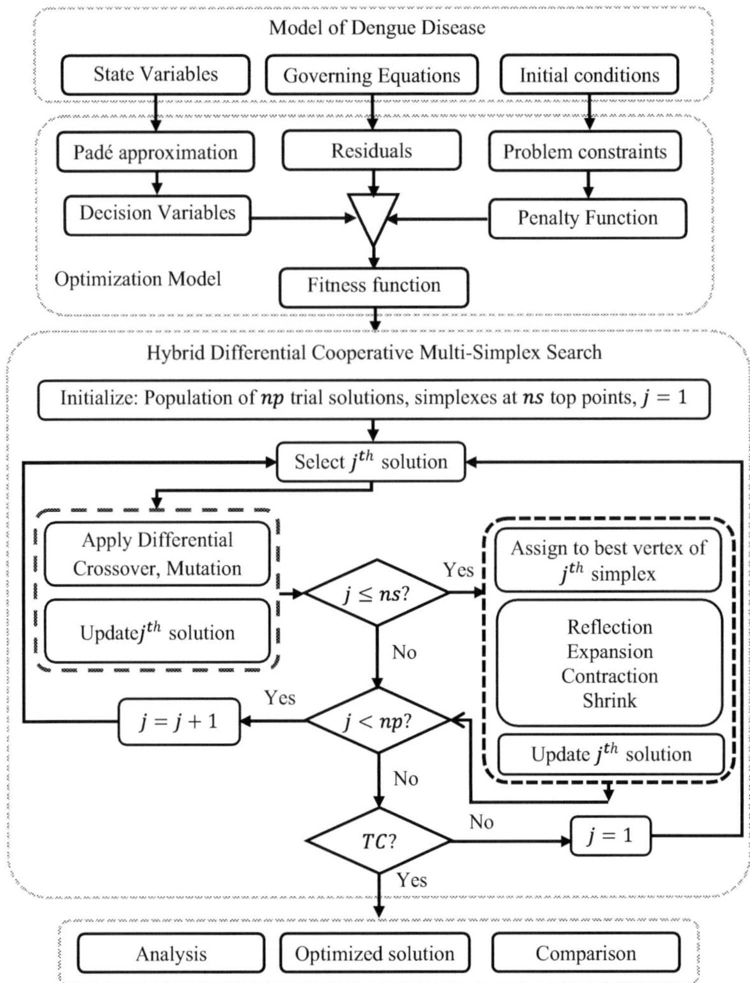

FIGURE 1.2 Flowchart of the proposed DCMP framework.

1.5 RESULTS AND DISCUSSIONS

The results have been produced by using various values of model parameters. Table 1.1 presents the real-life observations based on the values of model parameters.

The initial conditions are:

$$S_0 = 0.1,\ X_0 = 0.0001,\ I_0 = 0.0001,\ V_0 = 0.001, W_0 = 0.001.$$

To simulate the dengue model through the proposed DCMP scheme, the necessary algorithmic parameters are fixed as below:

$$Np = 50,\ ns = 5,\ T_{\max} = 3000,\ F = 0.5,\ CR = 0.9,\ n_1 = n_2 = 2.$$

TABLE 1.1
Parameters of Underlying Dengue Model

Parameter	Values Per Day
N_1	5,000
α_1	1/5
β_1	0.00005
μ_1	0.0000391
α_2	1/10
β_2	0.00008
μ_2	0.0000391
r	1/14
C (DFE)	3.00
C (EE)	300

The numerical solutions found by DCMP for various step lengths are shown in Figures 1.3–1.6. The values of step lengths used are 0.1, 1, and 10, respectively.

It can be observed from Figures 1.3, 1.4, 1.6, and 1.7 that convergence speed of the proposed DCMP scheme is extremely fast as compared to nonstandard finite difference scheme (NSFD) for smaller as well as larger step lengths. The phase portraits of infectious vs. susceptible human and vector populations presented in Figures 1.8 and 1.9 demonstrate that convergence of DCMP to steady states is almost linear, whereas NSFD has a long spiral like convergence path.

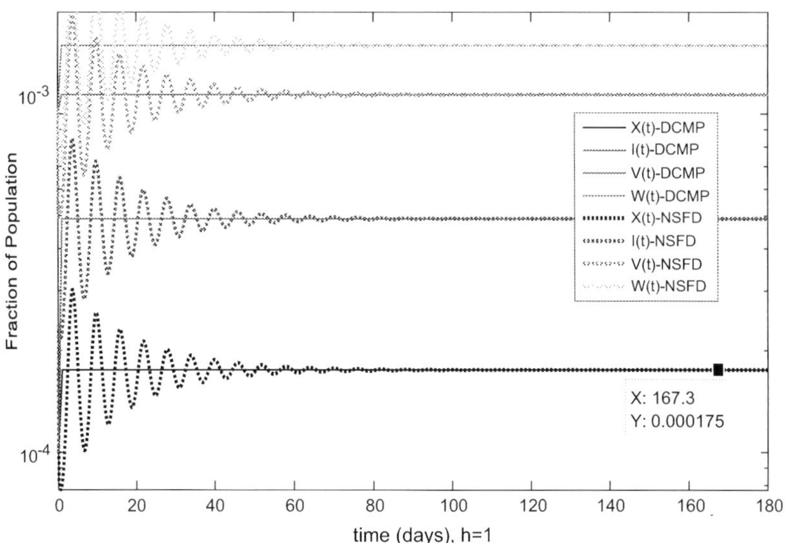

FIGURE 1.3 EE convergence curves.

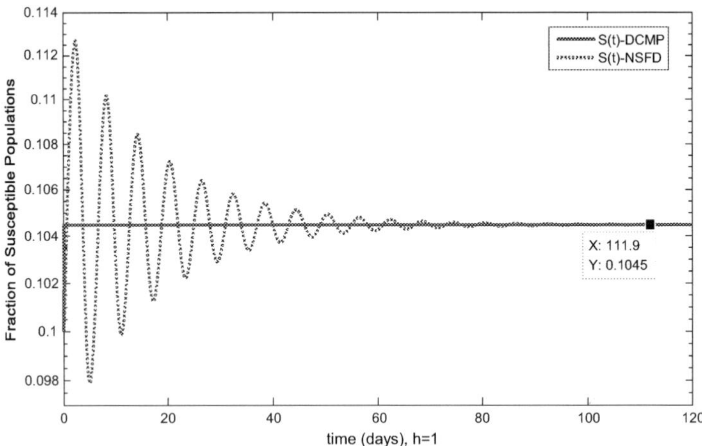

FIGURE 1.4 EE convergence curves for susceptible human population.

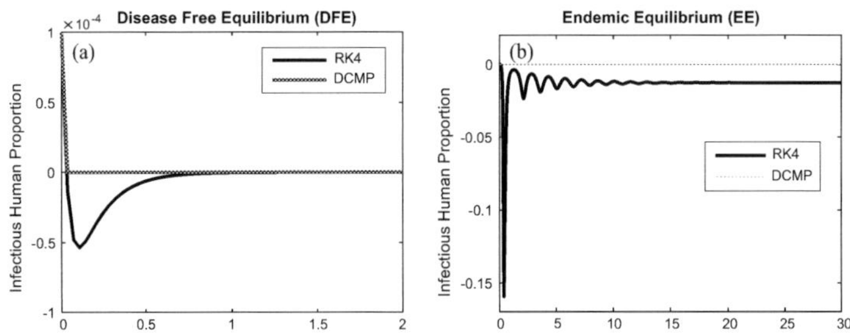

FIGURE 1.5 (a) RK4 does not preserve positivity for $h=13$ and (b) RK4 shows false convergence for $h=13$.

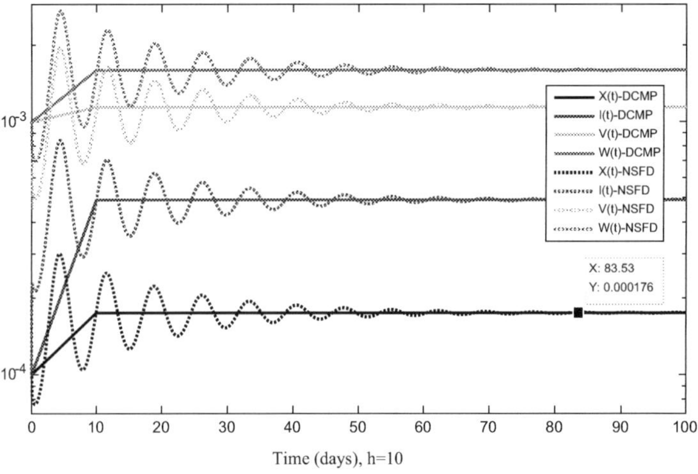

FIGURE 1.6 EE convergence curves of human and vector populations.

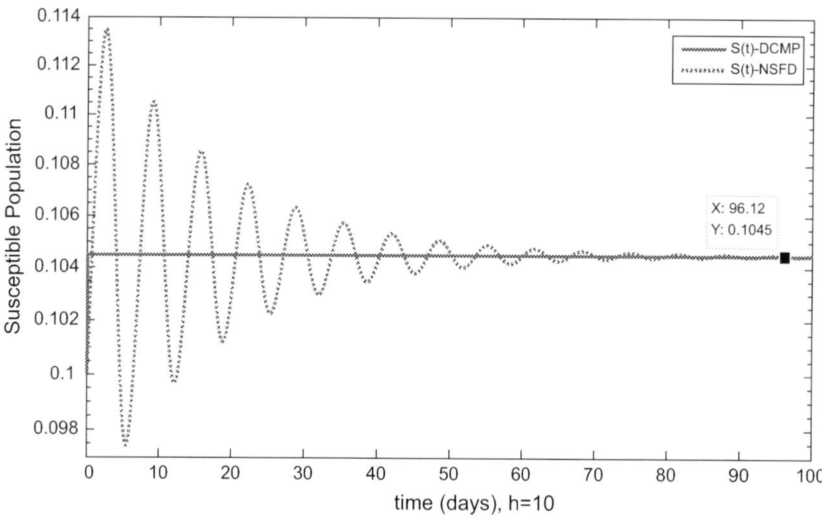

FIGURE 1.7 EE convergence history of susceptible human population for large step length.

Figure 1.5 depicts the divergence behavior of RK4 ($h=13$ years) for infectious human population proportions in both of the endemic equilibrium (EE) and disease free equilibrium (DFE) cases. On the other hand, the proposed DCMP scheme shows very fast convergences. Moreover, from Figures 1.3 to 1.7, it is evident that DCMP also preserves the positivity and boundedness of the solution.

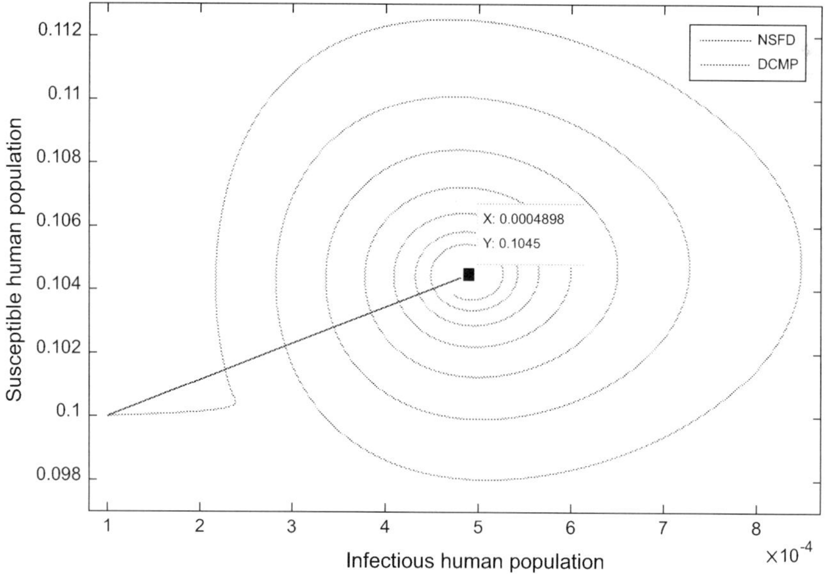

FIGURE 1.8 Phase portraits of DCMP and NSFD for human population.

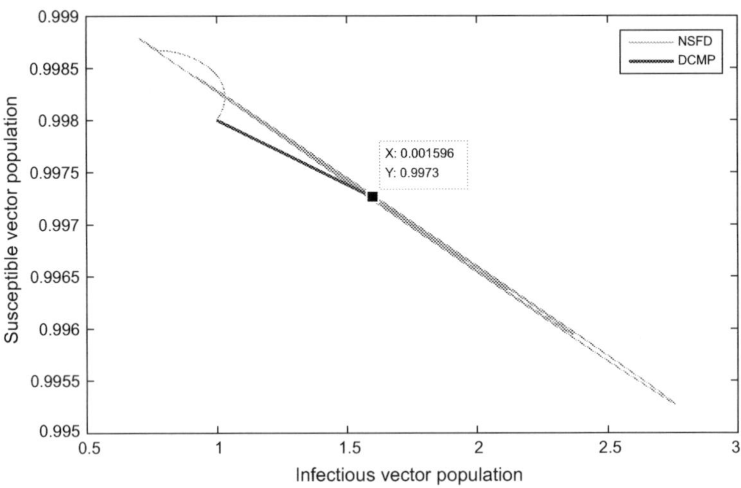

FIGURE 1.9 Phase portraits of DCMP and NSFD for vector population.

1.6 CONCLUSION

We have proposed a Padé approximation-based evolutionary computing framework for predicting and analyzing the dynamics of virus propagation of dengue disease involving virus incubation period. The proposed framework is comprised of Padè approximation, penalty function approach, NMSA, and DE for solving the underlying model numerically.

By analyzing the obtained facts and figures, the following remarks can be concluded:

- The evolutionary DCMP-based approach has successfully been developed and employed to solve the dengue disease transmission model involving virus incubation period.
- DCMP approximated state variables with high accuracy to meet the governing equations.
- The essential characteristics such as positivity, boundedness, and initial conditions of the real-world dynamical model were competently handled through the penalty function method.
- For various time steps, the DCMP produced highly accurate numerical solutions of the model that has no analytical solution so far.
- It is valuable to mention that convergence of the developed DCMP technique is in complete agreement with that produced by unconditionally convergent NSFD scheme but the convergence speed of DCMP is faster (please see Figures 1.3–1.9).
- Another superior aspect of the proposed DCMP scheme is the unconditional validity of the numerical solution generated by it for many other step lengths options, and hence, re-execution for the altered step length is prevented. This aspect straightforwardly reduces the computational efforts.

The DCMP method unconditionally converges to steady-state equilibriums regardless of the step length used. The developed framework also fits for several other nonlinear epidemiological models having no analytic solutions. It is important to note that the current implementations used Padé rational function of order (2, 2). Raising the order of Pade approximation rational, one can obtain more accurate numerical solutions with faster convergence. In the future, we intended to extend the proposed framework to solve fractional order nonlinear systems with a time delay factor.

REFERENCES

Alatas B (2017) Sports inspired computational intelligence algorithms for global optimization. *Arti Intel Rev*. https://doi.org/10.1007/s10462-017-9587-x.

Ali J, Saeed M, Chaudhry NA (2017) Low cost efficient remedial strategy for stagnated Nelder_Mead simplex method, *Pak J Sci* 69(1): 119–126.

Ali J, Saeed M, Chaudhry NA, Luqman M, Tabassum MF (2015) Artificial showering algorithm: A new meta-heuristic for unconstrained optimization. *Sci Int* (*Lahore*) 27(6): 4939–4942.

Ara A, Khan, NA, Razzaq, OA, Hameed, T, Raja, MAZ (2018) Wavelets optimization method for evaluation of fractional partial differential equations: an application to financial modelling. *Adv Differ Eq*. (8). https://doi.org/10.1186/s13662-017-1461-2.

Babaei M (2013) A general approach to approximate solutions of nonlinear differential equations using particle swarm optimization. *Appl Soft Comput* 13(7): 3354–3365.

Bojdi ZK, Ahmadi-Asl S, Aminataei A (2013) A new extended Padé approximation and its application. *Adv Num. Anal*. Article ID 263467. http://dx.doi.org/10.1155/2013/263467.

Cooley PC, Bartsch SM, Brown ST, Wheaton WD, Wagener DK, Lee BY (2016) Weekends as social distancing and their effect on the spread of influenza. *Comput Math Organ Theory* 22(1): 71–87. https://doi.org/10.1007/s10588-015-9198-5.

Goldberg DE (2005) *Genetic Algorithms in Search, Optimization, and Machine learning*. Pearson Education, Low Price Edition, Delhi.

Gubler DJ (1998) Dengue and dengue hemorrhagic fever. *Clin Microbio Rev* 11: 480–496.

Halstead SB (1998) Pathogenesis of dengue: Challenges to molecular biology. *Science* 239: 476–481.

Karaboga D, Basturk B (2007) A powerful and efficient algorithm for numerical function optimization: Artificial Bee Colony (ABC) algorithm. *J Glob Optim* 39: 459–471.

Karr CL, Wilson E (2003) A self-tuning evolutionary algorithm applied to an inverse partial differential equation. *Appl Intel* 19: 147–155.

Kennedy J, Eberhart R (1995) Particle Swarm Optimization. IEEE International Conference on Neural Networks, 1942–1948.

Lee ZY (2006) Method of bilaterally bounded to solution Blasius equation using particle swarm optimization. *Appl Math Comput* 179: 779–786.

McKinnon KIM (1998) Convergence of the Nelder-Mead simplex method to a non-stationary point. *SIAM J Optimiz* 9: 148–158.

Naowarat S, Rajabhat S (2011) Dynamical model for determining human susceptibility to dengue fever. *Am J Appl Sci* 8(11): 1101–1106.

Nelder JA, Mead R (1965) A simplex method for function minimization. *Comput J* 7(4): 308–313.

Padé H (1892) Sur la répresentation approchée d'une fonction par des fractions rationelles. *Ann Sci Ecole Normale* Sup 9 (Suppl.): 1–93.

Panagant N, Bureerat S (2014) Solving partial differential equations using a new differential evolution algorithm. *Math Probl Eng* (747490): 1–10.

Passino KM (2002) Bio mimicry of bacterial foraging for distribution optimization and control. *IEEE Control Syst.* 2(3): 52–67.

Pongsumpun P (2008) Mathematical model of Dengue disease with incubation period of virus. *World Acad Sci, Eng Technol* 44: 328–332.

Rafiq M, Ahmad MO (2015) Numerical modeling of dengue disease with incubation period of virus. *Pak J Eng & Appl Sci* 17: 19–29.

Rafiq M, Raza A, (2017) Numerical Modelling of Transmission Dynamics of Vector-Born Plant Pathogen. *Proceedings of 14th International Bhurban Conference on Applied Sciences and Technology IBCAST-2017*: 214–219.

Rafiq M, Raza A, Anayat A (2017) Numerical Modelling of Virus Transmission in a Computer Network. *Proceedings of 14th International Bhurban Conference on Applied Sciences and Technology IBCAST-2017*: 414–419.

Sadollah A, Eskander H, Bahreinejad A, Kim, JH (2015) Water Cycle algorithm with evaporation rate for solving constrained and unconstrained optimization problems. *Appl Soft Comput* 30: 58–71.

Storn R, Price K (1997) Differential evolution- a simple and efficient heuristic for global optimization over continuous spaces. *J Glob Optim* 11: 341–359.

Vijta M (2000) Some remarks on the Padé-approximations. *Proceedings of the 3rd TEMPUS-INTCOM Symposium*: 1–6.

Wang H, Wang W, Sun H, Rahnamayan S (2016) Firefly algorithm with random attraction. *Int J Bio-Inspir Comput* 8(1): 33–41.

World Health Organization (2012) *Dengue Bulletin* 36: 146–160.

Zafar Z, Rehan K, Mushtaq M, Rafiq M (2017) Numerical treatment for nonlinear brusselator chemical model. *J Differ Eq Appl* 23(3): 521–538.

2 Fuzzy-Genetic Approach to Epidemiology

Minakshi Biswas Hathiwala, Jignesh Pravin Chauhan, and Gautam Suresh Hathiwala
Marwadi University

CONTENTS

2.1 INTRODUCTION

The mechanism of world can be described using the models. Mathematical modeling deals with converting such models/systems into mathematical language, which includes concepts from algebra, geometry, differential equation, and topology, whereas its outcomes depend on the parameters considered to model that system. Mathematical modeling of the real-world problem plays an important role in many branches of science and technology. It is used for analyzing the behavior of system and understanding its mechanism. The problems that result in the necessity of studying complex control systems are extremely diverse and are generated by various branches of modern science and technology. Recently, connections between the mathematical sciences and biological sciences are increasing rapidly. New branches of mathematical sciences, including study of issues such as population growth models, epidemic models, and recently including study of genomes arising from the accumulation of DNA sequence data, have made biomathematics an interesting field.

In certain cases, the existing information that has a lack of precision or is qualitative in nature cannot be modeled efficiently by usual mathematical modeling approaches. Many natural processes cannot be described mathematically, or whose descriptions have much complexity to be of practical value, which has motivated mathematicians and researchers in fuzzy modeling and their identification techniques.

This chapter is an attempt to describe mathematically the functioning of fuzzy recombination of chromosomes. Chromosomal diseases happen when entire chromosome or enormous fragment of chromosome is either duplicated or missing or transformed. For instance, Down syndrome is a protuberant example of chromosomal irregularity. Single-gene disorder happens when transformation occurs in the gene, affecting the working of other genes, e.g., sickle cell anemia. Mutation in multiple genes results in multifactorial disorders, frequently accompanied by environmental causes. High blood pressure, arthritis, diabetes, and obesity are few of the multifactorial disorders. Mutations in nonchromosomal DNA, located in mitochondria, result in mitochondrial disorders and can influence other parts of the body, including muscles, veins, or brain. Gene plays a significant role in communicable diseases such as tuberculosis and AIDS. Moreover, it also plays a role in noncommunicable diseases such as diabetes and cancer. This section presents a brief overview about the role of genetics in few leading diseases, which weight down human population worldwide.

2.2 GENETIC EPIDEMIOLOGY AND TOPOLOGY

Conventional genomic analysis emphasizing on the genetic factor is responsible for precise phenotypes, whereas traditional study of epidemics is concerned about ecological causes and associated risk factors for individuals [1]. Genetic epidemiology is a combination of both the factors, namely, the role of genetic factors and their association with environmental factors, as far as the disease in human population is concerned. The study of genetic variation at the molecular level promises to contribute to the understanding of the etiology and pathogenesis of major chronic diseases that appear to have a genetic component, such as coronary heart diseases, cancer, and birth defects [2].

Much of genetic and epidemiological analysis involves determining the relationship between disease and exposure to risk factors, and whether a candidate exposure condition impacts the probability that an individual will have a disease diagnosis. The statistical tests to determine the probability are usually more complicated with multiple exposure variables such as genetic, behavioral, or medical conditions. The massive growth in genetic technology and its ever-expanding register of human genes are further adding to the increasing complexity [3]. One approach to identify risk exposures, and relationships between diseases and condition that are described, is simply to identify whether some patterns occur more frequently than expected. This implies that it becomes possible to explore the logical relationship and identify equality among pattern members. The mathematical structure called topological spaces provides the required platform for the discussion of nearness or sameness of such pattern through neighborhoods. Topology provides a general framework for analyzing structures or data with the advantage of being able to extract information from a large collection while not depending on the choice of any threshold value.

2.3 UNEQUAL CROSSOVER

The genetic data of living organisms is reserved in deoxy ribonucleic acid abbreviated as DNA. Each DNA molecule is packed in a thick-like structure called chromosome. The chromosomes differ in length from 10^5 base pairs in yeast to 10^8 base pairs in human [4]. A chromosome comprises gene blocks of DNA. Each protein is encoded by a gene. Alleles are different versions of the same gene. The whole collection of genetic material, i.e., all chromosomes, is known as a genome. In living organisms, a genome usually consists of homologous chromosomes. One of each homologous pair of chromosomes originates from the mother, while the other originates from the father.

During meiosis, two homologous chromosomes cross over and contribute to reassortment of genetic variation. In this process, both chromosomes break due to the pressure of mutual attraction. Further, these broken ends of chromosomes rejoin to the original chromosomes or can cross over to join the homologous original chromosomes [4]. The physical interchange of chromosomes results in recombination. Recombination generates a new combination of alleles at each generation in diploid organisms. Due to recombination, the new copy of homologous chromosomes may have different alleles. Thus, by the exchange of respective segments between the homologs, recombinant chromosomes can be generated, and these chromosomes are different from the original parental one (Figure 2.1).

Sometimes, it is possible that during meiosis, the breakage of two chromosomes occurs unequally. During this process, the phase length of the original chromosome changes, and in the new copy of chromosomes, one gets longer and the other gets shorter. This is often referred as an unequal crossover. Unequal crossover results because recombination occurs between two sites that are similar in nature, but not aligned precisely. When such an event occurs, the number of repeats increases in one

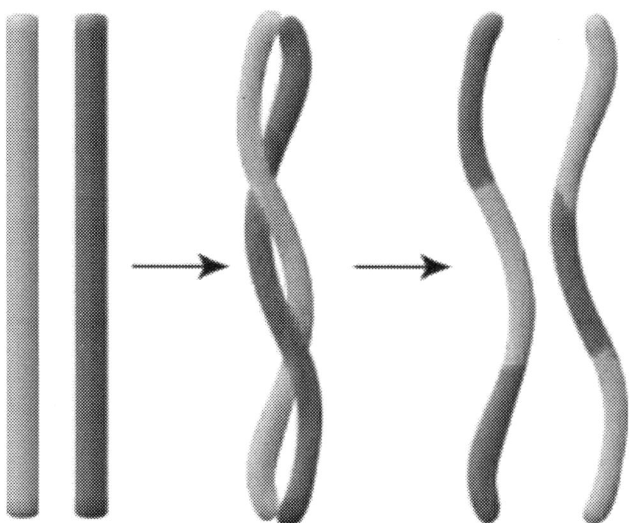

FIGURE 2.1 Recombination.

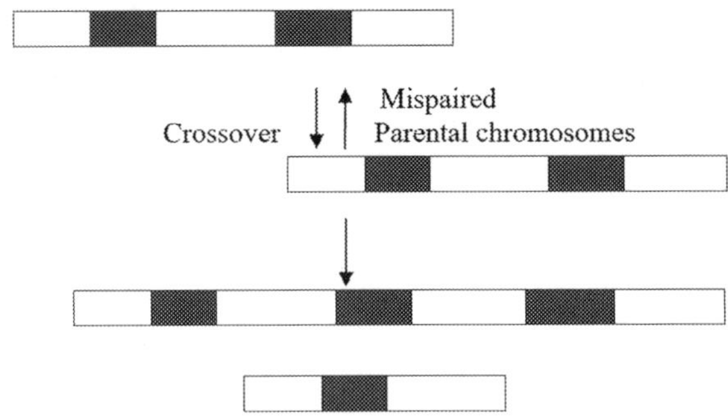

FIGURE 2.2 Unequal crossover.

chromosome, whereas that decreases in the other. In consequence, one chromosome in the new copy of homologous pair has a deletion and the other has an insertion (Figure 2.2).

2.4 MATHEMATICAL BACKGROUND

2.4.1 Fuzzy Sets

Mathematically, problems can be defined as the concept of functions and sets, which results in a rigid representation of problems that can be modeled using the concepts of fuzzy [5]. The fuzzy theory has great influence as it brings to our notice the very existence of uncertainty. Fuzzy theory can be treated to be a meaningful tool for treating such uncertainty, although it may not cover all uncertainties. Because of its large possibility, fuzzy theory is useful to an organization and human systems. An ordinary or a crisp set is a well-defined collection of finite, countable, or uncountable objects. A set is represented by the uppercase letters, and the objects in it are represented by the lowercase letters. If A is a set, then an object or element x in it is represented by $x \in A$. A characteristic function can be used to describe and represent a set. A characteristic function of A is mapping χ_A from A to the set $\{0,1\}$ that takes the value 1 if x is in A; otherwise, 0. However, in case of "fuzzy," the relation $\chi_A(x)$ of belonging to between x and A is not only "0 or 1" but also has a grade of membership, which may assume any real value between 0 and 1. The concept of fuzzy sets provides us a new way of observing and investigating the relationship between sets and their elements other than the traditional way of black or white. It says that there exists other possibilities besides the relation of *belonging to* with *not belonging to* between an element and a set. Thus, basically, fuzzy sets are the classes of an object with membership grades ranging between 0 and 1.

A fuzzy set F of a non-empty set U is defined through a mapping,

$$\mu_F : U \to [0,1].$$

This function μ_F is said to be the membership function, and this maps U to a membership space M. The value $\mu_F(x)$ is then said to be the membership grade of x in F. For simplicity, both the fuzzy set F and its membership function μ_F are usually denoted by F. If M has only the two points 0 and 1, F is crisp (or non-fuzzy) set and μ_F is the same as the characteristic function of the crisp set F. The range set of the membership function is a bounded set of real numbers, and its members are nonnegative.

For a non-empty set U, $I^U = \{F: U \rightarrow [0,1]\}$.

The members of I^U are said to be fuzzy subsets of U. 0_U and 1_U are the functions on U that takes the value equal to 0 and 1, respectively, for each value of a in U.

Let F and G be two fuzzy sets whose standard intersection, union, and complement, i.e., $F \cap G$, $F \cup G$, and F^c, are defined for each $a \in U$ by the following equations [6]:

$$(F \cap G)(a) = \min[F(a), G(a)]$$

$$(F \cup G)(a) = \max[F(a), G(a)]$$

$$F^c(a) = 1 - F(a)$$

$$F \subseteq G \text{ if } F(a) \leq G(a).$$

Note: "min" and "max" are used in place of infimum and supremum, respectively, for a finite collection of fuzzy sets.

For every fuzzy subset $F \in I^U$, *support* of F is given by

$$\text{supp}(F) = \{a \in U: F(a) > 0\}.$$

A fuzzy point on U is a fuzzy set

$$P_y^x(a) = \begin{cases} x, \text{ if } a = y \\ 0, \text{ if } a \neq y \end{cases}.$$

2.4.2 Fuzzy Pretopology

In mathematics, general topology, viz., point set topology or ordinary topology, deals with the characteristics of a space that are preserved under continuous distortions, such as stretching and bending, but not tearing or gluing. General topology, which is based on the crisp set, establishes the foundational aspects of other branches of topology. Pretopological spaces are the generalization of topological spaces. Two elements may be close to a third element via some relation; however, there is not sufficient structure to say which one of them is closer or nearer. Such a space is called a topological space. It has a suitable structure to embrace the concept of boundary. If we remove the underlying behavior of boundary from this structure, the weakest notion of nearness is uncovered, and thus, we obtain a pretopological space [7]. It is customary to define topology on a set through a class of open sets or a class of closed sets [8]. However, the concept of topology on a set can also be presented through operators such as closures, neighborhoods, and interiors instead of the conventional class of open or closed sets [9].

Fuzzy sets can be used in an extensive range of structures such as topological spaces, groups, rings, algebras, ideals, and vector spaces. They can also be applied in quantum particle physics and control theory. C.L. Chang [10] was the first to propose the concept of fuzzy topological spaces. He used fuzzy sets instead of crisp sets in the definition of point set topology and redefined the theory of ordinary topological spaces. Fuzzy topology is defined by extending ordinary topology to fuzzy setting, and the theory of ordinary topology is a special case of it. Although ordinary topology can be generalized to fuzzy topology, fuzzy topology has its own remarkable characteristics. It can magnify our interpretation of some structures in classical mathematics. Besides that, it provides a new way of observing significant results of classical mathematics [5]. For our convenience, we will now term the "topology" based on crisp sets to be crisp topology. Just like crisp topology, it is possible to define fuzzy pretopology by means of operators discussed below.

Definition 2.1 [11]

For a non-empty set U, a fuzzy pretopology on U is defined by an application α of I^U into I^U, which verifies the properties $\alpha(0_U)=0_U$ and $\alpha(F)\supseteq F$ for each $F\in I^U$. Then, (U,α) can be referred as a fuzzy pretopological space, or in short fpts.

It is possible that α may also satisfy some properties [11], such as

1. $\alpha(F)\supseteq\alpha(G)$, where $F\supseteq G$, and for each $F, G\in I^U$. In this case, (U,α) is said to be fpts of type I.
2. $\alpha(F\cup G)=\alpha(F)\cup\alpha(G)$ for each $F, G\in I^U$. In this case, (U,α) is said to be fpts of type D.
3. $\alpha(\alpha(F))=\alpha^2(F)=\alpha(F)$ for each $F\in I^U$. In this case, (U,α) is said to be fpts of type S.

If (U,α) is fpts of type I, D, and S, then it becomes a fuzzy topological space. Here, α will be called a closure operator in fpts (U,α).

Definition 2.2 [11]

By considering the fuzzy pretopological space (U,α), the interior is a function $i_\alpha: I^U\to I^U$ defined as $i_\alpha(F)=\left(\alpha\left(F^c\right)\right)^c$, where $F\in I^U$.

Then, the above properties becomes

1. $i_\alpha(0_U)=0_U$.
2. $i_\alpha(F)\subseteq F$ for each $F\in I^U$.
3. $i_\alpha(F)\subseteq i_\alpha(G)$ for each $F,G\in I^U$ so that $F\subseteq G$.

4. $i_\alpha(F \cap G) = i_\alpha(F) \cap i_\alpha(G)$ for each $F, G \in I^U$.
5. $i_\alpha^{\,2}(F) = i_\alpha(F)$ for each $F \in I^U$.

Definition 2.3 [11]

A class of fuzzy preneighborhoods of the point a in U is a collection $\mathcal{B}(a)$ of fuzzy subsets V, which satisfy $V(a) = 1$.

Definition 2.4 [11]

Let φ: $I^U \to [0,1]$ be a function. Then, φ can be referred as the degree of non-vacuity if the following conditions are fulfilled:

i. $\varphi(0_U) = 0$.
ii. $\varphi(F) = 1$ if there is a a in U such that $F(a) = 1$.
iii. $F \supseteq G$ implies $\varphi(F) \geq \varphi(G)$.

We see that $\varphi(F) = \sup_{a \in U} F(a)$.

If F is a fuzzy set of a non-empty set U, then we can construct a function α: $I^U \to I^U$ defined as

$$\{\alpha(F)\}(a) = \inf_{V \in \mathcal{B}(a)} \varphi(V \cap F).$$

This is a type I *closure* operator on U. This also relates fuzzy preneighborhoods to closure operator.

For our convenience, we sometimes denote $\alpha(F)$ by $\overline{F}$.

Definition 2.5 [12]

Let U be a non-empty set. Let $\tilde{U}$ be the collection of all fuzzy points of U. Let $\tilde{d}$ be a function on $\tilde{U}$, which takes the real values and verifies the following properties:

i. $\tilde{d}\left(P_a^r, P_b^s\right) \geq 0$.
ii. $\tilde{d}\left(P_a^r, P_b^s\right) = \tilde{d}\left(P_b^s, P_a^r\right)$.
iii. $\tilde{d}\left(P_a^r, P_b^s\right) = 0$ implies $P_a^r = P_b^s$.
iv. $\tilde{d}\left(P_a^r, P_b^s\right) \leq \tilde{d}\left(P_a^r, P_c^t\right) + \tilde{d}\left(P_b^s, P_c^t\right)$.
v. $\tilde{d}\left(P_a^r, P_b^s\right) = \tilde{d}\left(P_a^s, P_b^r\right)$ for all a, b, and c in U.

Then, $\tilde{d}$ is a classical metric on $\tilde{U}$ that satisfies an additional property (v), and it is known as fuzzy metric defined on U.

Definition 2.6 [12]

Let $\tilde{d}$ be a fuzzy metric defined on U. We define for any a, b in U and $\varepsilon > 0$

$$B\left(P_a^r, \varepsilon\right) = \left\{P_b^s : \tilde{d}\left(P_a^r, P_b^s\right) < \varepsilon\right\}$$

$$B'\left(P_a^r, \varepsilon\right) = \left\{P_b^s : \tilde{d}\left(P_a^r, P_b^s\right) \leq \varepsilon\right\}.$$

The first set is called an open ball, whereas the second set is called a closed ball—both centered at a with radius ε. These sets are treated as fuzzy sets of U whose corresponding membership functions are given by

$$B\left(P_a^r, \varepsilon\right)(b) = \sup\left\{s : \tilde{d}\left(P_a^r, P_b^s\right) <\right.$$

$$B'\left(P_a^r, \varepsilon\right)(b) = \sup\left\{s : \tilde{d}\left(P_a^r, P_b^s\right) \leq \varepsilon\right\}.$$

Thus, $\tilde{d}$ gives a fuzzy topology on U defined by the neighborhood basis $\mathcal{B}(a) = \left\{B\left(P_a^r, \varepsilon\right) : \varepsilon > 0\right\}$ for each $a \in U$.

Definition 2.7 [12]

A fuzzy weakly metrizable space is a fpts U, in which there lies a fuzzy metric $\tilde{d}$ and set Q, $Q' \subset (0, \infty)$ such that

$$\mathcal{B}(a) = \left\{B\left(P_a^r, \varepsilon\right) : \varepsilon \in Q\right\} \cup \left\{B'\left(P_a^r, \varepsilon\right) : \varepsilon \in Q'\right\}$$

is a fuzzy preneighborhood basis of U, where $B\left(P_a^r, \varepsilon\right)$ and $B'\left(P_b^r, \varepsilon\right)$ are the open and closed balls, respectively.

In crisp topology, separation axioms express how rich the population of open sets are. More precisely, each of them tells how tightly the distinct points or disjoint subsets can be wrapped in an open set. The following definitions are some of the separation axioms in fuzzy pretopology.

Definition 2.8 [12]

A T_0-fuzzy space is a fuzzy pretopological space U in which for all $a \neq b$, there exists $F \in \mathcal{B}(b)$ such that $a \notin F$, i.e., $F(a) = 0$, or there exists $G \in \mathcal{B}(a)$ such that $b \notin G$, i.e., $G(b) = 0$.

Definition 2.9 [12]

A T_1-fuzzy space is a fuzzy pretopological space U in which for all $a \neq b$, there exists $F \in \mathcal{B}(b)$ such that $a \notin F$, i.e., $F(a) = 0$.

Definition 2.10

A T_2-fuzzy space is a fuzzy pretopological space U in which for all $a \neq b$, there exists $F \in \mathcal{B}(a)$ and $G \in \mathcal{B}(b)$ such that $F \cap G = 0_U$.

It is clear from the definition that $T_2 \Rightarrow T_1 \Rightarrow T_0$.

Definition 2.11 [12]

A R_0-fuzzy space is a fuzzy pretopological space U in which for $a \in \overline{\{b\}}$ implies $b \in \overline{\{a\}}$, i.e., $\overline{\{b\}}(a) > 0 \Rightarrow \overline{\{a\}}(b) > 0$ for each a, b in U. Note that the fuzzy subset $\{a\}$ is such that $\{a\}(a) = 1$; otherwise, 0.

Theorem 2.1 [12]

A fuzzy weakly metrizable space is a R_0-fuzzy space.

Theorem 2.2 [12]

A fuzzy pretopological space U is $T_1 \Leftrightarrow \overline{\{a\}} = \{a\} \forall a \in U$.

Theorem 2.3

A fuzzy pretopological space U is $T_1 \Leftrightarrow U$ is both R_0 and T_0.

We know that finite objects are easy to handle and so they are considered as the well-behaved ones. In crisp topology, the property of compactness is not exactly finiteness, but it behaves a lot in that manner. Compactness tells how firmly a set is packed. The following definitions discuss the compactness in fuzzy pretopology.

For a type I fuzzy pretopological space (U, α), we have the following terms.

Definition 2.12 [11]

U is said to be 1-compact if and only if for every collection $A_{j|j \in J}$ of fuzzy subsets of U that satisfies $\cap_{j \in j_0} A_j \neq 0_U \, \forall J_0 \subseteq J$, where J_0 is finite, we have $\cap_{j \in J} \alpha(A_j) \neq 0_U$.

Definition 2.13 [13]

U is said to be 1-Lindelof if and only if for every collection $A_{j|j \in J}$ of fuzzy subsets of U that satisfies $\cap_{j \in J_0} A_j \neq 0_U$, where $J_0 \subseteq J$ and J_0 is countable, we have $\cap_{j \in J} \alpha(A_j) \neq 0_U$.

Definition 2.14 [13]

U is said to be countable 1-compact if and only if for every collection $A_{j|j \in J}$ of fuzzy subsets of U that satisfies $\cap_{j \in J_0} A_j \neq 0_U$, where $J_0 \subseteq J$ and J_0 is finite, we have $\cap_{j \in J} i_\alpha \left\{\alpha\left(A_j\right)\right\} \neq 0_U$.

Definition 2.15 [13]

U is said to be almost 1-compact if and only if for every collection $A_{j|j \in J}$ of fuzzy subsets of U that satisfies $\cap_{j \in J_0} i_\alpha\left(A_j\right) \neq 0_U$, where $J_0 \subseteq J$ and J_0 is finite, we have $\cap_{j \in J} \alpha\left(A_j\right) \neq 0_U$.

Definition 2.16 [13]

U is said to be nearly 1-compact if and only if for every collection $A_{j|j \in J}$ of fuzzy subsets of U that satisfies $\cap_{j \in J_0} i_\alpha\left(A_j\right) \neq 0_U$, where $J_0 \subseteq J$ and J_0 is finite, we have $\cap_{j \in J} i_\alpha \left\{\alpha\left(A_j\right)\right\} \neq 0_U$.

1-compact $\Rightarrow$ nearly 1-compact $\Rightarrow$ almost 1-compact.

In crisp topology, connectedness is referred as one of the principal topological properties that are used to differentiate topological spaces. This property can be generalized with the help of fuzzy sets as follows.

Definition 2.17 [14]

Let (U, α) be a type I fuzzy pretopological space. Then,

i. U is strongly fuzzy-connected if for every $F \in I^U, F \neq 0_U, \alpha(F) = 1_U$.
ii. U is one-sided fuzzy-connected if for every $A \in I^U, F \neq 0_U, \alpha(F) = 1_U$ or for every $G \in I^U, G \neq 0_U$ if $G \subset \{\alpha(F)\}^c$, then $F \subset \alpha(G)$.
iii. U is hyper-fuzzy-connected if for every $F \in I^U, F \neq 0_U, \alpha(F) = 1_U$ or there exists $G \in I^U, G \neq 0_U$ if $G \subset \{\alpha(F)\}^c$, then $F \subset \alpha(G)$.
iv. U is apo-fuzzy-connected if for every $F \in I^U, F \neq 0_U, \alpha(F) = 1_U$ or for every $G \in I^U, G \neq 0_U$ if $G \subset \{\alpha(F)\}^c$, then $\alpha(F) \cap \alpha(G) \neq 0_U$.
v. U is fuzzy-connected if for every $F \in I^U, F \neq 0_U, \alpha(F) = 1_U$ or $\alpha\left[\{\alpha(F)\}^c\right] \cap \alpha(F) \neq 0_U$.

Theorem 2.4 [14]

For type I fuzzy pretopological space (U, α),

i. U is one-sided fuzzy-connected if it is strongly fuzzy-connected.

ii. U is hyper-fuzzy-connected and apo-fuzzy-connected if it is one-sided fuzzy-connected.
iii. U is fuzzy-connected if it is hyper-fuzzy-connected.
iv. U is fuzzy-connected if it is apo-fuzzy-connected.

2.5 FUZZY TOPOLOGICAL PROPERTIES OF RECOMBINATION SPACE

2.5.1 Mathematical Definition of Recombination Sets

Recombination spaces can be defined on the basis of concept of recombination functions $R{:}C \times C \rightarrow P(C)$ [15]. Consider a pair of parental chromosomes a and b, and the set of recombination $R(a, b)$ comprises all recombinant chromosomes that are formed by recombining a and b with the help of certain class of crossover operators. We observe the following given properties:

a. $\{a, b\} \subseteq R(a, b)$.
b. $R(a, b) = R(b, a)$.
c. $R(a, a) = \{a\}$.
d. $\forall c \in R(a, b)$, the inequality $|R(a, c)| \leq |R(a, b)|$ holds.

Condition (a) is just for notation purpose. Condition (b) signifies the need of simple symmetry. Condition (c) asserts that no new type of chromosome will be created by recombination of one single type of chromosomes. Lastly, condition (d) explains the topological significance of recombination. It fundamentally states that recombinant chromosomes are more likely to be of the parental kinds, than they (parental kinds) are to each other. This essentially asserts the notion that recombinants are blends of the two parental kinds. Additionally, this situation implies that the amount of similarity in two types of chromosomes is in inverse proportion to the number of types of recombinants that are created by recombination process. Properties (a) and (b) are satisfied by a generalized recombination structure. The appropriate recombination sets of homologous crossover, in addition, also satisfy (c) and (d). It appears obvious to infer $R(a,b)$ as neighborhood for each $b \in C$. By (a), we have $a \in R(a, b)$ for all a, b. Thus, the set of recombinants forms a neighborhood basis if and only if for each a, b, c, there is a d such that

$$R(a, d) \subseteq R(a, b) \cap R(a, c) \tag{2.1}$$

Generally, the above restriction is not always satisfied. However, if we take the set of recombinants as a sub-basis of the neighborhood filters, the coarsest pretopology is constructed in which the recombination sets are neighborhoods. In the case of genome set which is finite, there exists a smallest neighborhood $N(a)$, i.e., a minimal element of the neighborhood basis. This is generally true in Alexandroff spaces, the spaces in which the neighborhood filters have a finite basis. If C is finite, then vicinities can be extracted directly from the *(sub)-basis* of recombination sets as

$$N(a) = \cap_{b \in C} R(a,b) \tag{2.2}$$

If C is infinite, then the vicinity $N(a)$ defined in the above result (2.2) should not necessarily be the neighborhood of a. Result (2.2) defines neighborhoods if the size of the recombination sets $R(a, b)$ is bounded.

A closure operator on the set of genotypes is persuaded by recombination set through the closure operator

$$cl(M) = \cup_{(x,y) \in M \times M} R(x,y). \tag{2.3}$$

With this closure [15], the recombination space becomes a crisp pretopological space.

2.5.2 Unrestricted Unequal Crossover

In this model [16], an extreme form of unequal crossover, i.e., occurrence of crossover with equal possibility at all intergenic areas wherever possible along with the both sides of the gene cluster is assumed. Every occurrence of crossover incident yields a pair of recombinant chromosomes. Every trial of recombination results in chromosomes with distinct gene copies compared with the original one. Here, a represents a chromosome with certain number of gene copies as well as the number of gene copies on the chromosome. All the possible recombinants between chromosomes with a copies and b copies of genes form the recombination set $R(a, b)$, which is defined as

$$R(a, b) = \{0, 1, \ldots, (a + b)\},$$

where the smallest neighborhood is represented by $N(a) = R(a, 0) = \{0, 1 \ldots, a\}$.

It is also observed that $N(b) \subseteq N(a)$ if and only if $b \leq a$.

Hence in this form of unequal crossover, the neighborhood shares at least $\{0\}$ if not a much larger set. This may be the case when any two chromosomes have mismatched crossover of any number of gene positions and their recombinant is a chromosome with no gene copy, $a = 0$.

The recombination space in this model is a crisp topology. If we look at the separation axioms, it fails to be, T_1 or R_0. Even it is not weakly metrizable [16]. With the closure defined in 2.3, it was shown that the recombination space in this model is connected; however, it fails to be strongly connected [17].

2.5.3 Fuzzy Pretopology in a Recombination Space

The recombinants obtained from the recombination of chromosomes have different possible outcomes. Whenever recombinants are assigned arbitrary values, the newly proposed set can be viewed as a fuzzy set [12]. Consider the fuzzy set of recombinants μ_{ab} derived from a recombination of chromosome a with chromosome b. The set μ_{ab} is a collection of all possible recombinants obtained from chromosomes with

a and b numbers of gene copies but with variable values of membership. If C represents the set of chromosomes, then the recombinant set of the homologous pair (a, b) has the membership function $\mu_{ab}: C \rightarrow [0,1]$.

Here, the fuzzy subset μ_{ab} is considered as a fuzzy preneighborhood of a as well as of b for each $a, b \in C$, by considering, respectively, $\mu_{ab}(a) = 1$ and $\mu_{ab}(b) = 1$. The closure operator given in (2.4) defines the concept of pretopology on the set of chromosomes

$$cl(\mu)(a) = \inf_{V \in \mathcal{B}(a)} \varphi(V \cap \mu), \tag{2.4}$$

where μ denotes any fuzzy subset on C and $\mathcal{B}(a)$ denotes all preneighborhoods of a. The set of chromosomes C along with the fuzzy closure defined in (2.4) satisfies the basic axioms of the type I fuzzy pretopology, and so we have a fuzzy pretopology on the set of all chromosomes C [12]. In our text, we will refer this fuzzy pretopological space as fuzzy recombination space.

In the fuzzy pretopological model, it is considered that each element of crisp recombination set $R(a, b)$ has different possibilities of occurrence. In case of unrestricted unequal crossover model, if a is a chromosome, then the recombination set $R(a, 0)$ being the smallest neighborhood is contained in the support of all fuzzy recombination subsets of the form μ_{ab} on C.

2.5.4 Separation Properties

In the unrestricted unequal crossover model [16], the set $R(a, b)$ comprises every single recombinant possible among chromosomes with a copies and b copies of gene, i.e.,

$$R(a, b) = \{0,1, \ldots, (a+b)\},$$

where the smallest neighborhood is represented by $N(a) = R(a, 0) = \{0,1 \ldots, a\}$.

In fuzzy pretopological model, each element has different possibilities of occurrences. For $a < b$, the support of all preneighborhoods of b contains the elements of

$$R(b, 0) = \{0,1,\ldots, a, a+1,\ldots, b\}.$$

This set also contains a. Thus, we can conclude that every preneighborhood of b contains a. So the space is not T_1 [12]. Therefore, the space is not T_2. Since it is not R_0, the recombination space is not fuzzy weakly metrizable.

2.5.5 Lindelofness and Compactness

Here, we are going to discuss about the properties of 1-Lindelofness and 1-compactness of the fuzzy recombination space in the case of unrestricted unequal crossover model [18].

Let $A_{j|j \in J}$ represent the fuzzy subsets on the set of chromosomes C, such that

$$\cap_{j \in J_0} A_j \neq 0_C$$

for a countable subset J_0 of J.

Now, $\cap_{j\in J_0} A_j \neq 0_C$ will imply $\inf_{j\in J_0} A_j(a_0) \neq 0$ for some $a_0 \in C$.
For such a_0,

$$A_j(a_0) \neq 0 \text{ for every } j \in J_0.$$

This implies a_0 belongs to support of all A_j. Since the A_j's are fuzzy recombination sets, their support is non-empty. This is because the support of any fuzzy recombination set μ_{ab} at least contains the elements of the smallest neighborhood of the corresponding crisp recombination set $R(a, b)$. As discussed in Section 2.5.2, the neighborhoods share at least $\{0\}$ if not a much larger set, we can assure that there exists at least one such a_0.

If we choose $a_0 = 0$, then a_0 belongs to support of all A_j and

$$cl(A_j)(a_0) > A_j(a_0) > 0 \text{ for every } j \in J$$

This implies $cl(A_j)(a_0) \neq 0$ for every $j \in J$. This further gives $\inf_{j\in J} cl(A_j) \neq 0_C$, or $\cap_{j\in J} cl(A_j) \neq 0_C$.

This shows that the space is 1-Lindelof. The same is true if J_0 is the finite subset of J.

Hence, the space is also 1-compact. Consequently, it is almost 1-compact as well as nearly 1-compact.

2.5.6 Connectedness

Here, we are going to discuss about the properties of apo fuzzy connectedness and fuzzy connectedness in the case of unrestricted unequal crossover model in fuzzy recombination space [14].

Let $A \in I^C$ be a fuzzy recombination set such that $A \neq 0_C$ and $cl(A) \neq 1_C$, and let $B \in I^C$ be a fuzzy recombination set such that $B \neq 0_C$ and $B \subset [cl(A)]^c$. Now, $B(a_0) > 0$ if a_0 belongs to the smallest neighborhood of the corresponding crisp recombination set of B. Then,

$$B(a_0) < 1 - \{cl(A)(a_0)\} \text{ will give } cl(A)(a_0) < 1.$$

Thus, there exists a a_0 such that whenever $B \subset [cl(A)]^c$, $cl(A) = 1_C$ is not true, and vice versa. Now, if $a_0 \in N(0) = \{0\}$, then $B(a_0) > 0$ and $A(a_0) > 0$. This is possible because in this form of unequal crossover, the neighborhoods share at least $\{0\}$ if not a much larger set. Since the fuzzy recombination space is of type I fuzzy pretopological space,

$$cl(B)(a_0) > B(a_0) > 0 \text{ and } cl(A)(a_0) > A(a_0) > 0.$$

This gives $cl(A) \cap cl(B) \neq 0_C$. Hence, the space is apo-fuzzy-connected and consequently fuzzy-connected. Since strongly fuzzy-connected implies connectedness, in any case the fuzzy recombination space is fuzzy-connected.

2.6 CONCLUSION

As we have seen, with the aid of pretopology, we can have the structural analysis of genes. Unequal crossover leads to a mutation in genes [19]. Mutations bring genetic variations as they have adverse effects such as altering the product of genes or preventing them from functioning partially or completely. There are various models in an unequal crossover. In this chapter, we have investigated the properties of compactness and connectedness of the fuzzy recombination space in unrestricted unequal crossover model. It is observed that space is 1- compact and connected in this model. It is also observed that there is a possibility that space could be strongly connected in this model. Further, separation properties are also examined. Among all properties, the property of connectedness plays a dynamic role. To extract the connectivity in genetic development, i.e., the nearness of one population to another, pretopological connectivity gives an appropriate structure. The outcomes revealed in this chapter of unequal crossover replicate the connectivity of the fuzzy recombination space in genetic epidemiology.

REFERENCES

1. Permutt M.A., Wasson J, Cox N., Genetic epidemiology of diabetes. *Journal of Clinical Investigation*, 115(6), 2005, 1431–1439.
2. Khoury M.J., Beaty T.H., Cohen B.H., *Fundamentals of Genetic Epidemiology*, Oxford University Press, 11, New York, 1993.
3. Khoury M.J., Little J., Burke W., *Human Genome Epidemiology: A Scientific Foundation for Using Genetic Information to Improve Health and Prevent Disease*, Oxford University Press, New York, 2004.
4. Dokholyan M.V., Buldyrev S.V., Havlin S., Stanley H.E., Model of unequal chromosomal crossing over in DNA sequencing. *Physica A*, Elsevier publication, 249, 1998, 594–599.
5. Palaniappan, N., *Fuzzy Topology*. Narosa Publishing House, New Delhi, 2005.
6. Klir, G. J., Yuan, B, *Fuzzy Sets and Fuzzy Logic Theory and Application*. Prentice Hall of India Private Limited, Delhi, 2013.
7. Stadler B.M.R., Stadler P.F., Wagner G., Fontana W., The topology of possible: Formal spaces underlying patterns of evolutionary change. *Journal of Theoretical Biology*, 2001, 213, 241–274.
8. Munkers, J.M., *Topology*. Prentice-Hall of India Private Limited, Delhi, 2013.
9. Stadler B.M.R., Stadler P. F., Generalized topological spaces in evolutionary theory and combinational chemistry. *Journal of Chemical Information and Computer Sciences*, 42, 2001, 577–585.
10. Chang C.L., Fuzzy topological spaces. *Journal of Mathematical Analysis and Applications*, 24, 1968, 182–190.
11. Badard R., Fuzzy pretopological spaces and their representation. *Journal of Mathematical Analysis and Applications*, 81, 1981, 378–390.
12. Ali T., Phukan C. K., Incompatibility of metric structure in recombination space. *International Journal of Computer Applications*, 43(14), 2012, 1–6.
13. Mashour A.S, Ramadan A.A, Monsef M.E. Abd EL., A note on compactness in L-fuzzy pretopological spaces. *Rocky Mountain Journal of Mathematics*, 20(1), 1990, 199–208.
14. Biswas M., Chetia B.C., Connectedness in fuzzy recombination space. *Journal of Global Research in Mathematical Archive*, 2(3), 2014, 95–101.

15. Stadler P. F., Wagner G. P., Recombination induced hypergraphs: A new approach to mutation-recombination isomorphism. *Complexity*, 2, 1996, 37–43.
16. Stadler B.M.R., Stadler P. F., Shpak M., Wanger G. P., Recombination spaces, metrics, and pretopologies. *ZeitschriftfurPhysikalischeChemie*, 216, 2002, 217–234.
17. Phukan C.K., Connected pretopology in Recombination Space, *Theory in Biosciences*, Springer Nature, Switzerland AG,139, 2020, 145–151.
18. Biswas M., Chetia B.C., Phukan C.K., Lindelofness and compactness in fuzzy recombination space. *International Journal of Mathematical Archive*, 5(9), 2014, 33–37.
19. Kruger J., Vogel F., Population genetics of unequal crossing over. *Journal of Molecular Evolution*, Springer-Verlag, 4, 1975, 201–24.

3 Role of Mathematical Models in Physiology and Pathology

Saktipada Nanda and Biswadip Basu Mallik
IEM

CONTENTS

3.1 INTRODUCTION

Models are constructed to manipulate the real world in the way that they will suit our purpose best. Researchers generally adopt three types of models: mathematical, numerical, and experimental. It is an accepted fact that physical models based on experimental investigation are superior to the analytical models based on theoretical investigations. But the applicability of the physical ones has got limitations due to the problems faced during the measurement and collection of data relevant for the investigation of the physical problems such as brain injury and blood flow. On the other hand, mathematical models based on simple geometry (principally spherical and cylindrical) can be effectively used to explore a variety of causes related to brain injury and blood flow over a short period of time. The idea of modeling is that it constitutes a tool which we can manipulate in the way that will suit us best and find the way of manipulating the real world supposed to be represented by it.

Based on mathematical concepts and using mathematical languages and symbolisms, analytical models are constructed (or modified). Now these models are extensively applied in solving problems of human biological sciences.

The objective of this dissertation is to present a methodology for the design and construction of sophisticated mathematical models in the study of "head injury" and "blood flow" problems related to human physiology, hematology, and pathology. They are being extensively used in making predictions for the output when one or more parameters are varied in some specified range.

It is a fact that all the physical phenomena show nonlinear response. Therefore, for evaluating a phenomenon in the real world, the scientists should make their efforts to develop a mathematical model that exhibits the physical system or phenomenon's nearly exact behavior. Researchers generally use three types of models: mathematical (or analytical), numerical, and experimental. Now, the results obtained from the study of various branches of applied mathematics (in particular, dynamical system analysis) and the increasing use of high-speed electronics computers have gained importance in the construction of analytical models for formulation, analysis, and prediction of critical problems related to human physiological system for a variety of reasons.

Mathematical models are developed using a suitable symbolism to define the operation of a physical system. They are being extensively used in making predictions for the output when a particular parameter is varied experimentally. If a real-life problem involves a variation of one or more physical and/or rheological parameters in some admissible range, then analytical modeling is a must for investigation of the problem. Mathematical modeling becomes essential in the cases where experimental investigations are difficult to be carried out in order to obtain adequate data. The results obtained through the proper use of these techniques may have significant predictions on the nature and behavior of the complex investigations occurred in brain dysfunction and circulatory disorders. The refined form of the results might have some applications in the initial treatment of these diseases to identify the location and degree of severity of the damage. The results of these studies might have some applications for the initial treatment of the trauma in the remote areas where modern treatment technique is still a dream.

Though head injury means injury of any part of the cranial vault, brain trauma is the most vulnerable problem from pathogenic, surgical, and preventive considerations. So continuous analytical as well as experimental investigations on the mechanics of the cranial vault with particular emphasis on the brain mechanics have been made since World War II. The deformation of the human brain exceeding the equilibrium limit may lead to morbidity and disability. It is desired that more studies based on actual data and including more parameters will throw light in the construction and modification of sophisticated and secured well-protective systems for the human cranial vault.

It is now an established fact that irregular blood circulation leading to cardiovascular failure is the major risk factor for life of the human being. A methodical investigation on hematology, and rheological and mechanical properties of vessel walls may open a new insight in the protection and treatment of major problems of cardiology and vascular system. Investigations on the flow characteristics of blood through human artery/vein largely depend on modeling of the physical shape and nature of the artery/vein, and rheological behavior of the streaming blood.

In several disciplines of physical and life sciences, mathematical models are developed and sometimes modified using the suitable characteristics to define the function of a physical system. If a physical problem involves a variation of one or more physical and/or rheological parameters in an admissible range, then mathematical modeling has no alternative. Also, mathematical modeling becomes essential in the cases where experimental investigations are difficult to be carried out in order to obtain adequate data. In biomechanics, the results obtained through studies of different biological systems and biomaterials have led scientists and engineers to design and construct artificial organs for human body.

In the realm of medical and physiological sciences, the use of mathematical models is growing day-by-day on account of various reasons discussed earlier. So analytical models are extensively used in the study of the mechanics of the function of the human cranial system and blood flow as it may provide better understanding of the key parameters controlling the system. Also, the output of the dissertation may add some additional input in the health sector.

3.2 ROLE OF MATHEMATICAL MODELS IN THE STUDY OF BRAIN INJURY PROBLEMS

Mathematical models give confidence in the study of injury of the human cranial system due to the following reasons:

i. It is not possible to conduct experiments (analytical or experimental) on human head.
ii. No human being will voluntarily allow investigation on any part of the cranial vault.
iii. As human head is much superior to other animals so the results of the experiments done on the inferior animals fail to explain the potential injury mechanism of the human head subjected to any kind of severe direct impact or an impulsive load on the head or rotation of the head.
iv. The properties exhibited by a living being stop just after the brain death.
v. The geometry of the cranial vault does not resemble any standard geometric structure.
vi. No two persons can be found whose brain geometries are exactly similar. So more and more sophisticated mathematical models are being developed or old models are being improved to deal with a few problems related to brain injury mechanism, which are believed to be of much interest for researchers working in this field (Figure 3.1).

Under normal physiological condition, the reactions of the organisms to various stimuli (excitements) are transmitted by the sensory nerve to the brain for analysis and synthesis of that simulation. Then, the finest adjustment and integration of the activity of all systems, organs, and tissues follows on the basis of that result. As an example, when a mosquito simply sits or bites at any point of the body, the information is instantaneously transmitted to the brain through the sensory nerves. The brain analyzes the signal and commands the appropriate organ (hand or leg) to take necessary action (strike or kill the mosquito). When a human body is infected with virus or bacteria, the brain commands the temperature control system of the body to increase the temperature of the body so that those parasites may be killed through the natural process. Broadly speaking, *brain* controls all activities of a living human organism.

It is well known that craniocerebral trauma accounts for most of the sudden deaths due to vehicular or soccer accidents, and naturally, the problem of head injury

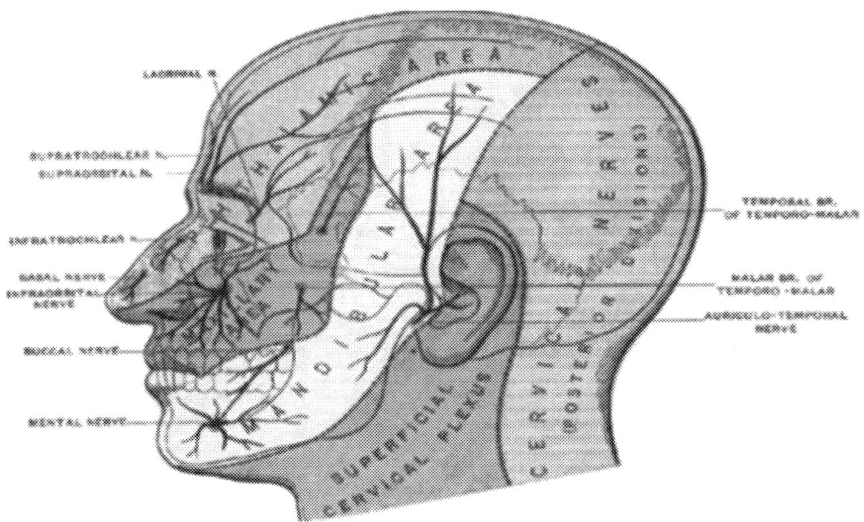

FIGURE 3.1 Basic structure of the human head.

should deserve scientific attention in both diagnosis and treatment of such trauma. Also, it deserves attention for drawing and giving shape for manufacture of the well-protected systems and devices for human head. In the broad perspective, head injury is a damage of any component of the head but injury of the brain attains the highest concern in the health sector as brain damage is the leading cause of death or morbidity in automotive accidents and fast games such as soccer and hockey.

Comprehensive studies have been conducted to understand the mechanics and reaction of the sensory organs owing to collision with some solid objects or biomechanical responses due to direct impact of the head with a projectile or a blow, or rotation of the head due to the effect of angular acceleration. Even then our knowledge about understanding of the direct cause of brain dysfunction and craniocerebral trauma is failing to explain potential injury mechanisms. As a result, the reactions of the human sensory nerves due to excitement cannot be drawn conclusively. Those pieces of information may supply necessary inputs for the health sector. It is therefore necessary to present a range of problems in the domain of biomechanics so that the basic mechanism producing brain damage may be explored. Also, the construction and analysis of mathematical models, and design of the protective instruments for cranial vault can proceed to a more sophisticated level. In most of the analytical studies associated with brain trauma, the spherical geometry and the layered structure of the cranial vault are considered.

Some of the models of human head used in the investigation of brain injury problems are shown in Figures 3.2 and 3.3.

The anatomical description and functional importance of the various components of the human head accompanied by the composition of the brain tissues and the functions performed by them may be relevant in this discussion.

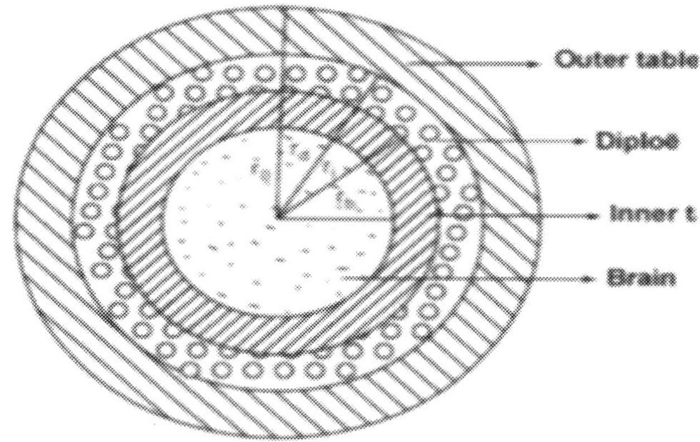

FIGURE 3.2 Schematic representation of the layered structure of the *cranium*

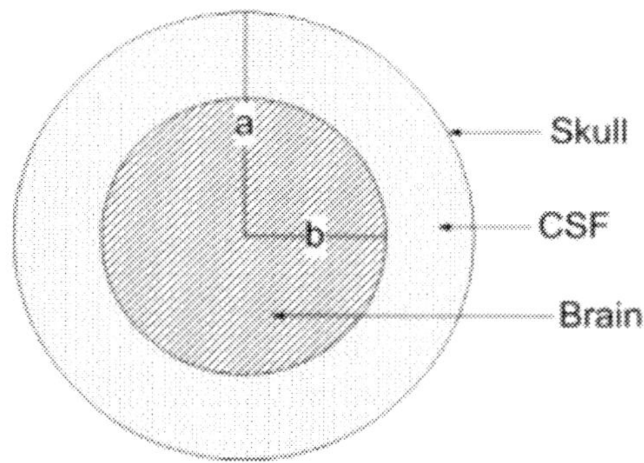

FIGURE 3.3 Simplified model of the composite structure of the brain, the CSF, and the skull.

Location of brain in the cranial vault (head) and its natural protection: The basic components of the human head that protect the brain from external hit are the scalp, the skull, the diploe, and the cerebrospinal fluid (CSF)—the brain being at the central location.

i. *Scalp*: It is the outermost cerebral covering varying in thickness from 6.4 to 12.7 mm. It appears to have a leathery consistency, and the outer surface is skin with hairy covering. Owing to its elastic property and hairy covering, it initially protects the brain from excess heat, humidity, and mild external blow.
ii. *Skull*: It completely encloses the brain except for the spinal cord opening. It is placed on the upper end of the vertebral column and acts as the bony protection of the brain. It is almost spheroidal in shape and has twofold structure, namely, the cranium and the face. The thickness of the skull varies from 9.6 to 12.7 mm for an adult human being and moves like a rigid body.

iii. *Diploe*: A highly vesicular layer separates the outer and the inner layers of eight solid bones forming the skull. It is a porous layer resembling a honeycomb, and the fluid substances filling the cavities are almost Newtonian viscous.
iv. *Cerebrospinal fluid (CSF)*: The CSF acts as a cushion between the hard cranium and the soft and delicate brain matters. The fluid is pressed out when intracranial pressure tends to rise, and more fluid is retained when the pressure tends to fall. Thus, mechanical injury to the brain is prevented by evenly distributing the change of pressure, if any. It is clear, colorless, nearly Newtonian fluid with a specific gravity of 1.004–1.007 containing protein, glucose, and inorganic salts (NaCl, KCl, etc.).
v. *Brain*: The egg-shaped brain is somewhat like a gel, although not as stiff. Its principal components are water (78%); esters (12%); protein (8%); and the remaining part containing small quantity of inorganic salts, soluble organic substances, and carbohydrates. It is divided texturally into white and gray matter, and its density is slightly more than that of water. The weight of an adult human brain ranges from 1100 to 1700 gms, and its capacity is approximately 1200 cc. The length and transverse diameter are about 165 and 140 mm, respectively. Experimental work on a variety of brain specimens establishes that brain exhibits a viscoelastic behavior (damping property).

In most of the analytical studies on brain injury, the components of the cranium are modeled as follows:

The geometry of the cranium is either spherical or spheroidal in shape, and is rigid in nature—the material behavior of the scalp tissues is viscoelastic.
The diploe is a porous layer resembling a honeycomb—the perforated portions are filled with some fluids modeled as Newtonian and viscous.
The CSF is a Newtonian fluid with a specific gravity nearly equal to water.
The egg-shaped brain tissues are viscoelastic.

Engin (1969) considered a mathematical model of the human head as a thin homogeneous elastic spherical shell filled with an inviscid compressible fluid. The dissipative (scattered and disintegrated) properties of the brain and the skull tissues are established following the outcomes of the experimental works done by McElhaney et al. (1970) and Jamison et al. (1968). So the improved models for the human cranium were developed and studied to investigate the cause, severity, and remedies of brain injury. Nanda (1988) adopted an improved model of the human head by a fluid-filled spherical shell—the fluid was homogeneous, compressible, and inviscid. An axisymmetric but local and radial impulsive force was applied to the system. Using integral transform method, the elastic behavior is studied and Bland's (1960) "correspondence principle" is applied to study the viscoelastic nature of the brain. The analytical expressions of normal stress components for axisymmetric torsionless motion and excess pressure of the fluid are obtained. Finally, using some numerical data representing relevant material constants and parameters involved in the study, the following pieces of information were obtained and exhibited through graphs:

i. Variation of radial skull displacement vs. time.
ii. The normal stress along the thickness of the shell at the time when the displacement along the radius attains the maximum value.
iii. The excess pressure of the fluid vs. radial distance. The severity of the brain damage in the vicinity was finally predicted.

It is now established that the medium and severe brain damage leading to cerebral concussion may occur due to the rotation of the head without any direct impact with rigid body or projectile. As no external injury can be detected so this is a problem of concern for the medical professionals. The cerebral concussion resulting from the angular acceleration was estimated theoretically by Misra and Roy (1988) after developing an appropriate finite difference technique. They paid due attention to the eccentricity of the skull by treating a fluid-filled prolate spheroidal shell as a model of the head. Misra and Chakraborty (1982) presented an analytical study on unrestricted vibration of the cranium paying due consideration on the skull thickness and the linearly viscoelastic behavior of the brain case as well as the gray matter (brain). The effect of an input rotational acceleration on the concussion level was analytically presented. In the analysis, a mathematical model was developed where the skull is treated as a rigid sphere and the physical nature of the CSF was Newtonian, inviscid but compressible possessing irrotational motion. The viscoelastic property of the brain was acknowledged in the analysis. The governing equations were written in the spherical coordinate system, and through the use of relevant boundary conditions, the solution of the problem was obtained in the elastic medium. Bland's "correspondence principle" was applied to get the viscoelastic solution for shearing strain components produced in the CSF present within the skull–brain system. It was established that the magnitude and the duration of the applied angular acceleration play a remarkable role in the prediction of the possibility of injury to the brain.

Nanda and Basu Mallik (2011) presented a model of the cranium where the porous layer (diploe) in the skull–brain system was taken into consideration. It was observed that consideration of porosity of the skull in the skull–brain system significantly affects the results.

Nanda et al. (2016) presented an improved model of the human head to estimate the severity of injury of the human brain when the cranium is subjected to a radial impulsive axisymmetric load. The analytical presentation was based on the response due to direct impact with a swiftly projected light body so that the effects due to rotation, if any, were negligible. The model adopted for head was a thin homogeneous viscoelastic hollow container (shell) filled with a homogeneous compressible fluid. The thin shell theory and Laplace transform technique were applied to have the analytical solution. Finally, radial skull displacement vs. time, normal stress distribution in different locations of the brain, and excess pressure generated in the skull–brain system were presented through graphs.

The investigation ultimately led to some remarkable conclusions:

The consideration of viscoelastic behavior of the skull tissues significantly affects the stress distribution in the vibrating skull–brain system.
Although the analysis was based on then shell theory, the obtained information may open a new era in the construction and study of more sophisticated models for human skull–brain system.
The results may predict the vicinity of severe brain trauma without surgical measures where improved imaging facility is not available.
The time zone at which the severe brain damage occurred may be predicted.

On the basis of the information, some initial remedial measures may be taken in a humble way so that the victims can be provided with some psychological assistance.

Nanda et al. (2017a) studied mathematically the result of acceleration produced due to the rotation of the head on the outset of cerebral attack where the brain case was modeled as a sphere. The skull was assumed to be a rigid spherical shell filled with an inviscid and compressible fluid (representing the CSF) and the gray matter (brain tissue) as linearly viscoelastic. The theoretical threshold of cerebral concussion was estimated. These incidents are regular for vehicular accidents and soccer victims in the football ground. For quantifying the analytical results obtained for shear strain and shear stress components in the brain, the obtained numerical values of physical and parametric quantities controlling the system were collected from various agencies working in the field. The computational works were performed through the development of the computer codes, and the obtained output was plotted using MATLAB® 8.5.

A number of pieces of relevant information were obtained from the analytical study.

An analytical model of the human cranium may analyze and present the variation of results when the key parameters controlling the system are varied in some specified range.
More refined results may be obtained if damping behavior of the brain tissues is given due consideration.
Results obtained through the study may be used in the design and construction of the protective instruments such as air cushion bags and helmets.

Human specimens obtained from surgical sources or hospitals are limited so a controlled database of information on head impacts should be collected from the high-risk zone areas where such tragedies occur frequently. In the USA model soccer competitions, the players covered their head with prescribed type of helmets, still head-to-head (or ground, or ball) collisions occur—some of them are so painful and serious that the victims require immediate hospitalization.

Basu Mallik and Nanda (2011) presented an analytical study on the effective use of American football helmets in mild traumatic brain injury (MTBI). The injury was instigated by accelerating the cranium. The effects on the head (in particular, brain) were investigated on the victims of the crash through mathematical modeling. The study explored a number of interesting conclusions.

As linear and rotational accelerations developed in the impact were paid due attention, so the outcomes of the investigation were more acceptable.

An important parameter involved in the reconstruction technique was "relative velocity" of the head with respect to impulsive load, and there might be an error in its measurement. In spite of the shortcoming, the method might add some inputs in the construction of protective gears for the heads of the soccer players or victims of the vehicular accidents.

These structures of protective gears are capable of absorbing the energy emitted and transmitted during collision or rotation of the head, and thus, serious damage of the sensory organs may be prevented. But the degree of impulse or acceleration transmitted to the hard exterior shell of the human head should not exceed the limit of fatal injury. It is expected that severe brain injuries due to the collision or rotation of the head in vehicular accidents or soccer games may be eliminated through further investigations involving more real-life data and parameters.

3.3 MATHEMATICAL MODELING FOR BLOOD FLOW IN HUMAN ARTERY/VEIN

After World War II, it was a common observation that craniocerebral trauma is the single major cause of death due to accidents in road or in fast games. Now researchers working in the domain of human anatomy, physiology, hematology, and pathology reported that arterial diseases based on the flow behavior of blood and mechanical disorder of the blood vessel walls contribute to more causalities in the advanced world. So more and more investigations (experimental and analytical) may be done to study the effect of high concentration of stresses in the arteries/veins that may lead to cardiovascular failure. Uncontrolled food habit (consumption of excess fatty, oily, and spicy food items), indifference and careless of physical exercise, and excess mental tension lead to the initiation and growth of the disease. The majority of the victims of the disease belong to the people in the economically creamy layer of the society. The number of victims of the disease is growing so fast that it seems to be a challenge to the human civilization.

The uncommon deposits of fats, cholesterol, and plaques in the inner lining of the narrow arteries/veins is technically defined as atherosclerosis (in the medical term "stenosis"). As a result, the arteries/veins become hard, stiff, and narrow, leading to the loss of their elastic property. Consequently, the usual flow of blood is restricted and sometimes blocked, leading to irregular and insufficient supply of blood to brain, heart, and other organs. The hematological disease "carotid circulatory disorders" is due to the development of constriction of blood flow in the artery. It plays a significant role in the occurrence of many severe cardiovascular diseases such as ischemia, stroke, and heart attack. The soft target of the disease are the older people. The exact reason behind the occurrence of the fatal disease is ambiguous, and the opinion of the investigators varies to some extent, but the rheological and physical nature of the blood together with the geometry of the stenosis might significantly contribute to the genesis and development of the hematological diseases. If the problem is neglected at the initial stage of attack and allowed to proceed to an advanced level, then the outcome might be an insufficient flow of blood to brain, heart, and other organs or rupture of the artery due to high concentration of pressure. It would be high risk for survival leading to morbidity or even death.

In accordance with the logic given for developing analytical models to study head injury problems, any investigations on blood flow irregularities in stenotic region, the rheological and physical nature of the blood together with the geometry of the stenosis were modeled mathematically by the researchers. The investigations on the circulation of blood in the human arterial chain made by the hematologists agreed that malfunctioning of the system is due to the change in dynamic nature of blood, pressure distribution, and the flow resistance (impedance) occurring as a consequence. So sophisticated and analytical models were devised on the characteristics of blood (Newtonian or non-Newtonian) and the stenotic geometry to explore modern approach on the cause and remedy of the hematological diseases (Figure 3.4).

A wide variety of analytical as well as experimental studies on streaming of blood through the branches of the arterial chain having a single or multiple stenosis were carried out by several investigators applying different blood models (Newtonian or non-Newtonian) and various geometry of the stenosis (polynomial, smooth cosine, or exponential). Results obtained from the experiments on blood confirmed that it is predominantly a suspension of erythrocytes (red cells) in plasma and may be better represented by non-Newtonian model when flowing through constricted arteries at weak shear rate, especially in the morbid condition. A good number of mathematicians, physicists, and medical professionals have contributed their experimental and theoretical research works on cause and remedies of restriction of blood flow and formation of undesirable growth in the constricted artery of the human body.

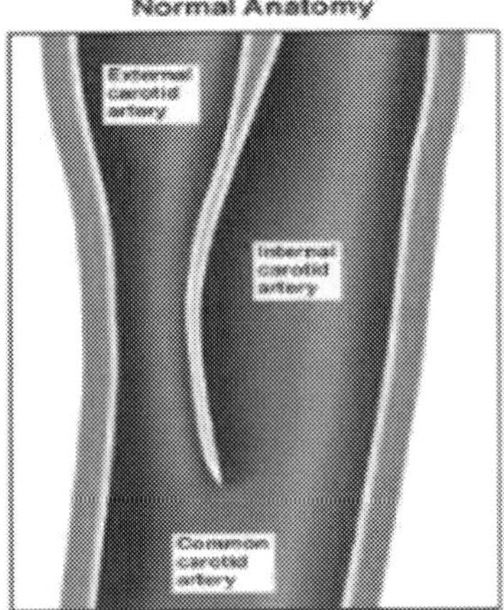

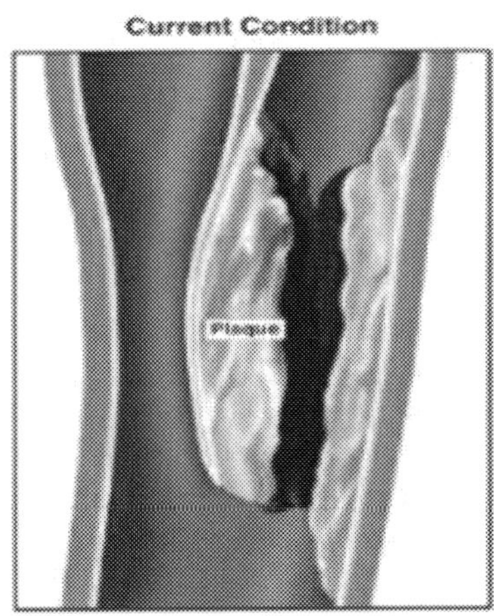

FIGURE 3.4 Presence of stenosis in the inner wall of the artery.

Their contributions as well as some experimental studies may be of help for the diagnosis and treatment of some of the related diseases. It is confirmed from the experimental works that blood is a suspension of red blood corpuscles (RBCs) in the plasma. It cannot be properly exhibited by Newtonian model where the fluid is considered homogeneous, single phase, and viscous, but may be described by non-Newtonian fluid model. Also, the growths within the inner wall of the artery may not be single, but in most of the cases, they are multiple or sometimes overlapped. So due attention should be given to the investigation where blood is represented by non-Newtonian fluid model and the geometry of the stenosis is complex.

A rigorous study of the previous literature on hematological problems due to uncommon growth within the artery and consequently irregularity in blood transmission or sometimes total obstruction led to constructing some fluid models for blood and standard geometrical models for the growth.

Most of the theoretical investigations are on discontinuity of the flow of blood in the narrow arteries constricted due to some abnormal growth in its inner lining and its consequences are as follows:

i. The non-Newtonian fluid model (Herschel–Bulkley (H-B), Casson, two-phase, Bingham plastic, power law) is accepted. Initially, the Newtonian fluid model was considered to represent blood.
ii. The geometries of the stenosis manifested in the artery were polynomial, exponential, or smooth cosine.
iii. The slip (velocity discontinuity at the wall of the artery) or no slip boundary conditions were imposed.

Some of the models of human artery/vein considered in the analysis of hematological problems are shown in Figures 3.5–3.9.

Many experimental and theoretical investigations were performed by several scholars to estimate the influence of the narrowing of the artery on the streaming behavior of blood applying different fluid models for blood and geometry of the stenotic growth. It is accepted that blood may be fairly closely represented by H-B fluid model—a non-Newtonian blood model at considerably low shear rates when flowing across a long hollow cylinder of diameter of 0.095 mm or less.

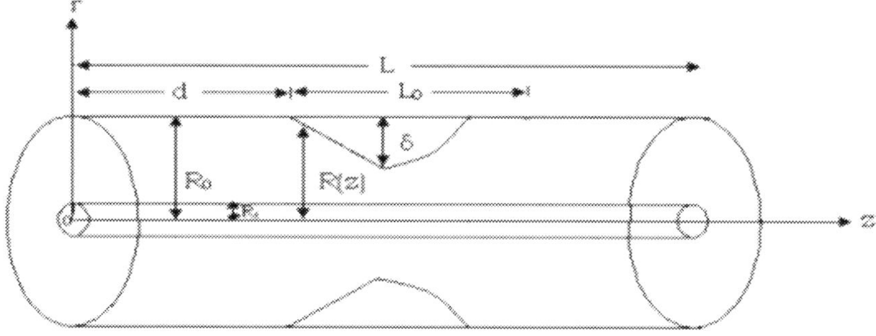

FIGURE 3.5 Geometry of a composite stenosis in a catheterized arterial segment.

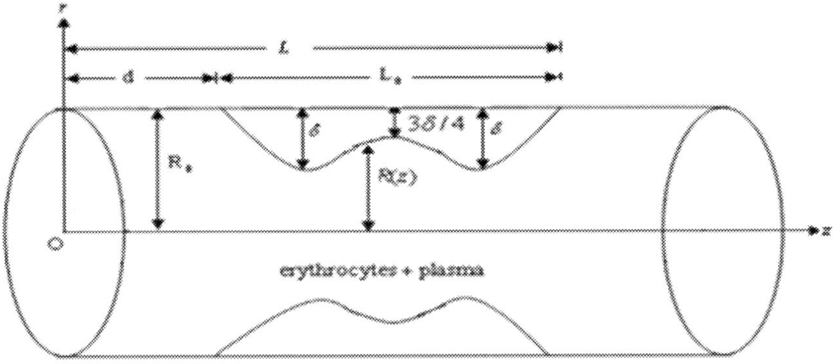

FIGURE 3.6 Geometry of an overlapping growth.

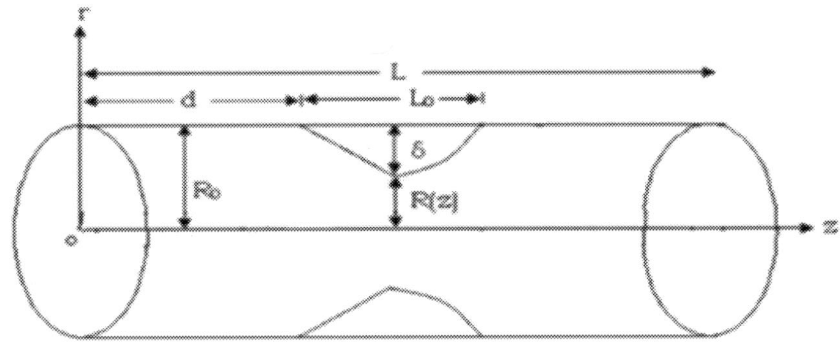

FIGURE 3.7 Geometry of a composite stenosis in an artery.

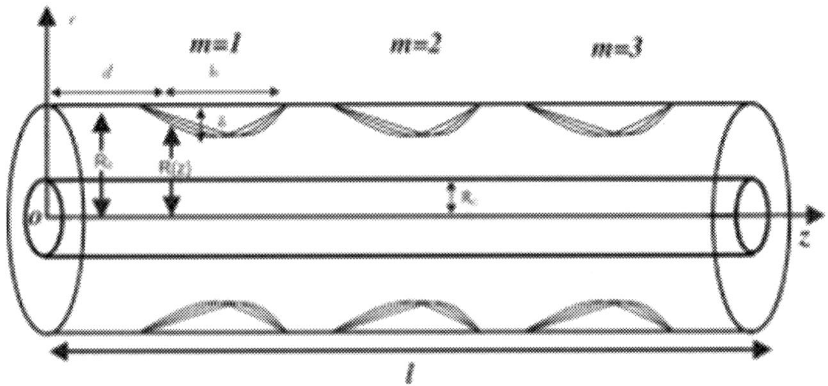

FIGURE 3.8 Geometry of axially symmetric but radially nonsymmetric multiple stenosis.

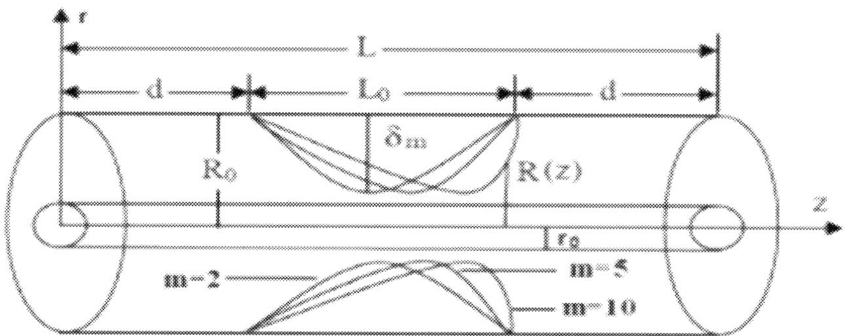

FIGURE 3.9 Geometry of the stenosed arterial segment.

A mathematical model was developed by Nanda and Bose (2012) for studying the blood flow through a narrow artery constricted by time-dependent multiple stenoses. The H-B fluid model of the blood was taken to consider the presence of suspended particles in the plasma. Both analytical and numerical techniques were exploited to get the solution of the problem. From the investigation, the results for the Newtonian model of the blood were also presented by the adjustment of the rheological parameters. The analysis provided a foresight for predicting which of the parameters have the dominating role on different flow characteristics—the skin friction, the impedance, and the volumetric flow rate at various locations of the morbid formation in the narrow artery.

Nanda and Basu Mallik (2012) developed a model of the human arterial system to study the characteristics of asymmetric and overlapping stenoses on the flow behavior of the streaming blood. In the analysis, the blood was modeled as a macroscopic two-phase fluid, while the abnormal growth manifested within the artery was represented by a polynomial model. The most important rheological parameter included in the study was hematocrit—the numerical measure of oxygen-carrying capacity of the blood—and the specified range for adult males is 41%–55% and for adult females is 37%–47%. The change in the resistance to flow with the variation of hematocrit (in its normal range) was presented through graphs.

More and more sophisticated catheters were developed and used in removing undesirable growths from the inner layer of the artery. Basu Mallik and Nanda (2012) presented an analytical model for examining the effect of axisymmetric flow of blood through a catheterized artery with multiple stenoses. In this model, the Newtonian fluid model of blood and the smooth cosine model of the growth were accepted. The length of the artery was taken large compared to its diameter so that special wall effects, if any, might be ignored. The motivation behind the modeling was to estimate the increased impedance and shear stress in a diseased artery during artery catheterization. It was noted that the catheter radius and height of the growth are the strong parameters affecting the flow qualitatively and quantitatively. The asymptotic nature of the impedance for different stenosis heights was an important observation of the investigation.

The analysis led to the following observations:

Size of the catheter should be cautiously chosen keeping in view of the stenosis height during medical treatment.
The results might have applications in detecting the range of the unusual growth, the critical location, and the severity of the ailment.

The model developed might have applications in the nonsurgical treatment of the obstruction on the movement of fluid due to the formation of undesirable growth in the arterial chain.

Basu Mallik et al. (2013) presented an analytical model for the transmission of streaming blood in an atherosclerotic human artery using the power law fluid model of blood and the smooth cosine model for the stenosis geometry. Following the outcome of some experimental observations, the presence of velocity discontinuity (slip) in the flow boundaries or in their immediate neighborhood was incorporated in some research. In the analysis, slip condition was also incorporated (which was neglected in many of the previous investigations) in the constricted wall of the human arteries/veins. The smooth cosine model for the stenosis geometry was incorporated in the analysis. Extensive qualitative and quantitative analyses were performed on the dominating factors responsible for the obstruction of blood transmission, and the consequences and the estimated results were exhibited through some graphs. As a special case, the corresponding results for a simplified model of blood—the Newtonian fluid model—were also presented in the study. As the pressure of the arterial segment might influence the hemodynamic behavior of blood, so the output had applications in reducing flow disorders in the human arterial system.

The investigation threw light in bringing out some interesting phenomena.

Physical parameters—length and height of the stenosis and flow variables—and apparent viscosity of blood and flow behavior index influenced the variation in resistance to flow substantially in both qualitative and quantitative measures.
The inclusion of the discontinuity of velocity condition at the constricted inner wall of the human arterial system and the viscosity parameter contributed a lot on the estimation of flow rate.
The investigation might present an opportunity for estimating the influence of flow variables and physical factors in different conditions of diseases related to blood flow.
The results obtained in the investigation might assist the medical practitioners in finding the rough location and the intensity of the ailment without any imaging.

The magnetic characteristics of blood are now justified due to the presence of hemoglobin (an iron compound exhibiting magnetic property) in the nonmagnetic plasma. The hematologists and chemists agree with the statement. The conductive flow induces voltage and current leading to a change in the streaming behavior of blood. Also, it was observed that the flow of blood in arteries may be controlled with the help of magnets. Nanda et al. (2013) proposed an analytical model to estimate the influence of stenosis on the streaming behavior of blood through a narrow contracted artery under the action of magnetic field. As the blood was flowing through narrow tubes, the Casson fluid model of blood was accepted in the analysis. The aggregates of proteins, fibrinogen, and globulin in an aqueous base plasma form a chain-like structure—known as *rouleaux*, which conducts as a plastic substance in the magnetic field. Under the physical condition, the non-Newtonian nature of the blood may be better represented by the Casson fluid model.

The study illuminates the salient points relevant to blood flow through the stenosed arteries.

In the investigation, the transmission of blood along the constricted (due to growth) artery was governed by the equation, including the magnetic intensity factor. So the outcome might add new inputs in the detection and treatment of the fatal disease.

MHD (magnetohydrodynamics) principles might have an application in reducing the excess flow of blood, and thus, rupture of the artery could be prevented.

Kumar (2015) presented a mathematical model to analyze the characteristics due to the inclusion of the elastic behavior of the arterial wall on the longitudinal velocity of blood considering the power law and H-B model of the blood. The flow was assumed axisymmetric and laminar and the arterial wall as elastic. The flow behaviors for H-B and power law fluid models of the blood were presented and compared for a standard measurement of the power index parameter ($n = 0.75$).

Nanda et al. (2017c) presented a mathematical model for estimating blood flow attributes transmitting along an elastic artery. The inclusion of the velocity discontinuity condition at the constricted inner wall of the human arterial system was given due weightage. Blood was represented by the power law fluid model, and the smooth cosine model for the stenosis shape was accepted in the analysis.

The theoretical analysis led to some relevant conclusions.

The consideration of elastic property of artery had a significant contribution in the flow behavior.

Rheological parameters such as fluid index parameter and consideration of slip velocity considerably altered the blood flow pattern.

Results obtained through the theoretical study (developing mathematical models for blood and stenosis) might have applications in the clinical treatment of the hematological patients.

Nanda et al. (2017b) developed an analytical model of the arterial segment to estimate the influence of multiple growth on transmitting characteristics of the blood along the narrow and constricted (diseased) artery. It was now established that stenosis could develop in series and might overlap. The nearly rigidity property of the vessel wall and Bingham plastic viscous fluid model representing blood were paid due consideration in the study. The geometry of the overlapped growth was expressed mathematically by an appropriate polynomial expression. An extensive quantitative analysis was performed through numerical computations on flow resistance and wall shear stress. Their variations with the change in the values of the key parameters in the admissible range were presented graphically.

A number of conclusions relevant to the health sector could be drawn from the output of the analysis of the mathematical model.

An analytical model might present an improved perception of the key material and physiological parameters that regulated the system.

The consideration of two-layered solid–fluid structure provided confidence in the analysis.

The output of the analytical investigation might have an initial support to the physicians in the diagnosis and treatment of the hematological problems.

3.4 CONCLUSION

Many mathematical models have already been developed to study different features of the problems of brain injury and blood flow in real life. Results obtained through these investigations would be utilized in the initial clinical treatment of the victims of the remote areas where modern medical facilities and improved imaging centers are not readily available. The inclusion of new physical and physiological parameters might lead to the development of more sophisticated mathematical models to recast the fatal problems more sincerely and to overcome the limitations attributed to the existing models. It seems to be a compulsion in the health sector to provide necessary support for building infrastructure for the theoretical research based on mathematical modeling.

It is expected that medical professionals will extend their valuable support by sharing data of the victims with the mathematicians and engineers so that the results might be more realistic and the output might benefit both the academicians and the medical professionals. The contribution of the investigators will be fruitful if the sufferers of the disease are provided with some initial support and the medical practitioners may get some preliminary idea about the fatal diseases.

REFERENCES

Basu Mallik, B. and Nanda, S. (2011). Effectiveness of American Football Helmets in Mild Traumatic Brain Injury. *IEMCON 2011* (5th–6th January, 2011)

Bland, D. R. (1960). *The Theory of Linear Viscosity*. Pergamon Press, Oxford.

Engin, A. E. and Liu, Y. K. (1970). Axisymmetric response of a fluid filled spherical shell in free vibrations. *J. Biomech.*, **3**, 11–22.

Jamison, C.B et al. (1968). Viscoelastic properties of soft tissues by discrete model characterization. *J. Biomech.*, **1**, 35–46.

Kumar S. (2015). Study of blood flow using Power law and HB non Newtonian fluid model through elastic artery. *International Conference on Frontiers in Mathematics* 2015, March 26–28, 2015 in Guwahati University. (Proceedings of ICFM 229–234).

McElhaney, J.H. et al. (1970). Mechanical properties of cranial bone. *J. Biomech.*, **3**, 495–512.

Basu Mallik, B. and Nanda, S.P. (2012). A math. Model of blood flow in a catherized artery with multiple stenoses. *IEM Int. J. Mgt. Tech.*, **2** (2), 73–77.

Basu Mallik, B. et al. (2013). A non-Newtonian fluid model for blood flow using Power law through an atherosclerotic arterial segment having slip velocity. *Int. J. Pharma, Chem Bio. Sci.*, **3** (3), 752–760.

Misra, J.S. and Chakraborty, S. (1982). A free-vibration analysis for the human cranial system. *J. Biomech.*, **15**, 635–645.

Misra, J.C. and Roy S. (1988) *Comp. Math. Appl. (UK)*, **16**, 247–266.

Nanda, S. and Bose R.K. (2012). A math. Model for blood flow through a narrow artery with multiple stenoses. *J. Appl. Math. Fluid Mech.*, **4** (3), 233–242.

Nanda, S.P. and Basu Mallik, B. (2011). Effect of an Axi-symmetric external load on the stress-field generated in a poroelastic spherical shell: A mathematical Model for head. *Int. J. Civil Eng*, **1** (1), 7–16.

Nanda, S.P. and Basu Mallik, B. (2012). A non-Newtonian two phase model for blood flow through arteries under stenotic condition. *Int. J. Pharma Bio Sci.*, **2** (2), 237–247.

Nanda, S.P. et al. (2013). Effect of magnetic field on flow behaviour of blood through a modeled atherosclerotic artery. *International Conference in Technology and Management Innovation in Computer and Communications in industry and academia, Conference Paper held in Kolkata on* 23 August, 2013.

Nanda, S. et al. (2016). An analytical study on the dynamic response of a fluid-filled spherical shell subjected to a local radial and axi-symmetric load. *Am. J. of Electron. Commun.*, **III** (1), 8–14.

Nanda, S. et al. (2017a). An analytical study on brain injury due to an input rotational acceleration on skull-brain system of the human head. *J. Chem Bio. Phy. Sci., Sec C*, **7** (1), 39–46.

Nanda, S. et al. (2017b). A study on Bingham plastic characteristics of blood flow through multiple overlapped stenosed arteries. *Saudi J. Eng. Technol*, **2** (9), 349–357.

Nanda, S. et al. (2017c). Study on the effect of non Newtonian nature of blood flowing through an elastic artery in slip condition, *J. Chem Bio. Phy. Sci., Sec C*, **7** (4), 934–942.

Yadav, S.S, and Kumar K. (2012). Bingham plastic characteristic of blood flow through a generalized artery with multiple stenoses. *Adv. Appl. Sci. Res.*, **3** (6), 3551–3557.

4 Machine-Learned Regression Assessment of the HIV Epidemiological Development in Asian Region

Rashmi Bhardwaj and Aashima Bangia
GGS Indraprastha University

CONTENTS

4.1 INTRODUCTION

HIV pervasiveness and incidence estimation in and around the developing countries, majorly in sub-Africa, have been demonstrated with statistically simulated models based first and foremost on the sentinel surveys consisting of pregnant women household ones. On the whole, a variation in this epidemiology depicts lesser innovative taints with a declined mortality for Saharan region of Africa. In 2000–2012 statistics, HIV in these African regions decreased to 50% in 2000 in comparison with 1 million lesser newly found HIV contaminations in 2012. Simultaneous increase in the estimation is via 20.8 million in 2000 towards around 20 million plus in 2012. This happened because of the largely improved survival rate through anti-retroviral therapy (ART). AIDS causing deaths dwindled from nearly 1.4 million in 2000 towards 1.2 million for the year 2012. Evidence suggested that the accessibility towards ART reduced the mortality rates plus contributed towards lower infected ones, resulting in slowly increasing rates in most regions, although there is an exemption of Angola, where newer contagions and deaths continued to rise (Figure 4.1).

Gene of transmission causative to the HIV epidemic in sub-Saharan Africa is unprotected intercourse. Risk of infection upsurges when there are multiple partners and consecutively sexually transmitted taint. Large magnitudes of new cases of infections would be accountable towards the long-term heterosexual relations. Amidst sub-Saharan couples, mostly at least one person is surely HIV positive, and at least two-thirds test immunized to be in the discordant relations. However, in Rwanda and Zambia, around 95% of newly found contagions are prevalent in individuals living with their partners. The extent to which new infecteds introduced in the long-term relationships with their other partners is still not known. Figure 4.2 clearly outlines the various possible transfer forms of the epidemiology through various agents.

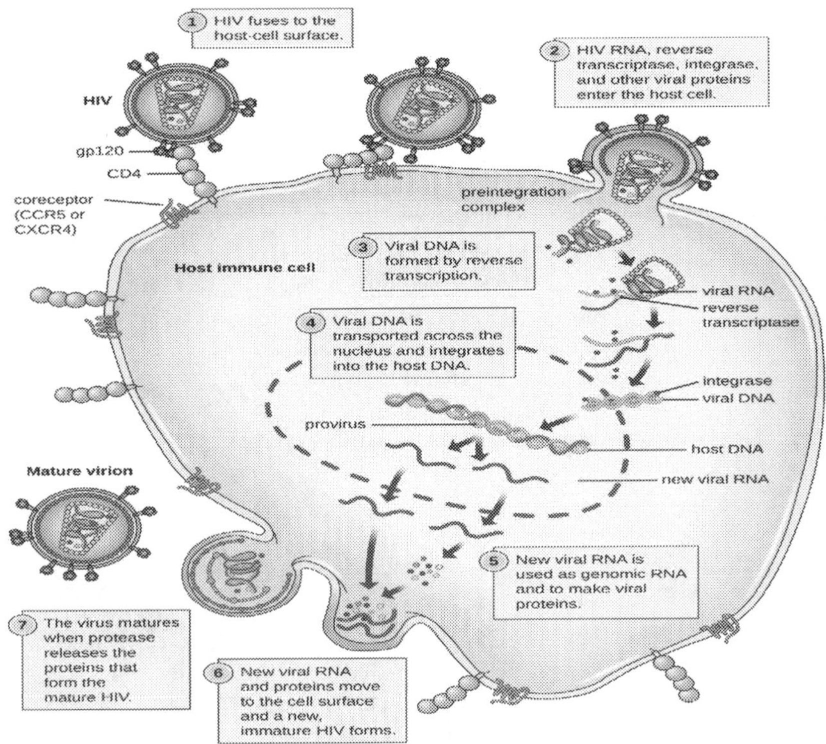

FIGURE 4.1 Detailed HIV virion changing the RNA with reverse transcriptase and other proteins to enter the host cell.

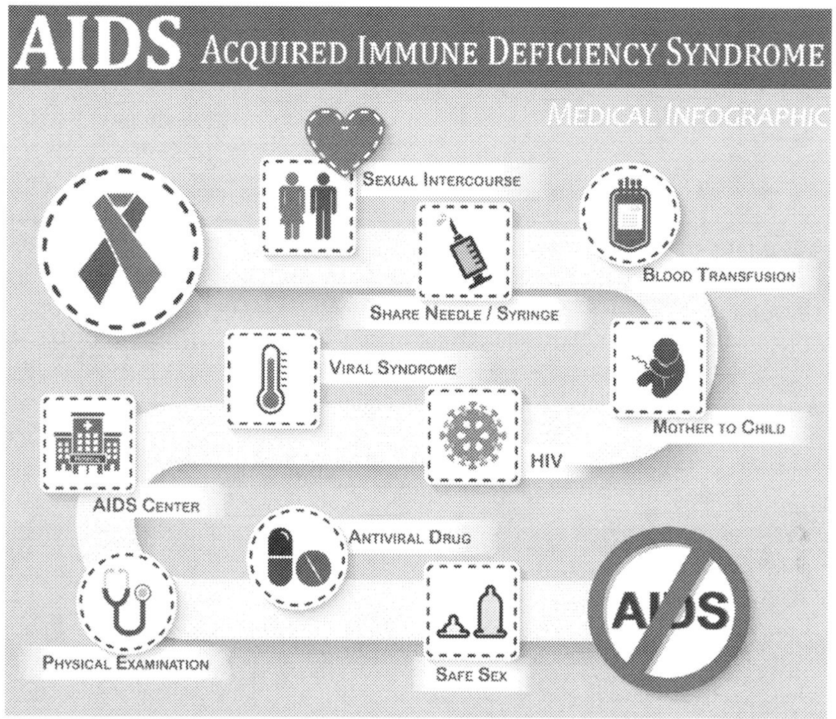

FIGURE 4.2 HIV-AIDS transmission, and causes and prevention.

Asia presently stands the second-largest HIV-infected region after Africa. As the trend in sub-Saharan region, occurrence is deteriorating in Asian region. Although, these regions involve necessary epidemiological-variances, essentially gets strenuous around significant populations. Even with bargained rates of HIV communication, numbers have increased significantly to assessed 4.8 million in 2012 from 3.8 million in 2000, chiefly for the reason that of survival. AIDS-related deaths in Asiatic regions have been gradually deteriorated during 2000s through roughly 26 million in year-2012 from that of 330 thousand in 2005. Though the accessibility towards antiretroviral improved, prevention towards mother-to-child transmission services is restricted, with <20% of pregnant women getting ART. Also, <60% of suitable public received ART in Asia in 2012. China and India have maximum incidences in Asia. The major way of transmission in China is heterosexual transmission; around 46.5% for assessed 78 million HIV-positive cases occur through heterosexual transmission. Injected drug utilizations are vital contributing factors in China and stand for nearly 28.4% infecteds. A subepidemic exclusive towards China happened among the former plasma donors, spread via tainted paraphernalia in the 1990s. Currently, India has the largest amount of infecteds, particularly in Asia (2.1 million infecteds). Overall expected prevalency of 0.3% in the age of 15–49 lesser than some countries in this asiatic areas, such as Cambodia with 0.8%, Myanmar around 0.6%, Vietnam with 0.4% and Malaysia having 0.4%. Laos counting on 0.3%), Nepal having 0.3%,

Philippines less than 0.1%, and Pakistan <0.1% (which has lower estimated prevalence rate as compared to India).

Within India, occurrence differs significantly: Areas infected with HIV prevalence are nearly four to five times as high in the southern part as in northern states. Mostly, HIV pervasiveness exists around sex workers in India, which seems to be dispersing from them towards their clients further to clients' other partners. Studies in India indicate that 50%–60% of clients of workers either married or had more than one relations. Exemption to this generality is in northeastern region wherever injected drug use (IDU) is the chief contributing factor.

One of the researches via a wide set of aims that consist of heart, lung, and blood disorders, and sleep morbidities plus various conditions is known to transpire together with HIV disease, including mental health issues and neurological illnesses. Their study strives to typify occurrence along with forecasters for the conditions with a sufficient statistical analysis. Broad inclusive criteria of re-enrollment might result in the recruitment of various population containing transgender people and drug injectors whose epidemiology may be dissimilar from the majority. It is important to spread awareness about HIV among these populations as the overall statistical strength of the cohort may perhaps get reduced when less people with the characteristics get enrolled.

Dramatic improvement to measure infection is the main sign of response to treatment, transmission risk, and risk for the progression. Existing ART schedules suppress HIV/AIDS, longevity of viral pools sustains, and thus, lifelong rehabilitation is a requisite to dodge its reactivation. The capability to precisely measure HIV viral load is critical towards tracking right size of viral reservoir and to assess plans towards HIV remission. Latest developments endorsed an enhanced measurement with accuracy of the viral reservoir that can identify single RNA/DNA molecules as well as nanotechnology to analyze one provirus at a time.

Bhardwaj and Bangia [1–6, 8, 9] discussed various methodologies for nonlinear analysis of the real-life case studies that involve atmospheric pollutants, meditating body, fuzzified PID controller, stock market trends, statistical HIV analysis etc. Bhardwaj [7] analyzed nonlinearities in environmental pollution. Bhardwaj and Datta [10] explored the consensus algo for the blockchains. Bangia *et al* [11] estimated the river quality with the help of wavelet conjuncted with AI. The global report [12] by UNAIDS stated all the facts and figures about the epidemic spread. Fettig et al. [13] described in detail the global epidemiology of HIV. Stengel [14] studied the mutation and control of immunodeficiency virus. Gao [15] analyzed computationally impulsive antiretroviral drug doses. Renshaw [16] modeled biological population in space and time. Samanta [17] described the extinction of nonautonomous HIV/AIDS model with time delay. Nosova and Romanyukha [18] discussed the mathematical model of infection transmission and dynamics of risk groups. Perelson and Nelson [19] analyzed the HIV-1 in vivo. Li and Wang [20] applied backward bifurcation on the infection with ART. Mhawej [21] controlled the infection and drug dosage. Shao [22] in his book explained in detail the pathogenesis. Wang [23] gave a delay-dependent model with HIV drug resistance. Elaiw [24] discussed a class of HIV with its properties. Narendra and Alan [25] also studied the dynamics of target cells. Leenheer and Smith [26] analyzed globally the virus dynamics. Perelson *et al*

[27] described the HIV-1 dynamics in vivo. Wodarz and Levy [28] studied the effects of different modes of viral spread on the dynamics of infected cells. Sharomia and Gumel [29] analyzed a multi-strain model of HIV with ART. Mishra [30] modified Chua attractor through differential operators.

None of the authors have studied machine-learned regression analysis of HIV model in the blood cells. In this paper, we studied the impact of time on the amount of uninfected cells, virus particles in the blood cells, and infected cells. Also, phase dynamics, intelligent regression procedures with their goodness of fit (GoF), and uncertainty of prediction towards the behavior of these cells are observed.

4.2 MATHEMATICAL DEVELOPMENT OF THE HIV EPIDEMIOLOGY

The mathematical model for the following three important components has been developed:

i. T defines the amount of uninfected blood cells.
ii. V defines the amount of virus particles in the blood cells.
iii. I defines the amount of infected blood cells.

The model helps to study the impact of HIV epidemic on these factors simultaneously with time and also to study how the change in one component influences the changes in other with respect to time. Some of the lemmas to transform the biological condition into a mathematical study are as follows:

Lemma 4.1: Contact rate is a time-varying function.
Lemma 4.2: Reverse Transcriptase Inhibitors (RTI) effect: Time-varying parameter, $0<\gamma<1$.
Lemma 4.3: Protease Inhibitors (PI) effect: Time-varying parameter, $0<\eta<1$.
Lemma 4.4: Transmission coefficient: Time-varying parameter, β.

All variable parameters T, V, and I are the positive constants as they can never stand negative. A mathematical development can be regulated based on the dynamism of HIV infecteds towards the study for spread with its treatment.

$$dT = (S - bT - (1-\gamma)\beta TV)dt \tag{4.1}$$

$$dV = ((1-\eta)NaI - cV)dt \tag{4.2}$$

$$dI = ((1-\gamma)\beta TV - aI)dt. \tag{4.3}$$

With the initial condition: $T(0)=T_0$, $V(0)=V_0$, $I(0)=I_0$,
where

T: amount of the uninfected target cells dT: rate for change of uninfected cells
S: constant source rate producing T cells b: cleared rate for the uninfected T cells.

Infection rate reduces from βTV to $(1-\gamma)\beta TV$.
V: amount of the virus particles in the blood cells
dV: rate of change of number of virus subdivisions produced by each infected cell
N: average number of the infective virus particles produced by the infected cell over their life span
a: cleared rate of infected cells
$(1-\eta)NaI$: virus subdivisions under the influence of PI drugs
c: rate at which viruses are cleared
I: amount of infected cells
dI: rate of change in infected cells.

4.2.1 Techniques Implemented for Analysis

4.2.1.1 Study of Phase Dynamics

For the dynamical systems, the phase diagram describes graphically the space where all of the conceivable positions of the structure can be characterized through every possible state corresponding to a unique point on phase diagram. Inside a phase space, all degrees of freedom (DOF), i.e., parameters of that particular system, are exemplified as the alignment of multidimensional space where one-dimensional system is referred as the phase line; two-dimensional space exists as phase plane. For each possible state of the system, there gets allowed a combination of values of system's considerations and the point is included in the multidimensional space. System's evolved phase with respect to time traces the pathway through the phase-space trajectory for that particular system via *high*-dimensional space.

4.2.1.2 Distribution Fitting

4.2.1.2.1 Beta-Distribution Fitting

Beta-distribution with four parameters indexed through two-shape parameters (P and Q), two parameters signifying minimum (A) and maximum (B). A and B are not estimated and are assumed to be the known parameters. Beta-density function is:

$$\beta_d(t \mid \alpha,\beta,\omega,\theta) = \frac{1}{B(\alpha,\beta)} \frac{(t-\omega)^{P-1}(\theta-t)^{Q-1}}{(\theta-\omega)^{P+Q-1}}, \alpha>0, \beta>0, \omega<t<\theta,$$

where

$$B(\alpha,\beta) = \frac{\Gamma(\alpha)\Gamma(\beta)}{\Gamma(\alpha+\beta)}$$

On transforming,

$$X = \frac{t-\omega}{\omega-\theta}$$

thus resulting in two-parameter beta-distribution. It is also known as the standardized form for beta-distribution whose density function is:

$$\beta_d(y \mid P,Q) = \frac{1}{B(P,Q)} y^{P-1}(1-y)^{Q-1}, P > 0, Q > 0, 0 < y < 1.$$

4.2.1.3 Goodness of Fit, Histogram, and Density Function

By grading GoF for various distributions, one can get impressions for whichever distribution is satisfactory and whichever is not. From cumulative distribution function (CDF), derive histogram and the probability density function (PDF).
Theorem: The measurement of discrepancies among observed and fitted values is regarded the deviation. For Poisson responses, deviances take this form:

$$D = 2\sum\left\{y_i \log\left(\frac{y_i}{\mu_i}\right) - (y_i - \mu_i)\right\}.$$

The first term is identical towards binomial deviance, demonstrating "twice a sum of observed times log of observed overfitted." The second term is the sum of differences between observed and fitted values, which is usually zero.

Lemma 4.1

For large sections, *distribution* of deviation is nearly chi-squared with *n-p* DOF, whenever *n* considers number of observations and *p* for number of parameters. Thus, deviance can be utilized directly towards testing GoF of this model.

4.2.1.4 Coefficient of Determination (R^2)

R-squared is the statistical measure of the closeness of data to look for fitted regression line. It is also known as the coefficient of determination and the coefficient of multidetermination for multiregressors. Description of *R*-squared is fairly simple; it is the percentage of retort variable having a variation, which is usually described through the regression.

R-squared lies between 0 and 1, where 0 indicates the model that describes none of the variability of this response data around the mean and 1 determines the prototype describing all of the variabilities around the response data of the mean.

4.2.1.2.2 Steps to Compute R-Squared

1. Take data points of observations of dependent and independent variables.
2. Find the line of best fit through the various regression models, whichever suit best.
3. Next, simulate the predicted values.
4. Now, subtract the actual observed values followed by squaring the outcomes.
5. Then, the summation process continues and equals the explained variation.
6. Further, to simulate the total variance, subtract the average value from the predicted ones.

7. Now, squaring the outcomes plus summation of the results.
8. Division of the first summation of explained variance by the simulated second sum of total variance.
9. Finally, subtract the so-called obtained values from one which is the R-squared value lying between 0 and 1.

$$R - \text{squared} = 1 - \left(\text{Explained variation / Total variation}\right)$$

4.2.1.5 Kolmogorov–Smirnov Test

The Kolmogorov–Smirnov (K-S) test is another GoF technique comparing the maximum distances among experimental and theoretical CDF.

Theorem: Maximum discrepancy between experimental and theoretical CDF is smaller than normally expected for the given sample, theoretical distribution, which is acceptable for modeling the underlying population considering a certain confidence level. H_0 *gets accepted at level-*α *that can be described in this equation:*

$$KS_{(d,d^{\tau})} < C(\alpha)\sqrt{\frac{d+d^{\tau}}{dd^{\tau}}},$$

where

$$KS_{(d,d^{\tau})} = \text{maximum discrepancy}$$

$$C(\alpha) = \text{critical value of } N \text{ sample size}$$

$$(\alpha) = \text{confidence level}$$

$$d = \text{number of elements on theoretical sample}$$

$$d^{\tau} = \text{number of elements on experimental sample.}$$

Lemma 4.2

Maximum discrepancy is defined by the following equation:

$$KS_{(d,d^{\tau})} = \sup_{x} \mid g_{\alpha,d}(x) - g_{\beta,d^{\tau}}(x),$$

where

$$g_{\alpha,d}(x) = \text{experimentally derived distribution function}$$

$$g_{\beta,d^{\tau}}(x) = \text{theoretically derived distribution function}$$

$$\sup = \text{supremum}$$

$$KS_{(d,d^{\tau})} = \text{maximum discrepancy.}$$

4.2.2 Analysis of Machine-Learned Regression Models

4.2.2.1 L-1 Norm Regression-Learned Model

This model approximates the median of the dependent variable and conditions for values of the independent variable. It can be compared to the least-squares model that involves estimating the mean of the dependent variable.

Theorem: L1 regression gives the regression plane that minimizes the sum of absolute residuals rather than the sum for the squared residuals. It performs the median regression by minimizing the sums of the absolute residuals.

Lemma 4.3

Extension of univariate-norm approximation for the conditional L1 functions via optimization of the piecewise linear objective function in the residuals. Plus it minimizes the sum of absolute residuals.

This was first familiarized by Boscovich around the eighteenth century. Asymptotic theory for L1-norm lapse relatively parallels the concept of univariate sample norms. Simulation for L1-norm lapse estimators can be verbalized as the linear programming problem can be further calculated via simplex/barrier methodologies. It has been used towards straight to estimate the models of upper medians for conditional distributions impartially more than deducing these relations for models on the basis of the conditional central tendency.

4.2.2.2 Logistic Regression-Learned Model

Logistic regression follows the path to derive multivariable composites to distinguish two or more groups. It does not include preventive statistical assumptions. It is used when the dependent/target variable is categorical. This regression provides a lot of benefits over statistical procedures. Interpretation of related importance for individual predictors stands upfront into this regression. Logistic regression function can also be used to simulate probability that individual belongs to one of these groups.

Logistic regression involves the procedures for investigating difficulties where there are more than one independent variable determining outcomes. In most of these studies, dependent variables are the dichotomous ones in which there appear only two conceivable outcomes.

Theorem: The regression searches for the best-fitting model for describing relations among dichotomous characteristics of dependent variable with a set of independent predictor variables. It produces coefficients plus the standard errors plus significance levels for the formulation towards the prediction of logit transformation of probability of presence of characteristics for interests:

$$\text{logit}(p) = b_0 + b_1X_1 + b_2X_2 + b_3X_3 + \cdots + b_iX_i,$$

where

p = probability of presence of characteristic of interest

bi = regression coefficient for X_i.

Lemma 4.4

The logit transformation is defined as the logged odds:

$$\text{Odds} = \frac{P}{1-P} = \frac{\text{Probability presence of characteristics}}{-\text{Probability absence of characteristics}}$$

$$\text{logit}(P) = Ln\left[\frac{P}{1-P}\right] = b_0 + b_i X_i,$$

where

$P =$ probability of an event occuring

$P/(1-\text{P}) =$ odds ratio

$\ln[P/(1-\text{P})] =$ logs odds ratio, i.e., logit.

An advantage of this regression is that it permits the examination of multi-explanatory variables through the addition of basic principles. The general equation is:

$$LR = \frac{1}{1+e^{-(\beta_0+\beta_1 X_1+\beta_2 X_2+\ldots.\beta_n X_n)}} = \frac{1}{1+e^{-(\beta_0+\sum \beta_i X_i)}},$$

where

$X_1, X_2, \ldots, X_n$ are the explanatory variables

$\beta_1, \beta_2, \ldots, \beta_n$ are the regression coefficients.

Mathematically, this regressor applies the maximum-likelihood estimation rather than the least-squares assessment procedure which can be used in the linear part. When this is applied for regression, it is a non-convex function for parameters. Gradient descent converges towards global minimum only when function is convex.

Logistic regression can be utilized towards classifying problems. Generally, logistic classifier can be customized for the linear combination for more than one feature value as argument of the sigmoidal function. Corresponding output for the sigmoidal function is the number 0 to 1. Middle value is deliberated as a threshold to establish whatever belongs towards class 1 plus towards class 0. Specifically, an input giving an output more than 0.5 can be taken belonging to class 1. Also, when outputs <0.5, the corresponding inputs are classified belonging towards class 0.

4.2.2.3 Poisson Regression-Learned Model

Poisson model is the statistical basis christened after French mathematician Siméon Denis Poisson. It is applied to model count variables. These models are best used for modeling occasions wherever outcomes are counts, particularly count data. Count to be articulated as the rate data, from number of times event occurs within the time frame to be denoted for raw count. It aids in analyzing both count and rate through allowing the determination of various explanatory variables (*X*-values) having an effect on the given response variable, *Y*-value.

Theorem: Let Poisson incidence rate μ can be expressed via the set of "k" regression variables. The expressions in relation to the magnitudes are:

$$\mu = t\exp\left(\beta_1 X_1 + \beta_2 X_2 + \cdots + \beta_k X_k\right),$$

where

$\mu -$ incidence rate

$$\beta_1 = \begin{cases} \text{intercept in the case,} & \text{when } X_1 \equiv 1 \\ \text{regression coefficients, otherwise} \end{cases}$$

$\beta_2,\ldots,\beta_k -$ regression coefficients having estimates as $b_1, b_2, \ldots, b_k$.

Theorem: Model probability of events y occurring within the precise time frame, considering y-incidences were unaffected through scheduling of previous occurrences, can be demonstrated mathematically as:

$$p_d\left(P = y \mid \mu\right) = \frac{\left(\mu_i t_i\right) y e^{-\mu_i t_i}}{y!}, \; y = 0,1,2\ldots,$$

where

$p_d(P = y \mid \mu) =$ probability of y events

$\mu =$ average number of times event may occur per unit of exposure

$$\mu_i = t_i\, \mu\left(x_i'\beta\right)$$
$$= t_i \exp\left(\beta_1 X_{1i} + \beta_2 X_{2i} + \cdots + \beta_k X_{ki}\right)$$

$y =$ events which can be failure, death, or existence.

Thus, for the provided set of values for regressor variables, the outputs follow the Poisson distribution.

4.2.3 Inferences of Multifaceted Epidemiological System

Using Eqs. (4.1)–(4.3), phase-space plot, time-series plot of each variable, and statistical characteristics have been computed. Also, machine-learned regressors have been plotted. Figure 4.3 depicts the phase space of the three variables *T*, *V*, and *I*. Distribution fitting via various distributions is simulated and can be concluded with the aid of *p*-values in Table 4.1. Table 4.2 gives the beta-distribution for four parameters for values, and standard errors for variables *T*, *V*, and *I*. Table 4.3 depicts the statistics of data values and parameters of three variables. Table 4.4 compares the K-S test outcomes for *T*, *V*, and *I*. Table 4.5 depicts the chi-squared test outcomes for

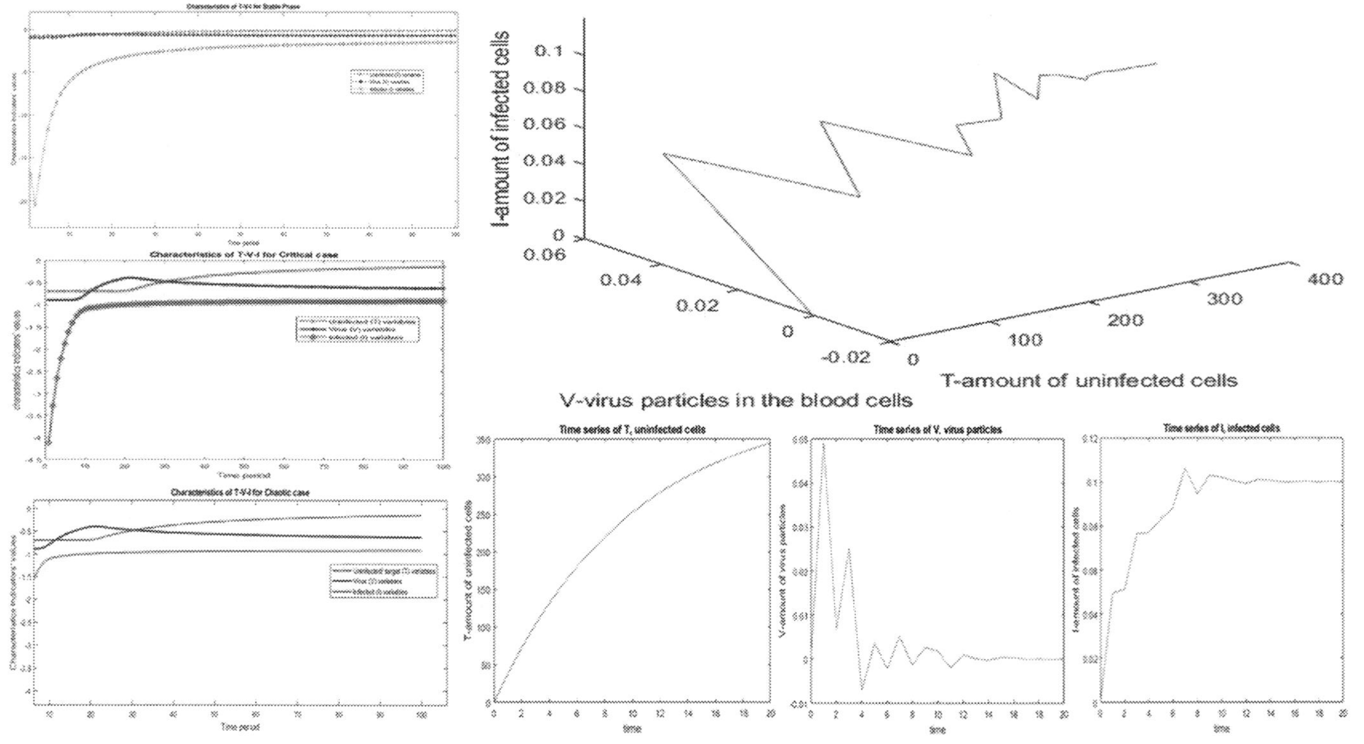

FIGURE 4.3 Phase dynamics and time series of the T, V, and I variables.

TABLE 4.1
Distribution Fitting Methods with Their *p*-Values

Distribution	*p*-Value
Beta4	**0.993**
Chi-square	0.000
Erlang	<0.0001
Exponential	0.081
Fisher–Tippett (1)	<0.0001
Fisher–Tippett (2)	0.045
Gamma (1)	<0.0001
Gamma (2)	0.416
Generalized Extreme Value (GEV)	0.016
Gumbel	<0.0001
Logistic	0.777
Normal	0.802
Normal (standard)	<0.0001
Student	<0.0001
Weibull (3)	0.955

TABLE 4.2
Beta-Distribution for Four Parameters for Values and Standard Errors for *T*, *V*, and *I*

Parameters	Values of *T*	Standard Errors of *T*	Values of *V*	Standard Errors of *V*	Values of *I*	Standard Errors of *I*
Alpha (α)	1.080	0.325	0.743	0.198	1.618	0.530
Beta (β)	0.670	0.181	1.964	0.626	0.555	0.144
ω	−3.478	25084.370	−0.008	0.000	−0.001	0.000
θ	347.839	0.000	0.050	0.000	0.107	0.000

TABLE 4.3
Statistics of Data Values and Parameters for *T*, *V*, and *I*

Statistics	Data Values for *T*	Parameters for *T*	Data Values for *V*	Parameters for *V*	Data Values for *I*	Parameters for *I*
Mean	223.678	213.336	0.004	0.008	0.087	0.080
Variance	10977.592	10606.255	0.000	0.000	0.001	0.001
Pearson's skewness	−0.655	−0.426	2.742	0.826	−2.078	−0.957
Pearson's kurtosis	−0.884	−1.048	7.113	−0.207	3.897	−0.051

TABLE 4.4
K-S Test-Outcomes for *T*, *V*, and *I*

K-S Test Outcomes (*T*)		K-S Test Outcomes (*V*)		K-S Test Outcomes (*I*)	
D	0.087	*D*	0.355	*D*	0.300
p-value	0.993	*p*-value	0.007	*p*-value	0.036
Alpha()	0.05	Alpha()	0.05	Alpha()	0.05

TABLE 4.5
Chi-squared Test Outcomes for *T*, *V*, and *I*

Chi-square	Test Outcomes (for *T*)	Test Outcomes (for *V*)	Test Outcomes (for *I*)
Chi-square (observed value)	1.094	36.375	19.551
Chi-square (critical value)	11.070	11.070	11.070
Degrees of freedom (DOF)	5	5	5
p-value (significance level)	0.955	<0.0001	0.002
Alpha()	0.05	0.05	0.05

all variables. Histograms for beta-distribution showing density vs. *T* can be seen in Figure 4.4a, density vs. *V* in Figure 4.4b, and density vs. *I* in Figure 4.4c. A comparison of GoF statistics showing DOF, -2log(likelihood), R^2 by McFadden, Cox and Snell, and Nagelkerke, Akaike Infromation criterion (AIC), Schwarz Bayesian Information Criterion (SBIC) for independent and full is given in Table 4.6 (Figure 4.5). Table 4.7 shows L1-norm regression results for *T*, *V*, and *I*. Figure 4.6a and b show the quantile process time and intercept of L1-norm regression for data of uninfected cells, respectively. Similarly, Figure 4.7a and b depict the quantile process time and intercept of L1-norm regression for data of virus cells, respectively. Figure 4.8a and b characterize the quantile process time and intercept of L1-norm regression for data of infected blood cells, respectively. Figure 4.9a–c are the regression trees of all variables *T*, *V*, and *I*. Figure 4.10 shows the factor analysis on the basis of variable axes F1 and F2. Figure 4.11 compares the eigenvalues, axes, and cumulative variability through the histogram in scree plot. Figure 4.12a and b show the logistic regressors (LR) of the targeted blood cells, and actual and predicted data of *T*, respectively. Figure 4.13a and b depict LR of the virus blood cells, and actual and predicted data of *V*, respectively. Figure 4.14a and b show LR of the infected blood cells, and actual and predicted data of *I*, respectively. Table 4.7 summarizes the factors of Poisson regression for all blood cells. Figure 4.15a–c show the training set and model of Poisson regression for *T*, *V*, and *I*. Figure 4.16 depicts the regression fit for all variables vs. time, which contains the data values, training set, and model plot for each one of them.

The distribution that fits best the data for the GoF test is the **beta**-distribution. For the uninfected, targeted, *T* blood cells, virus, *V* blood cells, and infected, *I* blood cells, statistics is estimated based on the input data and computed using the estimated parameters of Beta4 distribution.

For Poisson regression, the summarized information in Table 4.8 can be referred.

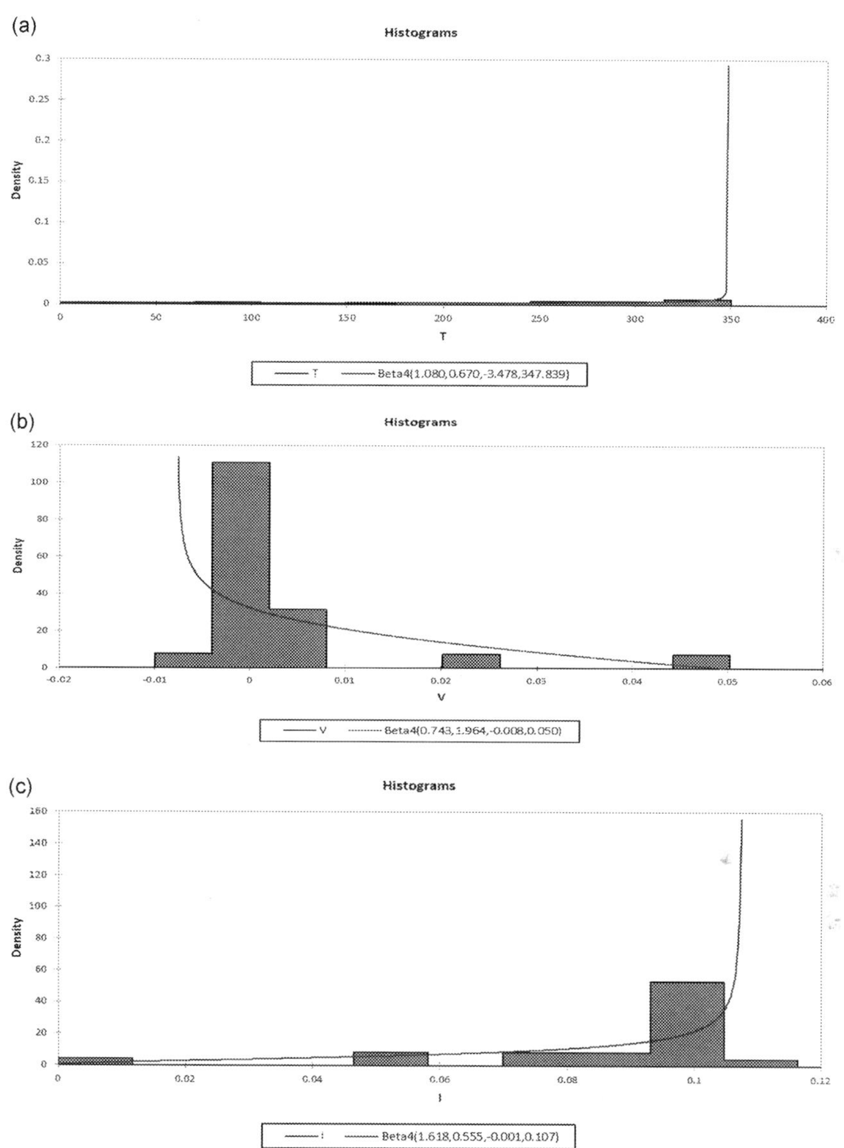

FIGURE 4.4 (a) Histograms for beta-distribution of density vs. uninfected cells, *T*. (b) Histograms for beta-distribution of density vs. virus-infected cells, *V*. (c) Histograms for beta-distribution of density vs. infected cells, *I*.

TABLE 4.6
Comparison of GoF Statistics for Variables *T*, *V*, and *I*

	GoF Statistics (Variable *T*)		GoF Statistics (Variable *V*)		GoF Statistics (Variable *I*):	
Statistic	**Independent**	**Full**	**Independent**	**Full**	**Independent**	**Full**
No. of Observations	21	21	21	21	21	21
Sum of weights	21.000	21.000	21.000	21.000	21.000	21.000
DOF	20	–19	20	–13	20	–17
-2 Log(likelihood)	127.870	7.781	118.506	22.578	125.097	13.050
R^2 (McFadden)	0.000	0.939	0.000	0.809	0.000	0.896
R^2 (Cox and Snell)	0.000	0.997	0.000	0.990	0.000	0.995
R^2 (Nagelkerke)	0.000	0.999	0.000	0.993	0.000	0.998
AIC	167.870	87.781	152.506	90.578	163.097	89.050
SBC	188.760	129.562	170.263	126.091	182.943	128.742
Iterations	0	9	0	9	0	9

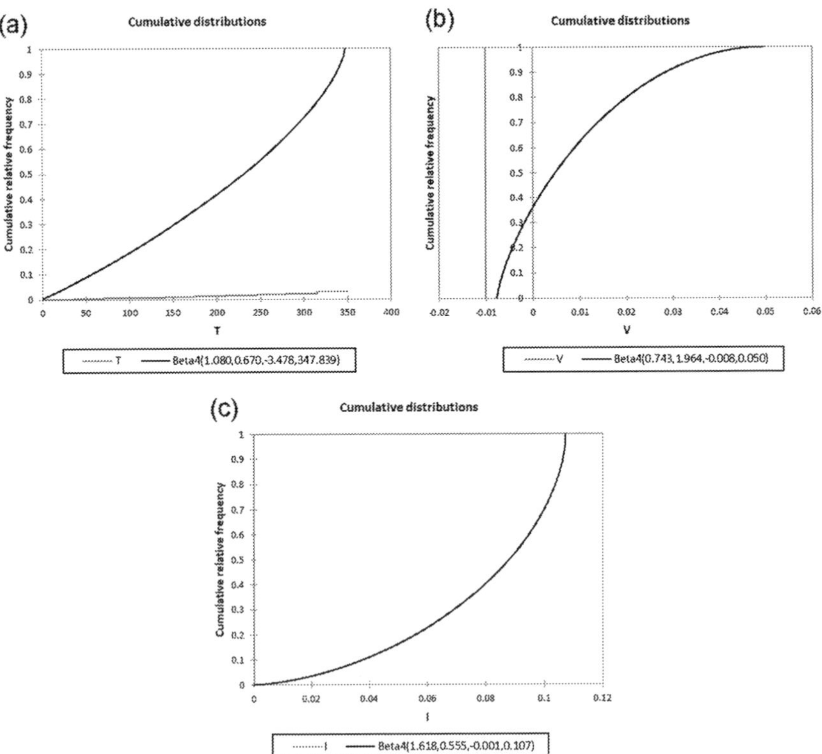

FIGURE 4.5 (a) Cumulative relative frequency vs *T*, (b) Cumulative relative frequency vs. *V*, and (c) Cumulative relative frequency vs. *I*.

TABLE 4.7
L1-Norm Regression Comparison for *T*, *V*, and *I* Variables

L1-Norm Regression/Results for Variable *T*			L1-Norm Regression/Results for Variable *V*			L1-Norm Regression/Results for Variable *I*		
Norms Process			Norms Process			Norms Process		
Norms	**Intercept**	**Time**	**Norms**	**Intercept**	**Time**	**Norms**	**Intercept**	**Time**
0.048	–17.220	17.220	0.048	–0.009	0.000	0.048	–0.005	0.005
0.095	–17.220	17.220	0.095	–0.005	0.000	0.095	–0.005	0.005
0.143	4.642	16.706	0.143	–0.003	0.000	0.143	0.042	0.003
0.190	4.642	16.706	0.190	–0.003	0.000	0.190	0.044	0.003
0.238	23.699	16.255	0.238	–0.002	0.000	0.238	0.043	0.003
0.286	23.699	16.255	0.286	0.000	0.000	0.286	0.043	0.003
0.333	40.128	15.864	0.333	0.000	0.000	0.333	0.068	0.002
0.381	40.128	15.864	0.381	0.000	0.000	0.381	0.070	0.002
0.429	54.076	15.529	0.429	0.002	0.000	0.429	0.069	0.002
0.476	54.076	15.529	0.476	0.004	0.000	0.476	0.069	0.002
0.524	65.672	15.249	0.524	0.005	0.000	0.524	0.074	0.002
0.571	65.672	15.249	0.571	0.006	0.000	0.571	0.074	0.002
0.619	75.019	15.022	0.619	0.000	0.000	0.619	0.078	0.002
0.667	82.203	14.847	0.667	0.000	0.000	0.667	0.084	0.001
0.714	82.203	14.847	0.714	0.000	0.000	0.714	0.084	0.001
0.762	87.288	14.723	0.762	0.009	0.000	0.762	0.100	0.000
0.810	87.288	14.723	0.810	0.009	0.000	0.810	0.103	0.000
0.857	80.861	15.437	0.857	0.032	–0.002	0.857	0.104	0.000
0.905	90.321	14.649	0.905	0.031	–0.001	0.905	0.106	0.000
0.952	82.958	15.385	0.952	0.054	–0.003	0.952	0.110	0.000

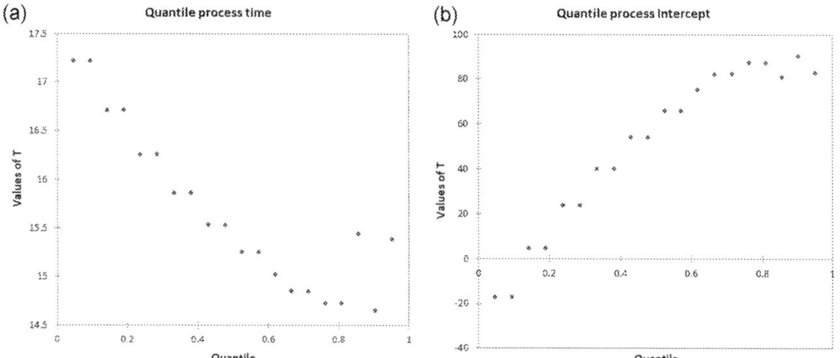

FIGURE 4.6 (a) L1-norm process intercept for the values of *T* and (b) L1-norm process time for the values of *T*.

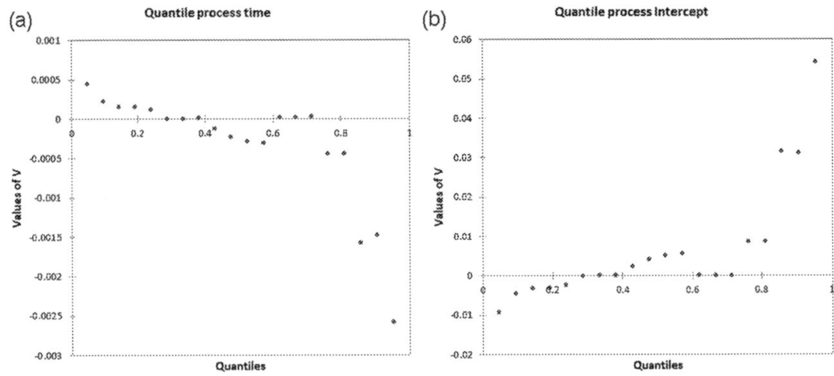

FIGURE 4.7 (a) L1-norm process intercept for the values of *V* and (b) L1-norm process time for the values of *V*.

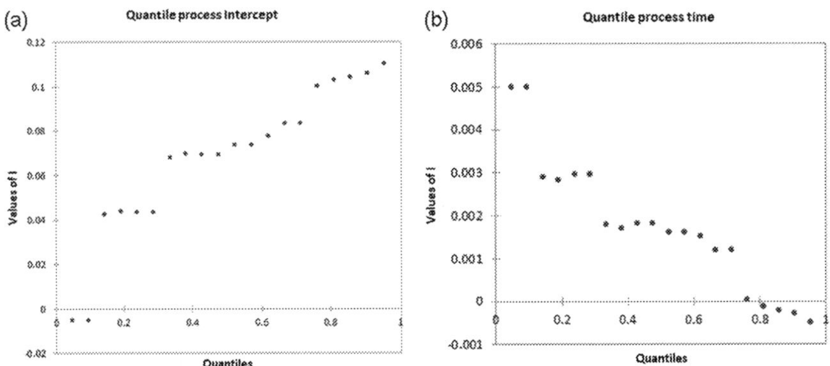

FIGURE 4.8 (a) L1-norm process intercept for the values of *I* and (b) L1-norm process time for the values of *I*.

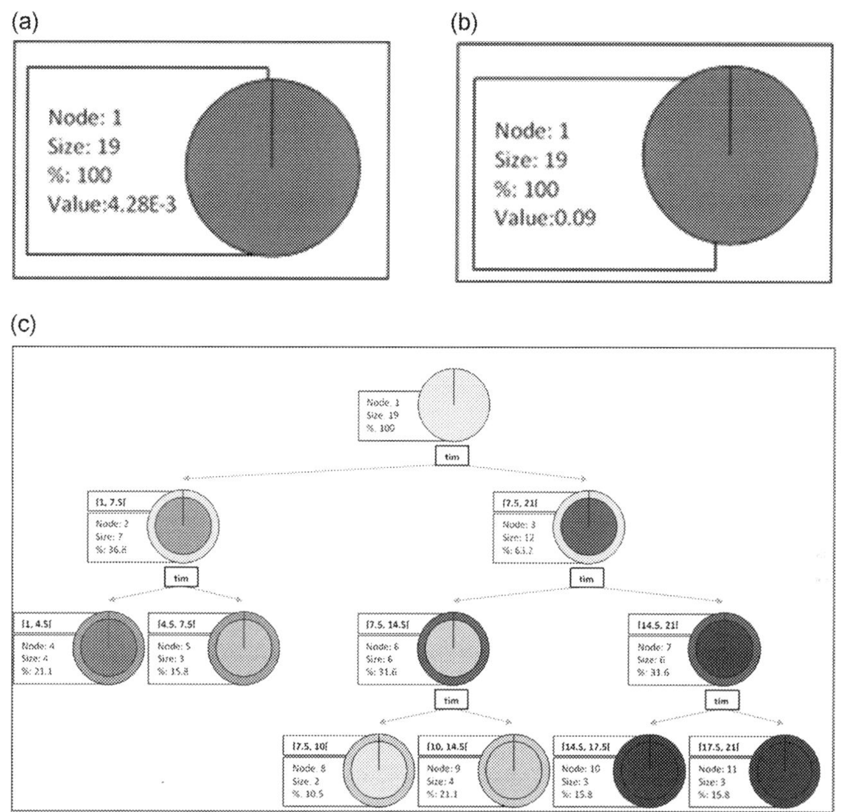

FIGURE 4.9 (a) Regression tree of variable *V*, (b) regression tree of variable *I*, and (c) regression tree of *T*.

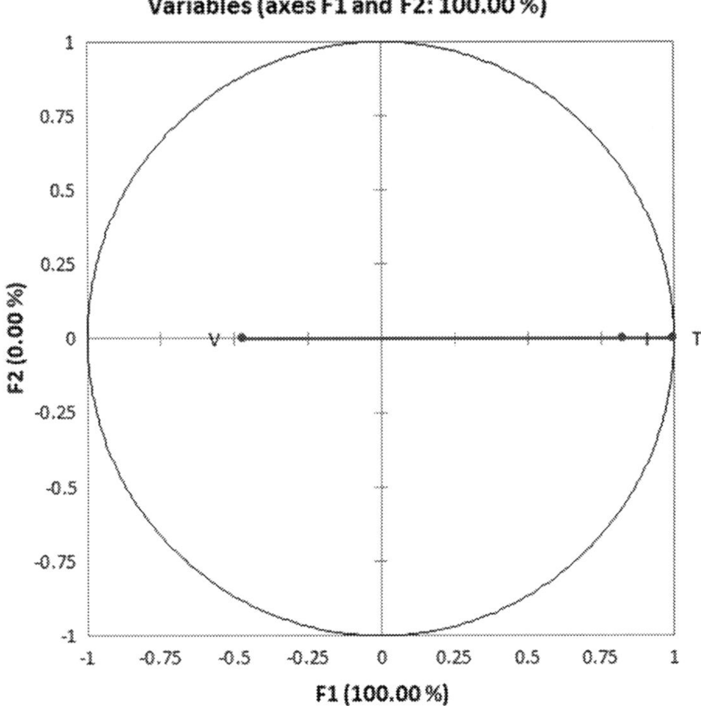

FIGURE 4.10 Factor analysis based on F1 and F2.

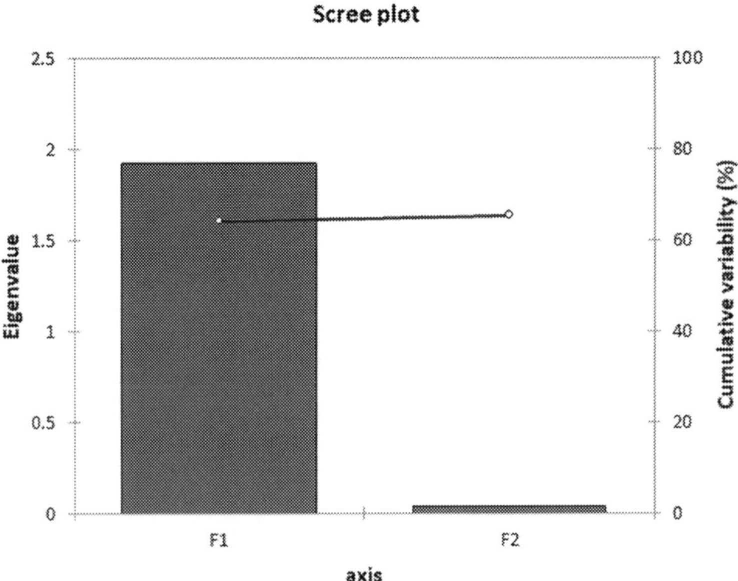

FIGURE 4.11 Factor analysis comparing eigenvalue, axes, and cumulative variability.

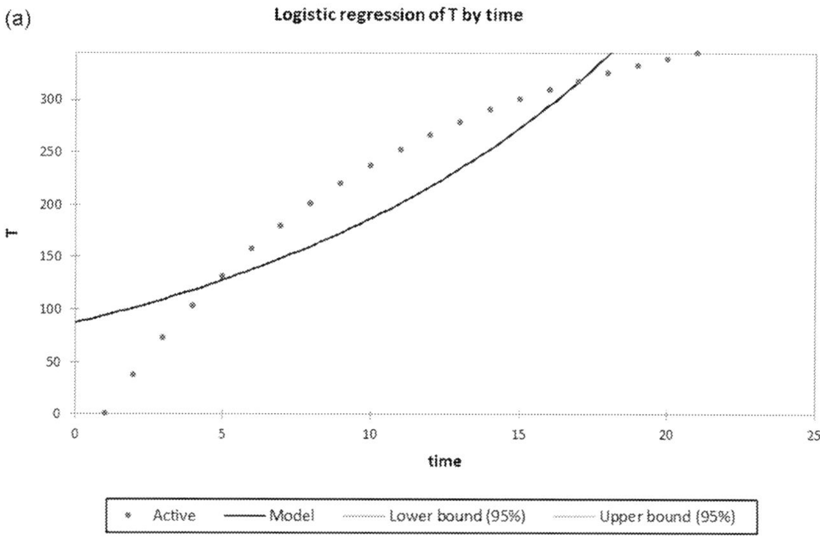

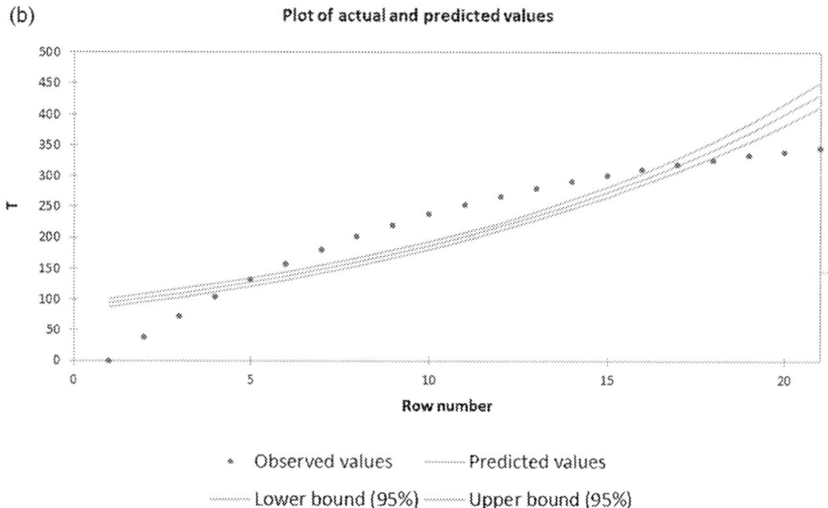

FIGURE 4.12 (a) LR of targeted blood cells (T) and (b) logistic plot of actual and predicted targeted blood cells (T).

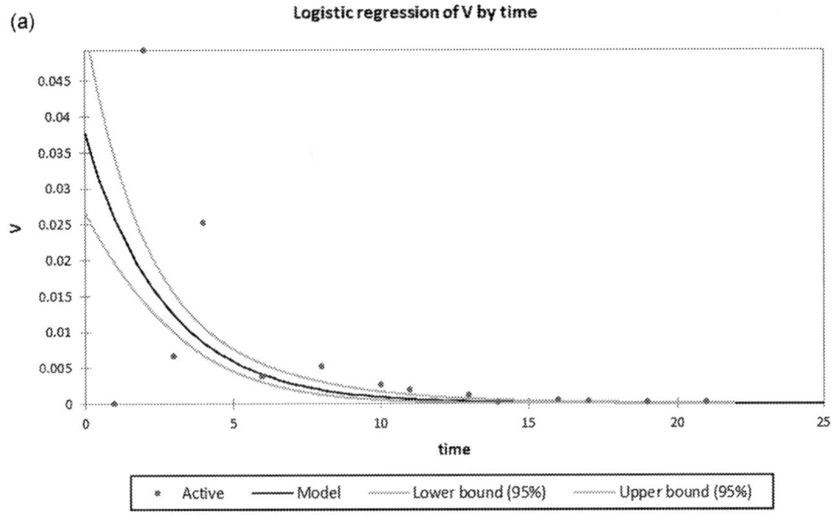

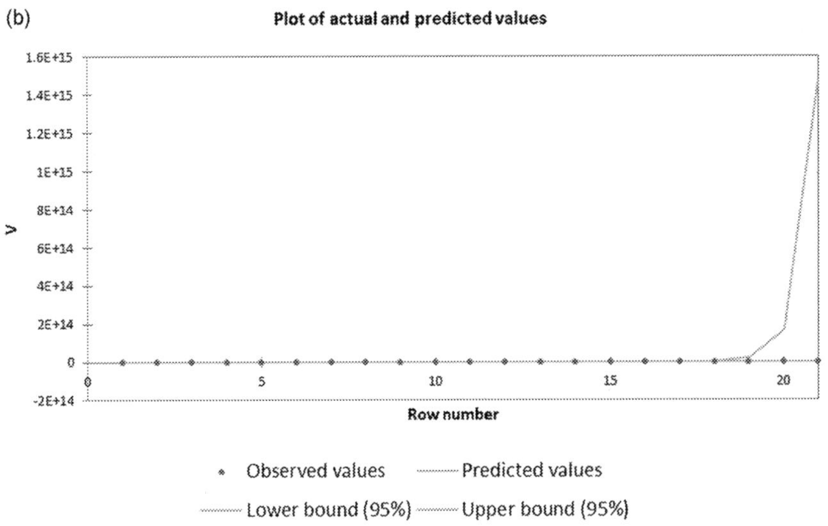

FIGURE 4.13 (a) LR of virus cells (V) and (b) logistic plot of actual and predicted virus blood cells (V).

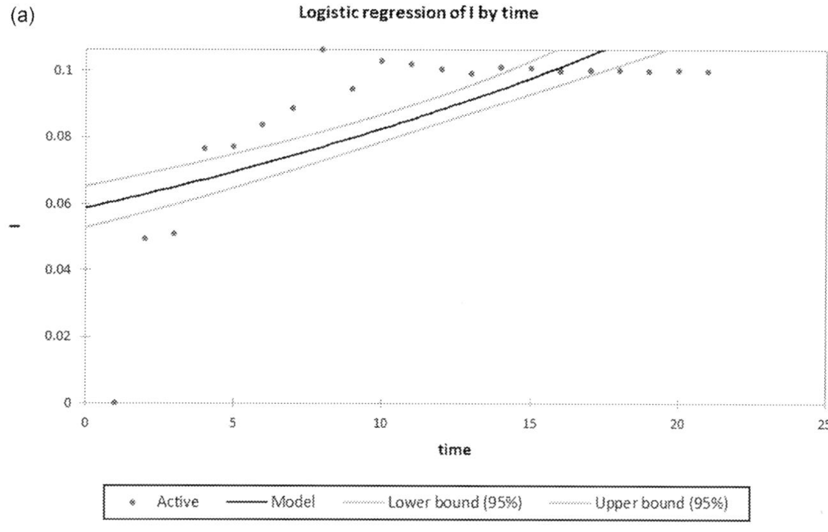

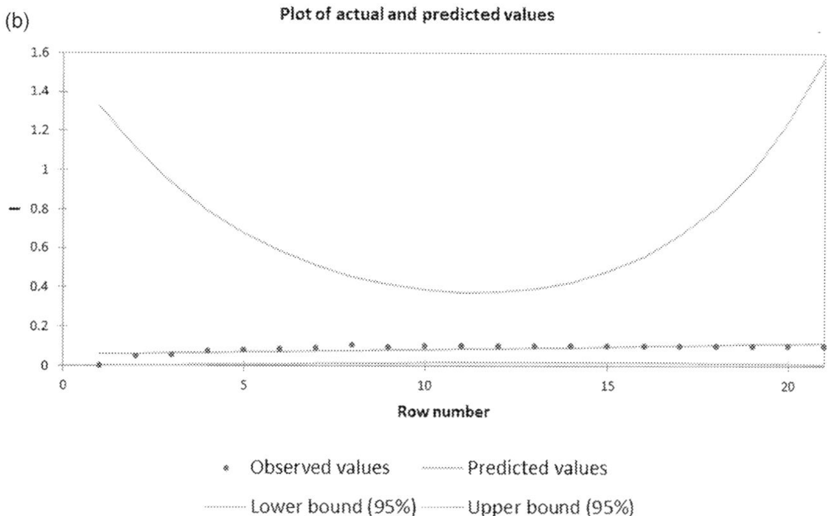

FIGURE 4.14 (a) LR of infected blood cells (I) and (b) logistic plot of actual and predicted infected blood cells (I).

TABLE 4.8
Summary for All *T-V-I* Variables

	Uninfected (*T*)	Virus cells (*V*)	Infected (*I*)
R^2	0.933	0.192	0.510
F	266.007	4.510	19.747
Pr>F	≪0.0001	0.047	0.000

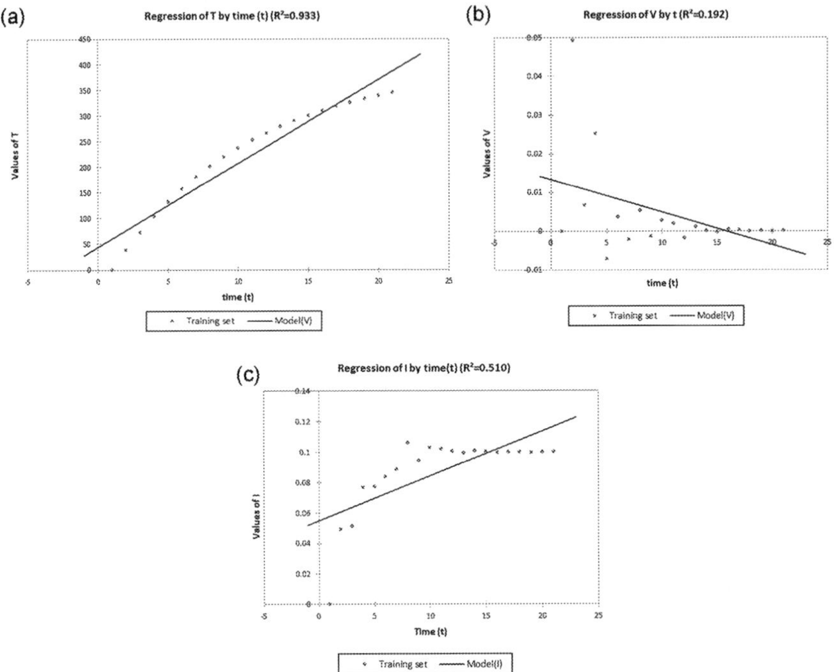

FIGURE 4.15 (a) Regressors of targeted blood cells (T), (b) regressors of virus blood cells (V), and (c) regressors of infected blood cells (I).

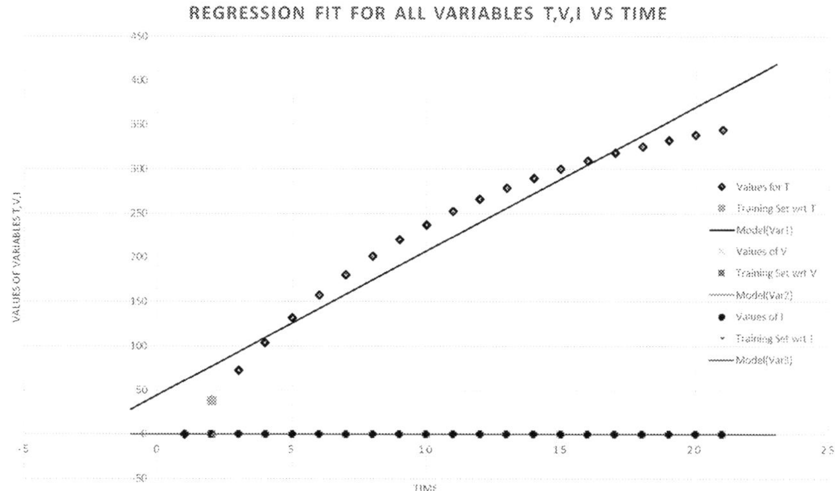

FIGURE 4.16 Regression fit of all three variables *T-V-I* with respect to time.

4.3 CONCLUSION

It can be concluded from time series of *V* that virus particles in the blood grow exponentially in the absence of any external factors (whether environmental or treatment related). Simultaneously, *I* (amount of infected cells) grows in the body causing a total failure of the immune system of the body, and *T* (amount of uninfected cells) will decrease with time. Range of *V* (virus particles) is very high as compared to *T* (uninfected) and *I* (infected cells) with time. *T* and *I* stabilize at a nonzero value after a period of time; there is an exponential increase in the number of virus particles. The anti-persistence behavior of the variables with time can be observed via the phase dynamical inferences. This implies that there will be a situation of chaos arising as the HIV virion progresses inside the body cells. Thus, it has been observed that HIV does not follow strictly the deterministic model, and it oscillates randomly about some norms so that equilibrium is never unconditionally fixed state. Therefore, the model with random perturbations aids to advance the acceptance of dynamical behavior, and with time, the growth of virus particles can be gradually decreased through these differential equations, and then the infection could tend towards zero level, which will after sometime be more or less not effective inside the body.

Though there exists geographical effect on this epidemic, higher ART coverage has deteriorated the prevalence in all regions. Increasing accessibility of ART pooled with particular preventive interventions in relevance to locals might diminish the occurrences further. It is imperative towards higher approach towards healthcare services for testing and treatment. Health establishments perhaps continuing the emphasis for an outreach towards recognizing HIV-positive ones, associating them to comprehensive-care. This would increase their quality of care, and also retain the affected-ones for treatment. Great accomplishments have been made for decreasing morbidity and mortality via first decade of 2000s, and continue emphasizing for the access to region-appropriate preventive scales.

ACKNOWLEDGMENT

Guru Gobind Singh Indraprastha University provided financial support and research facilities for the research.

REFERENCES

1. R. Bhardwaj, A. Bangia, Hybrid Fuzzified-PID controller for non-linear control surfaces for DC motor to improve the efficiency of electric battery driven vehicles. *Int. J. Recent Technol. Eng. (IJRTE)* 8(3) (2019) 2561–2568.
2. R. Bhardwaj, A. Bangia, Dynamic indicator for the prediction of atmospheric pollutants. *Asian J Water, Environ. Pollut.* 16(4) (2019) 39–50.
3. R. Bhardwaj, A. Bangia, Complexity dynamics of meditating body. *Indian J. Ind. Appl. Math.* 7(2) (2016) 106–116.
4. R. Bhardwaj, A. Bangia, Stock Market trend analysis during demonetization using soft-computing techniques. *2018 International Conference on Computing, Power and Communication Technologies (GUCON-2018)*, Galgotia University, Greater Noida. IEEEXplore Digital Library (2019), pp. 696–701.
5. R. Bhardwaj, A. Bangia, Statistical time series analysis of dynamics of HIV. *JNANABHA* Special Issue 48 (2018) 22–27.
6. R. Bhardwaj, A. Bangia, Neuro-Fuzzy analysis of demonetization on NSE. In J.C. Bansal, K.N. Das, A. Nagar, K. Deep, A.K. Ojha (Eds.) Advances in Intelligent Systems and Computing (AISC) (ISSN:2194–5357), Springer Proceedings: Soft Computing for Problem Solving (SocPros-2017), Springer, Singapore. (2019) pp. 853–861.
7. R. Bhardwaj. *Nonlinear Time Series Analysis of Environment Pollutants. Mathematical Modeling on Real World Problems: Interdisciplinary Studies in Applied Mathematics*, NOVA Publisher, New York. (2019) pp. 71–102.
8. R. Bhardwaj, A. Bangia. Dynamical forensic inference for Malware in IoT-based wireless transmissions. In K. Sharma, M. Makino, G. Shrivastava, B. Agarwal (Eds.) *Forensic Investigations and Risk Management in Mobile and Wireless Communications.* (2019) pp. 51–79.
9. R. Bhardwaj, A. Bangia. Data Driven Estimation of Novel COVID-19 Transmission Risks through Hybrid Soft-Computing Techniques. *Chaos, Soliton and Fractals.* (2020) doi:10.1016/j.chaos.2020.110152
10. R. Bhardwaj, D. Datta. Consensus Algorithm. 71, "Studies in Big Data" (2020) pp. 91–107. ISBN: 978-3-030-38676-4. Springer.
11. A. Bangia, R. Bhardwaj, K.V. Jayakumar. River water quality estimation through Artificial Intelligence Conjuncted with Wavelet Decomposition. 979. Numerical Optimization in Engineering and Sciences, (2020) pp. 107–123. Springer.
12. Global Report, UNAIDS report on the global AIDS epidemic 2013. UNAIDS Web site (2013). Available at: http://www.unaids.org/en/media/unaids/contentassets/documents/epidemiology/2013/gr2013/UNAIDS_Global_Report_2013_en.pdf.
13. J. Fettig, M. Swaminathan, C.S. Murrill, J.E. Kaplan, Global epidemiology of HIV. *Infect. Dis. Clin. North Am.* 28(3) (2014) 323–337.
14. R.F. Stengel, Mutation and control of the human immune-deficiency virus. *Math. Biosci.* 213 (2008) 93102.
15. T. Gao, W. Wang, X. Liu, Mathematical analysis of an HIV model with impulsive anti-retroviral drug doses. *Math. Comput. Simul.* 82 (2011) 653665.
16. E. Renshaw, *Modelling Biological Population in Space and Time*, Cambridge University Press, Cambridge (1995).

17. G.P. Samanta, Permanence and extinction of a nonautonomous HIV/AIDS epidemic model with distributed time delay. *Nonlinear Anal. Real. World Appl.* 12 (2011) 1163–1177.
18. E.A. Nosova, A.A. Romanyukha, Mathematical model of HIV-infection transmission and dynamics in the size of risk groups. *Math. Models Comput. Simul.* 5 (2013) 379–393.
19. A.S. Perelson, P.W. Nelson, Mathematical analysis of HIV-1 dynamics in vivo. *SIAM Rev.* 41 (1999) 3–44.
20. M.Y. Li, L. Wang, Backward bifurcation in a mathematical model for HIV infection in vivo with anti-retroviral treatment. *Nonlinear Anal. Real World Appl.* 17 (2014) 147–160.
21. M. Mhawej, C.H. Moog, F. Biafore, C. Brunet-Francois, Control of the HIV infection and drug dosage. *Biomed. Signal Process. Control* 5 (2010) 45–52.
22. Y.M. Shao, *HIV and the Pathogenesis of AIDS*, 3rd ed., Science Press, Beijing, China, (2010).
23. Y. Wang, F. Brauer, J. Wu, J.M. Heffernan, A delay-dependent model with HIV drug resistance during therapy. *J. Math. Anal. Appl.* 414 (2014) 514–531.
24. A.M. Elaiw, Global properties of a class of HIV models. *Nonlinear Anal. Real World Appl.* 11 (2010) 2253–2263.
25. M.D. Narendra, S.N. Alan, HIV dynamics with multiple infections of target cells. *Proc. Natl. Acad. Sci. U. S. A.* 102 (2005) 8198–8203.
26. P.D. Leenheer, H.L. Smith, Virus dynamics: A global analysis. *SIAM J. Appl. Math.* 63 (2003) 1313–1327.
27. A.S. Perelson, A. Neumann, M. Markowitz, J. Leonard, D. Ho, HIV-1dynamics in vivo: Virion clearance rate, infected cell life-span, and viral generation time. *Science* 271 (1996) 1582.
28. D. Wodarz, D.N. Levy, Effect of different modes of viral spread on the dynamics of multiply infected cells in human immune-deffciency virus infection. *J. R. Soc. Interface* 8 (2011) 289–300.
29. O. Sharomia, A.B. Gumel, Dynamical analysis of a multi-strain model of HIV in the presence of anti-retroviral drugs. *J. Biol. Dyn.* 2 (2008) 323–345.
30. J. Mishra. Modified Chua chaotic attractor with differential operators with non-singular kernels. *Chaos, Solitons Fractals* 125 (2019) 64–72.

5 Mathematical Modeling to Find the Potential Number of Ways to Distribute Certain Things to Certain Places in Medical Field

G. Mahadevan, S. Anuthiya, and M. Vimala Suganthi
Gandhigram Rural Institute – Deemed to be University

CONTENTS

5.1 INTRODUCTION

In Ref. [1–4], Duygu Vargor and Pinar Dundar first studied the concept of medium domination number of a graph "$\text{dom}(u,v)$ of G is the sum of number of u-v paths of length one and two of a graph G. The total number of vertices of a graph G dominates every pair of vertices $\text{TDV}(G) = \sum \text{dom}(u,v)$ for $u, v \in V(G)$. The medium domination number of a graph G is defined as $\text{MD}(G) = \dfrac{\text{TDT}(G)}{\binom{n}{2}}$."

Motivated by the above definition, *G*. Mahadevan et al. [5–10] introduced the concept of the extended medium domination number of a graph:edom (u,v) of G is the sum of number of $u-v$ paths of length ≤3 of a graph. The total number of vertices of a graph G dominates every pair of vertices $\text{ETDV}(G)=\sum \text{edom}(u,v)$ for $u,v \in V(G)$. The extended medium domination number of a graph G is defined as $\text{EMD}(G)=\frac{\text{ETDV}(G)}{\binom{n}{2}}$.

Motivated by the above definition, G. Mahadevan et al. [11] introduced the concept of double medium domination of a graph: $\text{DTwin}(u,v)$ is the sum of number of $u-v$ paths of length one two, three, and four. The total number of vertices dominates every pair of vertices SDTwin for $u,v \in V(G)$. In any simple graph G of n number of vertices, the double medium domination number of G is defined as $\text{DTD}(G)=\frac{\text{SDTwin}(G)}{\binom{n}{2}}$. We investigate the double twin domination number for some special type of graph.

The (m, n) tadpole graph, also called a dragon graph or kite graph, is the graph obtained by joining a cycle graph to a path with a bridge. The (m, n) lollipop graph is obtained by joining a complete graph to a path graph with a bridge. For $n \geq 3$, a closed helm graph CH_n is obtained from H_n by adding edges between the pendent vertices where supports in base wheel are adjacent. A uniform n-ply graph is a graph obtained from n distinct paths P_m, $m \geq 3$ by merging all the initial vertices to a vertex u and all the terminal vertices to a vertex v. The uniform n-ply graph is denoted by $P_n(u, v)$. $T_n(K_{1,m})$ is obtained by pasting the root vertex of $K_{(1,m)}$ to the nth vertex of the path in the triangular snake graph T.

Notation 5.1

DTwin (G): double twin domination number of a graph.
SDTwin (G): sum of double twin domination number of a graph.
DTD(G): double twin total domination number of a graph.

Definition 5.1

Let $G=(V,E)$ be a graph, and V and E be the vertex set and edge set, respectively. $\text{DTwin}(u,v)$ is the sum of number of u-v paths of length one, two, three, and four.

Definition 5.2

Let G be a graph. The total number of vertices dominates every pair of vertices $\text{SDTwin}(G)=\sum \text{DTwin}(u,v)$ for $u,v \in V(G)$.

Definition 5.3

In any simple graph G of n number of vertices, the double twin domination number of G is defined as $\text{DTD}(G) = \frac{\text{SDTwin}(G)}{\binom{n}{2}}$.

In Figure 5.1, DTwin (1, 2) = 1; DTwin (1, 3) = 2; DTwin (1, 4) = 3; DTwin (1, 5) = 2; DTwin (1, 6) = 2; DTwin (1, 7) = 3; DTwin (1, 8) = 2; DTwin (2, 3) = 3; DTwin (2, 4) = 3; DTwin (2, 5) = 5; DTwin (2, 6) = 3; DTwin (2, 7) = 4; DTwin (2, 8) = 3; DTwin (3, 4) = 3; DTwin (3, 5) = 3; DTwin (3, 6) = 3; DTwin (3, 7) = 4; DTwin (3, 8) = 3; DTwin (4, 5) = 2; DTwin (4, 6) = 2; DTwin (4, 7) = 3; DTwin (4, 8) = 3; DTwin (5, 6) = 2; DTwin (5, 7) = 3; DTwin (5, 8) = 4; DTwin (6, 7) = 3; DTwin (6, 8) = 4; DTwin (7, 8) = 3.

$$\text{SDTwin}(G) = 81; \quad \text{DTD}(G) = \frac{\text{SDTwin}(G)}{\binom{n}{2}} = \frac{81}{28}.$$

Observation 5.1

For any cycle graph C_n, $\text{SDTwin}(C_n) = 4n$ [11].

Observation 5.2

For any path P_n, $\text{SDTwin}(P_n) = 4n - 10$ [11].

Observation 5.3

For any star $(K_{1,n})$, $\text{SDTwin}(K_{1,n}) = \frac{n(n+1)}{2}$ [11].

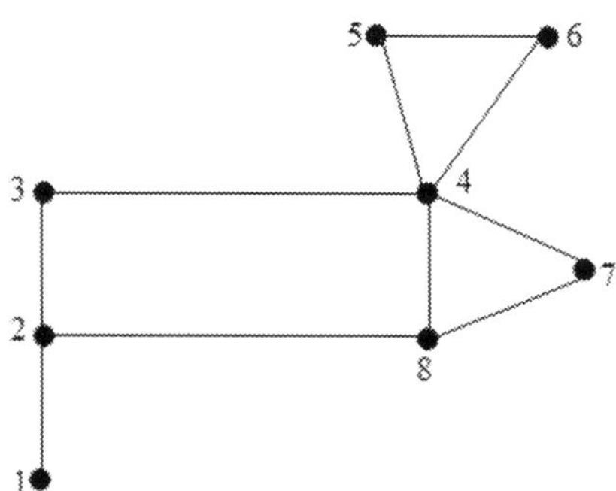

FIGURE 5.1 Example for Double Twin domination number of a graph.

5.2 REAL-LIFE APPLICATION OF MATHEMATICAL MODELING OF THE REAL-LIFE SITUATION USING OUR DOUBLE TWIN DOMINATION NUMBER OF A GRAPH IN MEDICAL FIELD

For practical mathematical modeling, we define $\mathrm{DTD}(G)=\frac{\mathrm{SDTwin}(G)}{\binom{n}{2}}$. According to the convenience, the availability of places and the accident zones, the ambulance services are arranged accordingly so that it could reach the nearby available hospital at a specified distance. Here, the ambulance driver must know all the possible ways to reach the hospital not only through short route but also through the other route. In this situation, it is vital to calculate how many number of ways to reach the hospital from the ambulance service spot.

This is nothing but the calculation of our newly introduced parameter called double twin domination number of a graph. This is explained in the following mathematical model.

Now consider the graph shown in Figure 5.2.

Let the vertices (or nodes) denote the places or junctions or hospitals or ambulance services points, etc. The edges denote the distance between (in terms of kilometers) one node and another node.

Consider the above graph theoretical model of the situations.

Here, H_i denotes hospitals, A_i denotes the ambulance services available spots, and P_i denotes some places between the hospitals and ambulance services available spots.

If any accident occurs in any places, it is necessary to arrange ambulance to the accident spot and then the affected patient should be transformed immediately using the ambulance to the nearby hospitals.

Now it is important that the driver should know the shortest path to reach the hospitals. In addition to that, not only the shortest path but also he must know all the possible ways to reach the hospital. The shortest path may have some obstacles such as very bad conditions of the roads, too many traffic jams, or it may be closed for security reasons. In that case, he can choose the other way to reach the hospital. Again, there is a possibility of obstacles in the second chosen path. In that case, again he must choose some other way to reach hospital.

Here in this model, we assume that the number of ways to reach the hospital from the ambulance service spot is four or less than four (i.e., the distance from the

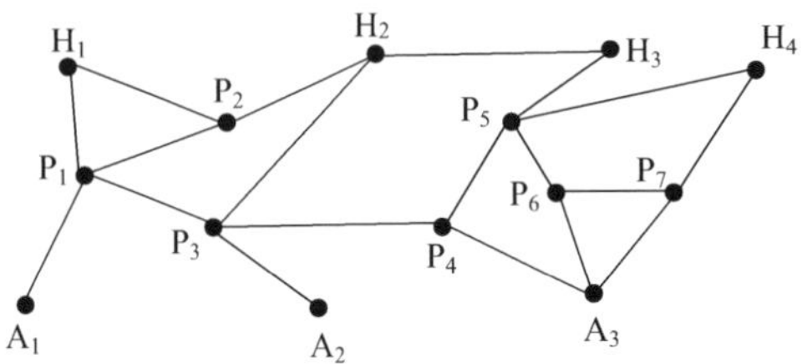

FIGURE 5.2 Example for real life application.

ambulance service spot to the hospital is four in length or less than four in length). Hence, we calculate all the possible ways to reach the hospital in less than or equal to four (in length). The above is nothing but the calculation of our newly introduced double twin domination number of a graph.

First, hospitals are constructed according to area's need, and then, survey is made on accident zones; an immediate ambulance service is constructed in those places or nearby places so that the distance between ambulance service and the hospital is four or less than four.

The pictorial representation is given in Figure 5.2. Suppose an accident occurs in the place P_1: Ambulance A_1 arrives to the place P_1 at distance 1 and can reach the hospital either in distance 3 or 2 or 1 according to the availability of places and pieces of equipment in hospitals H_1, H_2, and H_3.

Now, DTwin $(A_1, H_1) = 2$; DTwin $(A_1, H_2) = 3$; DTwin $(A_1, H_3) = 1$; DTwin $(A_1, H_4) = 0$; DTwin $(A_2, H_1) = 3$; DTwin $(A_2, H_2) = 2$; DTwin $(A_2, H_3) = 2$; DTwin $(A_2, H_4) = 1$; DTwin $(A_3, H_1) = 0$; DTwin $(A_3, H_2) = 2$; DTwin $(A_3, H_3) = 4$; DTwin $(A_3, H_4) = 3$.

$$\text{SDTwin}(G) = 23, \text{DTD}(G) = \frac{23}{21}.$$

As $\text{DTD}(G) = \frac{23}{21} = 1$, the shortest way to reach the hospital is 1, whereas the rest are the alternate ways.

These are the ways which are convenient and easy for the drivers in case of emergency or any block in the ways to transform the patient from the accident zone to the hospitals.

5.3 DOUBLE TWIN DOMINATION NUMBER OF DERIVED GRAPHS AND SPECIAL TYPE OF GRAPHS

Observation 5.4

If $G = T_{n.2}$, for $n \geq 5$, then $\text{DTD}(G) = \dfrac{4n+13}{\binom{n+2}{2}}$.

Observation 5.5

If $G = T_{n.3}$, for $n \geq 5$, then $\text{DTD}(G) = \dfrac{4n+18}{\binom{n+3}{2}}$.

Observation 5.6

If $G = T_{n.4}$, for $n \geq 5$, then $\text{DTD}(G) = \dfrac{4n+22}{\binom{n+4}{2}}$.

Theorem 5.1

If $G = T_{n,m}$, for $n \geq 5, m \geq 4$, then

$$\mathrm{DTD}(G) = \frac{4n+4m+6}{\binom{n+m}{2}}.$$

Proof:

Consider the tadpole graph $T_{n,m}$ with $n+m$ vertices. Let $T_{n,m}$ be a graph obtained by attaching the pendent vertex to one vertex of cycle C_n.

Let $\{a_1, a_2, a_3, \ldots, a_n\}$ be the vertices of the cycle C_n, and b_1 be the pendent vertex. Now the pendent vertex of a_1 is attached to b_1.

A cycle graph on m vertices and path graph on n vertices are connected with a bridge.

$$\mathrm{SDTwin}(G) = \text{for } \ u, vV(G).$$

$\mathrm{DTwin}(a_i, a_{i+1}) = 1$, for $i = 1$ to $n-1$. Therefore, $\sum_{i=1}^{n-1} \mathrm{DTwin}(a_i, a_{i+1}) = n-1$.

DTwin $(a_n, a_1) = 1$.

$\mathrm{DTwin}(a_i, a_{i+2}) = 1$, for $i = 1$ to $n-2$. Therefore, $\sum_{i=1}^{n-2} \mathrm{DTwin}(a_i, a_{i+2}) = n-2$.

DTwin (a_n, a_2) = DTwin $(a_{n-1}, a_1) = 1$.

$\mathrm{DTwin}(a_i, a_{i+3}) = 1$, for $i = 1$ to $n-3$. Therefore, $\sum_{i=1}^{n-3} \mathrm{DTwin}(a_i, a_{i+3}) = n-3$.

DTwin (a_n, a_3) = DTwin (a_{n-1}, a_2) = DTwin $(a_{n-2}, a_1) = 1$.

DTwin $(a_i, a_{i+4}) = 1$, for $i = 1$ to $n-4$. Therefore, $\sum_{i=1}^{n-4} \mathrm{DTwin}(a_i, a_{i+4}) = n-4$.

DTwin (a_n, a_4) = DTwin (a_{n-1}, a_3) = DTwin (a_{n-2}, a_2) = DTwin $(a_{n-2}, a_2) = 1$.

DTwin $(b_i, b_{i+1}) = 1$, for $i = 1$ to $n-1$. Therefore, $\sum_{i=1}^{n-1} \mathrm{DTwin}(b_i, b_{i+1}) = n-1$.

DTwin $(b_i, b_{i+2}) = 1$, for $i = 1$ to $n-2$. Therefore, $\sum_{i=1}^{n-2} \mathrm{DTwin}(b_i, b_{i+2}) = n-2$.

DTwin $(b_i, b_{i+3}) = 1$, for $i = 1$ to $n-3$. Therefore, $\sum_{i=1}^{n-3} \mathrm{DTwin}(b_i, b_{i+3}) = n-3$.

DTwin $(b_i, b_{i+4}) = 1$, for $i = 1$ to $n-4$. Therefore, $\sum_{i=1}^{n-4} \mathrm{DTwin}(b_i, b_{i+4}) = n-4$.

$\mathrm{DTwin}(b_1, a_{\pm i}) = 1$, for $i = 2$ to 4. Therefore, $\sum_{i=2}^{4} \mathrm{DTwin}(b_1, a_{\pm i}) = 6$.

$\text{DTwin}(b_2, a_{\pm i}) = 1$, for $i = 2,3$. Therefore, $\sum_{i=2}^{3} \text{DTwin}(b_2, a_{\pm i}) = 4$.

$\text{DTwin}(b_3, a_{\pm i}) = 1$, for $i = 2$. Therefore, $\sum_{i=2} \text{DTwin}(b_3, a_{\pm i}) = 2$.

$\text{DTwin}(a_1, b_i) = 1$, for $i = 1$ to 4. Therefore, $\sum_{i=1}^{4} \text{DTwin}(a_1, b_i) = 4$.

$$\text{SDTwin}(G) = n - 1 + 1 + n - 2 + 2 + n - 3 + 3 + n - 4 + 4 + m - 1 + 1$$
$$+ m - 2 + 2 + m - 3 + 3 + m - 4 + 4 + 6 + 4 + 2 + 4 = 4n + 4m + 6.$$

$$\text{DTD}(G) = \frac{\text{SDTwin}(G)}{\binom{n}{2}} = \frac{4n + 4m + 6}{\binom{n+m}{2}}.$$

Illustration. For the graph $T_{10,5}$
From Figure 5.3, DTwin $(a_1, a_2) = 1$; DTwin $(a_1, a_3) = 1$; DTwin $(a_1, a_4) = 1$; DTwin $(a_1, a_5) = 1$; DTwin $(a_1, a_7) = 1$; DTwin $(a_1, a_8) = 1$; DTwin $(a_1, a_9) = 1$; DTwin $(a_1, a_{10}) = 1$; DTwin $(a_1, b_1) = 1$; DTwin $(a_1, b_2) = 1$; DTwin $(a_1, b_3) = 1$; DTwin $(a_1, b_4) = 1$; DTwin $(a_2, a_3) = 1$; DTwin $(a_2, a_4) = 1$; DTwin $(a_2, a_5) = 1$; DTwin $(a_2, a_6) = 1$; DTwin $(a_2, a_8) = 1$; DTwin $(a_2, a_9) = 1$; DTwin $(a_2, a_{10}) = 1$; DTwin $(a_2, b_1) = 1$; DTwin $(a_2, b_2) = 1$; DTwin $(a_2, b_3) = 1$; DTwin $(a_3, a_4) = 1$; DTwin $(a_3, a_5) = 1$; DTwin $(a_3, a_6) = 1$; DTwin $(a_3, a_7) = 1$; DTwin $(a_3, a_9) = 1$; DTwin $(a_3, a_{10}) = 1$; DTwin $(a_3, b_1) = 1$; DTwin $(a_3, b_2) = 1$; DTwin $(a_4, a_5) = 1$; DTwin $(a_4, a_6) = 1$; DTwin $(a_4, a_7) = 1$; DTwin $(a_4, a_8) = 1$; DTwin $(a_4, a_{10}) = 1$; DTwin $(a_4, b_1) = 1$; DTwin $(a_5, a_6) = 1$; DTwin $(a_5, a_7) = 1$; DTwin $(a_5, a_8) = 1$; DTwin $(a_5, a_9) = 1$; DTwin $(a_5, a_{10}) = 1$; DTwin $(a_6, a_7) = 1$; DTwin $(a_6, a_8) = 1$; DTwin $(a_6, a_9) = 1$;DTwin $(a_6, a_{10}) = 1$; DTwin $(a_7, a_8) = 1$; DTwin $(a_7, a_9) = 1$; DTwin $(a_7, a_{10}) = 1$; DTwin $(a_8, a_9) = 1$; DTwin $(a_8, a_{10}) = 1$; DTwin $(a_8, b_1) = 1$; DTwin (a_9, a_{10}) =1; DTwin $(a_9, b_1) = 1$; DTwin $(a_9, b_2) = 1$; DTwin $(a_{10}, b_1) = 1$; DTwin $(a_{10}, b_2) = 1$; DTwin $(a_{10}, b_3) = 1$; DTwin $(b_1, b_2) = 1$; DTwin $(b_1, b_3) = 1$; DTwin $(b_1, b_4) = 1$; DTwin $(b_1, b_5) = 1$; DTwin $(b_2, b_3) = 1$; DTwin $(b_2, b_4) = 1$; DTwin $(b_2, b_5) = 1$; DTwin $(b_3, b_4) = 1$; DTwin $(b_3, b_5) = 1$; DTwin $(b_4, b_5) = 1$.

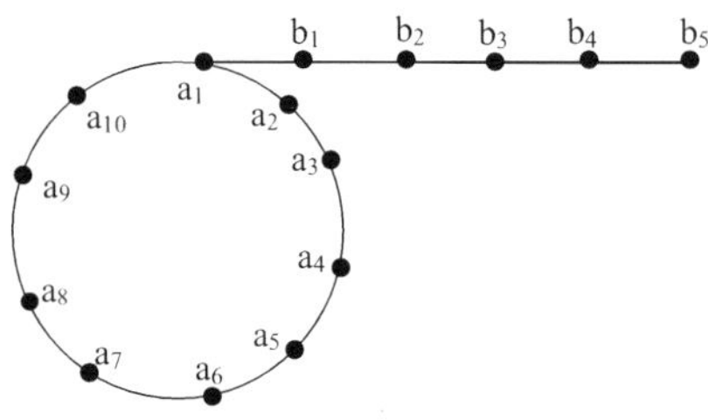

FIGURE 5.3 Tadpole graph.

$$\text{SDTwin}(G) = 66; \text{DTD}(G) = \frac{66}{105}$$

$$\text{SDTwin}(G) = 4n + 4m + 6$$

$$\text{SDTwin}(G) = 4(10) + 4(5) + 6 = 66$$

$$\text{DTD}(G) = \frac{4n + 4m + 6}{\binom{n+m}{2}}$$

$$\text{DTD}(G) = \frac{4(10) + 4(5) + 6}{\binom{10+5}{2}}$$

$$\text{DTD}(G) = \frac{66}{105}.$$

Observation 5.7

If $G = L_{m,1}$, for $m \geq 5$, $\text{DTD}(G) = \dfrac{2m^4 - 9m^3 + 10m^2 + 7m - 8}{2\binom{m+1}{2}}$.

Observation 5.8

If $G = L_{m,2}$, for $m \geq 5$, then $\text{DTD}(G) = \dfrac{2m^4 - 9m^3 + 12m^2 + 3m - 2}{2\binom{m+2}{2}}$.

Observation 5.9

If $G = L_{m,3}$, for $m \geq 5$, then $\text{DTD}(G) = \dfrac{2m^4 - 9m + 12m^2 + 5m + 2}{2\binom{m+3}{2}}$.

Observation 5.10

If $G = L_{m,4}$, for $m \geq 5$, then $\text{DTD}(G) = \dfrac{2m^4 - 9m^3 + 12m^2 + 5m + 10}{2\binom{m+4}{2}}$.

Theorem 5.2

If $G = L_{m.n}$, for $n \geq 5$; $m \geq 5$, then

$$\mathrm{DTD}(G) = \frac{2n^4 - 9n^3 + 12n^2 + 5n + 8m - 22}{2\begin{pmatrix} n+m \\ 2 \end{pmatrix}}.$$

Proof:

Consider the lollipop graph $L_{m,n}$ with $n + m$ vertices. Let $L_{m.n}$ be a graph obtained by attaching the pendent vertex to one vertex of complete K_m.

Let the $\{a_1, a_2, a_3, \ldots, a_n\}$ be the vertices of the complete K_m and b_n be the pendent vertex. Now the pendent vertex of a_1 is attached to $b_{1.}$

A complete graph on m vertices and path graph on n vertices are connected with a bridge.

$\mathrm{SDTwin}(G) =$ for $u, vV(G)$.

$\mathrm{DTwin}(a_1, a_i) = 2n^2 - 9n + 11$, for $i = 2$ to n. Therefore,

$$\sum_{i=2}^{n} \mathrm{DTwin}(a_1, a_i) = (n-1)(2n^2 - 9n + 11).$$

$\mathrm{DTwin}(a_1, a_i) = 2n^2 - 9n + 11$, for $i = 3$ to n. Therefore,

$$\sum_{i=3}^{n} \mathrm{DTwin}(a_2, a_i) = (n-2)(2n^2 - 9n + 11).$$

$\mathrm{DTwin}(a_3, a_i) = 2n^2 - 9n + 11$, for $i = 4$ to n. Therefore,

$$\sum_{i=4}^{n} \mathrm{DTwin}(a_3, a_i) = (n-3)(2n^2 - 9n + 11).$$

…………

$\mathrm{DTwin}(a_{n-1}, a_i) = 2n^2 - 9n + 11$, for $i = n$. Therefore,

$$\sum_{i=n} \mathrm{DTwin}(a_1, a_i) = (2n^2 - 9n + 11).$$

$\mathrm{DTwin}(a_1, b_i) = 1$, for $i = 1$ to 4. Therefore, $\sum_{i=1}^{4} \mathrm{DTwin}(a_1, b_i) = 4.$

$\mathrm{DTwin}(a_i, b_1) = n^2 - 4n + 5$, for $i = 2$ to n. Therefore,

$$\sum_{i=2}^{n} \mathrm{DTwin}(a_i, b_1) = (n-1)(n^2 - 4n + 5).$$

$\mathrm{DTwin}(a_i, b_2) = n - 1$, for $i = 2$ to n. Therefore, $\sum_{i=2}^{n} \mathrm{DTwin}(a_i, b_2) = (n-1)(n-1).$

$\mathrm{DTwin}(a_i, b_3) = 1$, for $i = 2$ to n. Therefore, $\sum_{i=2}^{n} \mathrm{DTwin}(a_i, b_3) = n - 1.$

$\mathrm{DTwin}(b_i, b_{i+1}) = 1$, for $i = 1$ to $m - 1$. Therefore, $\sum_{i=1}^{m-1} \mathrm{DTwin}(b_i, b_{i+1}) = m - 1.$

$\text{DTwin}(b_i, b_{i+2}) = 1$, for $i = 1$ to $m-2$. Therefore, $\sum_{i=1}^{m-2} \text{DTwin}(b_i, b_{i+2}) = m-2$.

$\text{DTwin}(b_i, b_{i+3}) = 1$, for $i = 1$ to $m-3$. Therefore, $\sum_{i=1}^{m-3} \text{DTwin}(b_i, b_{i+3}) = m-3$.

$\text{DTwin}(b_i, b_{i+4}) = 1$, for $i = 1$ to $m-4$. Therefore, $\sum_{i=1}^{m-4} \text{DTwin}(b_i, b_{i+4}) = m-4$.

$$\begin{aligned}\text{SDTwin}(G) = {} & (n-1)(2n^2 - 9n + 11) + (n-2)(2n^2 - 9n + 11) \\ & + \cdots + (2n^2 - 9n + 11) + 4 + (n-1)(n^2 - 4n + 5) \\ & + (n-1)(n-1) + (n-1) + (m-1) + (m-2) \\ & + (m-3) + (m-4)\end{aligned}$$

$$\text{SDTwin}(G) = \frac{2n^4 - 9n^3 + 12n^2 + 5n + 8m - 22}{2}$$

$$\text{DTD}(G) = \frac{2n^4 - 9n^3 + 12n^2 + 5n + 8m - 22}{2\binom{n+m}{2}}.$$

Illustration. For the graph $L_{5,5}$

From Figure 5.4, DTwin $(a_1, a_2) = 16$; DTwin $(a_1, a_3) = 16$; DTwin $(a_1, a_4) = 16$; DTwin $(a_1, a_5) = 16$; DTwin $(a_1, b_1) = 1$; DTwin $(a_1, b_2) = 1$; DTwin $(a_1, b_3) = 1$; DTwin $(a_1, b_4) = 1$; DTwin $(a_2, a_3) = 16$; DTwin $(a_2, a_4) = 16$; DTwin $(a_2, a_5) = 16$; DTwin $(a_2, b_1) = 10$; DTwin $(a_2, b_2) = 4$; DTwin $(a_2, b_3) = 1$; DTwin $(a_3, a_4) = 16$; DTwin $(a_3, a_5) = 16$; DTwin $(a_3, b_1) = 10$; DTwin $(a_3, b_2) = 4$; DTwin $(a_3, b_3) = 1$; DTwin $(a_4, a_5) = 16$; DTwin $(a_4, b_1) = 10$; DTwin $(a_4, b_2) = 4$; DTwin $(a_4, b_3) = 1$; DTwin $(a_5, b_1) = 10$; DTwin $(a_5, b_2) = 4$; DTwin $(a_5, b_3) = 1$; DTwin $(b_1, b_2) = 1$; DTwin $(b_1, b_3) = 1$; DTwin $(b_1, b_4) = 1$; DTwin $(b_1, b_5) = 1$; DTwin $(b_2, b_3) = 1$; DTwin $(b_2, b_4) = 1$; DTwin $(b_2, b_5) = 1$; DTwin $(b_3, b_4) = 1$; DTwin $(b_3, b_5) = 1$; DTwin $(b_4, b_5) = 1$.

$$\text{SDTwin}(G) = \frac{2n^4 - 9n^3 + 12n^2 + 5n + 8m - 22}{2}$$

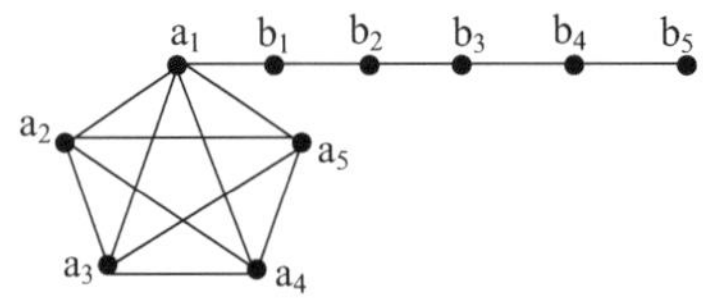

FIGURE 5.4 Lollipop graph.

$$\text{SDTwin}(G) = \frac{2(5)^4 - 9(5)^3 + 12(5)^2 + 5(5) + 8(5) - 22}{2}$$

$$\text{SDTwin}(G) = \frac{468}{2} = 234.$$

$$\text{DTD}(G) = \frac{2n^4 - 9n^3 + 12n^2 + 5n + 8m - 22}{2\binom{n+m}{2}}$$

$$\text{DTD}(G) = \frac{2(5)^4 - 9(5)^3 + 12(5)^2 + 5(5) + 8(5) - 22}{2\binom{5+5}{2}}$$

$$\text{DTD}(G) = \frac{234}{45}.$$

Theorem 5.3

If $G = CH_n$, where $n \geq 9$, then

$$\text{DTD}(G) = \frac{28n^2 - 56n}{2\binom{2n+1}{2}}.$$

Proof:

Consider the closed helm graph CH_n with $2n+1$ vertex. Let the root vertices be a, the inner cycle vertices be $b_1, b_2, \ldots b_n$ and the outer cycle vertices be $c_1, c_2, c_3, \ldots, c_n$.

Now, the vertices of b_1 are attached to c_1, b_2 to c_2, b_3 to $c_3 \ldots b_n$ to c_n, respectively.

$\text{SDTwin}(G) =$ for $u, vV(G)$.

$\text{DTwin}(a, b_i) = 9$, for $i = 1$ to n. Therefore, $\sum_{i=1}^{n} \text{DTwin}(a, b_i) = 9n$.

$\text{DTwin}(a, c_i) = 13$, for $i = 1$ to n. Therefore, $\sum_{i=1}^{n} \text{DTwin}(a, c_i) = 13n$.

$\text{DTwin}(b_i, b_{i+1}) = 7$, for $i = 1$ to $n-1$. Therefore, $\sum_{i=1}^{n-1} \text{DTwin}(b_i, b_{i+1}) = 7(n-1)$.

$$\text{DTwin}(b_n, b_1) = 7.$$

$\text{DTwin}(b_i, b_{i+2}) = 14$, for $i = 1$ to $n-2$. Therefore, $\sum_{i=1}^{n-2} \text{DTwin}(b_i, b_{i+2}) = 14(n-2)$.

$$\text{DTwin}(b_{n-1}, b_1) = 14; \text{DTwin}(b_n, b_2) = 14.$$

$$\text{DTwin}(b_i, b_{i+3}) = 14, \text{ for } i = 1 \text{ to } n-3. \text{ Therefore, } \sum_{i=1}^{n-3} \text{DTwin}(b_i, b_{i+3}) = 14(n-3).$$

$$\text{DTwin}(b_{n-2}, b_1) = 14; \text{DTwin}(b_{n-1}, b_2) = 14; \text{DTwin}(b_n, b_3) = 14.$$

$$\text{DTwin}(b_i, b_{i+4}) = 14, \text{ for } i = 1 \text{ to } n-4. \text{ Therefore, } \sum_{i=1}^{n-4} \text{DTwin}(b_i, b_{i+4}) = 14(n-4).$$

$$\text{DTwin}(b_{n-3}, b_1) = 14; \text{DTwin}(b_{n-2}, b_2) = 14; \text{DTwin}(b_{n-1}, b_3) = 14;$$

$$\text{DTwin}(b_n, b_4) = 14.$$

$$\text{DTwin}(b_1, b_i) = 13, \text{ for } i = 6 \text{ to } n-4. \text{ Therefore, } \sum_{i=6}^{n-4} \text{DTwin}(b_1, b_i) = 13(n-9).$$

$$\text{DTwin}(b_2, b_i) = 13, \text{ for } i = 7 \text{ to } n-3. \text{ Therefore, } \sum_{i=6}^{n-4} \text{DTwin}(b_2, b_i) = 13(n-9).$$

$$\text{DTwin}(b_3, b_i) = 13, \text{ for } i = 8 \text{ to } n-2. \text{ Therefore, } \sum_{i=8}^{n-2} \text{DTwin}(b_3, b_i) = 13(n-9).$$

$$\text{DTwin}(b_4, b_i) = 13, \text{ for } i = 9 \text{ to } n-1. \text{ Therefore, } \sum_{i=9}^{n-1} \text{DTwin}(b_4, b_i) = 13(n-9).$$

$$\text{DTwin}(b_5, b_i) = 13, \text{ for } i = 10 \text{ to } n. \text{ Therefore, } \sum_{i=10}^{n} \text{DTwin}(b_5, b_i) = 13(n-9).$$

$$\text{DTwin}(b_6, b_i) = 13, \text{ for } i = 11 \text{ to } n. \text{ Therefore, } \sum_{i=11}^{n} \text{DTwin}(b_5, b_i) = 13(n-10).$$

……

$$\text{DTwin}(b_{n-5}, b_n) = 13.$$

$$\text{DTwin}(b_i, c_i) = 5, \text{ for } i = 1 \text{ to } n. \text{ Therefore, } \sum_{i=1}^{n} \text{DTwin}(b_i, c_i) = 5n.$$

$$\text{DTwin}(b_i, c_{i+1}) = 7, \text{ for } i = 1 \text{ to } n-1. \text{ Therefore, } \sum_{i=1}^{n-1} \text{DTwin}(b_i, c_{i+1}) = 7(n-1).$$

$$\text{DTwin}(b_n, c_1) = 7.$$

$\text{DTwin}(b_i, c_{i-1}) = 7$, for $i = 2$ to n. Therefore, $\sum_{i=2}^{n} \text{DTwin}(b_i, c_{i-1}) = 7(n-1)$.

$$\text{DTwin}(b_1, c_n) = 7.$$

$\text{DTwin}(b_i, c_{i+2}) = 9$, for $i = 1$ to $n-2$. Therefore, $\sum_{i=1}^{n-2} \text{DTwin}(b_i, c_{i+2}) = 9(n-2)$.

$$\text{DTwin}(b_{n-1}, c_1) = 9;\ \text{DTwin}(b_n, c_2) = 9.$$

$\text{DTwin}(b_i, c_{i-2}) = 9$, for $i = 3$ to n. Therefore, $\sum_{i=3}^{n} \text{DTwin}(b_i, c_{i-2}) = 9(n-2)$.

$$\text{DTwin}(b_1, c_{n-1}) = 9;\ \text{DTwin}(b_2, c_n) = 9.$$

$\text{DTwin}(b_i, c_{i+3}) = 10$, for $i = 1$ to $n-3$. Therefore, $\sum_{i=1}^{n-3} \text{DTwin}(b_i, c_{i+3}) = 10(n-3)$.

$$\text{DTwin}(b_{n-2}, c_1) = 10;\ \text{DTwin}(b_{n-1}, c_2) = 10;\ \text{DTwin}(b_n, c_3) = 10.$$

$\text{DTwin}(b_i, c_{i-3}) = 10$, for $i = 4$ to n. Therefore, $\sum_{i=4}^{n} \text{DTwin}(b_i, c_{i-3}) = 10(n-3)$.

$$\text{DTwin}(b_1, c_{n-2}) = 10;\ \text{DTwin}(b_2, c_{n-1}) = 10;\ \text{DTwin}(b_3, c_n) = 10.$$

$\text{DTwin}(b_1, c_i) = 7$, for $i = 5$ to $n-3$. Therefore, $\sum_{i=5}^{n-3} \text{DTwin}(b_1, c_i) = 7(n-7)$.

$\text{DTwin}(b_2, c_i) = 7$, for $i = 6$ to $n-2$. Therefore, $\sum_{i=6}^{n-2} \text{DTwin}(b_2, c_i) = 7(n-7)$.

$\text{DTwin}(b_3, c_i) = 7$, for $i = 7$ to $n-1$. Therefore, $\sum_{i=7}^{n-1} \text{DTwin}(b_3, c_i) = 7(n-7)$.

$\text{DTwin}(b_4, c_i) = 7$, for $i = 8$ to n. Therefore, $\sum_{i=8}^{n} \text{DTwin}(b_4, c_i) = 7(n-7)$.

$\text{DTwin}(b_5, c_i) = 7$, for $i = 9$ to n. Therefore, $\sum_{i=5}^{n-3} \text{DTwin}(b_1, c_i) = 7(n-8)$.

……

$$\text{DTwin}(b_{n-4}, c_n) = 7.$$

$\text{DTwin}(c_1,b_i)=7$, for $i=5$ to $n-3$. Therefore, $\sum_{i=5}^{n-3}\text{DTwin}(c_1,b_i)=7(n-7)$.

$\text{DTwin}(c_2,b_i)=7$, for $i=6$ to $n-2$. Therefore, $\sum_{i=6}^{n-2}\text{DTwin}(c_2,b_i)=7(n-7)$.

$\text{DTwin}(c_3,b_i)=7$, for $i=7$ to $n-1$. Therefore, $\sum_{i=7}^{n-1}\text{DTwin}(c_3,b_i)=7(n-7)$.

$\text{DTwin}(c_4,b_i)=7$, for $i=8$ to n. Therefore, $\sum_{i=8}^{n}\text{DTwin}(c_4,b_i)=7(n-7)$.

$\text{DTwin}(c_5,b_i)=7$, for $i=9$ to n. Therefore, $\sum_{i=5}^{n-3}\text{DTwin}(c_1,b_i)=7(n-8)$.

………

$$\text{DTwin}(c_{n-4},b_n)=7.$$

$\text{DTwin}(c_i,c_{i+1})=3$, for $i=1$ to $n-1$. Therefore, $\sum_{i=1}^{n-1}\text{DTwin}(c_i,c_{i+1})=3(n-1)$.

$$\text{DTwin}(c_n,c_1)=3.$$

$\text{DTwin}(c_i,c_{i+2})=5$, for $i=1$ to $n-2$. Therefore, $\sum_{i=1}^{n-2}\text{DTwin}(c_i,c_{i+2})=5(n-2)$.

$$\text{DTwin}(c_{n-1},c_1)=5;\ \text{DTwin}(c_n,c_2)=5.$$

$\text{DTwin}(c_i,c_{i+3})=2$, for $i=1$ to $n-3$. Therefore, $\sum_{i=1}^{n-3}\text{DTwin}(c_i,c_{i+3})=2(n-3)$.

$$\text{DTwin}(c_{n-2},c_1)=2;\ \text{DTwin}(c_{n-1},c_2)=2;\ \text{DTwin}(c_n,c_3)=2.$$

$\text{DTwin}(c_i,c_{i+4})=2$, for $i=1$ to $n-4$. Therefore, $\sum_{i=1}^{n-4}\text{DTwin}(c_i,c_{i+3})=10(n-3)$.

$$\text{DTwin}(c_{n-3},c_1)=2;\ \text{DTwin}(c_{n-2},c_2)=2;\ \text{DTwin}(c_{n-1},c_3)=2;\ \text{DTwin}(c_n,c_4)=2.$$

$\text{DTwin}(c_1,c_i)=1$, for $i=6$ to $n-4$. Therefore, $\sum_{i=6}^{n-4}\text{DTwin}(c_1,c_i)=(n-9)$.

$\text{DTwin}(c_2,c_i)=1$, for $i=7$ to $n-3$. Therefore, $\sum_{i=6}^{n-4}\text{DTwin}(c_2,c_i)=(n-9)$.

$\text{DTwin}(c_3,c_i)=1$, for $i=8$ to $n-2$. Therefore, $\sum_{i=8}^{n-2}\text{DTwin}(c_3,c_i)=(n-9)$.

$\text{DTwin}(c_4,c_i)=1$, for $i=9$ to $n-1$. Therefore, $\sum_{i=9}^{n-1}\text{DTwin}(c_4,c_i)=(n-9)$.

$\text{DTwin}(c_5,c_i)=1$, for $i=10$ to n. Therefore, $\sum_{i=10}^{n}\text{DTwin}(c_5,c_i)=(n-9)$.

$\text{DTwin}(c_6,c_i)=1$, for $i=11$ to n. Therefore, $\sum_{i=11}^{n}\text{DTwin}(c_5,c_i)=(n-10)$.

……

$\text{DTwin}(c_{n-5},c_n)=1$.

$$\begin{aligned}\text{SDTwin}(G)&=9n+13n+7(n-1)+7+14(n-2)+14+14+14(n-3)+14+14\\&+14+14(n-4)+14+14+14+14+13(n-9)+13(n-9)\\&+13(n-9)+13(n-9)+[13(n-9)+13(n-10)+\cdots+13]+5n\\&+7(n-1)+7+7(n-1)+7+9(n-2)+9+9+9(n-2)+9+9\\&+10(n-3)+10+10+10+10(n-3)+10+10+10+7(n-7)\\&+7(n-7)+7(n-7)+[7(n-7)+7(n-8)+\cdots+7]+7(n-7)\\&+7(n-7)+7(n-7)+[7(n-7)+7(n-8)+\cdots+7]+3(n-1)+3\\&+5(n-2)+5+5+2(n-3)+2+2+2+2(n-4)+2+2+2+2\\&+(n-9)+(n-9)+(n-9)+(n-9)+[(n-9)+(n-10)+\cdots+1]\end{aligned}$$

$$\begin{aligned}\text{SDTwin}(G)&=9n+13n+7n+14n+14n+14n+13n-117+13n-117+13n-117\\&+13n-117+13n-117+13\left[\frac{(n-9)(n-8)}{2}\right]+5n+7n+7n+9n+9n\\&+10n+10n+7n-49+7n-49+7n-49+7\left[\frac{(n-7)(n-6)}{2}\right]+7n\\&-49+7n-49+7n-49+7\left[\frac{(n-7)(n-6)}{2}\right]+3n+5n+2n+2n+n\\&-9+n-9+n-9+n-9+\left[\frac{(n-9)(n-8)}{2}\right].\end{aligned}$$

$$\text{SDTwin}(G) = 238n - 798 + \left[\frac{\begin{array}{c}13n^2 - 221n + 936 + 7n^2 - 91n + 294 + 7n^2 - 91n \\ +294 + n^2 - 17n + 72\end{array}}{2}\right]$$

$$\text{SDTwin}(G) = \frac{28n^2 + 56n}{2}$$

$$\text{DTD}(G) = \frac{28n^2 + 56n}{2\binom{2n+1}{2}}.$$

Illustration. For the graph CH_{10}
From Figure 5.5, DTwin (1, 2) = 5; DTwin (1, 3) = 5; DTwin (1, 4) = 5; DTwin (1, 5) = 5; DTwin (1, 6) = 5; DTwin (1, 7) = 5; DTwin (1, 8) = 5; DTwin (1, 9) = 5; DTwin (1, 10) = 5; DTwin (1, 11) = 5; DTwin (1, 12) = 13; DTwin (1, 13) = 13; DTwin (1, 14) = 13; DTwin (1, 15) = 13; DTwin (1, 16) = 13; DTwin (1, 17) = 13; DTwin (1, 18) = 13; DTwin (1, 19) = 13; DTwin (1, 20) = 13; DTwin (1, 21) = 13; DTwin (2, 3) = 7; DTwin (2, 4) = 14; DTwin (2, 5) = 14; DTwin (2, 6) = 14; DTwin (2, 7) = 13; DTwin (2, 8) = 14; DTwin (2, 9) = 14; DTwin (2, 10) = 14; DTwin (2, 11) = 7; DTwin (2, 12) = 5; DTwin (2, 13) = 7; DTwin (2, 14) = 9; DTwin (2, 15) = 10; DTwin (2, 16) = 7; DTwin (2, 17) = 7; DTwin (2, 18) = 7; DTwin (2, 19) = 10; DTwin (2, 20) = 9; DTwin (2, 21) = 7; DTwin (3, 4) = 7; DTwin (3, 5) = 14; DTwin (3, 6) = 14; DTwin (3, 7) = 14; DTwin (3, 8) = 13; DTwin (3, 9) = 14; DTwin (3, 10) = 14; DTwin (3, 11) = 14; DTwin (3, 12) = 7; DTwin (3, 13) = 5; DTwin (3, 14) = 7; DTwin (3, 15) = 9; DTwin (3, 16) = 10; DTwin (3, 17) = 7; DTwin (3, 18) = 7; DTwin (3, 19) = 7; DTwin (3, 20) = 10; DTwin (3, 21) = 9; DTwin (4, 5) = 7; DTwin (4, 6) = 14; DTwin (4, 7) = 14; DTwin (4, 8) = 14; DTwin (4, 9) = 13; DTwin (4, 10) = 14; DTwin (4, 11) = 14; DTwin (4, 12) = 9; DTwin (4, 13) = 7; DTwin (4, 14) = 5; DTwin (4, 15) = 7; DTwin (4, 16) = 9; DTwin (4, 17) = 10; DTwin (4, 18) = 7; DTwin (4, 19) = 7; DTwin (4, 20) = 7; DTwin (4, 21) = 10; DTwin (5, 6) = 7; DTwin (5, 7) = 14; DTwin (5, 8) = 14; DTwin (5, 9) = 14; DTwin (5, 10) = 13;

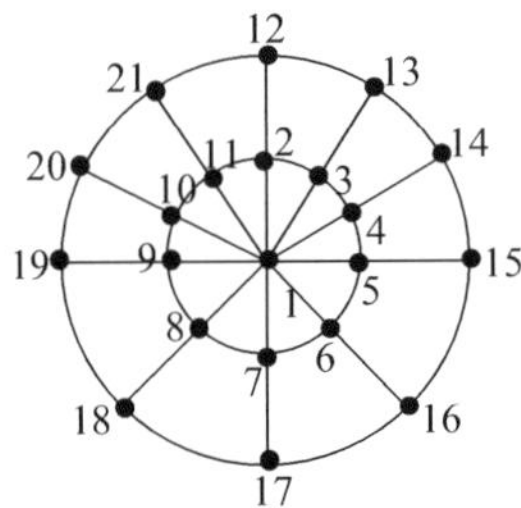

FIGURE 5.5 Closed helm graph.

DTwin (5, 11) = 14; DTwin (5, 12) = 10; DTwin (5, 13) = 9; DTwin (5, 14) = 7; DTwin (5, 15) = 5; DTwin (5, 16) = 7; DTwin (5, 17) = 9; DTwin (5, 18) = 10; DTwin (5, 19) = 7; DTwin (5, 20) = 7; DTwin (5, 21) = 7; DTwin (6, 7) = 7; DTwin (6, 8) = 14; DTwin (6, 9) = 14; DTwin (6, 10) = 14; DTwin (6, 11) = 13; DTwin (6, 12) = 7; DTwin (6, 13) = 10; DTwin (6, 14) = 9; DTwin (6, 15) = 7; DTwin (6, 16) = 5; DTwin (6, 17) = 7; DTwin (6, 18) = 9; DTwin (6, 19) = 10; DTwin (6, 20) = 7; DTwin (6, 21) = 7; DTwin (7, 8) = 7; DTwin (7, 9) = 14; DTwin (7, 10) = 14; DTwin (7, 11) = 14; DTwin (7, 12) = 7; DTwin (7, 13) = 7; DTwin (7, 14) = 10; DTwin (7, 15) = 9; DTwin (7, 16) = 7; DTwin (7, 17) = 5; DTwin (7, 18) = 7; DTwin (7, 19) = 9; DTwin (7, 20) = 10; DTwin (7, 21) = 7; DTwin (8, 9) = 7; DTwin (8, 10) = 14; DTwin (8, 11) = 14; DTwin (8, 12) = 7; DTwin (8, 13) = 7; DTwin (8, 14) = 7; DTwin (8, 15) = 10; DTwin (8, 16) = 9; DTwin (8, 17) = 7; DTwin (8, 18) = 5; DTwin (8, 19) = 7; DTwin (8, 20) = 9; DTwin (8, 21) = 10; DTwin (9, 10) = 7; DTwin (9, 11) = 14; DTwin (9, 12) = 10; DTwin (9, 13) = 7; DTwin (9, 14) = 7; DTwin (9, 15) = 7; DTwin (9, 16) = 10; DTwin (9, 17) = 9; DTwin (9, 18) = 7; DTwin (9, 19) = 5; DTwin (9, 20) = 7; DTwin (9, 21) = 9; DTwin (10, 11) = 7; DTwin (10, 12) = 9; DTwin (10, 13) = 10; DTwin (10, 14) = 7; DTwin (10, 15) = 7; DTwin (10, 16) = 7; DTwin (10, 17) = 10; DTwin (10, 18) = 9; DTwin (10, 19) = 7; DTwin (10, 20) = 5; DTwin (10, 21) = 7; DTwin (11, 12) = 7; DTwin (11, 13) = 9; DTwin (11, 14) = 10; DTwin (11, 15) = 7; DTwin (11, 16) = 7; DTwin (11, 17) = 7; DTwin (11, 18) = 10; DTwin (11, 19) = 9; DTwin (11, 20) = 7; DTwin (11, 21) = 5; DTwin (12, 13) = 3; DTwin (12, 14) = 5; DTwin (12, 15) = 2; DTwin (12, 16) = 2; DTwin (12, 17) = 1; DTwin (12, 18) = 2; DTwin (12, 19) = 2; DTwin (12, 20) = 5; DTwin (12, 21) = 3; DTwin (13, 14) = 3; DTwin (13, 15) = 5; DTwin (13, 16) = 2; DTwin (13, 17) = 2; DTwin (13, 18) = 1; DTwin (13, 19) = 2; DTwin (13, 20) = 2; DTwin (13, 21) = 5; DTwin (14, 15) = 3; DTwin (14, 16) = 5; DTwin (14, 17) = 2; DTwin (14, 18) = 2; DTwin (14, 19) = 1; DTwin (14, 20) = 2; DTwin (14, 21) = 2; DTwin (15, 16) = 3; DTwin (15, 17) = 5; DTwin (15, 18) = 2; DTwin (15, 19) = 2; DTwin (15, 20) = 1; DTwin (15, 21) = 2; DTwin (16, 17) = 3; DTwin (16, 18) = 5; DTwin (16, 19) = 2; DTwin (16, 20) = 2; DTwin (16, 21) = 1; DTwin (17, 18) = 3; DTwin (17, 19) = 5; DTwin (17, 20) = 2; DTwin (17, 21) = 2; DTwin (18, 19) = 3; DTwin (18, 20) = 5; DTwin (18, 21) = 2; DTwin (19, 20) = 3; DTwin (19, 21) = 5; DTwin (20, 21) = 3.

$$\mathrm{SDTwin}(G) = \frac{28n^2 + 56n}{2} = \frac{28(10)^2 + 56(10)}{2} = \frac{3{,}360}{2} = 1{,}680$$

$$\mathrm{DTD}(G) = \frac{28n^2 + 56n}{2\binom{2n+1}{2}} = \frac{1{,}680}{\binom{2(10)+1}{2}} = \frac{1{,}680}{210} = 8.$$

Theorem 5.4

If $G = K_{1,m}(C_n)$, then $\mathrm{DTD}(G) = \dfrac{13m^2 + 8\,nm + m}{2\binom{mn+1}{2}}$, where $n \geq 5, m \geq 2$.

Proof:

Consider the graph $G = K_{1,m}(C_n)$ with $mn+1$ vertex. Let $K_{1,m}(C_n)$ be the graph obtained by pasting m copies of cycle C_n to the pendent vertices of $K_{1,m}$. Let a be the root vertex of star graph and $a_{1i}, a_{2i}, \ldots, a_{ni}$ (where $= 1, 2, \ldots, m$) be the cycle vertices.

$\text{SDTwin}(G) = \sum \text{DTwin}(u,v)$ for $u, vV(G)$.

For any star graph, $\text{SDTwin}(K_{1,m}) = \frac{m(m+1)}{2}$.

For any cycle graph, $\text{SDTwin}(C_n) = 4n$; we have m copy of C_n.

$\text{SDTwin}(m \text{ copies of } C_n) = 4\,mn$.

$\text{DTwin}(a, a_{ij}) = 1$, for $i = 1$ to m; $j = 2$ to 4. Therefore, $\sum_{i=1}^{m} \text{DTwin}(a, a_{ij}) = 3m$.

$\text{DTwin}(a, a_{ij}) = 1$, for $i = 1$ to m; $j = n-2$ to n. Therefore, $\sum_{i=1}^{m} \text{DTwin}(a, a_{ij}) = 3m$.

$\text{DTwin}(a_{i1}, a_{jk}) = 1$, for $i = 1$ to $m-1$; $j = 2$ to m; $k = 2, 3$; $j > i$. Therefore,

$\sum_{i=1}^{m-1} \text{DTwin}(a_{i1}, a_{jk}) = 2\left[\frac{m(m-1)}{2}\right]$.

$\text{DTwin}(a_{i1}, a_{jk}) = 1$, for $i = 1$ to $m-1$; $j = 2$ to m; $k = n, n-1$; $j > i$. Therefore,

$\sum_{i=1}^{m-1} \text{DTwin}(a_{i1}, a_{jk}) = 2\left[\frac{m(m-1)}{2}\right]$.

$\text{DTwin}(a_{i2}, a_{jk}) = 1$, for $i = 1$ to $m-1$; $j = 2$ to m; $k = 1, 2, n$; $j > i$. Therefore,

$\sum_{i=1}^{m-1} \text{DTwin}(a_{i2}, a_{jk}) = 3\left[\frac{m(m-1)}{2}\right]$.

$\text{DTwin}(a_{i3}, a_{j1}) = 1$, for $i = 1$ to $m-1$; $j = 2$ to m; $j > i$. Therefore,

$\sum_{i=1}^{m-1} \text{DTwin}(a_{i3}, a_{j1}) = \left[\frac{m(m-1)}{2}\right]$.

$\text{DTwin}(a_{in}, a_{jk}) = 1$, for $i = 1$ to $m-1$; $j = 2$ to m; $k = 1, 2, n$; $j > i$. Therefore,

$\sum_{i=1}^{m-1} \text{DTwin}(a_{in}, a_{jk}) = 3\left[\frac{m(m-1)}{2}\right]$.

$\text{DTwin}(a_{i(n-1)}, a_{j1}) = 1$, for $i = 1$ to $m-1$; $j = 2$ to m; $j > i$. Therefore,

$\sum_{i=1}^{m-1} \text{DTwin}(a_{i(n-1)}, a_{j1}) = \left[\frac{m(m-1)}{2}\right]$.

$$\text{SDTwin}(G) = \frac{m(m+1)}{2} + 4nm + 3m + 3m + 2\left[\frac{m(m-1)}{2}\right] + 2\left[\frac{m(m-1)}{2}\right]$$
$$+ 3\left[\frac{m(m-1)}{2}\right] + \left[\frac{m(m-1)}{2}\right] + 3\left[\frac{m(m-1)}{2}\right] + \left[\frac{m(m-1)}{2}\right].$$

$$\text{SDTwin}(G)=\frac{m^2+m+8nm+12m+12m^2-12m}{2}=\frac{13m^2+8mn+m}{2}$$

$$\text{DTD}(G)=\frac{13m^2+8mn+m}{2\binom{mn+1}{2}}.$$

Illustration. For the graph $G = K_{1,4}(C_6)$

From Figure 5.6, DTwin (1, 2) = 1; DTwin (1, 3) = 1; DTwin (1, 4) = 1; DTwin (1, 5) = 2; DTwin (1, 6) = 1; DTwin (1, 7) = 1; DTwin (1, 8) = 1; DTwin (1, 9) = 1; DTwin (1, 10) = 1; DTwin (1, 11) = 2; DTwin (1, 12) = 1; DTwin (1, 13) = 1; DTwin (1, 14) = 1; DTwin (1, 15) = 1; DTwin (1, 16) = 1; DTwin (1, 17) = 2; DTwin (1, 18) = 1; DTwin (1, 19) = 1; DTwin (1, 20) = 1; DTwin (1, 21) = 1; DTwin (1, 22) = 1; DTwin (1, 23) = 2; DTwin (1, 24) = 1; DTwin (1, 25) = 1; DTwin (2, 3) = 1; DTwin (2, 4) = 2; DTwin (2, 5) = 2; DTwin (2, 6) = 2; DTwin (2, 7) = 1; DTwin (2, 8) = 1; DTwin (2, 9) = 1; DTwin (2, 10) = 1; DTwin (2, 12) = 1; DTwin (2, 13) = 1; DTwin (2, 14) = 1; DTwin (2, 15) = 1; DTwin (2, 16) = 1; DTwin (2, 18) = 1; DTwin (2, 19) = 1; DTwin (2, 20) = 1; DTwin (2, 21) = 1; DTwin (2, 22) = 1; DTwin (2, 24) = 1; DTwin (2, 25) = 1; DTwin (3, 4) = 1; DTwin (3, 5) = 2; DTwin (3, 6) = 2; DTwin (3, 7) = 2; DTwin (3, 8) = 1; DTwin (3, 9) = 1; DTwin (3, 13) = 1; DTwin (3, 14) = 1; DTwin (3, 15) = 1; DTwin (3, 19) = 1; DTwin (3, 20) = 1; DTwin (3, 21) = 1; DTwin (3, 25) = 1; DTwin (4, 5) = 1; DTwin (4, 6) = 2; DTwin (4, 7) = 2; DTwin (4, 8) = 1; DTwin (4, 14) = 1; DTwin (4, 20) = 1;

DTwin (5, 6) = 1; DTwin (5, 7) = 2; DTwin (6, 7) = 1; DTwin (6, 8) = 1; DTwin (6, 14) = 1; DTwin (6, 20) = 1; DTwin (7, 8) = 1; DTwin (7, 9) = 1; DTwin (7, 13) = 1; DTwin (7, 14) = 1; DTwin (7, 15) = 1; DTwin (7, 19) = 1; DTwin (7, 20) = 1; DTwin (7, 21) = 1; DTwin (7, 25) = 1; DTwin (8, 9) = 1; DTwin (8, 10) = 2; DTwin (8, 11) = 2; DTwin (8, 12) = 2; DTwin (8, 13) = 1; DTwin (8, 14) = 1; DTwin (8, 15) = 1; DTwin (8, 16) = 1; DTwin (8, 18) = 1; DTwin (8, 19) = 1; DTwin (8, 20) = 1; DTwin (8, 21) = 1; DTwin (8, 22) = 1; DTwin (8, 24) = 1; DTwin (8, 25) = 1; DTwin (9, 10) = 1; DTwin (9, 11) = 2; DTwin (9, 12) = 2; DTwin (9, 13) = 2; DTwin (9, 14) = 1; DTwin (9, 15) = 1; DTwin (9, 19) = 1; DTwin (9, 20) = 1; DTwin (9, 21) = 1; DTwin (9, 25) = 1; DTwin

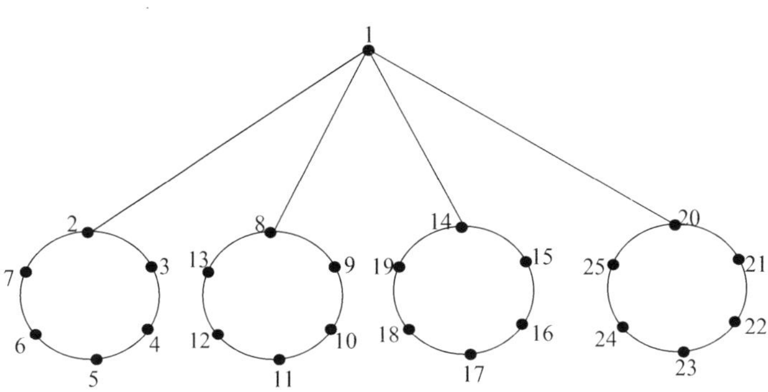

FIGURE 5.6 $K_{1,m}(C_n)$.

(10, 11) = 1; DTwin (10, 12) = 2; DTwin (10, 13) = 2; DTwin (10, 14) = 1; DTwin (10, 20) = 1; DTwin (11, 12) = 1; DTwin (11, 13) = 2; DTwin (12, 13) = 1; DTwin (12, 14) = 1; DTwin (12, 20) = 1; DTwin (13, 14) = 1; DTwin (13, 15) = 1; DTwin (13, 19) = 1; DTwin (13, 20) = 1; DTwin (13, 21) = 1; DTwin (13, 25) = 1; DTwin (14, 15) = 1; DTwin (14, 16) = 2; DTwin (14, 17) = 2; DTwin (14, 18) = 2; DTwin (14, 19) = 1; DTwin (14, 20) = 1; DTwin (14, 21) = 1; DTwin (14, 22) = 1; DTwin (14, 24) = 1; DTwin (14, 25) = 1; DTwin (15, 16) = 1; DTwin (15, 17) = 2; DTwin (15, 18) = 2; DTwin (15, 19) = 2; DTwin (15, 20) = 1; DTwin (15, 21) = 1; DTwin (15, 25) = 1; DTwin (16, 17) = 1; DTwin (16, 18) = 2; DTwin (16, 19) = 2; DTwin (16, 20) = 1; DTwin (17, 18) = 1; DTwin (17, 19) = 2; DTwin (18, 19) = 1; DTwin (18, 20) = 1; DTwin (19, 20) = 1; DTwin (19, 21) = 1; DTwin (19, 25) = 1; DTwin (20, 21) = 1; DTwin (20, 22) = 2; DTwin (20, 23) = 2; DTwin (20, 24) = 2; DTwin (20, 25) = 1; DTwin (21, 22) = 1; DTwin (21, 23) = 2; DTwin (21, 24) = 2; DTwin (21, 25) = 2; DTwin (22, 23) = 1; DTwin (22, 24) = 2; DTwin (22, 25) = 2; DTwin (23, 24) = 1; DTwin (23, 25) = 2; DTwin (24, 25) = 1.

$$\mathrm{SDTwin}(G) = \frac{13m^2 + 8mn + m}{2} = \frac{13(4)^2 + 8(4)(6) + (4)}{2} = \frac{404}{2} = 202.$$

$$\mathrm{DTD}(G) = \frac{13m^2 + 8mn + m}{2\binom{mn+1}{2}} = \frac{202}{\binom{(4)(6)+1}{2}} = \frac{202}{300}.$$

Theorem 5.5

If $G = P_n(u,v)$, then $\mathrm{DTD}(G) = \dfrac{12n^2 - 16n + 8nm}{2\binom{mn+2}{2}}$, where $n \ge 2, m \ge 2$.

Proof:

Consider the graph $G = P_n(u,v)$ with $mn+2$ vertexes. Let $\{b_{i1}, b_{i2}, \ldots, b_{im}\}$ be the vertices of the ith path for $i = 1$ to n.

Now join the initial vertices $\{b_{11}, b_{21}, \ldots, b_{n1}\}$ to the vertex u and join the terminal vertices $\{b_{1m}, b_{2m}, \ldots, b_{mn}\}$ to the vertex v.

$\mathrm{SDTwin}(G) = \sum \mathrm{DTwin}(u,v)$ for $u, vV(G)$.

For any path P_m, $\mathrm{SDTwin}(P_m) = 4m - 10$.

We have n copies of P_m, so $\mathrm{SDTwin}(nP_m) = n(4m - 10)$.

$\mathrm{DTwin}(u, b_{i1}) = 1$, for $i = 1$ to n. Therefore, $\sum_{i=1}^{n} \mathrm{DTwin}(u, b_{i1}) = n$.

$\mathrm{DTwin}(u, b_{i2}) = 1$, for $i = 1$ to n. Therefore, $\sum_{i=1}^{n} \mathrm{DTwin}(u, b_{i2}) = n$.

$\mathrm{DTwin}(u, b_{i3}) = 1$, for $i = 1$ to n. Therefore, $\sum_{i=1}^{n} \mathrm{DTwin}(u, b_{i3}) = n$.

$\text{DTwin}(u, b_{i4}) = 1$, for $i = 1$ to n. Therefore, $\sum_{i=1}^{n} \text{DTwin}(u, b_{i4}) = n$.

$\text{DTwin}(v, b_{in}) = 1$, for $i = 1$ to n. Therefore, $\sum_{i=1}^{n} \text{DTwin}(v, b_{in}) = n$.

$\text{DTwin}(v, b_{i(n-1)}) = 1$, for $i = 1$ to n. Therefore, $\sum_{i=1}^{n} \text{DTwin}(v, b_{i(n-1)}) = n$.

$\text{DTwin}(v, b_{i(n-2)}) = 1$, for $i = 1$ to n. Therefore, $\sum_{i=1}^{n} \text{DTwin}(v, b_{i(n-2)}) = n$.

$\text{DTwin}(v, b_{i(n-3)}) = 1$, for $i = 1$ to n. Therefore, $\sum_{i=1}^{n} \text{DTwin}(v, b_{i(n-3)}) = n$.

$\text{DTwin}(b_{i1}, b_{jk}) = 1$, for $i = 1$ to $n-1$; $j = 2$ to n; $k = 1,2,3$. Therefore,
$\sum_{i=1}^{n-1} \text{DTwin}(b_{i1}, b_{jk}) = 3\left[\frac{n(n-1)}{2}\right]$.

$\text{DTwin}(b_{i2}, b_{jk}) = 1$, for $i = 1$ to $n-1$; $j = 2$ to n; $k = 1,2$. Therefore,
$\sum_{i=1}^{n-1} \text{DTwin}(b_{i1}, b_{jk}) = 2\left[\frac{n(n-1)}{2}\right]$.

$\text{DTwin}(b_{i3}, b_{jk}) = 1$, for $i = 1$ to $n-1$; $j = 2$ to n; $k = 1$. Therefore,
$\sum_{i=1}^{n-1} \text{DTwin}(b_{i3}, b_{jk}) = \left[\frac{n(n-1)}{2}\right]$.

$\text{DTwin}(b_{im}, b_{jk}) = 1$, for $i = 1$ to $n-1$; $j = 2$ to n; $k = m, m-1, m-2$. Therefore,
$\sum_{i=1}^{n-1} \text{DTwin}(b_{im}, b_{jk}) = 3\left[\frac{n(n-1)}{2}\right]$.

$\text{DTwin}(b_{i(m-1)}, b_{jk}) = 1$, for $i = 1$ to $n-1$; $j = 2$ to n; $k = m, m-1$. Therefore,
$\sum_{i=1}^{n-1} \text{DTwin}(b_{i(m-1)}, b_{jk}) = 2\left[\frac{n(n-1)}{2}\right]$.

$\text{DTwin}(b_{i(m-2)}, b_{jk}) = 1$, for $i = 1$ to $n-1$; $j = 2$ to n; $k = m$. Therefore,
$\sum_{i=1}^{n-1} \text{DTwin}(b_{i(m-2)}, b_{jk}) = \left[\frac{n(n-1)}{2}\right]$.

$$\text{SDTwin}(G) = 4nm - 10n + n + n + n + n + n + n + 3\left[\frac{n(n-1)}{2}\right] + 2\left[\frac{n(n-1)}{2}\right] + \left[\frac{n(n-1)}{2}\right] + 3\left[\frac{n(n-1)}{2}\right] + 2\left[\frac{n(n-1)}{2}\right] + \left[\frac{n(n-1)}{2}\right].$$

$$\text{SDTwin}(G) = \frac{12n^2 + 8nm - 4n - 12n}{2}.$$

$$\text{SDTwin}(G) = \frac{12n^2 + 8nm - 16n}{2}$$

$$\text{DTD}(G) = \frac{12n^2 + 8nm - 4n - 12n}{2\binom{mn+2}{2}}.$$

Illustration. For the graph $P_4(u,v)$

From Figure 5.7, DTwin (1, 2) = 1; DTwin (1, 3) = 1; DTwin (1, 4) = 1; DTwin (1, 5) = 1; DTwin (1, 6) = 1; DTwin (1, 7) = 1; DTwin (1, 8) = 1; DTwin (1, 9) = 1; DTwin (1, 10) = 1; DTwin (1, 11) = 1; DTwin (1, 12) = 1; DTwin (1, 13) = 1; DTwin (2, 3) = 1; DTwin (2, 4) = 1; DTwin (2, 5) = 1; DTwin (2, 6) = 1; DTwin (2, 7) = 1; DTwin (2, 8) = 1; DTwin (2, 10) = 1; DTwin (2, 11) = 1; DTwin (2, 12) = 1; DTwin (2, 14) = 1; DTwin (3, 4) = 1; DTwin (3, 5) = 1; DTwin (3, 6) = 1; DTwin (3, 7) = 1; DTwin (3, 9) = 1; DTwin (3, 10) = 1; DTwin (3, 11) = 1; DTwin (3, 13) = 1; DTwin (3, 14) = 1; DTwin (4, 5) = 1; DTwin (4, 6) = 1; DTwin (4, 8) = 1; DTwin (4, 9) = 1; DTwin (4, 10) = 1; DTwin (4, 12) = 1; DTwin (4, 13) = 1; DTwin (4, 14) = 1; DTwin (5, 7) = 1; DTwin (5, 8) = 1; DTwin (5, 9) = 1; DTwin (5, 11) = 1; DTwin (5, 12) = 1; DTwin (5, 13) = 1; DTwin (5, 14) = 1; DTwin (6, 7) = 1; DTwin (6, 8) = 1; DTwin (6, 9) = 1; DTwin (6, 10) = 1; DTwin (6, 11) = 1; DTwin (6, 12) = 1; DTwin (6, 14) = 1; DTwin (7, 8) = 1; DTwin (7, 9) = 1; DTwin (7, 10) = 1; DTwin (7, 11) = 1; DTwin (7, 13) = 1; DTwin (7, 14) = 1; DTwin (8, 9) = 1; DTwin (8, 10) = 1; DTwin (8, 12) = 1; DTwin (8, 13) = 1; DTwin (8, 14) = 1; DTwin (9, 11) = 1; DTwin (9, 12) = 1; DTwin (9, 13) = 1; DTwin (9, 14) = 1; DTwin (10, 11) = 1; DTwin (10, 12) = 1; DTwin (10, 13) = 1; DTwin (10, 14) = 1; DTwin (11, 12) = 1; DTwin (11, 13) = 1; DTwin (11, 14) = 1; DTwin (12, 13) = 1; DTwin (12, 14) = 1; DTwin (13, 14) = 1.

$$\text{SDTwin}(G) = \frac{12n^2 + 8nm - 16n}{2} = \frac{12(3)^2 + 8(3)(4) - 16(4)}{2} = 78.$$

$$\text{DTD}(G) = \frac{12n^2 + 8nm - 4n - 12n}{2\binom{mn+2}{2}} = \frac{78}{91}.$$

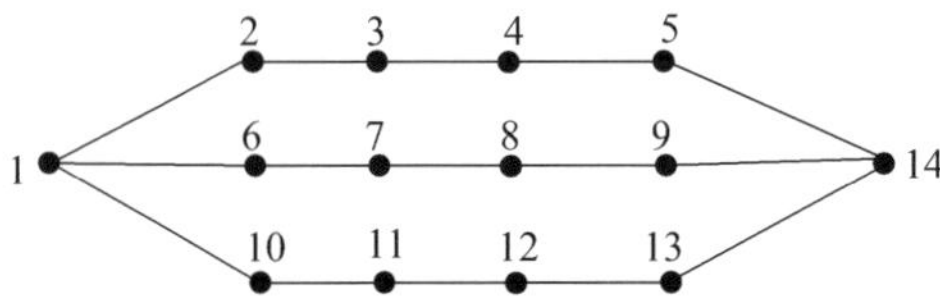

FIGURE 5.7 Uniform n-ply.

Theorem 5.6

If $G = T_n(K_{1.m})$, for $n \geq 3$ and $m \geq 2$, then

$$\text{DTD}(G) = \frac{m^2 + 25m + 84n - 204}{2\binom{2n-1+m}{2}}.$$

Proof:

$G = T_n(K_{1.m})$ is obtained by pasting the root vertex of $K_{(1.m)}$ to the nth vertex of the path in the triangular snake graph T_n. Hence, $T_n(K_{1.m})$ is the graph with $2n - 1 + \text{m}$ vertex.

Let $a_1, a_2, \ldots, a_n$ be the vertices of the path P_n, $b_1, b_2, \ldots, b_{n-1}$ be the vertices corresponding to their edges, and $a_1, a_2, \ldots, a_n$ be the pendent vertices of the star graph.

$$\text{SDTwin}(G) = \text{for } u, vV(G).$$

For any star $(K_{1.m})$, $\text{SDTwin}(G) = \dfrac{m(m+1)}{2}$.

$\text{DTwin}(a_i, a_{i+1}) = 2$, for $i = 1$ to $n-1$. Therefore. $\sum_{i=1}^{n-1} \text{DTwin}(a_i, a_{i+1}) = 2(n-1)$.

$\text{DTwin}(a_i, a_{i+2}) = 4$, for $i = 1$ to $n-2$. Therefore. $\sum_{i=1}^{n-2} \text{DTwin}(a_i, a_{i+2}) = 4(n-2)$.

$\text{DTwin}(a_i, a_{i+3}) = 4$, for $i = 1$ to $n-3$. Therefore, $\sum_{i=1}^{n-3} \text{DTwin}(a_i, a_{i+3}) = 4(n-3)$.

$\text{DTwin}(a_i, a_{i+4}) = 1$, for $i = 1$ to $n-4$. Therefore, $\sum_{i=1}^{n-4} \text{DTwin}(a_i, a_{i+4}) = (n-4)$.

$\text{DTwin}(a_i, b_i) = 2$, for $i = 1$ to $n-1$. Therefore, $\sum_{i=1}^{n-1} \text{DTwin}(a_i, b_i) = 2(n-1)$.

$\text{DTwin}(a_i, b_{i+1}) = 4$, for $i = 1$ to $n-2$. Therefore, $\sum_{i=1}^{n-2} \text{DTwin}(a_i, b_{i+1}) = 4(n-2)$.

$\text{DTwin}(a_i, b_{i+2}) = 4$, for $i = 1$ to $n-3$. Therefore, $\sum_{i=1}^{n-3} \text{DTwin}(a_i, b_{i+2}) = 4(n-3)$.

$\text{DTwin}(a_i, b_{i+3}) = 1$, for $i = 1$ to $n-4$. Therefore, $\sum_{i=1}^{n-4} \text{DTwin}(a_i, b_{i+3}) = (n-4)$.

$\text{DTwin}(a_i, b_{i-1}) = 2$, for $i = 2$ to n. Therefore, $\sum_{i=2}^{n} \text{DTwin}(a_i, b_{i-1}) = 2(n-1)$.

$\mathrm{DTwin}(a_i, b_{i-2}) = 4$, for $i = 3$ to n. Therefore, $\sum_{i=3}^{n} \mathrm{DTwin}(a_i, b_{i-2}) = 4(n-2)$.

$\mathrm{DTwin}(a_i, b_{i-3}) = 4$, for $i = 4$ to n. Therefore, $\sum_{i=4}^{n} \mathrm{DTwin}(a_i, b_{i-3}) = 4(n-3)$.

$\mathrm{DTwin}(a_i, b_{i-4}) = 1$, for $i = 5$ to n. Therefore, $\sum_{i=5}^{n} \mathrm{DTwin}(a_i, b_{i-4}) = (n-4)$.

$\mathrm{DTwin}(b_i, b_{i+1}) = 4$, for $i = 1$ to $n-2$. Therefore, $\sum_{i=1}^{n-2} \mathrm{DTwin}(b, b_{i+1}) = 4(n-2)$.

$\mathrm{DTwin}(b_i, b_{i+2}) = 4$, for $i = 1$ to $n-3$. Therefore, $\sum_{i=1}^{n-3} \mathrm{DTwin}(b_i, b_{i+2}) = 4(n-3)$.

$\mathrm{DTwin}(b_i, b_{i+3}) = 1$, for $i = 1$ to $n-4$. Therefore, $\sum_{i=1}^{n-4} \mathrm{DTwin}(b_i, b_{i+3}) = (n-4)$.

$\mathrm{DTwin}\left(a_{(n-3)}, c_j\right) = 1$, for $j = 1$ to m. Therefore, $\sum_{j=1}^{m} \mathrm{DTwin}\left(a_{(n-3)}, c_i\right) = m$.

$\mathrm{DTwin}\left(a_{(n-2)}, c_i\right) = 3$, for $j = 1$ to m. Therefore, $\sum_{j=1}^{m} \mathrm{DTwin}\left(a_{(n-2)}, c_i\right) = 3m$.

$\mathrm{DTwin}\left(a_{(n-1)}, c_i\right) = 2$, for $j = 1$ to m. Therefore, $\sum_{j=1}^{m} \mathrm{DTwin}\left(a_{(n-1)}, c_i\right) = 2m$.

$\mathrm{DTwin}\left(b_{(n-3)}, c_i\right) = 1$, for $j = 1$ to m. Therefore, $\sum_{j=1}^{m} \mathrm{DTwin}\left(b_{(n-3)}, c_i\right) = m$.

$\mathrm{DTwin}\left(b_{(n-2)}, c_i\right) = 3$, for $j = 1$ to m. Therefore, $\sum_{j=1}^{m} \mathrm{DTwin}\left(b_{(n-2)}, c_i\right) = 3m$.

$\mathrm{DTwin}\left(b_{(n-1)}, c_i\right) = 2$, for $j = 1$ to m. Therefore, $\sum_{j=1}^{m} \mathrm{DTwin}\left(b_{(n-1)}, c_i\right) = 2m$.

$$
\begin{aligned}
\mathrm{SDTwin}(G) = \frac{m(m+1)}{2} &+ 2(n-1) + 3(n-2) + 4(n-3) + (n-4) + 2(n-1) \\
&+ 4(n-2) + 4(n-3) + (n-4) + 2(n-1) + 4(n-2) \\
&+ 4(n-3) + (n-4) + 4(n-2) + 4(n-3) + (n-4) \\
&+ m + 2m + 3m + m + 2m + 3m.
\end{aligned}
$$

$$
\mathrm{SDTwin}(G) = \frac{m(m+1)}{2} + 42n - 102 + 12m.
$$

$$\text{SDTwin}(G) = \frac{m^2 + m + 84n - 204 + 24m}{2}.$$

$$\text{SDTwin}(G) = \frac{m^2 + 25m + 84n - 204}{2}$$

$$\text{DTD}(G) = \frac{m^2 + 25m + 84n - 204}{2\binom{2n-1+m}{2}}.$$

Illustration. For the graph $T_5(K_{1,5})$

From Figure 5.8, DTwin (1, 2) = 2; DTwin (1, 3) = 4; DTwin (1, 4) = 4; DTwin (1, 5) = 1; DTwin (1, 6) = 2; DTwin (1, 7) = 4; DTwin (1, 8) = 4; DTwin (1, 9) = 1; DTwin (2, 3) = 2; DTwin (2, 4) = 4; DTwin (2, 5) = 4; DTwin (2, 6) = 2; DTwin (2, 7) = 2; DTwin (2, 8) = 4; DTwin (2, 9) = 4; DTwin (2, 10) = 1; DTwin (2, 11) = 1; DTwin (2, 12) = 1; DTwin (2, 13) = 1; DTwin (2, 14) = 1; DTwin (3, 4) = 2; DTwin (3, 5) = 4; DTwin (3, 6) = 4; DTwin (3, 7) = 2; DTwin (3, 8) = 2; DTwin (3, 9) = 4; DTwin (3, 10) = 3; DTwin (3, 11) = 3; DTwin (3, 12) = 3; DTwin (3, 13) = 3; DTwin (3, 14) = 3; DTwin (4, 5) = 2; DTwin (4, 6) = 4; DTwin (4, 7) = 4; DTwin (4, 8) = 2; DTwin (4, 9) = 2; DTwin (4, 10) = 2; DTwin (4, 11) = 2; DTwin (4, 12) = 2; DTwin (4, 13) = 2; DTwin (4, 14) = 2; DTwin (5, 6) = 1; DTwin (5, 7) = 4; DTwin (5, 8) = 4; DTwin (5, 9) = 2; DTwin (5, 10) = 1; DTwin (5, 11) = 1; DTwin (5, 12) = 1; DTwin (5, 13) = 1; DTwin (5, 14) = 1; DTwin (6, 7) = 4; DTwin (6, 8) = 4; DTwin (6, 9) = 1; DTwin (7, 8) = 4; DTwin (7, 9) = 4; DTwin (7, 10) = 1; DTwin (7, 11) = 1; DTwin (7, 12) = 1; DTwin (7, 13) = 1; DTwin (7, 14) = 1; DTwin (8, 9) = 4; DTwin (8, 10) = 3; DTwin (8, 11) = 3; DTwin (8, 12) = 3; DTwin (8, 13) = 3; DTwin (8, 14) = 3; DTwin (9, 10) = 2; DTwin (9, 11) = 2; DTwin (9, 12) = 2; DTwin (9, 13) = 2; DTwin (9, 14) = 2; DTwin (10, 11) = 1; DTwin (10, 12) = 1; DTwin (10, 13) = 1; DTwin (10, 14) = 1; DTwin (11, 12) = 1; DTwin (11, 13) = 1; DTwin (11, 14) = 1; DTwin (12, 13) = 1; DTwin (12, 14) = 1; DTwin (13, 14) = 1.

$$\text{SDTwin}(G) = \frac{m^2 + 25m + 84n - 204}{2} = \frac{(5)^2 + 25(5) + 84(5) - 204}{2} = 183.$$

$$\text{DTD}(G) = \frac{m^2 + 25m + 84n - 204}{2\binom{2n-1+m}{2}} = \frac{183}{\binom{2(5)-1+(5)}{2}} = \frac{183}{91}.$$

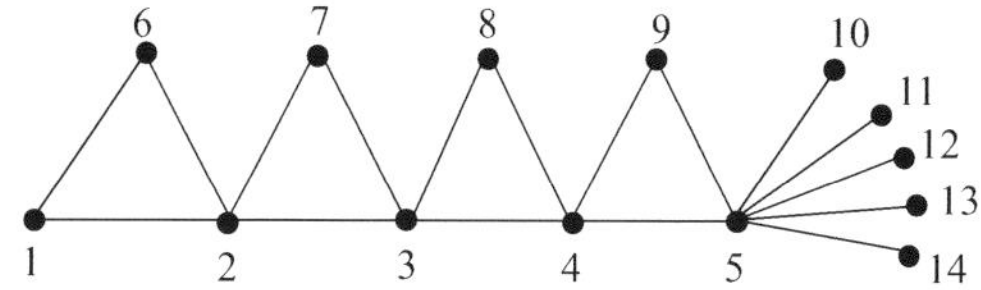

FIGURE 5.8 $T_n(K_{1,m})$.

5.4 CONCLUSION

In this chapter, we provided a mathematical modeling of the real-life situation using our double twin domination number of a graph in the medical field. More precisely, in case of emergency or any block in the roads, we suggested the ambulance driver the possible number of ways (not only shortest path) to transform the patient from the accident zone to the hospitals. The above model is nothing but finding our double twin domination number of a graph. We have investigated this parameter for some special types of graphs such as tadpole graph, lollipop graph, closed helm, uniform n-ply, and $K_{1,m}(C_n), T_n(K_{1,m})$.

REFERENCES

1. D. Vargor, P. Dundar, The medium domination number of a graph, *International Journal of Pure and Applied Mathematics*, 70(3), 297–306, 2011.
2. M. Ramachandran, N. Parvathi, The medium domination number of Jahangir graph J, nm, *Indian Journal of Science Technology*, 8(5), 400–406, 2015.
3. G. Mahadevan, V. Vijayalakshmi, C. Sivagnanam, Investigation of the medium domination number of some special types of graphs, *Australian Journal of Basic and Applied Sciences*, 9(35), 126–129, 2015.
4. G. Mahadevan, V. Vijayalakshmi, C. Sivagnanam, Medium domination number of J(m, n), Fn and S`(Bk, k), *Journal of Graph Labeling*, 2(2), 263–272, 2016.
5. G. Mahadevan, V. Vijayalakshmi, C. Sivagnanam, Extended medium domination number of a graph, *International Journal of Applied Engineering Research*, 10(92), 355–360, 2015.
6. G. Mahadevan, V. Vijayalakshmi, C. Sivagnanam, General result of extended medium domination number of a graph B(n, n), J(m, n) and Cn Θ K1,n, FACT *2016* - Proceedings of *4th* National Conference on Frontiers in Applied Sciences and Computer Technology, National Institute of Technology, Trichy, 4, 269–273, 2016.
7. G. Mahadevan, V. Vijayalakshmi, C. Sivagnanam, Some new results on extended medium domination number of few classes of graphs, *International Journal of Control Theory and Applications*, 9(10), 4241–4248, 2016.
8. G. Mahadevan, V. Vijayalakshmi, C. Sivagnanam, Extended medium domination number of some specialized types of trees and product graphs, *Asian Journal of Research in Social Sciences and Humanities*, 6(9), 1937–1953, 2016.
9. G. Mahadevan, V. Vijayalakshmi, C. Sivagnanam, Further results of extended medium domination number of some special types of graphs, *International Journal of Engineering Science, Advanced Computing and Bio-Technology*, 8(2), 55–74, 2017.
10. G. Mahadevan, V. Vijayalakshmi, Extended medium domination number of a Jahangir graph, *International Journal of Engineering and Applied Sciences*, 13(11), 8843–8846, 2018.
11. G. Mahadevan, S. Anuthiya, Double Twin Domination Number and Its Various Derived Graphs (Communicated).

6 Fractional SIRI Model with Delay in Context of the Generalized Liouville–Caputo Fractional Derivative

Ndolane Sene
Cheikh Anta Diop University

CONTENTS

6.1 INTRODUCTION

The applications of fractional calculus were extended in many domains: in fluids [18], in mechanics [1,5,18], in physics [15,20], in mathematical physic [1,19], in engineering [6], in mathematical modeling [1,5,6,12,17,24], in mathematical biology [14,23], and others. The importance of the fractional derivative is the fact that it is a derivative with memory. It is considered as the best derivative in modeling the dynamical systems. For recent applications of fractional calculus in the real-world problems, see Refs. [12,18,24]. Recently, the researchers have focused the epidemiological models as SIR model, as SIRI model, and as SIRS model according to the fractional derivatives. The investigations concern the stability of the free disease point, which corresponds to the disease that always dies out [11] and the stability of the endemic point, which is the persistence of the disease at this point when the disease is present. All the investigations are done according to the reproduction number denoted by $\mathcal{F}_0$. The studies are done according to the following conditions: $\mathcal{F}_0 = 1$, $\mathcal{F}_0 < 1$, and $\mathcal{F}_0 > 1$.

The literature of the epidemiological models in fractional calculus is very long. We summarize some of them in the next lines. Elazzouzi et al. [8] studied the asymptotic global stability of the fundamental points of the SIRI equation with delay. Wang and Wang [23] proved the stability of the trivial point for the SIR equation in the presence of saturated and treatment functions. Lahrouz et al. [11] proposed work on the SIRS model. Li and Bie [13] investigated the stability of the trivial points for the SIRS reaction-diffusion model with the Lyapunov function method. Rostamy and Mottaghi [16] investigated the stability of epidemics model admitting many fundamental points. Mouaouine et al. [14] proposed the stability analysis of the fractional SIR equation with an incidence rate represented by Liouville–Caputo derivative. Arafa et al. [4] investigated the fractional HIV infection model with drug therapy. And many others investigated the biology models. For more investigations, see Refs. [2,3].

In this paper, we investigate the fractional SIRI model using the generalized Liouville–Caputo derivative. We mainly consider the SIRI model with time delay. The SIRI model considered in this paper includes the delay in the saturated incidence rate function in the form $\frac{\vartheta SI(t-\tau)}{1+\nu S}$. We first prove the important properties as existence and uniqueness of the solution of our model. After that, we determine the reproduction number $\mathcal{R}_0$, the free fundamental point, and the endemic fundamental point. We study the global stability of the free fundamental point by a Lyapunov candidate function. Using the Lyapunov direct method, we study the stability of the endemic point of the proposed model too.

6.2 FRACTIONAL CALCULUS TOOLS AND STABILITY NOTIONS

In this section, we address the definitions and lemmas. We recall the fractional derivatives and the stability notions used in this new field. We remind the generalized fractional derivative and the recent fractional operators existing in fractional calculus.

Definition 6.1

Assuming that the function $h:[0,+\infty[\to\mathbb{R}$, we define the Liouville–Riemann integral of the function h as the following representation [9,10]:

$$\left(I_c^{\alpha,\rho}h\right)(t)=\frac{1}{\Pi(\alpha)}\int_0^t\left(\frac{t^\rho-s^\rho}{\rho}\right)^{\alpha-1}h(s)\frac{ds}{s^{1-\rho}}, \tag{6.1}$$

where $t>0, \alpha\in(0,1)$ is the order, and the function gamma is $\Pi(...)$.

Definition 6.2

Assuming that the function $h:[0,+\infty[\to\mathbb{R}$, we define the generalized Liouville–Caputo derivative of the function h as the following representation [9,10]:

$$\left(H_c^{\alpha,\rho}h\right)(t)=\frac{1}{\Pi(1-\alpha)}\int_0^t\left(\frac{t^\rho-s^\rho}{\rho}\right)^{-\alpha}h'(s)ds, \tag{6.2}$$

where $t>0, \alpha\in(0,1)$ is the order, and the function gamma is $\Pi(...)$.

Definition 6.3

Assuming that the function $h:[0,+\infty[\to\mathbb{R}$, we define the generalized derivative of the function h as the following representation [9]:

$$\left(H^{\alpha,\rho}h\right)(t)=\frac{1}{\Pi(1-\alpha)}\frac{d}{dt}\int_0^t\left(\frac{t^\rho-s^\rho}{\rho}\right)^{-\alpha}h(s)\frac{ds}{s^{1-\rho}}, \tag{6.3}$$

where $t>0, \alpha\in(0,1)$ is the order, and the function gamma is $\Pi(...)$.

Recently, new fractional derivatives appear in the literature, namely, the Fabrizio–Caputo derivative and the Baleanu–Abdon derivative.

Definition 6.4

Assuming that the function $h:[0,+\infty[\to\mathbb{R}$, we define the Fabrizio–Caputo derivative of the function h as the following representation [5]:

$$H_\alpha^{CF}h(t)=\frac{M(\alpha)}{1-\alpha}\int_0^t h'(s)\exp\left(-\frac{\alpha}{1-\alpha}(t-s)\right)ds, \tag{6.4}$$

where $t>0, \alpha\in(0,1)$ is the order, and the normalization function is $M(...)$.

Definition 6.5

Assuming that the function $h:[0,+\infty[\to\mathbb{R}$, we define the Dimitru–Abdon derivative of the function h as the following representation [5]:

$$H_\alpha^{ABC}h(t)=\frac{AB(\alpha)}{1-\alpha}\int_0^t h'(s)E_\alpha\left(-\frac{\alpha}{1-\alpha}(t-s)^\alpha\right)ds, \tag{6.5}$$

where $t>0, \alpha\in(0,1)$ is the order, and the normalization function is $AB(...)$.

The Simon Pierre Laplace transformation is used to solve the differential equations. We recall this transformation in the context of the fractional derivatives. We express the Pierre Laplace transformation of the generalized Liouville–Caputo derivative by the following equation:

$$\mathcal{L}_\rho\left\{\left(H_c^{\alpha,\rho}h\right)(t)\right\} = s^\alpha \mathcal{L}_\rho\left\{h(t)\right\} - s^{\alpha-1}h(0). \tag{6.6}$$

We recall the Pierre Laplace transformation of the function h by the following equation:

$$\mathcal{L}_\rho\left\{h(t)\right\}(s) = \int_0^\infty e^{-s\frac{t^\rho}{\rho}} h(t)\frac{dt}{t^{1-\rho}}. \tag{6.7}$$

Definition 6.6

We represent Mittag-Leffler function in our paper as the following equation [5]:

$$E_{\alpha,\beta}(t) = \sum_{j=0}^{\infty} \frac{t^j}{\Pi\left(\alpha j + \beta\right)}, \tag{6.8}$$

with $\alpha > 0$, $\beta \in \mathbb{R}$, and the variable t is into the set $\mathbb{C}$. We obtain the classical exponential when the condition $\alpha = \beta = 1$ is held.

In this chapter, we suppose the differential equation with Liouville–Caputo derivative expressed as the equation:

$$H_c^{\alpha,\rho}u = h(u,t), \tag{6.9}$$

where $u \in \mathbb{R}^n$ and f is Lipchitz continuous function $h: \mathbb{R}^n \times \mathbb{R}^+ \to \mathbb{R}^n$. We note by u^* the point which satisfies the equation $h\left(u^*\right) = 0$. We recall the definitions and lemmas; we will utilize them in our investigations.

Definition 6.7

The point u^* is stable when, for all $\epsilon > 0$, we can find $\delta = \delta(\epsilon)$ such that under the condition $u_0 < \delta$, we get the $u(t)$ of the differential Eq. (6.8) satisfying the relation $\left\|u(t) - u^*\right\| < \epsilon$ with $t > t_0$. The point x^* is asymptotically stable when the point is stable and obeys to [21,22]

$$\lim_{t\to+\infty} \left\|u(t) - u^*\right\| = 0.$$

We introduce the following lemma proved by Sene [21] for the generalized fractional derivative and fundamental in the stability analysis with the Lyapunov function method. See also Ref. [7].

Lemma 6.1

We consider the vectors $u \in \mathbb{R}^n$ which are differentiable. Under the assumption $t \geq 0$, we have the following condition [21]:

$$H_c^{\alpha,\rho}\left(\left[u-u^*\right]^T M\left[u-u^*\right]\right)\leq 2\left[u-u^*\right]^T MH_c^{\alpha,\rho}\left[u-u^*\right]\quad \alpha\in[0,1),\rho>0,$$

where $M\in\mathbb{R}^{n\times n}$ is a positive definite matrix, which is constant, square, and symmetric.

We introduce the second lemma in this work for the global asymptotic stability for the equilibrium points of the epidemiological models in the context of the Lyapunov direct method. We have the following lemma.

Lemma 6.2

We consider the vectors $u\in\mathbb{R}^n$ which are differentiable. Under the assumption $t\geq t_0$, we have the following condition [23]:

$$H_c^{\alpha,\rho}\left[u-u^*-u^*\ln\left(\frac{u}{u^*}\right)\right]\leq\left[1-\frac{u}{u^*}\right]H^{\alpha,\rho}u,\quad \text{with}\quad u^*\in\mathbb{R}^+. \tag{6.10}$$

6.3 PRESENTATION OF SIRI EPIDEMIC MODEL AND CHARACTERISTICS NUMBERS

This section addresses a class of diseases models with the generalized Liouville–Caputo derivative. The characteristic numbers as the reproduction number $\mathcal{F}_0$, and the trivial points for the fractional model will be determined. We suppose the fractional model is represented by the following equation:

$$H_c^{\alpha,\rho}S=\Delta-\mu S-\frac{\alpha SI(t-\tau)}{1+\nu S} \tag{6.11}$$

$$H_c^{\alpha,\rho}I=\frac{\alpha SI(t-\tau)}{1+\nu S}-(\mu+c+\gamma)I+\delta R \tag{6.12}$$

$$H_c^{\alpha,\rho}R=\gamma I-(\mu+\delta)R. \tag{6.13}$$

We make the following assumptions related to the initial boundary conditions:

$$S(\Theta)=\varphi_1(\Theta),\quad I(\Theta)=\varphi_2(\Theta),\quad R(\Theta)=\varphi_3(\Theta), \tag{6.14}$$

where $\theta\in[-h,0]$ and the vector function ϕ is defined as follows $\phi=(\phi_1,\phi_2,\phi_3)\in C^3([-h,0];\mathbb{R})$, where the set $\mathcal{C}([-h,0];\mathbb{R})$ contains all the functions which are continuous and mapping $[-h,0]$ into $\mathbb{R}$. In the SIRI model, S represents the value of susceptible individuals at t, I represents the value of infected individuals at t, and R represents the value of recovered individuals at t. The means of the parameters are summarized as follows: Δ is the recruitment rate, μ is the natural death rate, c is the death rate generated by infection, δ is the relapse rate, and α is the transmission rate. We represent the time delay by the parameter τ. In the beginning of this chapter, we prove the non-negativity and the boundness of the differential Eqs. (6.10)–(6.13) solutions. The procedure of the proof is classic and exists in many papers. We recall the procedure by applying it in our problem. For the proof of the positivity, we have the following reason. Let t_1 the solution of the set $\min\{t : S(t) = 0, \text{and } I(t) = 0\}$. We suppose $S(t_1) = 0$ and $S(t) < 0$ with $t \in [0, t_1]$. It follows that

$$H_c^{\alpha,\rho} S(t) > AS(t), \tag{6.15}$$

where the constant A satisfies the following relationship:

$$A = \min_{0 \le t \le t_1} \left\{ \frac{\Delta}{S} - \mu - \frac{\alpha I(t-\tau)}{1+\beta S} \right\}. \tag{6.16}$$

From Eq. (6.14), we know that we can find a continuous function satisfying the following relation:

$$H_c^{\alpha,\rho} S(t) = AS(t) + m(t). \tag{6.17}$$

Using the Simon Pierre Laplace transformation of the generalized Liouville–Caputo derivative, it is straightforward to see that the solution of Eq. (6.16) is given by the following relationship:

$$S(t) = S(0) E_\alpha \left(A \left(\frac{t^\rho}{\rho} \right)^\alpha \right) + \int_0^t \left(\frac{t^\rho - s^\rho}{\rho} \right)^{\alpha-1} E_{\alpha,\alpha} \left(A \left(\frac{t^\rho}{\rho} \right)^\alpha \right) ds. \tag{6.18}$$

From the above equation, we have the following relation:

$$S(t) \ge S(0) E_\alpha \left(A \left(\frac{t^\rho}{\rho} \right)^\alpha \right) \ge 0. \tag{6.19}$$

We repeat the same procedure to establish that the solution I is positive. We assume $I(t_1) = 0$ and $I(t) < 0$ with the condition $t \in [0, t_1]$. Then, we can find a constant B which satisfies the following equation:

$$H_c^{\alpha,\rho} I(t) > BI(t). \tag{6.20}$$

From Eq. (6.19), we know that we can find a continuous function n satisfying the equation:

$$H_c^{\alpha,\rho} I(t) = BI(t) + n(t). \tag{6.21}$$

Using the Simon Pierre Laplace transformation of the generalized Liouville–Caputo derivative, it is straightforward to see the solution of Eq. (6.20) is given by the following relationship:

$$I(t)=I(0)E_{\alpha}\left(B\left(\frac{t^{\rho}}{\rho}\right)^{\alpha}\right)+\int_{0}^{t}\left(\frac{t^{\rho}-s^{\rho}}{\rho}\right)^{\alpha-1}E_{\alpha.\alpha}\left(B\left(\frac{t^{\rho}}{\rho}\right)^{\alpha}\right)ds. \tag{6.22}$$

From the above equation, we have the following relation:

$$I(t)\geq I(0)E_{\alpha}\left(B\left(\frac{t^{\rho}}{\rho}\right)^{\alpha}\right)\geq 0. \tag{6.23}$$

Let's prove the positivity of the function R. We do the following reasoning. From Eq. (6.12), we have

$$\begin{aligned} H_{c}^{\alpha,\rho}R &= \gamma I-\mu R \\ &\geq -\mu R \end{aligned} . \tag{6.24}$$

Using the Simon Pierre Laplace transformation of the generalized Liouville–Caputo derivative, it is straightforward to see the solution of Eq. (6.12) satisfies the following relationship:

$$R(t)\geq R(0)E_{\alpha}\left(-\mu\left(\frac{t^{\rho}}{\rho}\right)^{\alpha}\right)\geq 0. \tag{6.25}$$

Finally, we conclude first $S\geq 0$, second $I\geq 0$, and at last $R\geq 0$. The last step consists of proving the boundedness of the function $S+I+R$. From the aforementioned relationships, we deduce the boundedness of all solutions of the SIRI model considered in this chapter. We have the following procedure to arrive at our end. Applying the generalized Liouville–Caputo derivative to $N=S+I+R$ and using Eqs. (6.10)–(6.12), we get the following relation:

$$H_{c}^{\alpha,\rho}N(t)=\Delta-\mu N(t)-cI(t). \tag{6.26}$$

Applying the Simon Pierre Laplace transformation of the generalized Liouville–Caputo derivative to Eq. (6.25), we obtain the following equation:

$$\begin{aligned} s^{\alpha}\bar{N}(s)-s^{\alpha-1}N(0) &= \frac{\Delta}{s}-\mu\bar{N}(s)-c\bar{I}(s) \\ \bar{N}(s) &= \frac{\Delta}{\mu}\left\{\frac{1}{s}-\frac{s^{\alpha-1}}{s^{\alpha}+\mu}\right\}+\frac{s^{\alpha-1}N(0)}{s^{\alpha}+\mu}-\frac{c\bar{I}(s)}{s^{\alpha}+\mu} \end{aligned} . \tag{6.27}$$

Applying the inverse of the Simon Pierre Laplace transformation of the generalized Liouville–Caputo derivative to Eq. (6.26), and neglecting the negative term, it follows the relationship

$$N(t) \leq \frac{\Delta}{\mu}\left[1 - E_{\alpha}\left(-\mu\left(\frac{t^{\rho}}{\rho}\right)^{\alpha}\right)\right] + N(0)E_{\alpha}\left(-\mu\left(\frac{t^{\rho}}{\rho}\right)^{\alpha}\right). \tag{6.28}$$

Equation (6.27) ensures the boundedness of all solutions for the SIRI model. Furthermore, we can observe that when the time converges to infinity, it holds the following relation:

$$N(t) \leq \frac{\Delta}{\mu}. \tag{6.29}$$

We are now ready to provide the equilibrium points of the SIRI model described by the generalized Liouville–Caputo derivative. Note that the fractional derivative has no impact on this step. The free point for the fractional SIRI model is

$$E_0 = \left(\frac{\Delta}{\mu}, 0, 0\right). \tag{6.30}$$

Before obtaining the endemic point, we get the reproduction number $\mathcal{F}_0$ of our model. Note we write Eqs. (6.10)–(6.12) in matrix form as the form:

$$H_c^{\alpha,\rho}\text{SIRI} = \Omega - \Gamma, \tag{6.31}$$

where $\text{SIRI} = (S, I, R)$, and the matrices Ω and Γ are represented in the following form:

$$\Omega = \begin{pmatrix} \dfrac{\alpha SI}{1+\nu S} \\ 0 \\ 0 \end{pmatrix} \quad \text{and} \quad \Gamma = \begin{pmatrix} (\mu + c + \gamma)I - \delta R \\ -\gamma I + (\mu + \delta)R \\ \dfrac{\alpha SI}{1+\nu S} - \Delta + \mu S \end{pmatrix}. \tag{6.32}$$

We represent the Jacobian matrix for the matrices Ω and Γ at the free point as the following form:

$$J\Omega(E_0) = \begin{pmatrix} \dfrac{\vartheta\Delta}{\mu + \nu\Delta} & 0 & 0 \\ 0 & 0 & 0 \\ 0 & 0 & 0 \end{pmatrix} \quad \text{and} \quad J\Gamma(E_0) = \begin{pmatrix} \mu + c + \gamma & -\Delta & 0 \\ -\gamma & \mu + \delta & 0 \\ \dfrac{\vartheta\Delta}{\mu + \nu\Delta} & 0 & \gamma \end{pmatrix}.$$

To calculate the reproduction number, we extract the following matrices from the above matrix:

$$F = \begin{pmatrix} \dfrac{\alpha\Delta}{\mu + v\Delta} & 0 \\ 0 & 0 \end{pmatrix} \quad \text{and} \quad V = \begin{pmatrix} \mu + c + \gamma & -\Delta \\ -\gamma & \mu + \delta \end{pmatrix}. \tag{6.33}$$

We obtain the reproduction number by applying the spectral radius to the matrix defined by FV^{-1}, which is given by

$$\mathcal{F}_0 = \rho\left(FV^{-1}\right) = \frac{\alpha\Delta(\mu+\delta)}{(\mu+\beta\Delta)\left[(\mu+c+\gamma)(\mu+\delta)-\gamma\delta\right]}. \tag{6.34}$$

Let prove the existence and uniqueness for endemic equilibrium denoted by $E_e = \left(S^*, I^*, R^*\right)$. It is the solution for the following equations:

$$D^{\alpha,\rho}S(t) = 0, \quad D^{\alpha,\rho}I(t) = 0, \quad D^{\alpha,\rho}R(t) = 0. \tag{6.35}$$

From Eq. (6.34), we have the following relationships:

$$0 = \Delta - \mu S^* - \frac{\alpha S^* I^*}{1 + vS^*}, \tag{6.36}$$

$$0 = \frac{\alpha S^* I^*}{1 + vS^*} - (\mu + c + \gamma)I^* + \delta R^*, \tag{6.37}$$

$$0 = \gamma I^* - (\mu + \delta)R^*. \tag{6.38}$$

We first assume the existence and uniqueness of I^*. We will prove its existence later. We first give the form of the endemic equilibrium point. From Eq. (6.37), we have

$$R^* = \frac{\gamma}{\mu + \delta} I^*. \tag{6.39}$$

Summing Eq. (6.35) and Eq. (6.36), we obtain the following relationship:

$$\Delta - \mu S^* - (\mu + c + \gamma)I^* + \delta R^* = 0. \tag{6.40}$$

From the aforementioned equation, we get the following relationship:

$$S^* = \frac{\Delta}{\mu} - \frac{\mu + c + \gamma - \dfrac{\delta\gamma}{\mu + \delta}}{\mu} I^*. \tag{6.41}$$

Note the existence and uniqueness of I^* are sufficient for the existence of S^* and R^*. We use the classical mean value theorem to prove it. Let the function m as the form

$$m(i) = \frac{\alpha\left[\frac{\Delta}{\mu} - \frac{\mu + c + \gamma - \frac{\delta\gamma}{\mu+\delta}}{\mu} i\right]}{1 + v\left[\frac{\Delta}{\mu} - \frac{\mu + c + \gamma - \frac{\delta\gamma}{\mu+\delta}}{\mu} i\right]} - (\mu + c + \gamma) + \frac{\gamma}{\mu + \delta}. \tag{6.42}$$

Note that at first from Eq. (6.41), we have the following relationship:

$$m\left(\frac{\mathcal{F}_0(\mu + v\Delta)}{\vartheta}\right) = -(\mu + c + \gamma) + \frac{\gamma}{\mu + \delta} < 0. \tag{6.43}$$

Note that at second from Eq. (6.41), we have the following relationship:

$$m(0) = \frac{\alpha\Delta}{\mu + v\Delta} - (\mu + c + \gamma) + \frac{\gamma}{\mu + \delta} = \frac{\alpha\Delta}{\mu + v\Delta}\left[1 - \frac{1}{\mathcal{F}_0}\right] > 0. \tag{6.44}$$

Furthermore, the function m is bijective, thus we can find a unique I^*, which is the solution of the equation $m(i) = 0$ and satisfies the following equation:

$$0 \leq I^* \leq \frac{\mathcal{F}_0(\mu + v\Delta)}{\alpha}. \tag{6.45}$$

Finally, we conclude the endemic equilibrium $E_e = \left(S^*, I^*, R^*\right)$ exists and is unique.

6.4 EXISTENCE AND UNIQUENESS OF THE SIRI MODEL WITH DELAY

In this section, we establish the solutions of all equations exist and are unique for our fractional model. This section justifies the investigation related to the solutions of the fractional SIRI epidemic model and the biological meaning of the model.

We begin with Eq. (6.10). Note that we can rewrite the fractional equation represented by Eq. (6.10) as the following form:

$$H_c^{\alpha,\rho} S = \Lambda(S). \tag{6.46}$$

We first prove the function Λ is Lipchitz continuous with Lipchitz constant, which we will determine. We have the following relationships for the above problem:

$$\|\Lambda(S_1)-\Lambda(S_2)\| = \left\|\Delta - \mu S_1 - \frac{\alpha S_1 I(t-\tau)}{1+\nu S_2}\right\|$$

$$= \left\|-\mu S_1 + \mu S_2 - \frac{\alpha S_1 I(t-\tau)}{1+\nu S_1} + \frac{\alpha S_1 I(t-\tau)}{1+\nu S_2}\right\|$$

$$\leq \mu\|S_1 - S_2\| + \alpha\|S_1 - S_2\|\left\|\frac{I(t-\tau)}{1+\nu S_1} + \frac{I(t-\tau)}{1+\nu S_2}\right\|. \tag{6.47}$$

We know the following relation:

$$\left\|\frac{I(t-\tau)}{1+\nu S_1} + \frac{I(t-\tau)}{1+\nu S_2}\right\| \leq \epsilon_1. \tag{6.48}$$

Substituting Eq. (6.47) into Eq. (6.46), we obtain the following relationship:

$$\|\Lambda(S_1)-\Lambda(S_2)\| \leq \epsilon\|S_1 - S_2\|, \tag{6.49}$$

where $\epsilon = \mu + \alpha\epsilon_1$. That is, the function Λ is Lipchitz continuous with Lipchitz constant ϵ.

The solution of Eq. (6.45) is obtained by applying the generalized fractional integral described in Definition 6.1; we have the following identity:

$$S(t) - S(\theta) = I^{\alpha,\rho}\Lambda(S). \tag{6.50}$$

The Picard operator is defined here by the following relationship:

$$PS(t) = S(\theta) + I^{\alpha,\rho}\Lambda(S). \tag{6.51}$$

We will prove the Picard operator is well defined by applying the Euclidean norm. We have the following equation:

$$\|PS(t) - S(\theta)\| = \left\|I^{\alpha,\rho}\Lambda(S)\right\|$$

$$\leq I^{\alpha,\rho}\|\Lambda(S)\|. \tag{6.52}$$

Using the fact Λ is Lipchitz continuous with Lipchitz constant ϵ, we obtain the following relationship:

$$\|PS(t) - S(\theta)\| \leq \frac{\rho^{1-\alpha}}{\Pi(\alpha)}\left[\frac{b^{\rho}}{\rho}\right]^{\alpha} m, \tag{6.53}$$

where $\Lambda(S) \leq m$. From Eq. (6.52), we get the function P, which is well definite. After that, we prove the operator P is a contraction, i.e.,

$$\begin{aligned}\|PS_1 - PS_2\| &= \left\|I^{\alpha,\rho}\left[\Lambda(S_1) - \Lambda(S_2)\right]\right\| \\ &\leq I^{\alpha,\rho}\left\|\Lambda(S_1) - \Lambda(S_2)\right\| \\ &\leq \left\|\Lambda(S_1) - \Lambda(S_2)\right\| I^{\alpha,\rho}(1). \end{aligned} \tag{6.54}$$

Using Eq. (6.48), we obtain the equation represented by

$$\|PS_1 - PS_2\| \leq \frac{\rho^{1-\alpha}}{\Pi(\alpha+1)}\left[\frac{b^{\rho}}{\rho}\right]^{\alpha} \epsilon\|S_1 - S_2\|. \tag{6.55}$$

From Eq. (6.54), we deduce P is a contraction when it holds the following relation:

$$\frac{\rho^{1-\alpha}}{\Pi(\alpha+1)}\left[\frac{b^{\rho}}{\rho}\right]^{\alpha} < \frac{1}{\epsilon}. \tag{6.56}$$

From the classical Banach fixed theorem, we conclude the equation $PS = S$ has unique solution.

We continue with Eq. (6.11). Note that we can rewrite the fractional equation represented by Eq. (6.11) as the following form:

$$H_c^{\alpha,\rho} I = \Omega(I). \tag{6.57}$$

We first prove the function Ω is Lipchitz continuous with Lipchitz constant, which we will determine. We have the following relationships for the above problem:

$$\begin{aligned}\left\|\Omega(S_1) - \Omega(S_2)\right\| &= \left\|\Delta - \mu S_1 - \frac{\alpha S_1 I(t-\tau)}{1+\nu S_1} - \Delta + \mu S_2 + \frac{\alpha S_2 I(t-\tau)}{1+\nu S_2}\right\| \\ &= \left\|-\mu S_1 + \mu S_2 - \frac{\alpha S_1 I(t-\tau)}{1+\nu S_1} + \frac{\alpha S_2 I(t-\tau)}{1+\nu S_2}\right\| \\ &\leq \mu\|S_1 - S_2\| + \alpha\|S_1 - S_2\|\left\|\frac{I(t-\tau)}{1+\nu S_1} + \frac{I(t-\tau)}{1+\nu S_2}\right\|. \end{aligned} \tag{6.58}$$

Or we know the following relationship:

$$\left\|\frac{I(t-\tau)}{1+\beta S_1} + \frac{I(t-\tau)}{1+\beta S_2}\right\| \leq \epsilon_1. \tag{6.59}$$

Substituting Eq. (6.58) into Eq. (6.57), we have the relation represented by

$$\left\|\Omega(S_1) - \Omega(S_2)\right\| \leq \epsilon\|S_1 - S_2\|, \tag{6.60}$$

where $\epsilon = \mu + \alpha\epsilon_1$ That is, the function Ω is Lipchitz continuous with Lipchitz constant ϵ.

The solution of Eq. (6.56) is obtained by applying the generalized fractional integral described in Definition 6.1, and we have the following identity:

$$S(t)-S(\theta)=I^{\alpha,\rho}\Omega(S). \tag{6.61}$$

The Picard operator is defined here by the following relationship:

$$PS(t)=S(\theta)+I^{\alpha,\rho}\Omega(S). \tag{6.62}$$

We will prove the Picard operator is well defined. Applying the Euclidean norm, we have the following relationship:

$$\begin{aligned}\|PS(t)-S(\theta)\| &= \left\|I^{\alpha,\rho}\Omega(S)\right\| \\ &\leq I^{\alpha,\rho}\|\Omega(S)\|.\end{aligned} \tag{6.63}$$

Using the fact Ω is Lipchitz continuous with Lipchitz constant ϵ, we obtain the following relation:

$$\|PS(t)-S(\theta)\|\leq\frac{\rho^{1-\alpha}}{\Pi(\alpha)}\left[\frac{b^{\rho}}{\rho}\right]^{\alpha}m, \tag{6.64}$$

where $\|\Omega(S)\|\leq m$. From Eq. (6.63), we get the function P which is well defined. After that, we prove the operator P is a contraction, i.e.,

$$\begin{aligned}\|PS_1-PS_2\| &= \left\|I^{\alpha,\rho}\left[\Omega(S_1)-\Omega(S_2)\right]\right\| \\ &\leq I^{\alpha,\rho}\left\|\Omega(S_1)-\Omega(S_2)\right\| \\ &\leq \left\|\Omega(S_1)-\Omega(S_2)\right\|I^{\alpha,\rho}(1).\end{aligned} \tag{6.65}$$

Using Eq. (6.59), we get the following relationship:

$$\|PS_1-PS_2\|\leq\frac{\rho^{1-\alpha}}{\Pi(\alpha+1)}\left[\frac{b^{\rho}}{\rho}\right]^{\alpha}\epsilon\|S_1-S_2\|. \tag{6.66}$$

From Eq. (6.65), we deduce P is a contraction when it holds the following relation:

$$\frac{\rho^{1-\alpha}}{\Pi(\alpha+1)}\left[\frac{b^{\rho}}{\rho}\right]^{\alpha}<\frac{1}{\epsilon}. \tag{6.67}$$

From the classical Banach fixed theorem, we conclude the existence of a unique solution for the equation $PS=S$.

We conclude with Eq. (6.12). Note that we write the fractional equation represented by Eq. (6.12) as the following form:

$$H_c^{\alpha,\rho} R = \Theta(R). \tag{6.68}$$

We first prove the function Θ is Lipchitz continuous with Lipchitz constant, which we will determine. We have the following relationships for the above problem:

$$\begin{aligned} \left\|\Theta(R_1) - \Theta(R_2)\right\| &= \left\|\gamma I - (\mu+\delta)R_1 - \gamma I + (\mu+\delta)R_2\right\| \\ &= \left\|-(\mu+\delta)R_1 + (\mu+\delta)R_2\right\| \\ &\leq (\mu+\delta)\left\|R_1 - R_2\right\|. \end{aligned} \tag{6.69}$$

After rearrangements, we obtain the following relationship:

$$\left\|\Theta(R_1) - \Theta(R_2)\right\| \leq (\mu+\delta)\left\|R_1 - R_2\right\|. \tag{6.70}$$

That is, the function Θ is Lipchitz continuous with Lipchitz constant $(\mu+\delta)$.

The solution of Eq. (6.67) is obtained by applying the generalized fractional integral described in Definition 6.1, and we have the following identity:

$$R(t) - R(\theta) = I^{\alpha,\rho}\Theta(R). \tag{6.71}$$

The Picard operator is defined here by the following relationship:

$$PR(t) = R(\theta) + I^{\alpha,\rho}\Theta(R). \tag{6.72}$$

We will prove the Picard operator is well defined by applying the Euclidean norm; we have the following relationship:

$$\begin{aligned} \left\|PR(t) - R(\theta)\right\| &= \left\|I^{\alpha,\rho}\Theta(R)\right\| \\ &\leq I^{\alpha,\rho}\left\|\Theta(R)\right\|. \end{aligned} \tag{6.73}$$

Using the fact Θ is Lipchitz continuous with Lipchitz constant $(\mu+\delta)$, we obtain the following relation:

$$\left\|PR(t) - R(\theta)\right\| \leq \frac{\rho^{1-\alpha}}{\Pi(\alpha)}\left[\frac{b^\rho}{\rho}\right]^\alpha m, \tag{6.74}$$

where $\left\|\Theta(R)\right\| \leq m$. From Eq. (6.73), we get the function P that is well defined. After that, we prove the operator P is a contraction, i.e.,

$$\begin{aligned} \left\|PR_1 - PR_2\right\| &= \left\|I^{\alpha,\rho}\left[\Theta(R_1) - \Theta(R_2)\right]\right\| \\ &\leq I^{\alpha,\rho}\left\|\Theta(R_1) - \Theta(R_2)\right\| \\ &\leq \left\|\Theta(R_1) - \Theta(R_2)\right\| I^{\alpha,\rho}(1). \end{aligned} \tag{6.75}$$

Using Eq. (6.69), we have the relation represented by

$$\|PR_1 - PR_2\| \le \frac{\rho^{1-\alpha}}{\Pi(\alpha+1)}\left[\frac{b^\rho}{\rho}\right]^\alpha (\mu+\delta)\|R_1 - R_2\|. \tag{6.76}$$

From Eq. (6.75), we deduce P is a contraction when it holds the following relationship:

$$\frac{\rho^{1-\alpha}}{\Pi(\alpha+1)}\left[\frac{b^\rho}{\rho}\right]^\alpha < \frac{1}{\mu+\delta}. \tag{6.77}$$

From the classical Banach fixed theorem, we conclude the solution of the equation $PR = R$ exists and is unique.

6.5 STABILITY OF THE SIRI EQUATION WITH DELAY

In this part, we discuss the global asymptotic stability of the free point E_0 and the endemic point E_e. We use the Lyapunov direct method. Note that Lemmas 6.1 and 6.2 are fundamental to our cases. Furthermore, from the boundedness of the solution, it follows from Eq. (6.28) the condition

$$S \le \frac{\Delta}{\mu}. \tag{6.78}$$

Equation (6.77) will be used to prove the negativity of the fractional derivative along with the trajectories.

We begin our investigation with the free point. Let the Lyapunov function be defined by

$$V(S,I,R) = S - S^* - S^*\ln\left(\frac{S}{S^*}\right) + I + \frac{\delta}{\mu+\delta}R + \theta I_t^{\alpha,\rho} I(t-\tau), \tag{6.79}$$

where $\kappa = \mu + c + \gamma$ and $\theta = \dfrac{\alpha\Delta}{\mu+\beta\Delta}$. The derivative of the Lyapunov function V is represented as

$$\begin{aligned} H_c^{\alpha,\rho} V &\le \left(1 - \frac{S^*}{S}\right) D^{\alpha,\rho} S + D^{\alpha,\rho} I + \frac{\delta}{\mu+\delta} D^{\alpha,\rho} R + \theta D^{\alpha,\rho} I_t^{\alpha,\rho} I(t-\tau) \\ &\le -\mu\frac{\left(S-S^*\right)^2}{S} + \frac{\alpha S^* I(t-\tau)}{1+vS} - \left[k - \frac{\gamma\delta}{\mu+\delta} + \theta\right] I - \theta I(t-\tau) \\ &\le -\mu\frac{\left(S-S^*\right)^2}{S} - \left[k - \frac{\gamma\delta}{\mu+\delta} + \theta\right] I - \frac{\alpha\Delta I(t-\tau)\left(S-S^*\right)}{\mu\left(1+vS^*\right)(1+vS)}. \end{aligned} \tag{6.80}$$

We can observe Eq. (6.79) is negative if and only if it holds the following identity:

$$\left[\mu + c + \gamma - \frac{\gamma\delta}{\mu + \delta}\right] + \frac{\alpha\Delta}{\mu + v\Delta} \geq 0. \tag{6.81}$$

We conclude the free point E_0 is asymptotically globally stable if and only if $\mathcal{F}_0 \leq 1$.

The second step of this section is to prove the endemic point E_1 is asymptotically globally stable if and only if the condition $\mathcal{F}_0 > 1$ is held. The proof and the calculations are not trivial. We use the Lyapunov function described by the following form:

$$V(S,I,R) = V_1(S,I,R) + V_2(S,I,R) + V_3(S,I,R), \tag{6.82}$$

where the first candidate function is in the form:

$$V_1(S,I,R) = S - S^* - S^*\ln\left(\frac{S}{S^*}\right) + I - I^* - I^*\ln\left(\frac{I}{I^*}\right), \tag{6.83}$$

the second candidate function is described in the following form:

$$V_2(S,I,R) = \kappa I_t^{\alpha,\rho}\left[I(t-\tau) - I^*\right], \tag{6.84}$$

and the third candidate function is represented by the following relationship:

$$V_3(S,I,R) = R - R^* - R^*\ln\left(\frac{R}{R^*}\right). \tag{6.85}$$

The generalized Liouville–Caputo derivative of the first function, along the trajectories of the differential Eqs. (6.10)–(6.11) and utilizing the relation in Eq. (6.9), is given in the following relationships:

$$\begin{aligned} H_c^{\alpha,\rho}V_1 &\leq \left[1 - \frac{S^*}{S}\right]H_c^{\alpha,\rho}S + \left[1 - \frac{I^*}{I}\right]H_c^{\alpha,\rho}I \\ &\leq \left[1 - \frac{S^*}{S}\right]\left[\Delta - \mu S - \frac{\alpha SI(t-\tau)}{1+vS}\right] \\ &+ \left[1 - \frac{I^*}{I}\right]\left[\frac{\alpha SI(t-\tau)}{1+vS} - (\mu + c + \gamma)I + \delta R\right] \end{aligned} \tag{6.86}$$

Under endemic equilibrium, we have the following equations:

$$\Delta = \mu S^* + \frac{\alpha S^* I^*}{1 + vS^*} \tag{6.87}$$

$$\frac{\alpha S^* I^*}{1 + vS^*} = (\mu + c + \gamma)I^* - \delta R^* \tag{6.88}$$

$$\gamma I^* = (\mu + \delta)R^*. \tag{6.89}$$

Substituting Eqs. (6.86)–(6.88) into Eq. (6.85), we have the following calculations:

$$D^{\alpha,\rho}V_1 \le -\mu\frac{\left(S-S^*\right)^2}{S} - \left[1-\frac{S^*}{S}\right]\left[\frac{\alpha SI(t-\tau)}{1+\nu S} - \frac{\alpha S^* I^*}{1+\nu S^*}\right]$$
$$+\left[1-\frac{I^*}{I}\right]\left[\frac{\alpha SI(t-\tau)}{1+\beta S} - \frac{\alpha S^* I^*}{1+\nu S^*}\right]$$
$$-\left(\mu+c+\gamma\right)\left[1-\frac{I^*}{I}\right]\left[I-I^*\right]+\delta\left[1-\frac{I^*}{I}\right]\left[R-R^*\right]. \tag{6.90}$$

The generalized Liouville–Caputo derivative of the second function and taking into account the delay, along the solutions of the differential Eqs. (6.10)–(6.12) and using relation in Eq. (6.8), is given in the following relationships:

$$D^{\alpha,\rho}V_2 = \kappa D^{\alpha,\rho} I_t^{\alpha,\rho}\left[I(t-\tau)-I^*\right]^2$$
$$= \kappa\left[I-I^*\right]^2 - \kappa\left[I(t-\tau)-I^*\right]^2. \tag{6.91}$$

We find Eq. (6.90) in Ref. [23]. The generalized Liouville–Caputo derivative of the third function, along the solutions of the differential Eqs. (6.10)–(6.12) and using relation in Eq. (6.9), is given in the following relationships:

$$D^{\alpha,\rho}V_3 \le \left[1-\frac{R^*}{R}\right]D^{\alpha,\rho}R$$
$$\le \left[1-\frac{R^*}{R}\right]\left[\gamma I-(\mu+\delta)R\right]. \tag{6.92}$$

Equation (6.91) under endemic point gives the following results:

$$D^{\alpha,\rho}V_3 \le \gamma\left[1-\frac{R^*}{R}\right]\left[I-I^*\right]-(\mu+\delta)\left[1-\frac{R^*}{R}\right]\left[R-R^*\right]. \tag{6.93}$$

Combining Eqs. (6.89), (6.90), (6.92) under the condition $\mathcal{R}_0 \le 1$, we arrive at the condition

$$D^{\alpha,\rho}V = D^{\alpha,\rho}V_1 + D^{\alpha,\rho}V_1 + D^{\alpha,\rho}V_3 \le 0. \tag{6.94}$$

Furthermore, when the conditions $S = S^*$, $I = I^*$, $R = R^*$, and $I(t-\tau) = I^*$, we obtain, using Eqs. (6.89), (6.90), and (6.92), the following relationship:

$$D^{\alpha,\rho}V = D^{\alpha,\rho}V_1 + D^{\alpha,\rho}V_1 + D^{\alpha,\rho}V_3 = 0. \tag{6.95}$$

Finally, we conclude under condition (6.93), the endemic–epidemic E_1 is asymptotically globally stable if and only if $\mathcal{F}_0 > 1$.

6.6 CONCLUSION

In this paper, we have analyzed the global stability of the free point and the endemic point for the SIRI equation. In our work, the incidence function has a delay. We have discussed the positivity and the boundedness of the solutions of the proposed model. We investigate the existence and uniqueness in the context of the Caputo–Liouville derivative, too. For the future investigation, we will focus on the analytical or numerical solutions.

REFERENCES

1. K.A. Abro, J.F. Gomez-Aguilar, A comparison of heat and mass transfer on a Walter's-B fluid via Caputo-Fabrizio versus Atangana-Baleanu fractional derivatives using the Fox-H function, *Euro. Phys. J. Plus*, **134**(3), 1–10 (2019).
2. C.N. Angstmann, B.I. Henry, A.V. McGann, A fractional-order infectivity SIR model, *Phys. A*, **452**, 86–93 (2016).
3. C.N. Angstmann, B.I. Henry, A.V. McGann, A fractional-order infectivity and recovery SIR model, *Fractal Fract.*, **1**, 11 (2017).
4. A.A.M. Arafa, S.Z. Rida, M. Khalil, A fractional-order model of HIV infection with drug therapy effect, *J. Egypt. Math. Soc.*, **22**(3), 538–543 (2013).
5. A. Atangana, D. Baleanu, New fractional derivatives with nonlocal and non-singular Kernel: Theory and application to heat transfer model, arXiv preprint arXiv:1602.03408 (2016).
6. A. Atangana, T. Mekkaoui, Trinition the complex number with two imaginary parts: Fractal, chaos and fractional calculus, *Chaos, Solitons Fractals*, **128**, 366–381 (2019).
7. N.A. Camacho, M.A.D. Mermoud, J.A. Gallegos, Lyapunov functions for fractional order systems, *Commun. Nonlinear Sci. Numer. Simul.*, **19**, 2951–2957 (2014).
8. A. Elazzouzi, A.L. Alaoui, M. Tilioua, D.F.M. Torres, Analysis of a SIRI epidemic model with distributed delay and relapse, *Stat. Opt. Inform. Comput.*, **7**(3), 545–557 (2019).
9. J. Fahd, T. Abdeljawad, D. Baleanu, On the generalized fractional derivatives and their Caputo modification, *J. Nonlinear Sci. Appl.*, **10**, 2607–2619 (2017).
10. J. Fahd, T. Abdeljawad, A modified Laplace transform for certain generalized fractional operators, *Results Nonlinear Anal.*, **2**, 88–98 (2018).
11. A. Lahrouz, L. Omari, D. Kiouach, Global analysis of a deterministic and stochastic nonlinear SIRS epidemic model, *Nonlinear Anal. Modell. Control*, **16**(1), 59–76 (2011).
12. K. Logeswari, C. Ravichandran, A new exploration on existence of fractional neutral integro-differential equations in the context of Atangana-Baleanu integral operator and non-linear Telegraph equation via fixed point method, *Chaos, Solitons Fractals*, **130**, 109439 (2020).
13. B. Li, Q. Bie, Long-time dynamics of an SIRS reaction-diffusion epidemic model, *J. Math. Anal. Appl.*, **475**, 1910–1926 (2019).
14. A. Mouaouine, A. Boukhouima, K. Hattaf, N. Yousfi, A fractional order SIR epidemic model with nonlinear incidence rate, *Adv. Differ. Equations*, **2018**, 160 (2018).
15. K.M. Owolabi, Numerical analysis and pattern formation process for space fractional superdiffusive systems, *Disct. Contin. Dyn. Syst. Ser. S*, **12**(3), 543–566 (2019).
16. D. Rostamy, E. Mottaghi, Stability analysis of a fractional-order epidemics model with multiple equilibriums, *Adv. Differ. Equations*, **2016**, 170, (2016).
17. K. Saad, D. Baleanu, A. Atangana, New fractional derivatives applied to the Korteweg–de Vries and Korteweg–de Vries–Burger's equations, *Comput. Appl. Math.*, **37**(6), 5203–5216 (2019).
18. N. Sene, Stokes' first problem for heated flat plate with Atangana-Baleanu fractional derivative, *Chaos, Solitons Fractals*, **117**, 68–75 (2018).

19. N. Sene, Integral balance methods for Stokes' first, equation described by the left generalized fractional derivative, *Physics*, **1**, 154–166 (2019).
20. N. Sene, Analytical solutions and numerical schemes of certain generalized fractional diffusion models, *Eur. Phys. J. Plus*, **134**, 199 (2019).
21. N. Sene, Stability analysis of the generalized fractional differential equations with and without exogenous inputs, *J. Nonlinear Sci. Appl.*, **12**, 562–572 (2019).
22. N. Sene, Global asymptotic stability of the fractional differential equations, *J. Nonlinear Sci. Appl.*, **13**, 171–175 (2020).
23. X. Wang, Z. Wang, Dynamic analysis of a delayed fractional-order SIR model with saturated incidence and treatment function, *Int. J. Bifurcation Chaos*, **28**(14), 1850180, (2018).
24. M. Yavuz, Characterization of two different fractional operators without singular Kernel, *Math. Model. Nat. Phen.*, **14**(3), 302 (2019).

7 Optimal Control of a Nipah Virus Transmission Model

Prabir Panja
Haldia Institute of Technology

Ranjan Kumar Jana
Sardar Vallabhbhai National Institute of Technology

CONTENTS

7.1 INTRODUCTION

Mathematical modeling can be helpful to study the dynamics and the control of an infectious disease. Experimentally, it is found that Nipah virus is a member of the genus Henipavirus. Nipah virus mainly spreads in SouthEast and South Asian region. Also, Nipah virus is becoming one of the most life-threatening disease. Goh et al. [1] reported the transmission of Nipah virus among the pig farmers in Malaysia. Transmission dynamics of Nipah virus at Bangladesh has been studied by Hsu et al. [2]. Again, Mondal et al. [3] investigated the optimal control analysis of a Nipah virus model. There exist very few mathematical models [4,5] of transmission and optimal control of Nipah virus in the literature. But the above-mentioned studies do not represent the actual dynamics of Nipah virus. In 2018, Nipah virus infection mainly spread in Kerala, India [6,7]. It is also found that Nipah virus mainly spreads through the bite of fruit bats in fruits and contamination of drinking water in Kerala, India.

Optimal control analysis of infectious disease through different possible control strategies has been studied by several researchers (Kar and Jana [8] and Lashari and Zaman [9], etc.). In 2016, Satterfield et al. [10] investigated about the development and research status of vaccines for Nipah virus. In 2017, Khan et al. [11] reported the optimal control analysis of pine wilt disease. In 2018, Khan et al. [12] investigated the impacts of media coverage campaign of hepatitis B disease transmission dynamics. Again, Khan et al. [13] studied the application of different types of control strategies to control pine wilt disease. Khan et al. [14] investigated the complex dynamics of Susceptible Exposed Infectious Recovered (SEIR) epidemic model. The infectious diseases cause several thousands of deaths worldwide, so the control of these diseases is very much necessary. Motivated by the above studies, we have investigated the optimal control analysis of Nipah virus transmission.

7.2 MODEL FORMULATION

According to Nipah virus transmission path, the possible routes of transmission for this outbreak include the consumption of fruits partially eaten by bats. When pig consumes these fruits, the Nipah virus is transmitted to the pig population. Then, the Nipah virus infects humans in two possible ways: (i) due to the contact of susceptible humans with infected pig and (ii) due to susceptible humans to infected human contact. Due to Nipah virus infection, pig population have been categorized into three subpopulations: (i) susceptible pig, (ii) infected pig, and (iii) recovered pig populations. Three types of human population, namely, (i) susceptible human, (ii) infected human, and (iii) recovered human, have been considered due to Nipah virus infection. Here, $V(t)$, $S_P(t)$, $I_P(t)$, $R_P(t)$, $S_H(t)$, $I_H(t)$, and $R_H(t)$ are the densities of Nipah virus, susceptible pig, infected pig, recovered pig, susceptible human, infected human, and recovered human, respectively. The eradication of the Nipah virus from human population is very much essential because it causes several thousands of deaths throughout the world. We have considered two control parameters, namely, vaccination for pig population and treatment for human population, for the eradication of Nipah virus. It is assumed that Nipah virus is present in the environment at a constant rate r_0. It is also considered r and d as the intrinsic growth rate and death rate of Nipah virus, respectively. It is assumed that the per-capita natural birth (or death) rate of pig population is μ_1. Also, here it is assumed that the size of pig population is N, i.e., $N = S_P + I_P + R_P$. Here, β_1, f_1, a, and u_1 are the transmission rate of Nipah virus from the susceptible pig to the infected pig, per-capita loss of immunity rate of recovered pig, effectiveness of vaccination control, and vaccination control parameter, respectively. It is assumed that some infected pig may be recovered due to natural immunity at a rate γ_1. Also, μ_2, μ_3, β_2, β_3, f_2, b, ρ, and u_2 are the recruitment rate of susceptible human, natural death rate of human, transmission rate of Nipah virus in human population through the contact of susceptible human with infected pig, transmission rate of Nipah virus in human population through the contact of susceptible human with infected human, per-capita loss of immunity rate of recovered human, effectiveness of treatment control parameter, constant related to recovered human and treatment control parameter, respectively. Again, $\in$ and γ_2 are the disease-related death rate of infected human due to Nipah virus infection and per-capita natural recovery rate of infected human, respectively. Keeping the above assumptions in mind, we have formulated a set of nonlinear differential equation as follows:

$$\left.\begin{aligned}
\frac{dV}{dt} &= r_0 + rV - dV \\
\frac{dS_P}{dt} &= \mu_1 N - \mu_1 S_P - \beta_1 V S_P + f_1 R_P - au_1 S_P \\
\frac{dI_P}{dt} &= \beta_1 V S_P - \mu_1 I_P - \gamma_1 I_P \\
\frac{dR_P}{dt} &= \gamma_1 I_P - \mu_1 R_P - f_1 R_P + au_1 S_P \\
\frac{dS_H}{dt} &= \mu_2 - \mu_3 S_H - \beta_2 S_H I_P - \beta_3 S_H I_H + f_2 R_H + bu_2(1-\rho) I_H \\
\frac{dI_H}{dt} &= \beta_2 S_H I_P + \beta_3 S_H I_H - (\varepsilon + \gamma_2 + \mu_3) I_H - bu_2 I_H \\
\frac{dR_H}{dt} &= \gamma_2 I_H - \mu_3 R_H - f_2 R_H + bu_2 \rho I_H
\end{aligned}\right\}, \tag{7.1}$$

which satisfies the initial conditions $V(0) > 0$, $S_P(0) > 0$, $I_P(0) > 0$, $R_P(0) > 0$, $S_H(0) > 0$, $I_H(0) > 0$, and $R_H(0) > 0$.

7.3 BOUNDEDNESS OF SOLUTIONS

In this section, we will prove the boundedness of solutions of our proposed model (7.1).

Theorem 7.1

All solutions of system (7.1) are bounded.

Proof: From the first equation of model (7.1), it is obtained that

$$\frac{dV}{dt} = r_0 + rV - dV.$$

Then by solving the above equation, it is obtained that $V = \frac{r_0}{d-r} + c_1 e^{(r-d)t}$. Now if $r - d < 0$ and $t \to \infty$, we have $V = \frac{r_0}{d-r}$.

Adding the second, third, and fourth equations of system (7.1), we have

$$\frac{d(S_P+I_P+R_P)}{dt}=\mu_1 N-\mu_1(S_P+I_P+R_P)$$

$$\frac{d(S_P+I_P+R_P)}{dt}+\mu_1(S_P+I_P+R_P)=\mu_1 N$$

By solving the above equation, it is obtained that

$$S_P+I_P+R_P=N+c_2e^{-\mu_1 t}.$$

As $t\to\infty$, we have $S_P+I_P+R_P=N$. So we have $S_P\le N, I_P\le N$ and $R_P\le N$. Again, by adding the fifth, sixth, and seventh equations of system (7.1), we have

$$\frac{d(S_H+I_H+R_P)}{dt}=\mu_2-\mu_3(S_H+I_H+R_P)-\in I_H$$

$$\frac{d(S_H+I_H+R_P)}{dt}+\mu_3(S_H+I_H+R_P)\le\mu_2$$

Then by solving the above equation using differential inequality [15], we have

$$S_H+I_H+R_H\le\frac{\mu_2}{\mu_3}+c_3e^{-\mu_3 t}.$$

As $t\to\infty$, we have $S_H+I_H+R_H\le\frac{\mu_2}{\mu_3}$. So it can be written as $S_H\le\frac{\mu_2}{\mu_3}, I_H\le\frac{\mu_2}{\mu_3}$ and $R_H\le\frac{\mu_2}{\mu_3}$.

Hence, all solutions of system (7.1) exist within the region

$$\sum=\left\{(V,S_P,I_P,R_P,S_H,I_H,R_H)\in R^7: V=\frac{r_0}{d-r},(S_P,I_P,R_P)\le N,(S_H,I_H,R_H)\le\frac{\mu_2}{\mu_3}\right\}.$$

7.4 EQUILIBRIUM POINTS AND BASIC REPRODUCTION NUMBER

Our proposed system (7.1) has the following possible equilibrium points:

i. The disease-free equilibrium point $E_0=\left(0,S_P^0,0,R_P^0,S_H^0,0,0\right)$, where

$$S_P^0=\frac{N(\mu_1+f_1)}{au_1+\mu_1+f_1}, R_P^0=\frac{au_1N}{au_1+\mu_1+f_1}\text{ and }S_H^0=\frac{\mu_2}{\mu_3}.$$

ii. Endemic equilibrium point $E^* = \left(V^*, S_P^*, I_P^*, R_P^*, S_H^*, I_H^*, R_H^*\right)$, where

$$V^* = \frac{r_0}{d-r},$$

$$S_P^* = \frac{(d-r)(\mu_1+\gamma_1)}{\beta_1 r_0} I_P^*,$$

$$I_P^* = \frac{\mu_1 N \beta_1 r_0}{(\mu_1+\gamma_1)[\mu_1(d-r)+\beta_1 r_0 + au_1(d-r) - f_1 au_1] - f_1\gamma_1\beta_1 r_0},$$

$$R_P^* = \left[\gamma_1 + \frac{au_1(d-r)(\mu_1+\gamma_1)}{\beta_1 r_0}\right] I_P^*,$$

$$S_H^* = \frac{[(\in + \gamma_2 + \mu_3) + bu_2] I_H^*}{\beta_2 I_P^* + \beta_3 I_H^*},$$

$$R_H^* = \frac{(\gamma_2 + bu_2\rho) I_H^*}{\mu_3 + f_2} \text{ and}$$

$$I_H^* = \frac{(\mu_3 + \beta_2 I_P^*) S_H^* - \mu_2}{bu_2(1-\rho) + \dfrac{f_2(\gamma_2 + bu_2\rho)}{\mu_3 + f_2} - \beta_3 S_H^*}$$

Now, we will determine the value of R_0 (basic reproduction number) of our proposed model. The concept of next-generation matrix [16] is used to obtain the value of (R_0). Let us define $\bar{x} = (V, I_P, I_H)$, which contains Nipah virus, infected pig, and infected human. Now, we will consider the following system:

$$\left.\begin{aligned} \frac{dV}{dt} &= r_0 + rV - dV \\ \frac{dI_P}{dt} &= \beta_1 V S_P - (\mu_1+\gamma_1) I_P \\ \frac{dI_H}{dt} &= \beta_2 S_H I_P + \beta_3 S_H I_H - (\in + \gamma_2 + \mu_3 + bu_2) I_H \end{aligned}\right\}. \tag{7.2}$$

Then, the matrices F and V are given by

$$F = \begin{pmatrix} 0 & 0 & 0 \\ \beta_1 S_P^0 & 0 & 0 \\ 0 & \beta_2 S_H^0 & \beta_3 S_H^0 \end{pmatrix}, V = \begin{pmatrix} d - r_0 & 0 & 0 \\ 0 & \mu_1 + \gamma_1 & 0 \\ 0 & 0 & \in + \gamma_2 + \mu_3 + bu_2 \end{pmatrix}.$$

Then,

$$V^{-1} = \begin{pmatrix} \frac{1}{d - r_0} & 0 & 0 \\ 0 & \frac{1}{\mu_1 + \gamma_1} & 0 \\ 0 & 0 & \frac{1}{\in + \gamma_2 + \mu_3 + bu_2} \end{pmatrix} \text{ and}$$

$$FV^{-1} = \begin{pmatrix} 0 & 0 & 0 \\ \frac{\beta_1 S_P^0}{d - r_0} & 0 & 0 \\ 0 & \frac{\beta_2 S_H^0}{\mu_1 + \gamma_1} & \frac{\beta_3 S_H^0}{\in + \gamma_2 + \mu_3 + bu_2} \end{pmatrix}$$

Therefore, the expression of R_0 is given by

$$R_0 = \text{ maximum eigen value of the matrix } FV^{-1}$$

$$= \frac{\beta_3 \mu_2}{\mu_3 \left(\in + \gamma_2 + \mu_3 + bu_2 \right)}.$$

7.5 STABILITY OF EQUILIBRIA

Here, stability of equilibrium points of our proposed model (7.1) have been investigated.

Theorem 7.2

Our proposed system (7.1) is locally asymptotically stable around E_0 if $r < d$ and $R_0 < 1$.

Proof: Let us consider a subsystem of system (7.1) as follows:

$$\left. \begin{aligned} \frac{dV}{dt} &= r_0 + rV - dV \\ \frac{dI_P}{dt} &= \beta_1 V S_P - \mu_1 I_P - \gamma_1 I_P \\ \frac{dI_H}{dt} &= \beta_2 S_H I_P + \beta_3 S_H I_H - \left(\in + \gamma_2 + \mu_3 \right) I_H - bu_2 I_H \end{aligned} \right\}. \tag{7.3}$$

Then, the variational matrix of model (7.3) at E_0 is given by

$$J = \begin{pmatrix} r-d & 0 & 0 \\ \beta_1 S_P^0 & -(\mu_1+\gamma_1) & 0 \\ 0 & \beta_2 S_H^0 & \beta_3 S_H^0 - (\in + \gamma_2 + \mu_3 + bu_2) \end{pmatrix}.$$

Then, the characteristic equation of the matrix J at E_0 is given by

$$(r-d-\lambda)(-\mu_1-\gamma_1-\lambda)\left(\beta_3 S_H^0 - \left(\in + \gamma_2 + \mu_3 + bu_2\right) - \lambda\right) = 0$$

The roots of the above equation are $\lambda = r-d, -\mu_1-\gamma_1, and\ \beta_3 S_H^0 - (\epsilon + \gamma_2 + \mu_3 + bu_2)$. So, our proposed system (7.1) is locally asymptotically stable around E_0 if $r < d$ and $\beta_3 S_H^0 - (\epsilon + \gamma_2 + \mu_3 + bu_2) < 0$, *i.e.*, $R_0 < 1$.

Theorem 7.3

Our proposed system (7.1) is locally asymptotically stable around E^* if $R_0 > 1$, $M_{11} < 0$, and $D_i > 0$, $i = 1, 2, \ldots, 6$ are given in the subsequent steps.
Proof: The variational matrix of model (7.1) is given by

$$J(E^*) = \begin{pmatrix} M_{11} & 0 & 0 & 0 & 0 & 0 & 0 \\ -M_{21} & -M_{22} & 0 & M_{24} & 0 & 0 & 0 \\ M_{21} & M_{32} & -M_{33} & 0 & 0 & 0 & 0 \\ 0 & M_{42} & M_{43} & -M_{44} & 0 & 0 & 0 \\ 0 & 0 & -M_{53} & 0 & -M_{55} & M_{56} & M_{57} \\ 0 & 0 & M_{53} & 0 & M_{65} & M_{66} & 0 \\ 0 & 0 & 0 & 0 & 0 & M_{76} & -M_{77} \end{pmatrix},$$

where

$$M_{11} = r-d, M_{21} = \beta_1 S_P, M_{22} = \mu_1 + \beta_1 V + au_1,$$

$$M_{24} = f_1, M_{32} = \beta_1 V, M_{33} = \mu_1 + \gamma_1, M_{42} = au_1,$$

$$M_{43} = \gamma_1, M_{44} = \mu_1 + f_1, M_{53} = \beta_2 S_H,$$

$$M_{55} = \mu_3 + \beta_2 I_P + \beta_3 I_H, M_{56} = bu_2(1-\rho) - \beta_3 S_H,$$

$$M_{57} = f_2, M_{65} = \beta_2 I_P + \beta_3 I_H,$$

$$M_{66} = \beta_3 S_H - (\in + \gamma_2 + \mu_3 + bu_2),$$

$$M_{76} = \gamma_2 + bu_2\rho, M_{77} = \mu_3 + f_2.$$

Then, the characteristic equation is given by

$$(M_{11}-\lambda)(\lambda^6+c_1\lambda^5+c_2\lambda^4+c_3\lambda^3+c_4\lambda^2+c_5\lambda+c_6)=0,$$

where c_i, $i = 1, 2, 3, 4, 5, 6$ are given in the appendix.

One root of the above characteristic equation is M_{11}. Now, the other roots of the above characteristic equation are negative real numbers or complex number with negative real numbers

$$D_1=c_1>0, D_2=\begin{vmatrix} c_1 & c_3 \\ 1 & c_2 \end{vmatrix}>0, D_3=\begin{vmatrix} c_1 & c_3 & c_5 \\ 1 & c_2 & c_4 \\ 0 & c_1 & c_3 \end{vmatrix}>0,$$

$$D_4=\begin{vmatrix} c_1 & c_3 & c_5 & 0 \\ 1 & c_2 & c_4 & 0 \\ 0 & c_1 & c_3 & c_5 \\ 0 & 1 & c_2 & c_4 \end{vmatrix}>0, D_5=\begin{vmatrix} c_1 & c_3 & c_5 & 0 & 0 \\ 1 & c_2 & c_4 & 0 & 0 \\ 0 & c_1 & c_3 & c_5 & 0 \\ 0 & 1 & c_2 & c_4 & 0 \\ 0 & 0 & c_1 & c_3 & c_5 \end{vmatrix}>0,$$

$$D_6=\begin{vmatrix} c_1 & c_3 & c_5 & 0 & 0 & 0 \\ 1 & c_2 & c_4 & 0 & 0 & 0 \\ 0 & c_1 & c_3 & c_5 & 0 & 0 \\ 0 & 1 & c_2 & c_4 & 0 & 0 \\ 0 & 0 & c_1 & c_3 & c_5 & 0 \\ 0 & 0 & 0 & 0 & 0 & c_6 \end{vmatrix}>0.$$

Hence the theorem.

7.6 OPTIMAL CONTROL OF NIPAH VIRUS MODEL

In this section, we have formulated the optimal control of Nipah virus transmission model. First, we have discussed about the existence of optimal control problem in the next subsection.

7.6.1 Existence

Theorem 7.4

An optimal control $\left(u_1^*, u_2^*\right)$ exists for the problem (7.4).

Proof: We will follow Theorem 4.1 and Corollary 4.1 in [17] to prove this theorem. Let $s(t, x, u)$ be the right-hand side of our proposed system (7.1), where $x = (V, S_P, I_P, R_P, S_H, I_H, R_H)$ and $u = (u_1, u_2)$. Now, we will verify the following conditions:

i. $|s(t,0,0)|\le D, |s_{\vec{x}}(t,\vec{x},\vec{u})|\le D(1+|\vec{u}|)$ and $|s_{\vec{u}}(t,\vec{x},\vec{u})|\le D$, where $s\in C^1$ (class of differential function) and D is a constant.

ii. The admissible set $\Im$ of all solutions to the system (7.1) with corresponding control in U_{ad} is non-empty.
iii. $s(t,\vec{x},\vec{u}) = a_1(t,\vec{x}) + a_2(t,\vec{x})\vec{u}$.
iv. The control set $\Gamma = [0,1]\times[0,1]$ is closed, convex, and compact.
v. The integrand of the objective functional is convex in Γ.

Now, we will verify the above conditions

$$s(t,\vec{x},\vec{u}) = \begin{pmatrix} r_0 + rV - dV \\ \mu_1 N - \mu_1 S_P - \beta_1 V S_P + f_1 R_P - a u_1 S_P \\ \beta_1 V S_P - \mu_1 I_P - \gamma_1 I_P \\ \gamma_1 I_P - \mu_1 R_P - f_1 R_P + a u_1 S_P \\ \mu_2 - \mu_3 S_H - \beta_2 S_H I_P - \beta_3 S_H I_H + f_2 R_H + b u_2 (1-\rho) I_H \\ \beta_2 S_H I_P + \beta_3 S_H I_H - (\epsilon + \gamma_2 + \mu_3) I_H - b u_2 I_H \\ \gamma_2 I_H - \mu_3 R_H - f_2 R_H + b u_2 \rho I_H \end{pmatrix}$$

So, we can say that $s(t,\vec{x},\vec{u}) \in C^1$ and $|\, s(t,0,0)\,| = \begin{pmatrix} r_0 \\ \mu_1 N \\ 0 \\ 0 \\ \mu_2 \\ 0 \\ 0 \end{pmatrix}$. Also, we have

$$|\, s_{\vec{x}}(t,\vec{x},\vec{u})\,| = \left\| \begin{pmatrix} r-d & 0 & 0 & 0 & 0 & 0 & 0 \\ -\beta_1 S_P & -\mu_1 - \beta_1 V - a u_1 & 0 & f_1 & 0 & 0 & 0 \\ \beta_1 S_P & \beta_1 V & -\mu_1 - \gamma_1 & 0 & 0 & 0 & 0 \\ 0 & a u_1 & \gamma_1 & -f_1 & 0 & 0 & 0 \\ 0 & 0 & -\beta_2 S_H & 0 & -\mu_3 - \beta_2 I_P - \beta_3 I_H & -\beta_3 S_H + b u_2 (1-\rho) & f_2 \\ 0 & 0 & \beta_2 S_H & 0 & \beta_2 I_P + \beta_3 I_H & \beta_3 S_H - (\epsilon + \gamma_2 + \mu_3) - b u_2 & 0 \\ 0 & 0 & 0 & 0 & 0 & \gamma_2 + b u_2 \rho & -f_2 \end{pmatrix} \right\|$$

and

$$|s_{\vec{u}}(t,\vec{x},\vec{u})| = \left\| \begin{pmatrix} 0 & 0 \\ -aS_P & 0 \\ 0 & 0 \\ aS_P & 0 \\ 0 & b(1-\rho)I_H \\ 0 & -bI_H \\ 0 & b\rho I_H \end{pmatrix} \right\|.$$

Since $V, S_P, I_P, R_P, S_H, I_H$, and R_H are bounded, there exists a constant D such that

$$|s(t,0,0)| \leq D, |s_{\vec{x}}(t,\vec{x},\vec{u})| \leq D(1+|\vec{u}|), |s_{\vec{u}}(t,\vec{x},\vec{u})| \leq D.$$

This proves that the condition (i) is satisfied.

$$s(t,\vec{x},\vec{u}) = \begin{pmatrix} r_0 + rV - dV \\ \mu_1 N - \mu_1 S_P - \beta_1 V S_P + f_1 R_P - au_1 S_P \\ \beta_1 V S_P - \mu_1 I_P - \gamma_1 I_P \\ \gamma_1 I_P - \mu_1 R_P - f_1 R_P + au_1 S_P \\ \mu_2 - \mu_3 S_H - \beta_2 S_H I_P - \beta_3 S_H I_H + f_2 R_H + bu_2(1-\rho) I_H \\ \beta_2 S_H I_P + \beta_3 S_H I_H - (\epsilon + \gamma_2 + \mu_3) I_H - bu_2 I_H \\ \gamma_2 I_H - \mu_3 R_H - f_2 R_H + bu_2 \rho I_H \end{pmatrix}$$

$$+ \begin{pmatrix} 0 & 0 \\ -aS_P & 0 \\ 0 & 0 \\ aS_P & 0 \\ 0 & b(1-\rho)I_H \\ 0 & -bI_H \\ 0 & b\rho I_H \end{pmatrix} \times \begin{pmatrix} u_1 \\ u_2 \end{pmatrix}.$$

Thus, condition (iii) is verified. From the definition, condition (iv) is obviously satisfied. We need to verifying (v), for completion of the proof.

To prove the convexity of $g(t,\vec{x},\vec{u})$, we need to prove that

$$(1-p)g(t,\vec{x},\vec{u}) + ph(t,\vec{x},\vec{v}) \geq g(t,\vec{x},(1-p)\vec{u} + p\vec{v}),$$

where $g(t,\vec{x},\vec{u}) = B_1 I_P + B_2 I_H + B_3 u_1^2 + B_4 u_2^2$ and $\vec{u}, \vec{v}$ are two control parameters with $p \in [0,1]$. So, we have

$$(1-p)g(t,\vec{x},\vec{u})+ph(t,\vec{x},\vec{v})=B_1I_P+B_2I_H+(1-p)[B_3u_1^2+B_4u_2^2]+q[B_3v_1^2+B_4v_2^2]$$

and

$$g(t,\vec{x},(1-p)\vec{u}+p\vec{v})=B_1I_P+B_2I_H+B_3[(1-p)u_1+pv_1]^2+B_4[(1-p)u_2+pv_2]^2.$$

Furthermore, we have

$$\begin{aligned}
&(1-p)g(t,\vec{x},\vec{u})+pg(t,\vec{x},\vec{v})-g(t,\vec{x},(1-p)\vec{u}+p\vec{v})\\
&=B_3\left\{[(1-p)u_1^2+pv_1^2]-[(1-p)u_1+pv_1]^2\right\}\\
&\quad+B_4\left\{[(1-p)u_2^2+pv_2^2]-[(1-p)u_2+pv_2]^2\right\}\\
&=B_3[\sqrt{q(1-q)}u_1-\sqrt{q(1-q)}v_1]^2+B_4[\sqrt{q(1-q)}u_2-\sqrt{q(1-q)}v_2]^2\\
&=B_3q(1-q)[u_1-v_1]^2+B_4q(1-q)[u_2-v_2]^2\geq 0
\end{aligned}$$

Hence the theorem.

7.6.2 Construction of Optimal Control Problem

Here, we will construct an optimal control problem in a specified time interval using Pontryagin's maximum principle [18]. We will try to reduce the number of humans from Nipah virus infection as well as to minimize the cost of control strategies. Now, we have constructed an objective functional in $[t_0, t_f]$ as follows:

$$J(u_1,u_2)=\min_{u_1,u_2}\int_{t_0}^{t_f}\left[B_1I_P+B_2I_H+B_3u_1^2+B_4u_2^2\right]dt, \tag{7.4}$$

subject to our proposed model (7.1). The state variables of (7.4) are $V(t)$, $S_P(t)$, $I_P(t)$, $R_P(t)$, $S_H(t)$, $I_H(t)$, and $R_H(t)$. Also, u_1 and u_2 represent vaccination and treatment control parameters, respectively. The parameters B_1 and B_2 are the balancing coefficients for infected pig and infected human, respectively. Also, the parameters B_3 and B_4 are the weight constraints for the control variables u_1 and u_2, respectively. To minimize the side effects and overdoses, the squares of u_1 and u_2 have been considered ([19,20]).

Our objective is to determine the optimal value of u_1 and u_2 such that

$$J(u_1^*,u_2^*)=\min_{u_1,u_2\in\odot}J(u_1,u_2), \tag{7.5}$$

where $\odot=\{(u_1,u_2):u_1,u_2$ are the measurable functions and $0\leq u_1\leq 1, 0\leq u_2\leq 1, t\in[0,t_f]\}$ shows the control set.

First, we define the Lagrangian as follows:

$$L=B_1I_P+B_2I_H+B_3u_1^2+B_4u_2^2.$$

Now, the Hamiltonian H can be defined as follows:

$$H(L,V,S_P,I_P,R_P,S_H,I_H,R_H,u_1,u_2,\lambda_1,\lambda_2,\lambda_3,\lambda_4,\lambda_5,\lambda_6,\lambda_7)$$
$$= L+\lambda_1\frac{dV}{dt}+\lambda_2\frac{dS_P}{dt}+\lambda_3\frac{dI_P}{dt}+\lambda_4\frac{dR_P}{dt}+\lambda_5\frac{dS_H}{dt}+\lambda_6\frac{dI_H}{dt}+\lambda_7\frac{dR_H}{dt} \tag{7.6}$$

$$\begin{aligned}
&= B_1I_P+B_2I_H+B_3u_1^2+B_4u_2^2+\lambda_1\left\{r_0+rV-dV\right\}\\
&\quad+\lambda_2\left\{\mu_1N-\mu_1S_P-\beta_1VS_P+f_1R_P-au_1S_P\right\}\\
&\quad+\lambda_3\left\{\beta_1VS_P-\mu_1I_P-\gamma_1I_P\right\}+\lambda_4\left\{\gamma_1I_P-\mu_1R_P-f_1R_P+au_1S_P\right\}\\
&\quad+\lambda_5\left\{\mu_2-\mu_3S_H-\beta_2S_HI_P-\beta_3S_HI_H+f_2R_H+bu_2(1-\rho)I_H\right\}\\
&\quad+\lambda_6\left\{\beta_2S_HI_P+\beta_3S_HI_H-(\varepsilon+\gamma_2+\mu_3)I_H-bu_2I_H\right\}\\
&\quad+\lambda_7\left\{\gamma_2I_H-\mu_3R_H-f_2R_H+bu_2\rho I_H\right\}
\end{aligned} \tag{7.7}$$

where λ_i for $i = 1, 2, 3, 4, 5, 6, 7$ are the adjoint variables. Also, the adjoint variables can be obtained by solving the following system of equations:

$$\frac{d\lambda_1}{dt}=-\frac{\partial H}{\partial V}=\beta_1S_P(\lambda_2-\lambda_3)-(r-d)\lambda_1$$

$$\frac{d\lambda_2}{dt}=-\frac{\partial H}{\partial S_P}=\mu_1\lambda_2+\beta_1V(\lambda_2-\lambda_3)+au_1(\lambda_2-\lambda_4)$$

$$\frac{d\lambda_3}{dt}=-\frac{\partial H}{\partial I_P}=-B_1+\lambda_3(\mu_1+\gamma_1)-\lambda_4\gamma_1+\beta_2S_H(\lambda_5-\lambda_6)$$

$$\frac{d\lambda_4}{dt}=-\frac{\partial H}{\partial R_P}=(\mu_1+f_1)\lambda_4-f_1\lambda_2$$

$$\frac{d\lambda_5}{dt}=-\frac{\partial H}{\partial S_H}=\mu_3\lambda_5+(\beta_2I_P+\beta_3I_H)(\lambda_5-\lambda_6)$$

$$\begin{aligned}\frac{d\lambda_6}{dt}=-\frac{\partial H}{\partial I_H}&=-B_2+\lambda_5(\beta_3S_H-bu_2(1-\rho))\\&\quad+\lambda_6(\Sigma+\gamma_2+\mu_3+bu_2-\beta_3S_H)-\lambda_7(\gamma_2+bu_2\rho)\end{aligned}$$

$$\frac{d\lambda_7}{dt}=-\frac{\partial H}{\partial R_H}=(\mu_3+f_2)\lambda_7-f_2\lambda_5$$

satisfying the transversality conditions $\lambda_i(t_f) = 0$ for $i = 1, 2, 3, 4, 5, 6, 7$.

Theorem 7.5

The optimal values of the control parameters u_I and u_2 are $u_1^* = \max\{0, \min(\overline{u}_1, 1)\}$, $u_2^* = \max\{0, \min(\overline{u}_2, 1)\}$, where $\overline{u}_1 = \dfrac{aS_P(\lambda_2 - \lambda_4)}{2B_3}$ and $\overline{u}_2 = \dfrac{I_H\left(b\lambda_6 - b(1-\rho)\lambda_5 - b_p\lambda_7\right)}{2B_4}$.

Proof: From the optimality conditions (i.e., $\dfrac{\partial H}{\partial u_1} = 0$ and $\dfrac{\partial H}{\partial u_2} = 0$), it is obtained that

$$u_1 = \frac{aS_P\left(\lambda_2 - \lambda_4\right)}{2B_3}(= \overline{u}_1) \text{ and } u_2 = \frac{I_H\left(b\lambda_6 - b(1-\rho)\lambda_5 - b\rho\lambda_7\right)}{2B_4}(= \overline{u}_2).$$

Also, it is known that the maximum and minimum values of the control parameters are 1 and 0, respectively. So, we have $u_I = 0$ if $\overline{u}_1 \leq 0$, $u_I = 1$ if $\overline{u}_1 \geq 1$, and $u_I = \overline{u}_1$ for rest of the time. The similar result holds for other control parameter u_2. So we can get the optimum value of J (see Eq. (7.4)) for (u_1^*, u_2^*). This completes the proof.

7.7 NUMERICAL SIMULATION

We have solved system (7.1) numerically using MATLAB® to get a better insight of the proposed model. Due to unavailability of real data related to the Nipah virus transmission, we have considered some hypothetical values of the parameters. The parametric values that have been considered for Figure 7.1 are $r_0 = 10$, $r = 0.01$, $d = 0.1$, $\mu_1 = 1/20$, $N = 350$, $\beta_1 = 0.3$, $f_1 = 0.4$, $\gamma_1 = 4.5$, $\mu_2 = 10$, $\mu_3 = 1/65$, $f_2 = 0.4$, $\beta_2 = 0.003$, $\beta_3 = 0.0002$, $E = 1/45$, $\gamma_2 = 4.5$, $u_1 = 0$, and $u_2 = 0$. Figure 7.1 represents the time-series diagrams of each of the respective population. From this figure, it is seen that the endemic equilibrium point is locally asymptotically stable. So it can be concluded that the Nipah virus infection exists in human population.

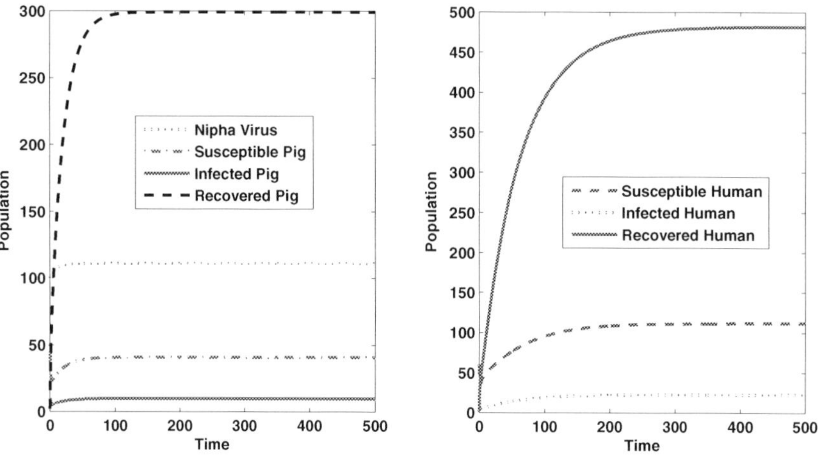

FIGURE 7.1 Stability of the endemic equilibrium point.

Using the same set of parametric values used in Figure 7.1 except $f_1 = 0.01$, $f_2 = 0.02$, $\beta_2 = 0.008$, $\beta_3 = 0.006$, $\gamma_1 = 0.1$, and $\gamma_2 = 0.1$, the solutions of system (7.1) are presented in Figure 7.2. From this figure, it is observed that if the value of β_1 (transmission rate of Nipah virus in pig population) gradually decreases, then the number of infected pig and also the number of infected human gradually decrease. Hence, it can be concluded that the transmission rate of Nipah virus in human population will be decreased if we can restrict the consumption of infected fruits (partially bite by fruit bats) by pig and human population.

Now using the same set of parametric values used in Figure 7.2, the change of number of infected human with the variation of β_2 is plotted in Figure 7.3. From this figure, it is observed that as the value of β_2 (transmission rate of Nipah virus from infected pig to human population) decreases, then the number of infected human gradually decreases. So it can be concluded that the Nipah virus transmission may be under control if we can restrict the use of infected pig as a food source for human.

Again, the change of number of infected human with the variation of β_3 (Nipah virus transmission rate from infected human to susceptible human) is shown in Figure 7.4. From this figure, it is observed that as the value of β_3 decreases, the number of infected human gradually decreases. So it can be concluded that the Nipah virus transmission rate in human population may be decreased if we can keep infected human far away from susceptible human.

Also, the change of infected pig and infected human with the variation of γ_1 (natural recovery rate of infected pig) is plotted in Figure 7.5. As the value of γ_1 increases, the number of infected human gradually decreases. Hence, it can be concluded that if the natural recovery rate of infected pig increases, then the Nipah virus transmission gradually decreases. Again, the change of infected human with the variation of γ_2 (natural recovery rate of infected human) is presented in Figure 7.6. This figure shows that as the value of γ_2 gradually increases, the number of infected human gradually

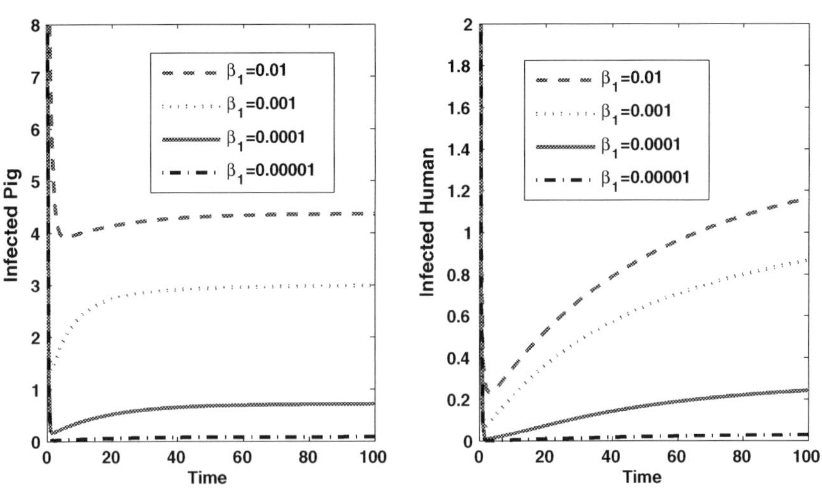

FIGURE 7.2 Change of infected pig and infected human with the variation of β_1.

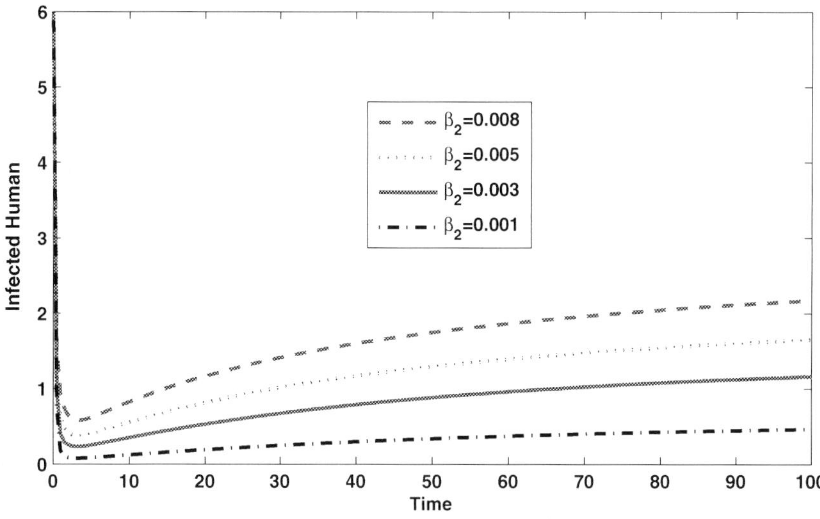

FIGURE 7.3 Change of infected human with the variation of β_2.

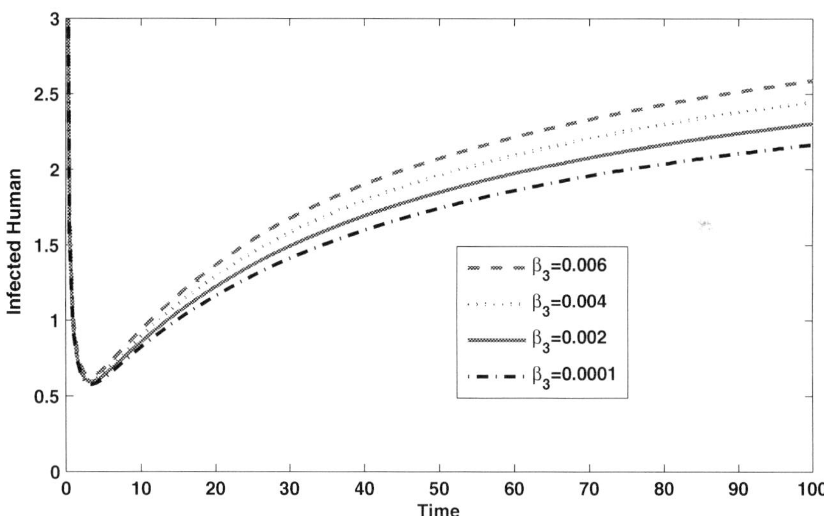

FIGURE 7.4 Change of infected human with the variation of β_3.

decreases. It can be said that if the natural recovery rate of infected human increases, then the Nipah virus transmission will be under control.

To solve Eq. (7.4), we have considered the parametric values as $r_0 = 10.0$, $r = 0.2$, $d = 0.3$, $\mu_1 = 0.005$, $N_1 = 5000$, $f_1 = 0.1$, $\gamma_1 = 4.5$, $\mu_2 = 10$, $\mu_3 = 1/65$, $f_2 = 0.2$, $E = 1/45$, $\gamma_2 = 4.5$, $\beta_1 = 0.0001$, $\beta_2 = 0.008$, $\beta_3 = 0.0001$, $b = 0.8$, $a = 0.6$, $\rho = 0.6$, $B_1 = 30.0$, $B_2 = 20.0$, $B_3 = 10.0$, and $B_4 = 5.0$. Also, we have considered the time interval 20 units and initial

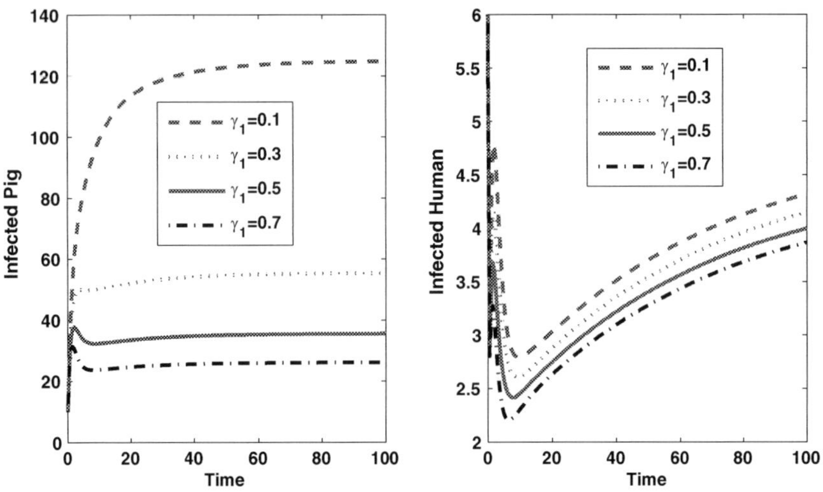

FIGURE 7.5 Change of infected pig and infected human with the variation of γ_1.

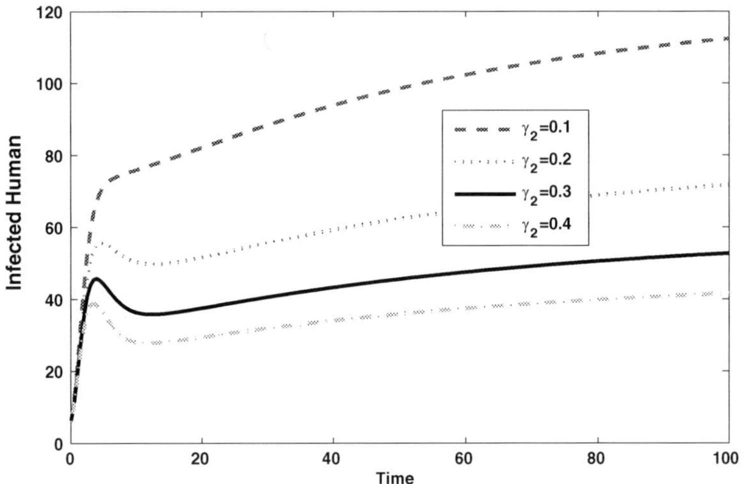

FIGURE 7.6 Change of infected human with the variation of γ_2.

population for Nipah virus, susceptible pig, infected pig, recovered pig, susceptible human, infected human, and recovered human as 500, 500, 200, 100, 600, 200 and 150, respectively. Then, the state variables and adjoint variables have been solved with the help of forward RK method and backward RK method, respectively.

Time evolution of susceptible pig (S_p) and recovered pig (R_p) with control and without control is plotted in Figure 7.7. From this figure, it is seen that in the presence of optimal vaccination and treatment control, the number of susceptible pig gradually decreases and that of recovered pig gradually increases. It can be concluded that due to vaccination of susceptible pig, the number of recovered pig increases. It may be helpful to decrease the Nipah virus transmission in pig population. Again, in the

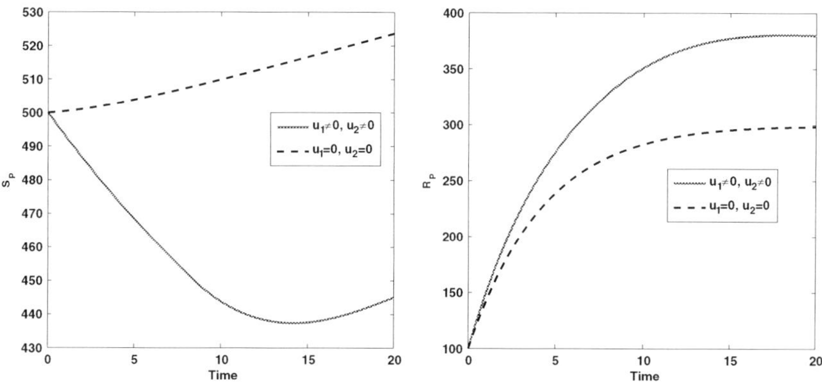

FIGURE 7.7 Change of susceptible pig and recovered pig population with and without control.

presence and absence of vaccination and treatment control, the variation of infected human (I_H) is shown in Figure 7.8. From this figure, it is observed that the number of infected human gradually decreases due to the presence of vaccination and treatment control. Hence, it can be concluded that vaccination and treatment control may reduce the number of infected human. Again, the optimal vaccination control parameter u_1 and optimal treatment control parameter u_2 are plotted with respect to time in Figure 7.9. From this figure, it is seen that to eradicate the Nipah virus infection from human population, we should continue the vaccination program up to 9 units of time with full effort, and then it will be decreased with respect to time and ultimately becomes zero at the end of 20 units of time. Again, we should continue the treatment on infected human with full effort up to 17 units of time, and then it will be decreased with respect to time and ultimately becomes zero at the end of 20 units of time.

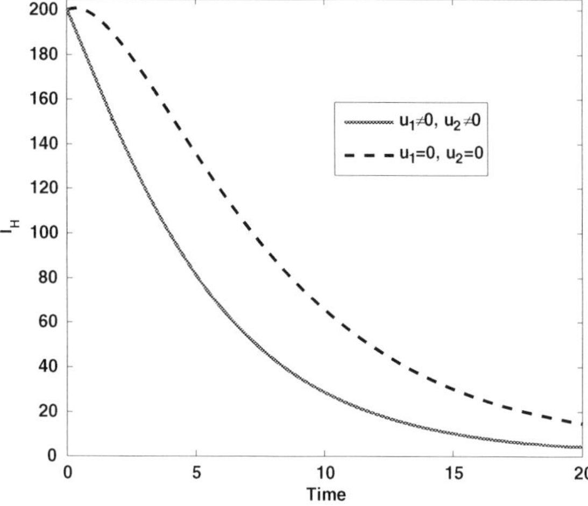

FIGURE 7.8 Change of infected human population with and without control.

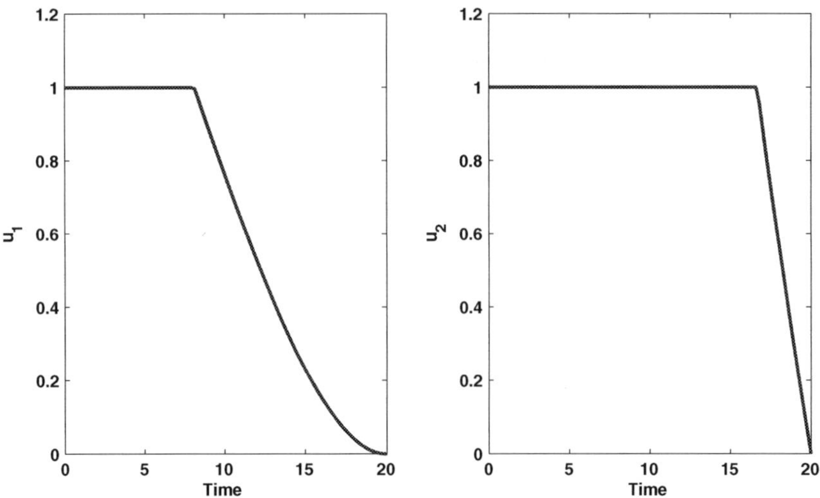

FIGURE 7.9 Vaccination and treatment controls.

7.8 CONCLUSION

In this chapter, we have studied the transmission dynamics of Nipah virus in pig population and human population. Nipah virus is transmitted to the human population through eating of fruits partially consumed by bats and using water from wells infested by bat. Bats are known to drink toddy that is collected in open containers, and occasionally urinate in it, which makes it contaminated with the virus. To reduce the Nipah virus from human population, two control strategies, namely, vaccination and treatment, have been used. From the theoretical and numerical results, the following conclusions have been drawn:

i. It is observed that the Nipah virus transmission in pig population may be decreased if we can restrict the consumption of fruits partially consumed by bats.

ii. It is seen that the Nipah virus transmission in human population may be under control if we can restrict the use of infected pig as a food source for human. Also, it may be under control if we culture the pig far away from the locality of human.

iii. Again, it can be concluded that the Nipah virus infection may be removed or disappeared from human population if we can restrict the contact between susceptible human and infected human.

iv. If the natural recovery rate of infected pig and infected human gradually increases, then the Nipah virus transmission will be under control.

v. To reduce Nipah virus transmission in human population, combined effects of vaccination and treatment controls must be taken into consideration concurrently.

APPENDIX

$$c_1 = M_{22} + M_{33} + M_{44} + M_{55} - M_{66} + M_{77}$$

$$c_2 = M_{22}M_{33} + (M_{22} + M_{33})\{M_{44}M_{55} - M_{66}M_{77} + (M_{44} + M_{55})(M_{77} - M_{66})\}$$
$$- M_{24}M_{42} - M_{56}M_{65}$$

$$c_3 = M_{44}M_{55}(M_{77} - M_{66}) - M_{66}M_{77}(M_{44} + M_{55})$$
$$+ (M_{22} + M_{33})\{M_{44}M_{55} - M_{66}M_{77} + (M_{44} + M_{55})(M_{77} - M_{66})\}$$
$$- M_{56}M_{65}\{M_{22} + M_{33} + M_{44} + M_{77}\} - M_{57}M_{65}M_{76} + M_{24}M_{42}(M_{66} - M_{55})$$
$$- M_{24}M_{42}M_{77} - M_{24}M_{42}M_{33} - M_{24}M_{32}M_{43}$$

$$c_4 = M_{22}M_{33}\{M_{44}M_{55} - M_{66}M_{77} + (M_{44} + M_{55})(M_{77} - M_{66})\} - M_{44}M_{55}M_{66}M_{77}$$
$$(M_{22} + M_{33})\{M_{44}M_{55}(M_{77} - M_{66}) - M_{66}M_{77}(M_{44} + M_{55})\}$$
$$-M_{56}M_{65}\{M_{22}M_{33} + M_{44}M_{77} + (M_{22} + M_{33})(M_{44} + M_{77})\}$$
$$-M_{57}M_{65}M_{76}(M_{22} + M_{33} + M_{44})$$
$$+M_{24}M_{42}M_{55}M_{66} + M_{24}M_{42}M_{77}(M_{66} - M_{55})$$
$$+M_{24}M_{33}M_{42}(M_{66} - M_{55}) - M_{24}M_{33}M_{42}M_{77} + M_{24}M_{42}M_{56}M_{65}$$
$$+M_{24}M_{32}M_{43}(M_{66} - M_{55} - M_{77})$$

$$c_5 = -M_{44}M_{55}M_{66}M_{77}(M_{22} + M_{33})$$
$$+ M_{22}M_{33}\{M_{44}M_{55}(M_{77} - M_{66}) - M_{66}M_{77}(M_{44} + M_{55})\}$$
$$- M_{56}M_{65}\{M_{44}M_{77}(M_{22} + M_{33}) + M_{22}M_{33}(M_{44} + M_{77})\}$$
$$- M_{57}M_{65}M_{76}\{M_{22}M_{33} + M_{44}(M_{22} + M_{33})\} + M_{24}M_{42}M_{55}M_{66}M_{77}$$
$$+ M_{24}M_{42}M_{33}M_{55}M_{66} + M_{24}M_{42}M_{33}M_{77}(M_{66} - M_{55})$$
$$+ M_{42}M_{56}M_{65}M_{77} + M_{33}M_{42}M_{56}M_{65} + M_{42}M_{57}M_{65}M_{76}$$

$$c_6 = M_{24}M_{42}M_{33}M_{55}M_{66}M_{77} + M_{33}M_{42}M_{56}M_{65}M_{77} + M_{33}M_{42}M_{57}M_{65}M_{76}$$
$$+ M_{24}M_{32}M_{43}M_{55}M_{66}M_{77} + M_{24}M_{32}M_{43}M_{56}M_{65}M_{77} + M_{24}M_{32}M_{43}M_{57}M_{65}M_{76}$$
$$- M_{22}M_{33}M_{44}M_{55}M_{66}M_{77} - M_{22}M_{33}M_{44}M_{56}M_{65}M_{77} - M_{22}M_{33}M_{44}M_{57}M_{65}M_{76}$$

REFERENCES

1. K.J. Goh, C.T. Tan, N.K. Chew et al. Clinical features of Nipah virus encephalitis among pig farmers in Malaysia. *N. Engl. J. Med.* 342(2000) 1229–1235.
2. V.P. Hsu, M.J. Hossain, U.D. Parashar. Nipah virus encephalitis reemergence, *Bangladesh. Emerg. Infect. Dis.* 10(2004)2082–2087.
3. M.K. Mondal, M. Hanif, M.H.A. Biswas. A mathematical analysis for controlling the spread of Nipah virus infection. *Int. J. Model. Simul.* 2017, https://doi.org/10.1080/02286203.2017.1320820.
4. M.H.A. Biswas. Model and control strategy of the deadly Nipah virus (NiV) infections in Bangladesh. *Res. Rev. Biosci.* 6(2012) 370–377.
5. M.H.A. Biswas. Optimal control of Nipah virus (NiV) infections: A Bangladesh scenario. *J. Pure. Appl. Math.* 12(2014) 77–104.
6. C.N.N. Manveena Suri. 10 confirmed dead from Nipah virus outbreak in India (22 May 2018).
7. Nipah virus outbreak: Death toll rises to 14 in Kerala, two more cases identified. *Hindustan Times*, 27 May 2018.
8. T.K. Kar, S. Jana. A theoretical study on mathematical modelling of an infectious disease with application of optimal control. *BioSystems.* 111(2013) 37–50.
9. A. Lashari, G. Zaman. Optimal control of a vector borne disease with horizontal transmission. *Nonlinear Anal. RWA.* 13(2012) 203–212.
10. B.A. Satterfield, B.E. Dawes, G.N. Milligan. Status of vaccine research and development of vaccines for Nipah virus. *Vaccine.* 34(2016) 2971–2975.
11. M.A. Khan, K. Ali, E. Bonyah, K. O. Okosun, S. Islam, A. Khan. Mathematical modeling and stability analysis of Pine wilt disease with optimal control. *Sci. Reports.* 7 3115, DOI:10.1038/s41598-017-03179-w.
12. M.A. Khan, S. Islam, G. Zaman. Media coverage campaign in Hepatitis B transmission model. *Appl. Math. Comput.* 331(2018) 378–393.
13. M.A. Khan, R. Khan, Y. Khan, S. Islam. A mathematical analysis of Pine wilt disease with variable population size and optimal control strategies. *Chaos. Solit. Fract.* 108(2018) 205–217.
14. M.A. Khan, Y. Khan, S. Islam. Complex dynamics of an SEIR epidemic model with saturated incidence rate and treatment. *Physica A.* 493(2018) 210–227.
15. G. Birkhoff, G.C. Rota. *Ordinray Differential Equations.* Ginn: Boston, MA (1982).
16. P. Driessche, J. Watmough. Reproduction numbers and sub-threshold endemic equilibria for compartmental models of disease transmission. *Math. Biosci.* 180(2002) 29–48.
17. W.H. Fleming, R.W. Rishel. *Deterministic and Stochastic Optimal Control.* Springer-Verlag: New York (1975).
18. L.S. Pontryagin, V.G. Boltyanskii, R.V. Gamkrelidze, E.F. Mishchenko. *The Mathematical Theory of Optimal Processes.* Wiley: New York (1962).
19. H.R. Joshi. Optimal control of an HIV immunology model. *Optimal. Con. Appl. Method.* 23(2002) 199–213.
20. S. Lenhart, J.T. Workman. *Optimal Control Applied to Biological Models. Mathematical and Computational Biology Series.* Chapman & Hall/CRC Press: London (2007).

8 Application of Eternal Domination in Epidemiology

G. Mahadevan and T. Ponnuchamy
Gandhigram Rural Institute – Deemed to be University

Selvam Avadayappan
VHNSN College

Jyoti Mishra
Gyan Ganga Institute of Technology and Sciences

CONTENTS

8.1 INTRODUCTION

In this chapter, choose a simple, connected graph only, denoted by $G(V, E)$, where V is a vertex set and E is an edge set. It has p vertices and q edges. We consider a cycle C_m, a path P_m, a complete graph K_m and a complete bipartite graph $K_{m.n}$. In the mathematical field of graph theory, a prism graph on n vertices denoted by Y_n, $n \geq 3$ is defined by Cartesian product of a cycle with a single edge. The ladder graph can be obtained as the Cartesian product of two paths: one of which has only one edge and it is denoted by L_n, $n \geq 1$: the friendship graph F_n can be constructed by joining n copies of the cycle C_3 with a common vertex: a fan graph $F_{m,\,n}$ is defined as the graph join $\overline{K}_m + P_n$, where $\overline{K}_m$ is the empty graph on m vertices: a star S_k is the complete bipartite graph $K_{1,k}$. The wheel graph W_N is a star graph, join the pendent vertices by cycle [1–4]. The helm graph is denoted by H_n. In[5], W. Goddard et. al., introduced the concept of eternal domination in graphs. The dominating set S_0 (subset of a vertex set) of the graph G is said to be an eternal dominating set, if for any sequence $v_1, v_2, v_3, \ldots v_k$ of vertices, there exists a sequence of vertices $u_1, u_2, u_3, \ldots u_k$ with $u_1 \in S_{i-1}$ and u_i equal to or adjacent tov_i, such that each set $S_i = S_{i-1} - \{u_i\} \cup \{v_i\}$ is the dominating set in G [5]. The eternal domination number of G is denoted by $\gamma_\infty(G)$ and defined by $\gamma_\infty(G) = \min\{|S| : S \text{ is an eternal dominating set}\}$. We investigate this number for standard graphs and product-related graphs. Also we find the application for this concept to epidemiology.

8.2 EPIDEMIOLOGY

Day by day, many diseases are identified in the medical field. Some diseases will be cured with the help of medicine and treatment. But some diseases cannot be cured. This is the main problem in the medical field. How to cure or control this type of disease is termed as epidemiology. The concept of eternal domination in graph theory plays a vital role in epidemiology [6–12]. i.e. how to cure the disease in a particular area as well as how to control the disease within that area (i.e. to stop the disease spread to other unaffected areas).

8.2.1 Real-Life Application in the Concept of Eternal Domination in Epidemiology

If a disease or some infections are identified in the particular country, government will take all possible steps to control the disease from spreading to other places in the globe. In order to control the disease, medical camps are arranged in the dominating area of the country so that all the places could seek the immediate treatment. If the percentage of infection is more in some places other than the dominated areas, then to save the people a medical camp which is already located in the dominating area is moved to its neighboring area (i.e. either the area will be the most infected area or its very nearby area is infected). The minimum and efficient medical camps have to be arranged so that all the places are monitored by the doctors to give their best treatments to the infected peoples. In addition, there must be possible ways to transfer the medical camp to the neighboring places so that even after the transfer, it remains as a dominating area. This is obtained by using the eternal domination number of the pictorial representation of the country.

Following is the mathematical model for the eternal domination number in epidemiology.

Consider the country (Figure 8.1) which contains six areas. Suppose a disease (such as corona, AIDS) is spread in these places. A proper medicine or treatments are not still found to cure the disease, but it can be controlled from spreading. First, we chose the dominating areas in the country, which are v_1, v_2, v_6, v_4, and we arrange the immediate camp in those areas.

If the area v_3 is infected most when compared with the dominating areas, a medical camp should be arranged in this area so we move the medical camp v_6 to its neighboring infected area v_3 so that the medical camps are again in the dominated areas of the country.

Similar, arrangements are made when we see the other most infected areas. This process is repeated until the disease is controlled. This can be identified by the location of the medical camp that is after all the movements of medical camps if we get v_1, v_2, v_6, v_4 as the places where the medicals camps located currently. v_1, v_2, v_6, and v_4 are the places which have been chosen as the dominating places initially. If we get the dominating places which are chosen initially, then the disease is almost controlled. This is the concept of eternal domination in graph theory.

Here $V(G)=\{v_1, v_2, v_3, v_4, v_5, v_6\}$ is a vertex set in G. Consider the set $S_0=\{v_1, v_2, v_6, v_4\}$. Here, $N[S_0]=V(G)$. This gives that S_0 is a dominating set in G. Now, $S_1=S_0-\{v_6\}\cup\{v_5\}$ is a dominating set in G. Also $S_2=S_1-\{v_2\}\cup\{v_3\}$, $S_3=S_2-\{v_5\}\cup\{v_6\}$, and $S_4=S_3-\{v_3\}\cup\{v_2\}=S_0$ are all dominating sets in G. Therefore S_0 is an eternal dominating set in G, which is minimum. This gives that $\gamma_\infty(G)=4$.

8.3 ETERNAL DOMINATION NUMBER OF STANDARD GRAPHS

Eternal domination number for some standard type of graphs has been discussed in Ref. [13]. In this section, we have found the eternal domination number for prism graph, ladder graph, friendship graph, fan graph, and helm graph.

Theorem: 8.1

For any Prism graph, Y_n, $n\geq 3$, $\gamma_\infty(Y_n)=n$.

Proof: Consider the prism graph Y_n, $n\geq 3$. Here $V(Y_n)=\{u_1, u_2, u_3,.. u_n, v_1, v_2, v_3,.. v_n\}$ has *2n* vertices. Consider the set $S=\{u_1, u_2, u_3,.. u_n\}$.

Claim: S is an eternal dominating set in Y_n.

Here $N[S]=V(Y_n)$. This gives that S is a dominating set in Y_n. Now $S_1=S-\{u_1\}\cup\{v_1\}$ is also a dominating set in Y_n. $S_2=S_1-\{u_2\}\cup\{v_2\}$ is also dominating set in Y_n. Proceeding in this way we find $S_i=S_{i-1}-\{u_i\}\cup\{v_i\}$, $1\leq i\leq k$ for suitable k such that $S_k=S$. All the sets S_i, $1\leq i\leq k$ are dominating sets in Y_n. This implies S is an eternal dominating set in Y_n. Therefore

$$\gamma_\infty(Y_n)\leq n \tag{8.1}$$

Claim: $\gamma_\infty(Y_n)\not< n$

Consider the set S' with $S'|<n$. That is $S'|\leq n-1$. Now suppose $S'|=n-1$. That is $S'=\{u_1, u_2, u_3,.. u_{n-1}\}$. And the vertex v_n is not adjacent to any vertex in S', which implies S' is not a dominating set in Y_n. Therefore S' is not an eternal dominating set in S'. Therefore $\gamma_\infty(Y_n)\neq n-1$. Also any set with cardinality less then *n*–1 is not an eternal dominating set in Y_n. This gives that $\gamma_\infty(Y_n)\not< n-1$ which implies $\gamma_\infty(Y_n)>n-1$, which implies that

$$\gamma_\infty(Y_n)\geq n. \tag{8.2}$$

From (8.1) and (8.2) $\gamma_\infty(Y_n)=n$.

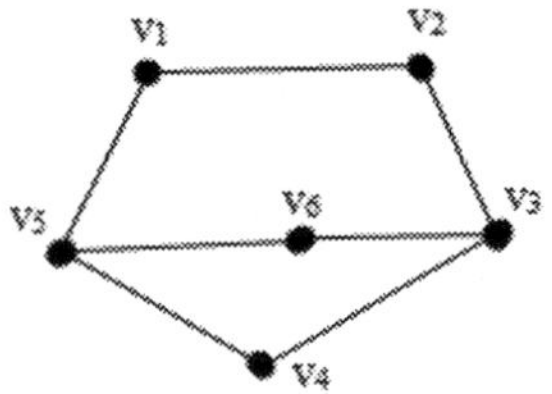

FIGURE 8.1 The graph whose eternal Domination number is 4.

Example 8.1

Consider the Prism graph Y_3

Here $V(Y_3)=\{u_1, u_2, u_3, v_1, v_2, v_3\}$ is a vertex set in Y_3. Consider the set $S_0=\{u_1, u_2, u_3\}$. Here $N[S_0]=V(Y_3)$. This gives that S_0 is a dominating set in Y_3. Now $S_1=S_0-\{u_1\}\cup\{v_1\}$ is a dominating set in Y_3. Also $S_2=S_1-\{v_1\}\cup\{v_2\}$, $S_3=S_2-\{v_2\}\cup\{v_3\}$, $S_4=S_3-\{v_3\}\cup\{v_1\}$, and $S_5=S_4-\{v_1\}\cup\{u_1\}=S_0$ are all dominating set in Y_3. Therefore S_0 is an eternal dominating set in Y_3, which is minimum. This gives that $\gamma_\infty(Y_3)=3$.

Also $\gamma_\infty(Y_4)=4$.

$$\gamma_\infty(Y_7)=7.$$

Theorem: 8.2

For any Ladder graph L_n, $n\ge 4$, $\gamma_\infty(L_n)=n+1$.

Proof: Consider the ladder graph L_n, $n\ge 2$. Here $V(L_n)=\{u_1, u_2, u_3,.. u_n, v_1, v_2, v_3,.. v_n\}$ has $2n$ vertices. Consider the set $S=\{u_1, u_2, u_3,.. u_n, v_1\}$.

Claim: S is an eternal dominating set in L_n.

Here $N[S]=V(L_n)$. This gives that S is a dominating set in L_n. Now $S_1=S-\{v_1\}\cup\{v_2\}$ is also a dominating set in L_n. In this way we find $S_i=S_{i-1}-\{v_i\}\cup\{v_{i+1}\}$, $S_j=S_{j-1}-\{v_i\}\cup\{v_{i-1}\}$, $1\le i\le n$, $n\le j\le k$ for suitable k such that $S_k=S$. All the sets S_i, S_j, $1\le i\le n$, $n\le j\le k$ are dominating sets in L_n. This implies S is an eternal dominating set in L_n. Therefore

$$\gamma_\infty(L_n)\le n+1. \tag{8.3}$$

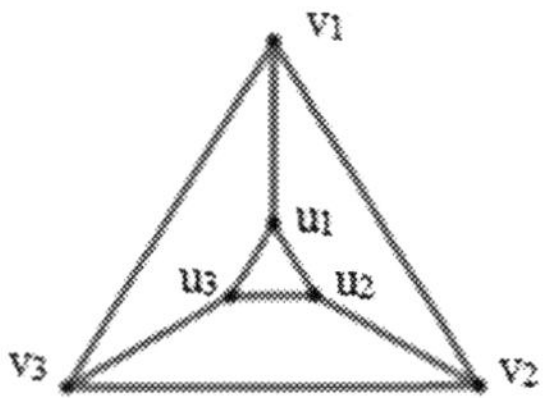

FIGURE 8.2 The Prism Graph.

Claim: $\gamma_\infty(L_n) \nless n+1$.

Consider the set S' with $S'| < n+1$. That is $S'| \leq n$. Now suppose $S'| = n$. That is $S' = \{u_1, u_2, u_3,.. u_n\}$. And every vertex in $V - S'$ is adjacent to some vertices in S', which implies S' is a dominating set in L_n. Now $S_1' = S' - \{u_n\} \cup \{v_1\}$ is a dominating set. In this way, $S_2' = S_1' - \{v_1\} \cup \{v_2\}$ is a dominating set, but $S_3' = S_2' - \{v_2\} \cup \{v_3\}$ is not a dominating set because the vertex v_1 is not adjacent to any vertex in S_3'. Therefore S' is not an eternal dominating set in L_n. Therefore $\gamma_\infty(L_n) \neq n$. Also, any set with cardinality less than n is not an eternal dominating set in L_n. This gives that $\gamma_\infty(L_n) \nless n$, which implies $\gamma_\infty(L_n) > n$, which implies that

$$\gamma_\infty(L_n) \geq n+1. \tag{8.4}$$

From (8.3) and (8.4) $\gamma_\infty(L_n) = n+1$.

Note 8.1 Consider the Ladder graphs L_2 and L_3. Here the eternal domination numbers are 2 and 3 respectively. That is $\gamma_\infty(L_n) = n \neq n+1$, $n = 2,3$.

Theorem: 8.3

For any Wheel graph W_N, $N = n+1$, $n \geq 4$, $\gamma_\infty(W_N) = N-3 = n-2$.

Proof: Consider the wheel graph W_N, $n \geq 3$. Here, $V(W_N) = \{u, v_1, v_2, v_3,.. v_n\}$ has *n+1* vertices. Consider the set $S = \{u, v_1, v_2, v_3, \ldots v_{n-3}\}$.

Claim: S is an eternal dominating set in W_N.

Here $N[S] = V(W_N)$. This gives that S is a dominating set in W_N. Now $S_1 = S - \{v_{n-3}\} \cup \{v_{n-2}\}$ is also a dominating set in W_N. In this way we find $S_i = S_{i-1} - \{v_{i+1}\} \cup \{v_i\}$, $1 \leq i \leq k$ for suitable k such that $S_k = S$. All the sets S_i, $1 \leq i \leq k$ are the dominating sets in W_N. This implies S is an eternal dominating set in W_N. Therefore

$$\gamma_\infty(W_N) \leq n-2 \tag{8.5}$$

Claim: $\gamma_\infty(W_n) \nless n-2$.

Consider the set S' with $S'| < n-2$. That is $S'| \leq n-3$. Now suppose $S'| = n-3$. That is $S' = \{u, v_1, v_2, v_3,.. v_{n-4}\}$. And every vertex in $V - S'$ is adjacent to some vertices in S' which implies S' is a dominating set in W_N. Now $S_1' = S' - \{u\} \cup \{v_{n-3}\}$ is not a dominating set because the vertex v_{n-1} is not adjacent to any vertex in S_1'. Therefore S' is not an eternal dominating set in W_N. Therefore $\gamma_\infty(W_N) \neq n-3$. Also any set with cardinality less then n-3is not an eternal dominating set in W_N. This gives that $\gamma_\infty(W_n) \nless n-3$ which implies $\gamma_\infty(W_N) > n-3$, which implies that

$$\gamma_\infty(W_N) \geq n-2. \tag{8.6}$$

From (8.5) and (8.6) $\gamma_\infty(W_N) = n-2 = N-3$.

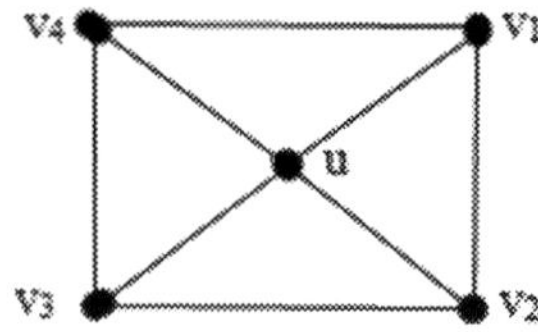

FIGURE 8.3 The Wheel Graph.

Example 8.2

Consider the wheel graph W_5 (Figure 8.3).

Here $V(W_5)=\{u, v_1, v_2, v_3, v_4\}$ is a vertex set in W_5. Consider the set $S_0=\{u, v_1\}$. Here $N[S_0]=V(W_5)$. This gives that S_0 is a dominating set in W_5. Now $S_1=S_0-\{v_1\}\cup\{v_2\}$ is a dominating set in W_5. Also $S_2=S_1-\{v_2\}\cup\{v_3\}$, $S_3=S_2-\{v_3\}\cup\{v_4\}$, $S_4=S_3-\{v_4\}\cup\{v_1\}=S_0$ are all dominating sets in W_5. Therefore S_0 is an eternal dominating set in W_5, which is minimum. This gives that $\gamma_\infty(W_5)=2=5-3$.

Also $\gamma_\infty(W_6)=3=6-3$.

$$\gamma_\infty(W_{10})=7=10-3.$$

Note 8.2 Consider the wheel graph W_4. Here the eternal domination number is 1. That is $\gamma_\infty(W_4)=1$. Since $W_4\cong K_4$.

Theorem: 8.4

For any friendship graph F_n, $n\geq 2, \gamma_\infty(F_n)=n$.

Proof: Consider the Friendship graph $F_n, n\geq 3$. Here $V(F_n)=\{u, v_1, v_2, v_3,\ldots v_n, u_1, u_2, u_3,\ldots u_n\}$ has *2n+1* vertices. Consider the set $S=\{v_1, v_2, v_3,\ldots v_n\}$.

Claim: S is an eternal dominating set in F_n.

Here $N[S]=V(F_n)$. This gives that S is a dominating set in F_n. Now $S_1=S-\{v_1\}\cup\{u_1\}$ is also a dominating set in F_n. In this way we find $S_i=S_{i-1}-\{v_i\}\cup\{u_i\}$, $S_{j+1}=S_j-\{u_i\}\cup\{v_i\}$, $1\leq i\leq n$, $n\leq j\leq k$ for suitable k such that $S_k=S$. All the sets S_i, S_j $1\leq i\leq n, n\leq j\leq k$ $S_i=S_j$ if $i=j=n$. are the dominating sets in F_n. This implies S is an eternal dominating set in F_n. Therefore

$$\gamma_\infty(F_n)\leq n. \tag{8.7}$$

Claim: $\gamma_\infty(F_n)\not< n$.

Consider the set $S'$with $|S'|<n$. That is $|S'|\leq n-1$. Now suppose $|S'|=n-1$. That is $S'=\{u, v_1, v_2, v_3, v_4, v_5,\ldots v_{n-2}\}$. And every vertex in $V-S'$ is adjacent to some vertices in S', which implies that S' is a dominating set in F_n. Now $S_1'=S'-\{u\}\cup\{u_1\}$ is not a dominating set because the vertex v_{n-1} is not adjacent to any vertex in S_1'. Therefore S' is not an eternal dominating set in F_n. Therefore $\gamma_\infty(F_n)\neq n-1$. Also any

set with cardinality less than *n-1* is not an eternal dominating set in F_n. This gives that $\gamma_\infty(F_n) \not< n-1$, which implies $\gamma_\infty(F_n) > n-1$, which implies that

$$\gamma_\infty(F_n) \geq n. \tag{8.8}$$

From (8.7) and (8.8), $\gamma_\infty(F_n) = n$.

Example 8.3

Consider the friendship graph F_5 (Figure 8.4).
Here $V(F_5) = \{u, v_1, v_2, v_3, v_4, v_5, u_1, u_2, u_3, u_4, u_5\}$ is a vertex set in F_5. Consider the set $S_0 = \{v_1, v_2, v_3, v_4, v_5\}$. Here $N[S_0] = V(F_5)$. This gives that S_0 is a dominating set in F_5. Now $S_1 = S_0 - \{v_1\} \cup \{u_1\}$ is a dominating set in F_5. Also $S_2 = S_1 - \{v_2\} \cup \{u_2\}$, $S_3 = S_2 - \{v_3\} \cup \{u_3\}$, $S_7 = S_6 - \{u_2\} \cup \{v_2\}$, $S_8 = S_7 - \{u_3\} \cup \{v_3\}$, $S_9 = S_8 - \{u_4\} \cup \{v_4\}$, and $S_{10} = S_9 - \{u_5\} \cup \{v_5\} = S_0$ are all dominating sets in F_5. Therefore S_0 is an eternal dominating set in F_5, which is minimum. This gives that $\gamma_\infty(F_5) = 5$.

Also $\gamma_\infty(F_6) = 6$.

$$\gamma_\infty(F_{10}) = 10.$$

Theorem 8.5

For any Fan graph $F_{m,n}$, $\gamma_\infty(F_{m,n}) = n-1$, $m \geq n \geq 3$.

Proof: Consider the Fangraph $F_{m,n}, m \geq n \geq 3$. Here $V(F_{m,n}) = \{v_1, v_2, v_3, .. v_n, u_1, u_2, u_3, .. u_m\}$ has $m+n$ vertices. Consider the set $S = \{v_1, v_2, v_3, \ldots v_{n-1}\}$.

Claim: S is an eternal dominating set in $F_{m,n}$.

Here $N[S] = V(F_{m,n})$. This gives that S is a dominating set in $F_{m,n}$. Now $S_1 = S - \{v_1\} \cup \{u_1\}$ is also a dominating set in $F_{m,n}$. In this way we find $S_i = S_{i-1} - \{v_i\} \cup \{u_i\}$, $S_{j+1} = S_j - \{u_i\} \cup \{v_i\}$, $1 \leq i \leq n$, $n \leq j \leq k$ for suitable k such

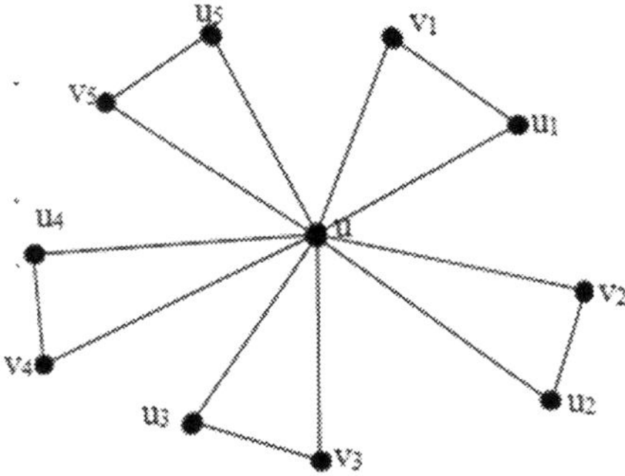

FIGURE 8.4 The Friendship Graph.

that $S_k = S$. All the sets S_i, S_j $1 \le i \le n$, $n \le j \le k$ & $S_i = S_j$ if $i = j = n$. are dominating sets in $F_{m,\,n}$. This implies S is an eternal dominating set in $F_{m,\,n}$. Therefore

$$\gamma_{\infty}\left(F_{m,\,n}\right) \le n-1. \tag{8.9}$$

Claim: $\gamma_{\infty}\left(F_{m,n}\right) \nless n-1$.

Consider the set S' with $|S'| < n-1$. That is $|S'| \le n-2$. Now suppose $|S'| = n-2$. That is $S' = \{v_1,\, v_2,\, v_3,\, v_4,\, v_5, \ldots\, v_{n-2}\}$. And the vertex v_n is not adjacent to any vertex in S'. This gives that S' is not a dominating set in $F_{m,\,n}$. Therefore S' is not an eternal dominating set in $F_{m,\,n}$. Therefore $\gamma_{\infty}\left(F_{m,\,n}\right) \ne n-2$. Also any set with cardinality less then *n-2* is not an eternal dominating set in $F_{m,\,n}$. This gives that $\gamma_{\infty}\left(F_{m,n}\right) \nless n-2$, which implies $\gamma_{\infty}\left(F_{m,\,n}\right) > n-2$, which implies that

$$\gamma_{\infty}\left(F_{m,\,n}\right) \ge n-1. \tag{8.10}$$

From (8.9) and (8.10) $\gamma_{\infty}\left(F_{m,\,n}\right) = n-1$.

Example 8.4

Consider the fan graph $F_{3,3}$ (Figure 8.5).
Here $V\left(F_{3,3}\right) = \{v_1,\, v_2,\, v_3,\, u_1,\, u_2,\, u_3\}$ is a vertex set in $F_{3,3}$. Consider the set $S_0 = \{v_1,\, v_2\}$. Here $N[S_0] = V\left(F_{3,3}\right)$. This gives that S_0 is a dominating set in $F_{3,3}$. Now $S_1 = S_0 - \{v_2\} \cup \{v_3\}$ is a dominating set in $F_{3,3}$. Also $S_2 = S_1 - \{v_1\} \cup \{u_1\}$, $S_3 = S_2 - \{v_3\} \cup \{u_3\}$, $S_4 = S_3 - \{u_1\} \cup \{u_2\}$, $S_5 = S_4 - \{u_2\} \cup \{v_1\}$, $S_6 = S_5 - \{u_3\} \cup \{v_2\} = S_0$ are all dominating sets in $F_{3,3}$. Therefore S_0 is an eternal dominating set in $F_{3,3}$ which is minimum. This gives that $\gamma_{\infty}\left(F_{3,3}\right) = 2$.

Also, $\gamma_{\infty}\left(F_{3,4}\right) = 3$.

$$\gamma_{\infty}\left(F_{7,6}\right) = 5.$$

Theorem: 8.6

For any Helm graph H_n, F_n, $\gamma_{\infty}\left(H_n\right) = n+1$.

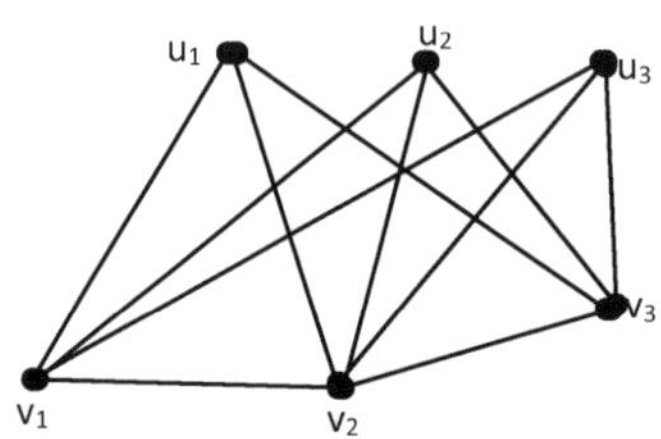

FIGURE 8.5 The Fan Graph.

Proof: Consider the Helmgraph $H_n, n \geq 3$. Here, $V(H_n) = \{u, v_1, v_2, v_3, .. v_n, u_1, u_2, u_3, .. u_n\}$ has *2n+1* vertices. Consider the set $S = \{u, v_1, v_2, v_3, ... v_n\}$.

Claim: S is an eternal dominating set in H_n.

Here $N[S] = V(H_n)$. This gives that S is a dominating set in H_n. Now $S_1 = S - \{v_1\} \cup \{u_1\}$ is also a dominating set in H_n. In this way we find $S_i = S_{i-1} - \{v_i\} \cup \{u_i\}$, $S_{j+1} = S_j - \{u_i\} \cup \{v_i\}$ $1 \leq i \leq n$, $n \leq j \leq k$ for suitable k such that $S_k = S$. All the sets S_i, S_j $1 \leq i \leq n$, $n \leq j \leq k$ & $S_i = S_j$ if $i = j = n$. are dominating sets in H_n. This implies S is an eternal dominating set in H_n. Therefore

$$\gamma_\infty(H_n) \leq n + 1. \tag{8.11}$$

Claim: $\gamma_\infty(H_n) \not< n + 1$.

Consider the set S' with $S'| < n + 1$. That is $S'| \leq n$. Now suppose $S'| = n$. That is $S' = \{v_1, v_2, v_3, v_4, v_5, ... v_n\}$. Also, every vertex in $V - S'$ is adjacent to some vertices in S', which implies S' is a dominating set in H_n. Now $S_1' = S' - \{v_n\} \cup \{u\}$ is not a dominating set because the vertex u_n is not adjacent to any vertex in S_1'. Therefore S' is not an eternal dominating set in H_n. Therefore $\gamma_\infty(H_n) \neq n$. Also any set with cardinality less then n is not an eternal dominating set in H_n. This gives that $\gamma_\infty(H_n) \not< n$, which implies $\gamma_\infty(H_n) > n$, which implies that

$$\gamma_\infty(H_n) \geq n + 1 \tag{8.12}$$

From (8.11) and (8.12), $\gamma_\infty(H_n) = n + 1$.

Example 8.5

Consider the Helm graph H_4 (Figure 8.6).

Here $V(H_4) = \{u, v_1, v_2, v_3, v_4, u_1, u_2, u_3, u_4\}$ is a vertex set in H_4. Consider the set $S_0 = \{u, v_1, v_2, v_3, v_4, \}$. Here $N[S_0] = V(H_4)$. This gives that S_0 is a dominating set in H_4. Now $S_1 = S_0 - \{v_1\} \cup \{u_1\}$ is a dominating set in H_4. Also $S_2 = S_1 - \{v_2\} \cup \{u_2\}$, $S_3 = S_2 - \{v_3\} \cup \{u_3\}$, $S_4 = S_3 - \{v_4\} \cup \{u_4\}$,

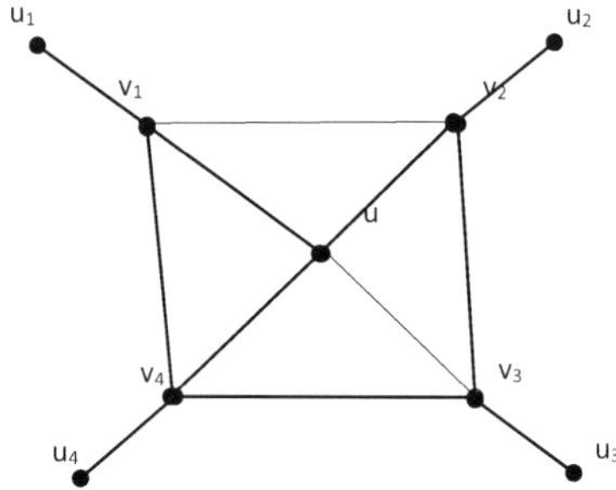

FIGURE 8.6 The Helm Graph.

$S_5 = S_4 - \{u_4\} \cup \{v_4\}$, $S_6 = S_5 - \{u_1\} \cup \{v_1\}$, $S_7 = S_6 - \{u_2\} \cup \{v_2\}$, $S_8 = S_7 - \{u_3\} \cup \{v_3\} = S_0$ are all dominating sets in H_4. Therefore S_0 is an eternal dominating set in H_4, which is minimum. This gives that $\gamma_\infty(H_4) = 5$.

Also $\gamma_\infty(H_6) = 7$.

$$\gamma_\infty(H_{10}) = 11.$$

8.4 ETERNAL DOMINATION NUMBER OF SOME PRODUCT RELATED GRAPHS

Lexicographic product of paths: In this section, we find the eternal domination number of the lexicographic product of paths. We recall the existing definition of lexicographic product of graphs: The lexicographic product of the graphs G and H is denoted by $G \circ H$, whose vertex set is $V(G) \times V(H)$. Two vertices (g, h) and (g', h') are adjacent in $G \circ H$ if $gg' \in E(G)$ (or) $g = g'$ and $hh' \in E(H)$. That is $V(G \circ H) = \{(g, h) : g \in V(G) \text{ and } h \in V(H)\}$ $E(G \circ H) = \{(g, h)(g', h') : gg' \in E(G)$ (or) $g = g'$ and $hh' \in E(H)\}$. The number of vertices in the lexicographic product of $G \circ H$ is $|V(G)||V(H)|$.

Theorem 8.7

For any $n \geq 3$, we have $\gamma_\infty(P_2 \circ P_n) = n - 1$.

Proof: Let P_2 and P_n be the paths on 2 vertices and n vertices, respectively. The lexicographic product of P_2 and P_n is denoted by $P_2 \circ P_n$ and has $2n$ vertices. Here $V(P_2 \circ P_n) = \{ (u_i, v_j) / 1 \leq i \leq 2, 1 \leq j \leq n\}$.

Consider the set $S = \{(u_1, v_k), 1 \leq k \leq n-1\}$. Then $|S| = n - 1$.

Claim: S is an eternal dominating set in $P_2 \circ P_n$.

Here $N[S] = V(P_2 \circ P_n)$. This gives that S is a dominating set in $P_2 \circ P_n$. Now $S_{1,1} = S - \{u_1, v_1\} \cup \{u_2, v_1\}$ is also a dominating set in $P_2 \circ P_n$. In this way we find $S_{i,j} = S_{i,j} - \{u_i, v_j\} \cup \{u_i, v_{j-1}\}, 1 \leq i \leq 2, 1 \leq j \leq n-1, S_{2,n} = S$. All the sets $S_{i,j}$ are the dominating sets in $P_2 \circ P_n$. This implies that S is an eternal dominating set in $P_2 \circ P_n$. Therefore

$$\gamma_\infty(P_2 \circ P_n) \leq n - 1 . \tag{8.13}$$

Claim: $\gamma_\infty(P_2 \circ P_n) \not< n - 1$.

Consider the set S' with $|S'| < n - 1$. That is $|S'| \leq n - 2$. Now suppose $|S'| = n - 2$. That is $S' = \{u_1, v_j, 1 \leq j \leq n - 2\}$. And the vertex u_1, v_n is not adjacent to any vertex in S'. This gives that S' is not a dominating set in $P_2 \circ P_n$. Therefore S' is not an eternal dominating set in $P_2 \circ P_n$. Therefore $\gamma_\infty(P_2 \circ P_n) \neq n - 2$. Also any set with cardinality less then *n-2* is not an eternal dominating set in $P_2 \circ P_n$. This gives that $\gamma_\infty(P_2 \circ P_n) \nless n-2$, which implies $\gamma_\infty(P_2 \circ P_n) > n-2$, which implies that

$$\gamma_\infty(P_2 \circ P_n) \geq n-1. \tag{8.14}$$

From (8.13) and (8.14) $\gamma_\infty(P_2 \circ P_n) = n-1$.

Theorem 8.8

For any $n \geq m \geq 3$, we have $\gamma_\infty(P_m \circ P_n) = m-1$.

Proof: Let P_m and P_n be the paths on m vertices and n vertices, respectively. The lexicographic product of P_m and P_n is denoted by $P_m \circ P_n$ and has mn vertices. Here $V(P_m \circ P_n) = \{ (u_i, v_j) / 1 \leq i \leq m, 1 \leq j \leq n \}$.

Consider the set $S = \{(u_i, v_1), 1 \leq i \leq m-1\}$. Then $|S| = m-1$.

Claim: S is an eternal dominating set in $P_m \circ P_n$.

Here $N[S] = V(P_m \circ P_n)$. This gives that S is a dominating set in $P_m \circ P_n$. Now $S_{1.1} = S - \{u_1, v_1\} \cup \{u_2, v_1\}$ is also a dominating set in $P_m \circ P_n$. In this way we find $S_{i,j} = S_{i,j} - \{u_i, v_j\} \cup \{u_i, v_{j-1}\}$, $1 \leq i \leq m$, $1 \leq j \leq n$, $S_{m.n} = S$. All the sets $S_{i.j}$ are the dominating sets in $P_m \circ P_n$. This implies S is an eternal dominating set in $P_m \circ P_n$. Therefore

$$\gamma_\infty(P_m \circ P_n) \leq m-1 . \tag{8.15}$$

Claim: $\gamma_\infty(P_m \circ P_n) \nless m-1$.

Consider the set S' with $|S'| < m-1$. That is $|S'| \leq m-2$. Now suppose $|S'| = m-2$. That is $S' = \{u_i, v_1, 1 \leq i \leq m-2\}$. And the vertex u_m, v_1 is not adjacent to any vertex in S'. This gives that S' is not a dominating set in $P_m \circ P_n$. Therefore S' is not an eternal dominating set in $P_m \circ P_n$. Therefore $\gamma_\infty(P_m \circ P_n) \neq m-2$. Also any set with cardinality less then *m-2* is not an eternal dominating set in $P_m \circ P_n$. This gives that $\gamma_\infty(P_m \circ P_n) \nless m-2$ which implies $\gamma_\infty(P_m \circ P_n) > m-2$, which implies that

$$\gamma_\infty(P_m \circ P_n) \geq m-1 \tag{8.16}$$

From (8.15) and (8.16) $\gamma_\infty(P_m \circ P_n) = m-1$.

Example 8.6

Consider the graphs (Figures 8.7–8.9).

P2: u1 u2

FIGURE 8.7 Path P_2.

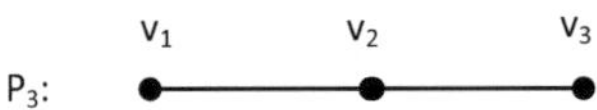

FIGURE 8.8 Path P_3.

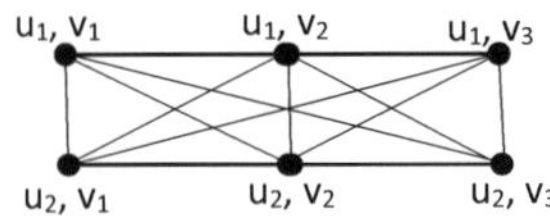

FIGURE 8.9 The lexicographic product $P_2 \circ P_3$.

Then the lexicographic product $P_2 \circ P_3$

Here $V(P_2 \circ P_3) = \{(u_i, v_j) : 1 \le i \le 2, 1 \le j \le 3\}$. Consider $S = \{(u_1, v_1), (u_1, v_2)\}$ be the subset of $V(P_2 \circ P_3)$. Then $N[S] = V(P_2 \circ P_3)$. Therefore S is a dominating set. Now $S_1 = S - (u_1, v_1) \cup (u_2, v_1)$, $S_2 = S_1 - (u_2, v_1) \cup (u_2, v_2)$, $S_3 = S_4 - (u_2, v_2) \cup (u_2, v_3)$, and $S_4 = S_3 - (u_2, v_3) \cup (u_1, v_1) = S$. Each set S_i is a dominating set in $P_2 \circ P_3$, which is minimum. Therefore $\gamma_\infty(P_2 \circ P_3) = 2$.

Similarly,

$$\gamma_\infty(P_4 \circ P_5) = 3.$$

$$\gamma_\infty(P_7 \circ P_{10}) = 6.$$

***Lexicographic product of cycles*:** In this section, we find the eternal domination number of the Lexicographic product of cycles.

Theorem 8.9

For any $n \ge 3$, we have $\gamma_\infty(C_3 \circ C_n) = 3$.

Proof: Let C_3 and C_n be the cycles on *3* vertices and n vertices, respectively. The lexicographic product of C_3 and C_n is denoted by $C_3 \circ C_n$ and has $3n$ vertices. Here $V(C_3 \circ C_n) = \{(u_i, v_j) / 1 \le i \le 3, 1 \le j \le n\}$.

Consider the set $S = \{(u_i, v_1), 1 \le i \le 2\}$. Then $|S| = 3$.

Claim: S is an eternal dominating set in $C_3 \circ C_n$.

Here $N[S] = V(C_3 \circ C_n)$. This gives that S is a dominating set in $C_3 \circ C_n$. Now $S_{1,1} = S - \{u_1, v_1\} \cup \{u_2, v_2\}$ is also a dominating set in $C_3 \circ C_n$. In this way we find $S_{i,j} = S_{i,j} - \{u_i, v_j\} \cup \{u_i, v_{j-1}\}, 1 \le i \le 3, 1 \le j \le n, S_{3,n} = S$. All the sets $S_{i,j}$ are dominating sets in $C_3 \circ C_n$. This implies S is an eternal dominating set in $C_3 \circ C_n$. Therefore

$$\gamma_\infty(C_3 \circ C_n) \le 2. \tag{8.17}$$

Claim: $\gamma_\infty(C_3 \circ C_n) \not< 2$.

Consider the set $S' = \{u_1, v_1\}$. Here $|S'| = 1$. And the vertex u_3, v_2 is not adjacent to any vertex in S'. This gives that S' is not a dominating set in $C_3 \circ C_n$. Therefore S'

is not an eternal dominating set in $C_3 \circ C_n$. Therefore $\gamma_\infty(C_3 \circ C_n) \neq 1$. This gives that $\gamma_\infty(C_3 \circ P_n) \not< 1$, which implies $\gamma_\infty(C_3 \circ C_n) > 1$, which implies that

$$\gamma_\infty(C_3 \circ P_n) \geq 2 \tag{8.18}$$

From (8.17) and (8.18) $\gamma_\infty(C_3 \circ C_n) = 2$.

Theorem 8.10

For any $n \geq m \geq 4$, we have $\gamma_\infty(C_m \circ C_n) = m - 1$.

Proof: Let C_m and C_n be the cycles on m vertices and n vertices, respectively. The lexicographic product of C_m and C_n is denoted by $C_m \circ C_n$ and has mn vertices. Here $V(C_m \circ C_n) = \{(u_i, v_j) / 1 \leq i \leq m, 1 \leq j \leq n\}$.

Consider the set $S = \{(u_i, v_1), 1 \leq i \leq m-1\}$. Then $S| = m - 1$.

Claim: S is an eternal dominating set in $C_m \circ C_n$.

Here $N[S] = V(C_m \circ C_n)$. This gives that S is a dominating set in $C_m \circ C_n$. Now $S_{1.1} = S - \{u_1, v_1\} \cup \{u_2, v_1\}$ is also a dominating set in $C_m \circ C_n$. In this way we find $S_{i,j} = S_{i,j} - \{u_i, v_j\} \cup \{u_i, v_{j-1}\}, 1 \leq i \leq m, 1 \leq j \leq n, S_{m,n} = S$. All the sets $S_{i,j}$ are the dominating sets in $C_m \circ C_n$. This implies S is an eternal dominating set in $C_m \circ C_n$. Therefore

$$\gamma_\infty(C_m \circ C_n) \leq m - 1 . \tag{8.19}$$

Claim: $\gamma_\infty(C_m \circ C_n) \not< m - 1$.

Consider the set $S'$with $S'| < m - 1$. That is $S'| \leq m - 2$. Now suppose $S'| = m - 2$. That is, $S' = \{u_i, v_1, 1 \leq i \leq m - 2\}$. And the vertices $u_m, v_j, 2 \leq j \leq n$ are not adjacent to any vertex in S'. This implies that S' is not a dominating set in $C_m \circ C_n$. Therefore S' is not an eternal dominating set in $C_m \circ C_n$. Therefore $\gamma_\infty(C_m \circ C_n) \neq m - 2$. Also any set with cardinality less then *m-2* is not an eternal dominating set in $C_m \circ C_n$. This gives that $\gamma_\infty(C_m \circ C_n) \not< m - 2$ which implies $\gamma_\infty(C_m \circ C_n) > m - 2$, which implies that

$$\gamma_\infty(C_m \circ C_n) \geq m - 1 \tag{8.20}$$

From (8.19) and (8.20), $\gamma_\infty(C_m \circ C_n) = m - 1$.

Example 8.7

Consider the graphs (Figures 8.10–8.12).

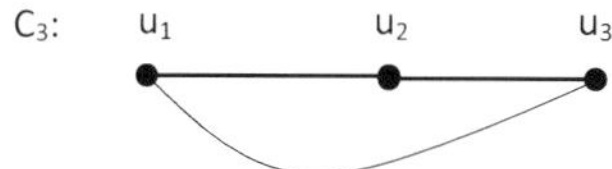

FIGURE 8.10 Cycle C_3.

FIGURE 8.11 Cycle C_3.

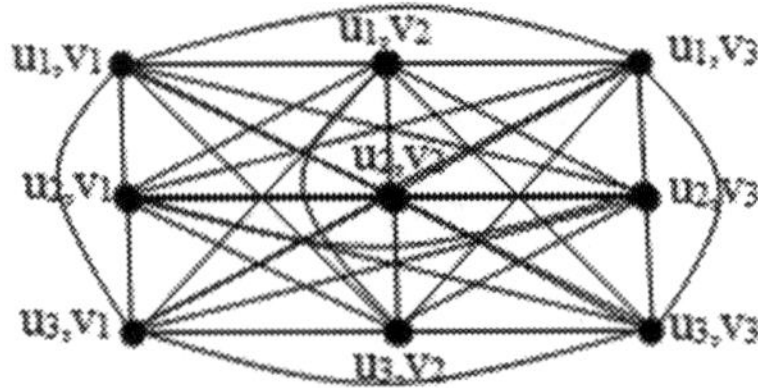

FIGURE 8.12 The lexicographic product $C_3 \circ C_3$.

Then the lexicographic product $C_3 \circ C_3$

Here $V(C_3 \circ C_3) = \{(u_i, v_j) : 1 \le i \le 3, 1 \le j \le 3\}$.

Let $S = \{(u_1, v_1), (u_1, v_2), (u_1, v_3)\}$ be the subset of $V(C_3 \circ C_3)$. Here $N[S] = V(C_3 \circ C_3)$. Therefore S is a dominating set. Now $S_1 = S - (u_1, v_1) \cup (u_2, v_1)$, $S_2 = S_1 - (u_2, v_1) \cup (u_2, v_2)$, $S_3 = S_4 - (u_2, v_2) \cup (u_2, v_3)$, $S_4 = S_3 - (u_2, v_3) \cup (u_3, v_3)$, $S_5 = S_4 - (u_3, v_3) \cup (u_3, v_2)$, $S_6 = S_5 - (u_3, v_2) \cup (u_3, v_1)$ and $S_7 = S_6 - (u_3, v_1) \cup (u_1, v_1) = S$. Each set S_i is a dominating set in $C_3 \circ C_3$, which is minimum. Therefore $\gamma_\infty(C_3 \circ C_3) = 2$.

Similarly,

$$\gamma_\infty(C_4 \circ C_5) = 3$$

$$\gamma_\infty(C_7 \circ C_{10}) = 6.$$

Lexicographic product of paths and cycles: In this section we find the Triple connected eternal domination number of the Lexicographic product of paths and cycles.

Theorem 8.11

For any $n \ge 3$, we have $\gamma_\infty(P_2 \circ C_n) = 2$.

Proof: Let P_2 and C_n be the path and cycle on *2* vertices and n vertices, respectively. The lexicographic product of P_2 and C_n is denoted by $P_2 \circ C_n$ has *2n* vertices. Here $V(P_2 \circ C_n) = \{ (u_i, v_j) / 1 \le i \le 2, 1 \le j \le n\}$.

Consider the set $S = \{(u_1, v_1), (u_2, v_1) \}$. Then $|S| = 2$.

Claim: S is an eternal dominating set in $P_2 \circ C_n$.

Here $N[S] = V(P_2 \circ C_n)$. This gives that S is a dominating set in $P_2 \circ C_n$. Now $S_{1,1} = S - \{u_1, v_1\} \cup \{u_1, v_2\}$ is also a dominating set in $P_2 \circ C_n$. In this way we find $S_{i,j} = S_{i,j} - \{u_i, v_j\} \cup \{u_i, v_{j-1}\}, 1 \le i \le 2, 1 \le j \le n-1, S_{2,n} = S$. All the sets $S_{i,j}$ are the dominating sets in $P_2 \circ C_n$. This implies S is an eternal dominating set in $P_2 \circ C_n$. Therefore

$$\gamma_\infty(P_2 \circ C_n) \le n - 1. \tag{8.21}$$

Claim: $\gamma_{\infty}(P_2 \circ C_n) \not< 2$.

Consider the set $S' = \{u_1, v_1\}$. Here $|S'| = 1$. And the vertex (u_1, v_n) in $V - S'$ is not adjacent to any vertex in S' which implies that S' is not a dominating set in $P_2 \circ C_n$. Therefore S' is not an eternal dominating set in $P_2 \circ C_n$. Therefore $\gamma_{\infty}(P_2 \circ C_n) \neq 1$. This gives that $\gamma_{\infty}(P_2 \circ C_n) > 1$, which implies that

$$\gamma_{tc.\,\infty}(P_2 \circ C_n) \geq 2 \tag{8.22}$$

From (8.21) and (8.22), $\gamma_{tc.\,\infty}(P_2 \circ C_n) = 2$.

Example 8.8

Consider the graphs (Figures 8.13–8.15).

Then the lexicographic product $P_2 \circ C_3$

Here $V(P_2 \circ C_3) = \{(u_i, v_j) : 1 \leq i \leq 2, 1 \leq j \leq 3\}$.

Let $S = \{(u_1, v_1), (u_2, v_1)\}$ be the subset of $V(P_2 \circ C_3)$. Here $N[S] = V(P_2 \circ C_3)$. Therefore S is a dominating set. Now $S_1 = S - (u_1, v_1) \cup (u_1, v_2)$, $S_2 = S_1 - (u_2, v_1) \cup (u_2, v_2)$, $S_3 = S_4 - (u_2, v_2) \cup (u_2, v_3)$, and $S_4 = S_3 - (u_2, v_3) \cup (u_2, v_1)$. Each set S_i is the dominating set in $P_2 \circ C_3$, which is minimum. Therefore $\gamma_{\infty}(P_2 \circ C_3) = 3$.

Similarly,

$$\gamma_{\infty}(P_2 \circ C_5) = 2$$

$$\gamma_{\infty}(P_2 \circ C_{10}) = 2.$$

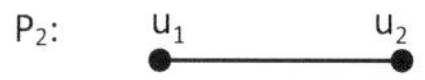

P2: u1 u2

FIGURE 8.13 Path P_2.

FIGURE 8.14 Cycle C_3.

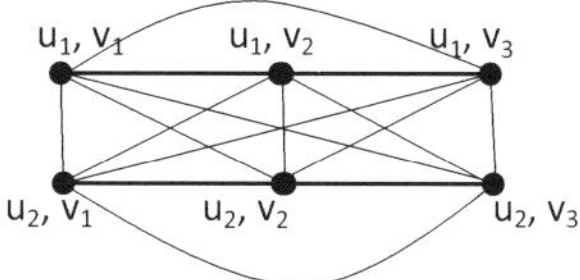

FIGURE 8.15 The lexicographic product $P_2 \circ C_3$.

Theorem 8.12

For any $n > m \geq 3$, we have $\gamma_{tc,\infty}(P_m \circ C_n) = m - 1$.

Proof: Let P_m and C_n be the path and cycle on m vertices and n vertices respectively. The lexicographic product of P_m and C_n is denoted by $P_m \circ C_n$ and has mn vertices. Here $V(P_m \circ C_n) = \{(u_i, v_j) / 1 \leq i \leq m, 1 \leq j \leq n\}$.

Consider the set $S = \{(u_i, v_1), 1 \leq i \leq m-1\}$. Then $|S| = m - 1$.

Claim: S is an eternal dominating set in $P_m \circ C_n$.

Here $N[S] = V(P_m \circ C_n)$. This gives that S is a dominating set in $P_m \circ C_n$. Now $S_{1,1} = S - \{u_1, v_1\} \cup \{u_2, v_1\}$ is also a dominating set in $P_m \circ C_n$. In this way we find $S_{i,j} = S_{i,j} - \{u_i, v_j\} \cup \{u_i, v_{j-1}\}, 1 \leq i \leq m, 1 \leq j \leq n, S_{m,n} = S$. All the sets $S_{i,j}$ are the dominating sets in $P_m \circ C_n$. This implies S is an eternal dominating set in $P_m \circ C_n$. Therefore

$$\gamma_\infty(P_m \circ C_n) \leq m - 1. \tag{8.23}$$

Claim: $\gamma_\infty(P_m \circ C_n) \nless m - 1$.

Consider the set $S'$with $|S'| < m - 1$. That is $|S'| \leq m - 2$. Now suppose $|S'| = m - 2$. That is $S' = \{u_i, v_1, 1 \leq i \leq m - 2\}$. And the vertex (u_m, v_1) in $V - S'$ is not adjacent to any vertex in S', which implies S' is a not dominating set in $P_m \circ C_n$. This gives that S' is not an eternal dominating set in $P_m \circ C_n$. Therefore $\gamma_\infty(P_m \circ C_n) \neq m - 2$. Also any set with cardinality less then *m-2* is not an eternal dominating set in $P_m \circ C_n$. This gives that $\gamma_\infty(P_m \circ C_n) \nless m - 2$ which implies $\gamma_{tc,\infty}(P_m \circ C_n) > m - 2$, which implies that

$$\gamma_\infty(P_m \circ C_n) \geq m - 1 \tag{8.24}$$

From (8.23) and (8.24), $\gamma_{tc,\infty}(P_m \circ C_n) = m - 1$.

Example 8.9

Consider the graphs (Figures 8.16–8.18).

FIGURE 8.16 Path P_3.

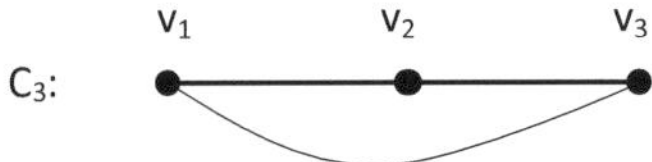

FIGURE 8.17 Cycle C_3.

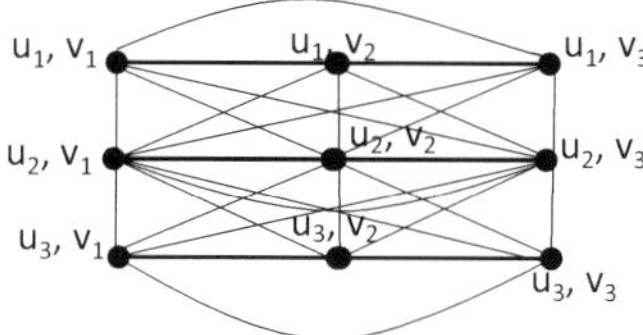

FIGURE 8.18 The lexicographic product $P_3 \circ C_3$.

Then the lexicographic product $P_3 \circ C_3$.

Here $V(P_3 \circ C_3) = \{(u_i, v_j) : 1 \le i \le 3, 1 \le j \le 3\}$.

Let $S = \{(u_1, v_1), (u_2, v_1)\}$ be the subset of $V(P_3 \circ C_3)$. Here $N[S] = V(P_3 \circ C_3)$. Therefore S is a dominating set. Now $S_1 = S - (u_1, v_1) \cup (u_1, v_2)$, $S_2 = S_1 - (u_2, v_1) \cup (u_2, v_2)$, $S_3 = S_4 - (u_2, v_2) \cup (u_2, v_3)$, $S_4 = S_3 - (u_2, v_3) \cup (u_3, v_3)$, $S_5 = S_4 - (u_3, v_3) \cup (u_3, v_2)$, $S_6 = S_5 - (u_3, v_2) \cup (u_3, v_1)$, and $S_7 = S_6 - (u_3, v_1) \cup (u_1, v_1)$. Each set S_i is triple connected dominating set in $P_3 \circ C_3$, which is minimum. Therefore $\gamma_\infty(P_3 \circ C_3) = 2$.

Similarly,

$$\gamma_\infty(P_4 \circ C_5) = 3$$

$$\gamma_\infty(P_7 \circ C_7) = 6.$$

Corona product of paths: We recall the existing definition of Corona product of graphs: The corona product of graphs G and H is denoted by *G[H]*, defined by one copy of *G* and $V(G)|$ copies of *H*. Here the ith vertex in *G* is adjacent to every vertex in the ith copy of *H*. In this section we find the eternal domination number of corona product of Paths.

Theorem 8.13

For any $n > 3$, we have $\gamma_{tc.\infty}(P_2[P_n]) = 2n - 2$.

Proof: Let P_2 and P_n be the paths on *2* vertices and n vertices, respectively. The corona product of P_2 and P_n is denoted by $P_2[P_n]$ has *2n+2* vertices. Here $V(P_2[P_n]) = \{(u_i), (v_{i,j}) / 1 \le i \le 2, 1 \le j \le n\}$.

Consider the set $S = \{(u_i), (v_{i,k}), 1 \le i \le 2, 1 \le k \le n-2\}$. Then $S| = 2n - 2$.

Claim: S is an eternal dominating set in $P_2[P_n]$.

Here $N[S] = V(P_2[P_n])$. This gives that S is a dominating set in $P_2[P_n]$. Now $S_{1.1} = S - \{v_{1,n-2}\} \cup \{v_{1,n-1}\}$ is also a dominating set in $P_2[P_n]$. In this way we find $S_{i,j} = S_{i,j} - \{v_{i,j-1}\} \cup \{v_{i,j}\}$, $1 \le i \le 2$, $1 \le j \le n-1$, $S_{2.n} = S$. All the sets $S_{i,j}$ are dominating sets in $P_2[P_n]$. This implies S is an eternal dominating set in $P_2[P_n]$. Therefore

$$\gamma_\infty\left(P_2\left[P_n\right]\right) \leq 2n-2. \tag{8.25}$$

Claim: $\gamma_\infty\left(P_2\left[P_n\right]\right) \nless 2n-2$.

Consider the set S' with $|S'| < 2n-2$. That is, $|S'| \leq 2n-3$. Now suppose $|S'| = 2n-3$. That is, $S' = \{v_{i,\,j},\ 1 \leq j \leq n-2\} - \{v_{1,n-2}\}$. And the vertex $v_{1,n}$ is not adjacent to any vertex in S'. This gives that S' is not a dominating set in $P_2\left[P_n\right]$. Therefore, S' is not an eternal dominating set in $P_2\left[P_n\right]$. Therefore $\gamma_\infty\left(P_2\left[P_n\right]\right) \neq 2n-3$. Also any set with cardinality less then $2n$–3 is not an eternal dominating set in $P_2\left[P_n\right]$. This gives that $\gamma_\infty\left(P_2\left[P_n\right]\right) \nless 2n-3$ which implies $\gamma_\infty\left(P_2\left[P_n\right]\right) > 2n-3$, which implies that

$$\gamma_\infty\left(P_2\left[P_n\right]\right) \geq 2n-2 \tag{8.26}$$

From (8.25) and (8.26), $\gamma_\infty\left(P_2\left[P_n\right]\right) = 2n-2$.

Note 8.3 The eternal domination number of $P_2\left[P_3\right]$ is 4.

Theorem 8.14

For any $n \geq m \geq 3$, we have $\gamma_\infty\left(P_m\left[P_n\right]\right) = m(n-1)$.

Proof: Let P_m and P_n be the paths on m vertices and n vertices respectively. The corona product of P_m and P_n is denoted by $P_m\left[P_n\right]$ has $m(n+1)$ vertices. Here $V\left(P_m\left[P_n\right]\right) = \{(u_i),\ (v_{i,\,j})\,/\,1 \leq i \leq m,\ 1 \leq j \leq n\}$.

Consider the set $S = \{(u_i),\ (v_{i,\,k}),\ 1 \leq i \leq m,\ 1 \leq k \leq n-2\}$. Then $|S| = m(n-1)$.

Claim: S is an eternal dominating set in $P_m\left[P_n\right]$.

Here $N[S] = V\left(P_m\left[P_n\right]\right)$. This gives that S is a dominating set in $P_m\left[P_n\right]$. Now $S_{1,1} = S - \{v_{1,n-2}\} \cup \{v_{1,n-1}\}$ is also a dominating set in $P_m\left[P_n\right]$. In this way, we find $S_{i,\,j} = S_{i,\,j} - \{v_{i,\,j-1}\} \cup \{v_{i,\,j}\},\ 1 \leq i \leq m,\ 1 \leq j \leq n-1, S_{m,\,n} = S$. All the sets $S_{i,j}$ are dominating sets in $P_m\left[P_n\right]$. This implies S is an eternal dominating set in $P_m\left[P_n\right]$. Therefore

$$\gamma_\infty\left(P_m\left[P_n\right]\right) \leq m(n-1)\ . \tag{8.27}$$

Claim: $\gamma_\infty\left(P_m\left[P_n\right]\right) \nless m(n-1)$.

Consider the set S' with $|S'| < m(n-1)$. That is, $|S'| \leq m(n-1)-1$. Now suppose $|S'| = m(n-1)-1$. That is, $S' = \{v_{i,\,j},\ 1 \leq i \leq m,\ 1 \leq j \leq n-2\} - \{v_{1,n-2}\}$. And the vertex $v_{1,n}$ is not adjacent to any vertex in S'. This gives that S' is not a dominating set in $P_m\left[P_n\right]$. Therefore, S' is not an eternal dominating set in $P_m\left[P_n\right]$. Therefore $\gamma_\infty\left(P_m\left[P_n\right]\right) \neq m(n-1)-1$. Also any set with cardinality less then $m(n-1)-1$ is not an eternal dominating set in $P_m\left[P_n\right]$. This gives that $\gamma_\infty\left(P_2\left[P_n\right]\right) \nless (n-1)-1$, which implies $\gamma_\infty\left(P_m\left[P_n\right]\right) > m(n-1)-1$, which implies that

$$\gamma_\infty\left(P_2\left[P_n\right]\right) \geq m(n-1) \tag{8.28}$$

From (8.27) and (8.28), $\gamma_\infty\left(P_m\left[P_n\right]\right) = m(n-1)$.

P2: u1 u2

FIGURE 8.19 Path P_2.

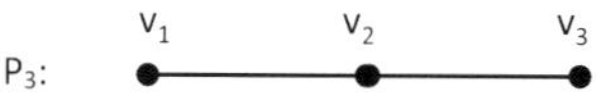

FIGURE 8.20 Path P_3.

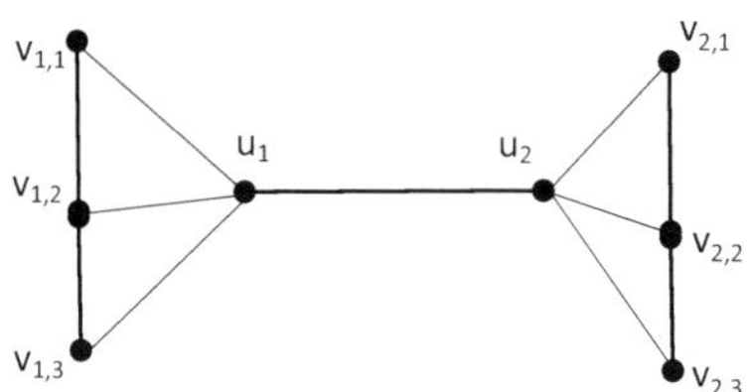

FIGURE 8.21 The Corona product P_2 [P_3].

Example 8.10

Consider the graphs (Figures 8.19–8.21).

Then the Corona product $P_2[P_{3.}]$.

Here $V(P_2[P_3]) = \{(u_i), (v_{i,\,j}) : 1 \le i \le 2, 1 \le j \le 3\}$.

Let $S = \{(v_{i,\,k})1 \le i \le 2,\ 1 \le k \le 2\}$ be the subset of $V(P_2 \circ P_3)$. Here $N[S] = V(P_2[P_3])$. Therefore S is a dominating set. Now $S_1 = S - (v_{1.2}) \cup (v_{1.3})$, $S_2 = S_1 - (v_{2.2}) \cup (v_{2.3})$, $S_3 = S_2 - (v_{1.3}) \cup (v_{1.2})$, $S_4 = S_3 - (v_{2.3}) \cup (v_{2.2}) = S$. Each set S_i is a dominating set in $P_2[P_3]$, which is minimum. Therefore $\gamma_\infty(P_2[P_3]) = 4$.

Similarly,

$$\gamma_\infty(P_4[P_5]) = 16$$

$$\gamma_\infty(P_7[P_{10}]) = 63.$$

Corona product of cycles: In this section we find the eternal domination number of corona product of Cycles.

Theorem 8.15

For any $n \ge 3$, we have $\gamma_{tc.\,\infty}(C_3[C_n]) = 3n - 6$.

Proof: Let C_3 and C_n be the cycles on *3* vertices and n vertices respectively. The corona product of C_3 and C_n is denoted by $C_3[C_n]$ has *3n+3* vertices. Here $V(C_3[C_n]) = \{(u_i), (v_{i.\,j}) / 1 \le i \le 3,\ 1 \le j \le n\}$.

Consider the set $S=\{(v_{i,k}), 1\le i\le 3, 1\le k\le n-2\}$. Then $|S|=3n-6$.

Claim: S is an eternal dominating set in $C_3[C_n]$.

Here $N[S]=V(C_3[C_n])$. This gives that S is a dominating set in $C_3[C_n]$. Now $S_{1,1}=S-\{v_{1,n-2}\}\cup\{v_{1,n-1}\}$ is also a dominating set in $C_3[C_n]$. In this way, we find $S_{i,j}=S_{i,j}-\{v_{i,j-1}\}\cup\{v_{i,j}\}, 1\le i\le 2, 1\le j\le n-1, S_{2,n}=S$. All the sets $S_{i,j}$ are dominating sets in $C_3[C_n]$. This implies S is an eternal dominating set in $C_3[C_n]$. Therefore,

$$\gamma_\infty(C_3[C_n])\le 3n-6\ . \tag{8.29}$$

Claim: $\gamma_\infty(C_3[C_n])\nless 3n-6$.

Consider the set S' with $|S'|<3n-6$. That is $|S'|\le 3n-5$. Now suppose $|S'|=3n-5$. That is, $S'=\{v_{i,j}, 1\le j\le n-2\}-\{v_{1,n-2}\}$. And the vertex $v_{1,n-1}$ is not adjacent to any vertex in S'. This gives that S' is not a dominating set in $C_3[C_n]$. Therefore S' is not an eternal dominating set in $C_3[C_n]$. Therefore $\gamma_\infty(C_3[C_n])\ne 3n-5$. Also any set with cardinality less then $3n$-5 is not an eternal dominating set in $C_3[C_n]$. This gives that $\gamma_\infty(C_3[C_n])\nless 3n-5$, which implies $\gamma_\infty(C_3[C_n])>3n-5$, which implies that

$$\gamma_\infty(C_3[C_n])\ge 3n-6 \tag{8.30}$$

From (8.29) and (8.30), $\gamma_\infty(C_3[C_n])=3n-6$.

Theorem 8.16

For any $n\ge m\ge 4$, we have $\gamma_\infty(C_m[C_n])=m(n-2)$.

Proof: Let C_m and C_n be the cycles on m vertices and n vertices respectively. The corona product of C_m and C_n is denoted by $C_m[C_n]$ has $m(n+1)$ vertices. Here $V(C_m[C_n])=\{(u_i),(v_{i,j})/1\le i\le 3, 1\le j\le n\}$.

Consider the set $S=\{(v_{i,k}), 1\le i\le m, 1\le k\le n-2\}$. Then $|S|=m(n-2)$.

Claim: S is an eternal dominating set in $C_m[C_n]$.

Here $[S]=V(C_m[C_n])$. This gives that S is a dominating set in $C_m[C_n]$. Now $S_{1,1}=S-\{v_{1,n-2}\}\cup\{v_{1,n-1}\}$ is also a dominating set in $C_m[C_n]$. In this way, we find $S_{i,j}=S_{i,j}-\{v_{i,j-1}\}\cup\{v_{i,j}\}, 1\le i\le m, 1\le j\le n-1, S_{2,n}=S$. All the sets $S_{i,j}$ are dominating sets in $C_m[C_n]$. This implies S is an eternal dominating set in $C_m[C_n]$. Therefore

$$\gamma_\infty(C_m[C_n])\le m(n-2). \tag{8.31}$$

Claim: $\gamma_\infty(C_m[C_n])\nless m(n-2)$.

Consider the set $S'$with $|S'|<m(n-2)$. That is, $|S'|\le m(n-2)-1$. Now suppose $|S'|=m(n-2)-1$. That is $S'=\{v_{i,j}, 1\le j\le n-2\}-\{v_{1,n-2}\}$. Since the vertex

$v_{1,n-1}$ is not adjacent to any vertex in S'. This gives that S' is not a dominating set in $C_m\left[C_n\right]$. Therefore S' is not an eternal dominating set in $C_m\left[C_n\right]$. Therefore $\gamma_\infty\left(C_m\left[C_n\right]\right) \neq m(n-2)-1$. Also any set with cardinality less then $m(n-2)-1$ is not an eternal dominating set in $C_m\left[C_n\right]$. This gives that $\gamma_\infty\left(C_m\left[C_n\right]\right) \not< m(n-2)-1$, which implies $\gamma_\infty\left(C_m\left[C_n\right]\right) > m(n-2)-1$, which implies that

$$\gamma_\infty\left(C_m\left[C_n\right]\right) \geq m \tag{8.32}$$

From (8.31) and (8.32), $\gamma_\infty\left(C_m\left[C_n\right]\right) = m(n-2)$.

Example 8.11

Consider the graphs (Figures 8.22–8.24).

Then the corona product $C_3\left[C_3\right]$.

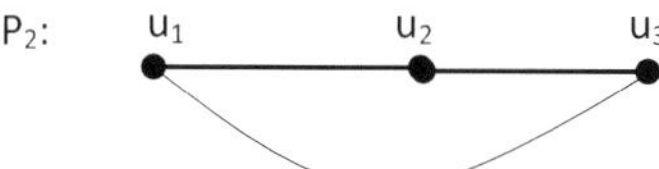

FIGURE 8.22 Cycle C_3.

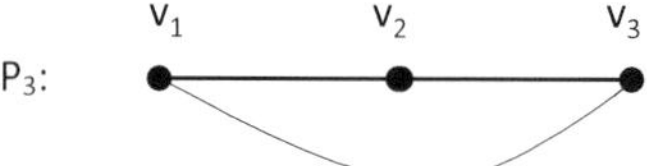

FIGURE 8.23 Cycle: C_3.

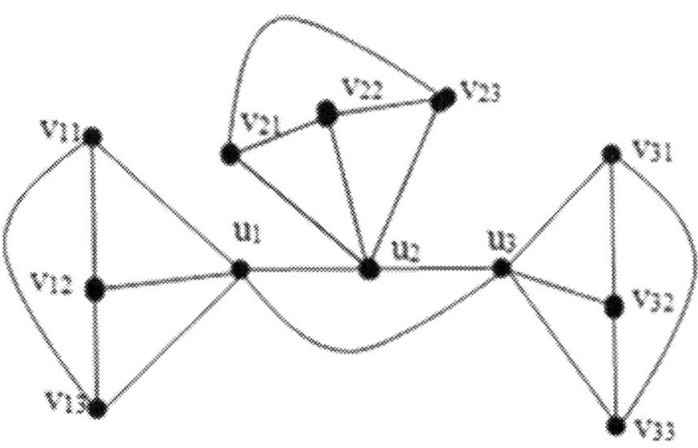

FIGURE 8.24 The Corona product C_3 $[C_3]$.

Here $V(C_3[C_3]) = \{(u_i), (v_{i,j}) : 1 \le i \le 3, 1 \le j \le 3\}$.

Here $V(C_3[C_3]) = \{(u_i), (v_{i,j}) : 1 \le i \le 3, 1 \le j \le 3\}$.

Let $S = \{(v_{i,k}) 1 \le i \le 3, 1 \le k \le 2\}$ be the subset of $V(C_3[C_3])$. Here $N[S] = V(C_3[C_3])$. Therefore S is a dominating set. Now $S_1 = S - (v_{1,2}) \cup (v_{1,3})$, $S_2 = S_1 - (v_{2,2}) \cup (v_{2,3})$, $S_3 = S_2 - (v_{1,3}) \cup (v_{1,2})$, $S_4 = S_3 - (v_{2,3}) \cup (v_{2,2}) = S$. Each set S_i is a dominating set in $C_3[C_3]$, which is minimum. Therefore $\gamma_\infty(C_3[C_3]) = 6$.

Similarly,

$$\gamma_\infty(C_4[C_5]) = 12$$

$$\gamma_\infty(C_7[C_{10}]) = 56.$$

Corona product of paths and cycles: In this section we find the eternal domination number of corona product of paths and cycles.

Theorem 8.17

For any $n \ge 3$, we have $\gamma_{tc,\infty}(P_2[C_n]) = 2n - 4$.

Proof: Let P_2 and C_n be the path and cycle on 2 vertices and n vertices respectively. The corona product of P_2 and C_n is denoted by $P_2[C_n]$ has $2n+2$ vertices. Here $V(P_2[C_n]) = \{(u_i), (v_{i,j}) / 1 \le i \le 2, 1 \le j \le n\}$.

Consider the set $S = \{(v_{i,k}), 1 \le i \le 2, 1 \le k \le n-2\}$. Then $|S| = 2n - 4$.

Claim: S is an eternal dominating set in $P_2[C_n]$.

Here $N[S] = V(P_2[C_n])$. This gives that S is a dominating set in $P_2[C_n]$. Now $S_{1,1} = S - \{v_{1,n-2}\} \cup \{v_{1,n-1}\}$ is also a dominating set in $P_2[C_n]$. In this way we find $S_{i,j} = S_{i,j} - \{v_{i,j-1}\} \cup \{v_{i,j}\}$, $1 \le i \le 2$, $1 \le j \le n-1$, $S_{2,n} = S$. All the sets $S_{i,j}$ are dominating sets in $P_2[C_n]$. This implies S is an eternal dominating set in $P_2[C_n]$. Therefore

$$\gamma_\infty(P_2[C_n]) \le 2n - 4. \tag{8.33}$$

Claim: $\gamma_\infty(P_2[C_n]) \not< 2n - 4$.

Consider the set S' with $|S'| < 2n-4$. That is $|S'| \le 2n-3$. Now suppose $|S'| = 2n-3$. That is, $S' = \{v_{i,j}, 1 \le j \le n-2\} - \{v_{1,n-2}\}$. And the vertex $v_{1,n-1}$ is not adjacent to any vertex in S'. This gives that S' is not a dominating set in $P_2[C_n]$. Therefore, S' is not an eternal dominating set in $P_2[C_n]$. Therefore $\gamma_\infty(P_2[C_n]) \ne 2n-3$. Also any set with cardinality less then $2n$-3 is not an eternal dominating set in $P_2[C_n]$. This gives that $\gamma_\infty(P_2[C_n]) \not< 2n - 3$, which implies

$$\gamma_\infty(P_2[C_n]) > 2n - 3 \Rightarrow \gamma_\infty(P_2[C_n]) \ge 2n - 4. \tag{8.34}$$

From (8.33) and (8.34), $\gamma_\infty(P_2[C_n]) = 2n - 4$.

Theorem 8.18

For any $n \geq m \geq 3$, we have $\gamma_\infty\left(P_m\left[C_n\right]\right) = m(n-2)$.

Proof: Let P_m and C_n be the path and cycle on m vertices and n vertices respectively. The corona product of P_m and C_n is denoted by $P_m\left[C_n\right]$ has $m(n+1)$ vertices. Here $V\left(P_m\left[C_n\right]\right) = \left\{(u_i), (v_{i,j}) / 1 \leq i \leq m, 1 \leq j \leq n\right\}$.

Consider the set $S = \left\{\left(v_{i,k}\right), 1 \leq i \leq m, 1 \leq k \leq n-2\right\}$. Then $S| = m(n-2)$.

Claim: S is an eternal dominating set in $P_m\left[C_n\right]$.

Here $N[S] = V\left(P_m\left[C_n\right]\right)$. This gives that S is a dominating set in $P_m\left[C_n\right]$. Now $S_{1.1} = S - \left\{v_{1.n-2}\right\} \cup \left\{v_{1.n-1}\right\}$ is also a dominating set in $P_m\left[C_n\right]$. In this way we find $S_{i,j} = S_{i,j} - \left\{v_{i,j-1}\right\} \cup \left\{v_{i,j}\right\}, 1 \leq i \leq m, 1 \leq j \leq n-1, S_{2,n} = S$. All the sets $S_{i,j}$ are dominating sets in $P_m\left[C_n\right]$. This implies S is an eternal dominating set in $P_m\left[C_n\right]$. Therefore

$$\gamma_\infty\left(P_m\left[C_n\right]\right) \leq m(n-2) \quad . \tag{8.35}$$

Claim: $\gamma_\infty\left(P_m\left[C_n\right]\right) \not< m(n-2)$.

Consider the set $S'$with $S'| < m(n-2)$. That is $S'| \leq m(n-2)-1$. Now suppose $S'| = m(n-2)-1$. That is $S' = \left\{v_{i,j}, 1 \leq j \leq n-2\right\} - \left\{v_{1.n-2}\right\}$. And the vertex $v_{1.n-1}$ is not adjacent to any vertex in S'. This gives that S' is not a dominating set in $P_m\left[C_n\right]$. Therefore, S' is not an eternal dominating set in $P_m\left[C_n\right]$. Therefore $\gamma_\infty\left(P_m\left[C_n\right]\right) \neq m(n-2)-1$. Also any set with cardinality less then $m(n-2)-1$ is not an eternal dominating set in $P_m\left[C_n\right]$. This gives that $\gamma_\infty\left(P_m\left[C_n\right]\right) \not< m(n-2)-1$, which implies $\gamma_\infty\left(P_m\left[C_n\right]\right) > m(n-2)-1$, which implies that

$$\gamma_\infty\left(P_m\left[C_n\right]\right) \geq m \tag{8.36}$$

From (8.35) and (8.36), $\gamma_\infty\left(P_m\left[C_n\right]\right) = m(n-2)$.

Example 8.12

Consider the graphs (Figures 8.25–8.27).

Then the Corona product $P_2\left[C_3\right]$.

FIGURE 8.25 Path P_2.

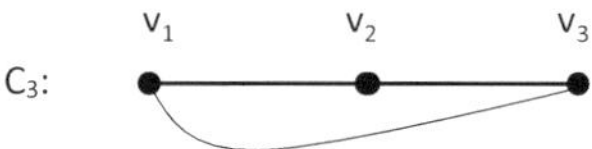

FIGURE 8.26 Cycle C_3.

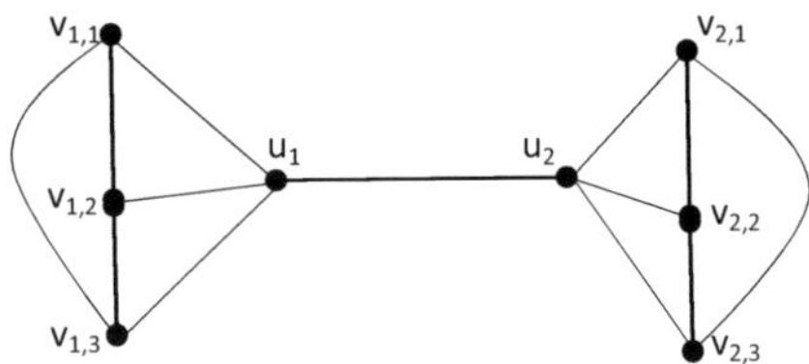

FIGURE 8.27 The Corona product $P_2\,[C_3]$.

Here $V\left(P_2\left[C_3\right]\right)=\left\{\left(u_i\right),\left(v_{i,\,j}\right):1\leq i\leq 2,1\leq j\leq 3\right\}$.

Let $S=\left\{\left(v_{i,\,1}\right)1\leq i\leq 2,\right\}$ be the subset of $V\left(P_2[C_3]\right)$. Here $N[S]=V\left(P_2\left[C_3\right]\right)$. Therefore S is a dominating set. Now $S_1=S-\left(v_{1,1}\right)\cup\left(v_{1,3}\right)$, $S_2=S_1-\left(v_{2,1}\right)\cup\left(v_{2,3}\right)$, $S_3=S_2-\left(v_{1,3}\right)\cup\left(v_{1,2}\right), S_4=S_3-\left(v_{2,3}\right)\cup\left(v_{2,2}\right)=S$. Each set S_i is a dominating set in $P_2\left[C_3\right]$, which is minimum. Therefore $\gamma_\infty\left(P_2\left[C_3\right]\right)=2$.

Similarly,

$$\gamma_\infty\left(P_4\left[C_5\right]\right)=12$$

$$\gamma_\infty\left(P_7\left[C_{10}\right]\right)=56.$$

ACKNOWLEDGMENT

The research work was supported by DSA (Departmental special assistance) Gandhigram Rural Institute—Deemed to be university—Gandhigram, under University Grants Commission, New Delhi.

REFERENCES

1. F. Harary. "*Graph Theory*". Addison Wesley Publishing Company pvt Ltd: Boston, MA, (1969).
2. G. Mahadevan, T. Ponnuchamy and A. Selvam "Non-split perfect triple connected domination number on different product of paths". *International Journal of Mathematical Combinatoricspp*: 118–131 (2018).
3. J.A. Bondy and U.S.R. Moorthy, "*Graph Theory with Applications*" The Macmillan Press Ltd: Great Britain (1976).
4. J. L. Gross and J. Yellen, "Handbook of graph theory" CRC Press Ltd: Boca Raton, FL (2003).
5. S. Finbow and M.E. Messinger, "Eternal domination on 3 X n Grid graphs". *Australasian Journal of Combinatorics* 61(2): 156–174 (2015).
6. J. Mishra. "Unified fractional calculus results related with ¯H function". *Journal of International Academy of Physical Sciences* 20 (3): 185–195 (2016).
7. J. Mishra, and N. Pandey. "An integral involving general class of polynomials with I-function". *International Journal of Scientific and Research Publications*: 204 (2013).
8. J. Mishra. "Fractional hyper-chaotic model with no equilibrium". *Chaos, Solitons & Fractals* 116: 43–53 (2018).

9. J. Mishra. "A remark on fractional differential equation involving I-function". *The European Physical Journal Plus* 133 (2): 36 (2018).
10. J. Mishra. "Modified Chua chaotic attractor with differential operators with non-singular kernels". *Chaos, Solitons & Fractals* 125: 64–72 (2019).
11. J. Mishra. "Analysis of the Fitzhugh Nagumo model with a new numerical scheme". *Discrete & Continuous Dynamical Systems-S*: 781 (2018).
12. J. Mishra. "Numerical analysis of a chaotic model with fractional differential operators: From Caputo to Atangana–Baleanu". *Methods of Mathematical Modelling: Fractional Differential Equations*: 167 (2019).
13. W. Goddard, S.M. Hedetniemi and S.T. Hedetniemi, "Eternal security in graphs". *Journal of Combinatorial Mathematics and Combinatorial Computing*: 52 (2008).

9 Numerical Analysis of Coupled Time-Fractional Differential Equations Arising in Epidemiological Models

Manish Goyal
GLA University

Amit Prakash
National Institute of Technology

Shivangi Gupta
GLA University

CONTENTS

9.1 INTRODUCTION

Fractional calculus [1] is important in fluid dynamic traffic, electromagnetic theory, dielectric polarization, neurophysiology, electrochemical processes, plasma physics, viscoelasticity, bioinformatics, colored noise, control theory, etc. We usually deal with the real-world practices of arbitrary order. Fractional-order differential equations (FDEs) accomplish the systems with a memory effect. The nonlocal property is the main benefit of working with them in physical models. It signifies that the future system state is dependent on the former states also. Thus, the models having fractional-order derivatives adhere to the reality. The study of coupled FDE [2–4] can be found in engineering; biomechanics; solid-state physics; modeling of electromagnetic, gravitational problems; and mathematical modeling of epidemiological models such as Lassa hemorrhagic fever, cholera model, malaria model, SIR model, Ebola fever model, and rotavirus model. In [5], the solution of an epidemic model given by the coupled differential equation is found by Rafei et al.

Now, consider a system of nonlinear coupled time-fractional differential equations as

$$D_t^{\alpha} u(x, t) + S_1(u, v) + Q_1(u, v) = g_1(x, t),$$

$$D_t^{\alpha} v(x, t) + S_2(u, v) + Q_2(u, v) = g_2(x, t), \; (0 < \alpha \leq 1) \tag{9.1}$$

with initial settings,

$$u(x, 0) = f_1(x), \; v(x, 0) = f_2(x). \tag{9.2}$$

Here, S_1, S_2; Q_1, and Q_2 are the linear and nonlinear operators, respectively. $g_1(x, t)$ and $g_2(x, t)$ are the source terms. u and v are the differentiable functions of space variable x and time variable t. D_t^{α} is Caputo fractional operator [6–8] of order that permits conditions to comprise in modeling a problem. The time-fractional equations [9,10] illustrate the particle motion taking memory in time. The memory-dependent derivative shows the inflection of system memory. FDEs are also used for the simulation of electrophysiologic activity of single cardiomyocyte. The phenomena of electrical activity in heart and the transmission dynamics of epidemic are impacted by memory factor in time. So fractional-order modeling is applicable for such kind of systems. The study of Eq. (9.1) is therefore significant. It motivated us to solve Eq. (9.1) by reliable HPTM (homotopy perturbation method via the Laplace transform).

Homotopy perturbation technique (HPM) was initiated by He [11–21], and it is used [22–24] in solving important nonlinear problems in various fields. HPM has several restrictions: Solution has a chain of small parameters that has problems as the majority of nonlinear problems have no trifling parameter. Later, Ji-huan He recommended the creation of an equation of homotopy with auxiliary term that disappears absolutely if parameter p is either 0 or 1. But such methods have some other limitations like massive computation with more time consumption. So they necessitate to be linked with a transform operator. Hybrid methods using integral transforms [25–28] are useful to treasure some solution of nonlinear FDE. HPTM is a collective form of HPM and transform of Laplace. In [29], Goyal et al. found solutions to the coupled FDE using HPTM. The reliability of solution procedure of a nonlinear equation is an important characteristic rather than modeling dimensions of equations [30–35].

Our aim is to get the solution of Eq. (9.1) by HPTM and match the results with the existing techniques. The paper is presented as follows. An introduction is given in Section 9.1. In Section 9.2, basic definitions of arbitrary derivative in Caputo sense and Mittag–Leffler function are given. In Section 9.3, the plan of HPTM is shown.

The convergence of bounded and positive series solution is discussed along with the execution of HPTM on Eq. (9.1) on two test examples. In Section 9.4, we discuss the results using figures and tables. In Section 9.5, the results are reviewed to find final inferences.

9.2 PRELIMINARIES

Definition 9.1 [6]

Areal function $h(\chi)$, $\chi > 0$ is in space

a. C_ζ, $\zeta \in \mathbb{R}$ if there exists a real number q (>ζ), s.t. $h(\chi) = \chi^q h_1(\chi)$, $h_1(\chi) \in C[0,\infty)$.
   Clearly, $C_\zeta \subset C_\gamma$ if $\gamma \le \zeta$.
b. C_ζ^m, $m \in \mathbb{N} \cup \{0\}$ if $h^{(m)} \in C_\zeta$

Definition 9.2 [6]

Arbitrary-order Caputo derivative [7] of $h(t)$; $h(t) \in C_{-1}^m$, $m \in \mathbb{N} \cup \{0\}$ is:

$$D_t^\beta h(t) = \begin{cases} I^{m-\beta} h^{(m)}(t),\ m-1 < \beta < m,\ m \in N, \\ \dfrac{d^m}{dt^m} h(t),\ \beta = m, \end{cases}$$

a. [8] $I_t^\zeta h(x, t) = \dfrac{1}{\Gamma\zeta} \displaystyle\int_0^t (t-s)^{\zeta-1} h(x, s) ds;\ \zeta, t > 0.$

b. [8] $D_\tau^\nu V(x, \tau) = I_\tau^{m-\nu} \dfrac{\partial^m V(x, \tau)}{\partial \tau^m},\ m-1 < \nu \le m.$

c. [8] $D_t^\zeta I_t^\zeta h(t) = h(t),\ m-1 < \zeta \le m,\ m \in \mathbb{N}.$

d. [8] $I_t^\zeta D_t^\zeta h(t) = h(t) - \displaystyle\sum_{k=1}^{m-1} h^{(k)}(0^+) \frac{t^k}{k!},\ m-1 < \zeta \le m,\ m \in \mathbb{N}.$

e. [8] $I^\beta t^\alpha = \dfrac{\Gamma(\alpha+1)}{\Gamma(\beta+\alpha+1)} t^{\beta+\alpha}.$

Definition 9.3 [6]

Mittag–Leffler function is

$E_\zeta(z) = \displaystyle\sum_{m=0}^{\infty} \frac{z^m}{\Gamma(1+\zeta m)},\ \zeta > 0, z \in C,$ valid in the entire complex plane.

Definition 9.4 [7]

Laplace transform (LT) of arbitrary-order derivative in Caputo sense is

$$L\left[D^{\alpha}g(t)\right]=p^{\alpha}F(p)-\sum_{k=0}^{n-1}p^{\alpha-k-1}g^{(k)}(0),\ n-1<\alpha\leq n.$$

9.3 BASIC PLAN OF HPTM FOR COUPLED FDE IN EPIDEMIC MODEL

To elucidate the procedure of HPTM, we ponder over a model given by Eq. (9.1).

Taking LT on Eq. (9.1) and using the initial conditions (9.2), we get

$$\left.\begin{aligned} u(p,t)&=p^{-1}h_1(x)-p^{-\alpha}L\left[g_1(x,t)\right]+p^{-\alpha}L\left[S_1(u,v)+Q_1(u,v)\right],\\ v(p,t)&=p^{-1}h_2(x)-p^{-\alpha}L\left[g_2(x,t)\right]+p^{-\alpha}L\left[S_2(u,v)+Q_2(u,v)\right].\end{aligned}\right\} \quad (9.3)$$

Taking inverse transform,

$$\left.\begin{aligned} u(x,t)&=G_1(x,t)+L^{-1}\left[p^{-\alpha}L\{S_1(u,v)+Q_1(u,v)\}\right],\\ v(x,t)&=G_2(x,t)+L^{-1}\left[p^{-\alpha}L\{S_2(u,v)+Q_2(u,v)\}\right].\end{aligned}\right\} \quad (9.4)$$

Here, $G_1(x,t)$ and $G_2(x,t)$ have come from source term using the initial values.

We apply HPM, and it is supposed that

$$\left.\begin{aligned} u_n(x,t)&=\sum_{n=0}^{\infty}p^n u_n(x,t),\\ v_n(x,t)&=\sum_{n=0}^{\infty}p^n v_n(x,t).\end{aligned}\right\} \quad (9.5)$$

It is to be noted that p is the homotopy parameter and $p\in[0,1]$.

Nonlinear terms are put as:

$$\left.\begin{aligned} Nu(x,t)&=\sum_{n=0}^{\infty}p^n H_n(u),\\ Nu(x,t)&=\sum_{n=0}^{\infty}p^n H_n'(v),\end{aligned}\right\} \quad (9.6)$$

where H_n and H'_n are He's polynomials of $u_0, u_1, u_2, \ldots, u_n$ and $v_0, v_1, v_2, \ldots, v_n$, respectively. They are calculated by the formulae:

$$\left.\begin{aligned} H_n(u_0,u_1,u_2,\ldots) &= \frac{1}{n!}\frac{\partial^n}{\partial p^n}\left[N\left(\sum_{n=0}^{\infty} p^i u_i\right)\right]_{p=0}, \quad n=0,1,2,\ldots \\ H'_n(v_0,v_1,v_2,\ldots) &= \frac{1}{n!}\frac{\partial^n}{\partial p^n}\left[N\left(\sum_{n=0}^{\infty} p^i v_i\right)\right]_{p=0}, \quad n=0,1,2,\ldots \end{aligned}\right\} \tag{9.7}$$

Substituting Eqs. (9.5) and (9.6) into Eq. (9.4), we get

$$\left.\begin{aligned} \sum_{n=0}^{\infty} p^n u_n(x,t) &= G_1(x,t) + pL^{-1}\left[p^{-\alpha} L\{S_1(u,v) + Q_1(u,v)\}\right], \\ \sum_{n=0}^{\infty} p^n v_n(x,t) &= G_2(x,t) + pL^{-1}\left[p^{-\alpha} L\{S_2(u,v) + Q_2(u,v)\}\right]. \end{aligned}\right\} \tag{9.8}$$

This is a union of LT and HPM with He's polynomials.

We equate the coefficients of various powers of p and get

$$\left.\begin{aligned} p^0 &: u_0(x,t) = G_1(x,t), \\ & v_0(x,t) = G_2(x,t), \end{aligned}\right\}$$

$$\left.\begin{aligned} p^n &: u_n(x,t) = L^{-1}\left[p^{-\alpha} L\{S_1(u_{n-1}, v_{n-1}) + H_{n-1}(u,v)\}\right], n > 0,\ n \in \mathbb{N}, \\ & v_n(x,t) = L^{-1}\left[p^{-\alpha} L\{S_2(u_{n-1}, v_{n-1}) + H'_{n-1}(u,v)\}\right], n > 0,\ n \in \mathbb{N}. \end{aligned}\right\}$$

Persisting this way, the ensuing components may also be attained.

The solution is

$$\left.\begin{aligned} u(x,t) &= \lim_{p \to 1} \sum_{n=0}^{\infty} p^n u_n(x,t), \\ v(x,t) &= \lim_{p \to 1} \sum_{n=0}^{\infty} p^n v_n(x,t). \end{aligned}\right\} \tag{9.9}$$

9.3.1 Convergence Analysis

Now, the conditions for convergence of the applied scheme are presented. The series (9) is convergent [36] in most cases, but some views were stated by Ji-huan He [11] for gaining rate of convergence on nonlinear operator.

1. Second derivative of Nu, Nv in respect to u, v should be less since the parameter may be fairly large, i.e., $p \to 1$.
2. Norm of $L^{-1}\dfrac{\partial N}{\partial u}$ and $L^{-1}\dfrac{\partial N}{\partial v}$ should be smaller than 1 for the series to converge.

Theorem 9.1 [37]

Let W and X be the Banach spaces. Let $N:W \rightarrow X$ be a nonlinear contraction mapping, i.e., for any $\vartheta, \tilde{\vartheta} \in W; \left\| N(\vartheta) - N\left(\tilde{\vartheta}\right) \right\| \leq \chi \left\| \vartheta - \tilde{\vartheta} \right\|, 0 < \chi < 1,$

Then by Banach theorem, taking fixed point u, i.e., $N(u) = u$

The sequence made by HPM is

$$V_n = N(V_{n-1}), \ V_{n-1} = \sum_{i=0}^{n-1} u_i, \ n = 1,2,3,\ldots,$$

and let $V_0 = v_0 = u_0 \in B_r(u)$, where $B_r(u) = \left\{ u^* \in W \mid \left\| u^* - u \right\| < r \right\}$.

Then, we have the following statements:

i. $\left\| V_n - u \right\| \leq \chi^n \left\| v_0 - u \right\|$,
ii. $V_n \in B_r(u)$,
iii. $\lim_{n \to \infty} V_n = u$.

9.3.2 Implementation of HPTM

To show the applicability of the presented scheme, we give few examples.

Example 9.1

Take an inhomogeneous time-fractional coupled differential equation,

$$\left. \begin{array}{l} D_t^\alpha u(x, t) - v_x(x, t) - u(x, t) + v(x, t) = -2, \\ D_t^\alpha v(x, t) + u_x(x, t) - u(x, t) + v(x, t) = -2, \end{array} \right\} \quad (0 < \alpha \leq 1, t \geq 0) \qquad (9.10)$$

With conditions,

$$u(x, 0) = 1 + e^x, \ v(x, 0) = -1 + e^x. \qquad (9.11)$$

When $\alpha = 1$, the solution of Eq. (9.10) is

$$u = 1 + e^{x+t}, \ v = -1 + e^{x-t}. \qquad (9.12)$$

Taking LT on Eq. (9.10), we get

$$\left. \begin{array}{l} u(p, t) = \dfrac{1 + e^x}{p} - \dfrac{2}{p^{\alpha+1}} + p^{-\alpha} L\left(D_x v(x, t) + u(x, t) - v(x, t) \right), \\ v(p, t) = \dfrac{-1 + e^x}{p} - \dfrac{2}{p^{\alpha+1}} + p^{-\alpha} L\left(-D_x u(x, t) + u(x, t) - v(x, t) \right) \end{array} \right\}. \qquad (9.13)$$

Taking inverse LT, we get

$$\left.\begin{aligned} u(x,t) &= 1+e^x - \frac{2t^\alpha}{\Gamma(1+\alpha)} + L^{-1}\left[p^{-\alpha}L\left(D_x v(x,t)+u(x,t)-v(x,t)\right)\right], \\ v(x,t) &= -1+e^x - \frac{2t^\alpha}{\Gamma(1+\alpha)} + L^{-1}\left[p^{-\alpha}L\left(-D_x u(x,t)+u(x,t)-v(x,t)\right)\right] \end{aligned}\right\}. \tag{9.14}$$

Applying HPM on Eq. (9.14), we get

$$\left.\begin{aligned} \sum_{n=0}^{\infty} p^n u_n(x,t) &= 1+e^x - \frac{2t^\alpha}{\Gamma(1+\alpha)} + pL^{-1}\left[p^{-\alpha}L\left(D_x v(x,t)+u(x,t)-v(x,t)\right)\right], \\ \sum_{n=0}^{\infty} p^n v_n(x,t) &= -1+e^x - \frac{2t^\alpha}{\Gamma(1+\alpha)} + pL^{-1}\left[p^{-\alpha}L\left(-D_x u(x,t)+u(x,t)-v(x,t)\right)\right] \end{aligned}\right\}. \tag{9.15}$$

We equate the coefficients of various powers of p in Eqs. (9.15) and get

$$p^0: u_0(x,t) = 1+e^x - \frac{2t^\alpha}{\Gamma(1+\alpha)},\ v_0(x,t) = -1+e^x - \frac{2t^\alpha}{\Gamma(1+\alpha)},$$

$$p^1: u_1(x,t) = \left(2+e^x\right)\frac{t^\alpha}{\Gamma(1+\alpha)},\ v_1(x,t) = \left(2-e^x\right)\frac{t^\alpha}{\Gamma(1+\alpha)},$$

$$p^2: u_2(x,t) = \frac{e^x t^{2\alpha}}{\Gamma(1+2\alpha)},\ v_2(x,t) = \frac{e^x t^{2\alpha}}{\Gamma(1+2\alpha)},$$

$$p^3: u_3(x,t) = \frac{e^x t^{3\alpha}}{\Gamma(1+3\alpha)},\ v_3(x,t) = \frac{-e^x t^{3\alpha}}{\Gamma(1+3\alpha)},$$

and so on.

Now, the solution is

$$\left.\begin{aligned} u(x,t) &= \lim_{p\to 1}\sum_{n=0}^{\infty} p^n u_n(x,t) = 1+e^x\sum_{k=0}^{\infty}\frac{t^{k\alpha}}{\Gamma(1+k\alpha)} = 1+e^x E_\alpha\left(t^\alpha\right). \\ v(x,t) &= \lim_{p\to 1}\sum_{n=0}^{\infty} p^n v_n(x,t) = -1+e^x\sum_{k=0}^{\infty}\frac{(-1)^k t^{k\alpha}}{\Gamma(1+k\alpha)} = -1+e^x E_\alpha\left(-t^\alpha\right) \end{aligned}\right\}. \tag{9.16}$$

Here, $E_\alpha(z)$ is the one-parameter Mittag–Leffler function [6]. By theorem for non-linear mapping N, the sufficient condition for convergence of HPTM is strictly contraction N.

So, we write

$$\left\|u_0 - u\right\| = \left\|1-e^t\right\|,$$

$$\|U_1 - u\| = \|u_0 + u_1 - u\|$$

$$\left\| = 1 + \frac{t^\alpha}{\Gamma(1+\alpha)} - e^t \right\| \le \|1 - e^t\| \left\| 1 + \frac{t^\alpha}{(1-e^t)\Gamma(1+\alpha)} \right\|.$$

Since, for any $t \in [0,1]$, $0 < \alpha \le 1$, we have

$$\left\| 1 + \frac{t^\alpha}{(1-e^t)\Gamma(1+\alpha)} \right\| \le \chi = 0.418 < 1,$$

Hence,

$$\|U_1 - u\| \le \chi \|1 - e^t\|$$

$$= \chi \|u_0 - u,\|$$

$$\|U_2 - u\| = \left\| 1 + \frac{t^\alpha}{\Gamma(1+\alpha)} + \frac{t^{2\alpha}}{\Gamma(1+2\alpha)} - e^t, \right\|$$

$$\le \|1 - e^t 1\| + \left\| \frac{\frac{t^\alpha}{\Gamma(1+\alpha)} + \frac{t^{2\alpha}}{\Gamma(1+2\alpha)}}{(1-e^t)} \right\|.$$

But, for any $t \in [0, 1]$, $0 < \alpha \le 1$,

$$\left\| 1 + \frac{\frac{t^\alpha}{\Gamma(1+\alpha)} + \frac{t^{2\alpha}}{\Gamma(1+2\alpha)}}{(1-e^t)} \right\| \le 0.127 < \chi.$$

Thus,

$$\|U_2 - u\| \le \chi^2 \|u_0 - u,\|$$

$$\|U_3 - u\| = \left\| 1 + \frac{t^\alpha}{\Gamma(1+\alpha)} + \frac{t^{2\alpha}}{\Gamma(1+2\alpha)} + \frac{t^{3\alpha}}{\Gamma(1+3\alpha)} - e^t \right\|,$$

$$\le \|1 - e^t\| \left\| 1 + \frac{\frac{t^\alpha}{\Gamma(1+\alpha)} + \frac{t^{2\alpha}}{\Gamma(1+2\alpha)} + \frac{t^{3\alpha}}{\Gamma(1+3\alpha)}}{(1-e^t)} \right\|.$$

But for any $t \in [0, 1]$, $0 < \alpha \le 1$,

$$\left\| 1 + \frac{\dfrac{t^{\alpha}}{\Gamma(1+\alpha)} + \dfrac{t^{2\alpha}}{\Gamma(1+2\alpha)} + \dfrac{t^{3\alpha}}{\Gamma(1+3\alpha)}}{\left(1-e^{t}\right)} \right\| \le 0.030 < \chi,$$

Thus,

$$\|U_3 - u\| \le \chi^3 \|u_0 - u\|,$$

$$\|U_n - u\| \le \|\chi^n u_0 - u\|.$$

Therefore,

$$\lim_{n\to\infty} \|U_n - u\| \le \lim_{n\to\infty} \chi^n \|u_0 - u\| = 0,$$

i.e.,

$$u(x, t) = 1 + e^{x} \sum_{k=0}^{\infty} \frac{t^{k\alpha}}{\Gamma(1+k\alpha)},$$

which reduces to the solution $u(x, t) = 1 + e^{x+t}$ at $\alpha = 1$.

Similarly,

$$\|v_0 - v\| = \left\|1 - e^{-t}\right\|,$$

$$\|V_1 - v\| = \|v_0 + v_1 - v\|$$

$$\left\| = 1 - \frac{t^{\alpha}}{\Gamma(1+\alpha)} - e^{-t}, \right\|$$

$$\le \left\|1 - e^{-t}\right\| \left\| 1 - \frac{t^{\alpha}}{\left(1-e^{-t}\right)\Gamma(1+\alpha)} \right\|.$$

Since, for any $t \in [0, 1]$, $0 < \alpha \le 1$, we have

$$\left\| 1 - \frac{t^{\alpha}}{\left(1-e^{-t}\right)\Gamma(1+\alpha)} \right\| \le \chi = 0.5819 < 1,$$

Hence,

$$\|V_1 - v\| \le \chi \left\|1 - e^{-t}\right\| = \chi \|v_0 - v,\|$$

$$\|V_2 - v\| = \left\| 1 - \frac{t^{\alpha}}{\Gamma(1+\alpha)} + \frac{t^{2\alpha}}{\Gamma(1+2\alpha)} - e^{-t} \right\|,$$

$$\leq \left\|1-e^{-t}\right\| \left\| 1-\frac{\dfrac{t^{\alpha}}{\Gamma(1+\alpha)}-\dfrac{t^{2\alpha}}{\Gamma(1+2\alpha)}}{\left(1-e^{-t}\right)} \right\|.$$

But for any $t \in [0,1]$, $0<\alpha \leq 1$,

$$\left\| 1-\frac{\dfrac{t^{\alpha}}{\Gamma(1+\alpha)}-\dfrac{t^{2\alpha}}{\Gamma(1+2\alpha)}}{\left(1-e^{-t}\right)} \right\| \leq 0.2091 < \chi,$$

Thus,

$$\|V_2 - v\| \leq \chi^2 \|v_0 - v\|,$$

$$\|V_3 - v\| = \left\| 1-\frac{t^{\alpha}}{\Gamma(1+\alpha)}+\frac{t^{2\alpha}}{\Gamma(1+2\alpha)}-\frac{t^{3\alpha}}{\Gamma(1+3\alpha)}-e^{-t} \right\|,$$

$$\leq \left\|1-e^{-t}\right\| \left\| 1-\frac{\dfrac{t^{\alpha}}{\Gamma(1+\alpha)}-\dfrac{t^{2\alpha}}{\Gamma(1+2\alpha)}+\dfrac{t^{3\alpha}}{\Gamma(1+3\alpha)}}{\left(1-e^{-t}\right)} \right\|.$$

But for any $t \in [0,1]$, $0<\alpha \leq 1$,

$$\left\| 1-\frac{\dfrac{t^{\alpha}}{\Gamma(1+\alpha)}-\dfrac{t^{2\alpha}}{\Gamma(1+2\alpha)}+\dfrac{t^{3\alpha}}{\Gamma(1+3\alpha)}}{\left(1-e^{-t}\right)} \right\| \leq 0.0546 < \chi,$$

Thus,

$$\|V_3 - v\| \leq \chi^3 \|v_0 - v\|,$$

$$\|V_n - v\| \leq \chi^n \|v_0 - v\|.$$

Therefore,

$$\lim_{n\to\infty} V_n - v \leq \lim_{n\to\infty} \chi^n v_0 - v = 0,$$

i.e.,

$$v(x,t) = -1 + e^{x} \sum_{k=0}^{\infty} \frac{(-1)^k t^{k\alpha}}{\Gamma(1+k\alpha)},$$

which reduces to the solution $v(x,t) = -1 + e^{x-t}$ at $\alpha = 1$.

It approves that the HPTM used for solving Eq. (9.10) gives convergent, bounded, and positive solution. It is noted that the result completely agrees with other methods [34].

Example 9.2

Consider another system of inhomogeneous nonlinear FDE,

$$\left.\begin{aligned} D_t^{\alpha}u(x,t)+v(x,t)\,D_x u(x,t)+u(x,t)=1,\\ D_t^{\alpha}v(x,t)-u(x,t)D_x v(x,t)-v(x,t)=1, \end{aligned}\right\}\quad (0<\alpha\le 1,\ t\ge 0) \tag{9.17}$$

With conditions,

$$u(x,0)=e^{x},\ v(x,\ 0)=e^{-x}. \tag{9.18}$$

The solution of Eq. (9.17) when $\alpha=1$ is

$$u(x,t)=e^{x-t},\ v(x,t)=e^{-x+t}. \tag{9.19}$$

Taking LT on Eq. (9.17), we get

$$\left.\begin{aligned} u(p,t)=\frac{e^{x}}{p}+\frac{1}{p^{\alpha+1}}+p^{-\alpha}L\big(-v(x,t)D_x u(x,t)-u(x,t)\big),\\ v(p,t)=\frac{e^{-x}}{p}+\frac{1}{p^{\alpha+1}}+p^{-\alpha}L\big(u(x,t)D_x v(x,t)+v(x,t)\big). \end{aligned}\right\} \tag{9.20}$$

Taking inverse LT, we get

$$\left.\begin{aligned} u(x,t)=e^{x}+\frac{t^{\alpha}}{\Gamma(1+\alpha)}-L^{-1}\left[p^{-\alpha}L\big(v(x,t)D_x u(x,t)+u(x,t)\big)\right],\\ v(x,t)=e^{-x}+\frac{t^{\alpha}}{\Gamma(1+\alpha)}+L^{-1}\left[p^{-\alpha}L\big(u(x,t)D_x v(x,t)+v(x,t)\big)\right]. \end{aligned}\right\} \tag{9.21}$$

Applying HPM on Eq. (9.21), we get

$$\left.\begin{aligned} \Sigma p^{n}u_n(x,t)=e^{x}+\frac{t^{\alpha}}{\Gamma(1+\alpha)}-pL^{-1}\left[p^{-\alpha}L\big(v(x,t)D_x u(x,t)+u(x,t)\big)\right],\\ \Sigma p^{n}v_n(x,t)=e^{-x}+\frac{t^{\alpha}}{\Gamma(1+\alpha)}+pL^{-1}\left[p^{-\alpha}L\big(u(x,t)D_x v(x,t)+v(x,t)\big)\right]. \end{aligned}\right\} \tag{9.22}$$

We equate coefficients of various powers of p in Eq. (9.22) and get

$$p^{0}: u_0(x,t)=-e^{x}\left[\frac{t^{\alpha}}{\Gamma(1+\alpha)}\right], v_0(x,t)=-e^{x}\left[\frac{t^{\alpha}}{\Gamma(1+\alpha)}\right],\Bigg\}$$

$$p^1: u_1(x,t) = -e^x\left[\frac{t^\alpha}{\Gamma(1+\alpha)}+\frac{t^{2\alpha}}{\Gamma(1+2\alpha)}\right]-\left[\frac{t^\alpha}{\Gamma(1+\alpha)}+\frac{t^{2\alpha}}{\Gamma(1+2\alpha)}\right],$$

$$v_1(x,t) = e^{-x}\left[\frac{t^\alpha}{\Gamma(1+\alpha)}-\frac{t^{2\alpha}}{\Gamma(1+2\alpha)}\right]-\left[\frac{t^\alpha}{\Gamma(1+\alpha)}+\frac{t^{2\alpha}}{\Gamma(1+2\alpha)}\right],$$

$$p^2: u_2(x,t) = e^x\left[\frac{t^{2\alpha}}{\Gamma(1+2\alpha)}+(1-c_1)\frac{t^{3\alpha}}{\Gamma(1+3\alpha)}+c_2\frac{t^{5\alpha}}{\Gamma(1+5\alpha)}\right]$$
$$+\frac{t^{2\alpha}}{\Gamma(1+2\alpha)}+(1+c_1)\frac{t^{3\alpha}}{\Gamma(1+3\alpha)}-c_2\frac{t^{5\alpha}}{\Gamma(1+5\alpha)}$$

$$v_2(x,t) = e^{-x}\left[\frac{t^{2\alpha}}{\Gamma(1+2\alpha)}-(1-c_1)\frac{t^{3\alpha}}{\Gamma(1+3\alpha)}-c_2\frac{t^{5\alpha}}{\Gamma(1+5\alpha)}\right]$$
$$-\frac{t^{2\alpha}}{\Gamma(1+2\alpha)}+(1+c_1)\frac{t^{3\alpha}}{\Gamma(1+3\alpha)}-c_2\frac{t^{5\alpha}}{\Gamma(1+5\alpha)},$$

$$p^2: u_3(x,t) = -\frac{t^{3\alpha}}{\Gamma(1+3\alpha)}-(1+c_1)\frac{t^{4\alpha}}{\Gamma(1+4\alpha)}-c_2\frac{t^{5\alpha}}{\Gamma(1+5\alpha)}+c_2\frac{t^{6\alpha}}{\Gamma(1+6\alpha)}$$
$$-\left(1-c_1^2\right)c_3\frac{t^{7\alpha}}{\Gamma(1+7\alpha)}+2(1-c_1)\,c_2c_4\frac{t^{9\alpha}}{\Gamma(1+9\alpha)}+c_2^2c_5\frac{t^{11\alpha}}{\Gamma(1+11\alpha)}$$
$$-e^x\left[\frac{t^{3\alpha}}{\Gamma(1+3\alpha)}+(1-c_1)\frac{t^{4\alpha}}{\Gamma(1+4\alpha)}-c_2\frac{t^{5\alpha}}{\Gamma(1+5\alpha)}\right.$$
$$-(2c_1c_6+c_2)\frac{t^{6\alpha}}{\Gamma(1+6\alpha)}+\left(1-c_1^2\right)c_3\frac{t^{7\alpha}}{\Gamma(1+7\alpha)}-2c_2c_7\frac{t^{8\alpha}}{\Gamma(1+8\alpha)}$$
$$\left.+2c_1c_2c_4\frac{t^{9\alpha}}{\Gamma(1+9\alpha)}-c_2^2c_5\frac{t^{11\alpha}}{\Gamma(1+11\alpha)}\right]$$

$$v_3(x,t) = -\frac{t^{3\alpha}}{\Gamma(1+3\alpha)}+(1+c_1)\frac{t^{4\alpha}}{\Gamma(1+4\alpha)}-c_2\frac{t^{5\alpha}}{\Gamma(1+5\alpha)}-c_2\frac{t^{6\alpha}}{\Gamma(1+6\alpha)}$$
$$+(1-c_1)^2c_3\frac{t^{7\alpha}}{\Gamma(1+7\alpha)}+2(1-c_1)c_2c_4\frac{t^{9\alpha}}{\Gamma(1+9\alpha)}+c_2^2c_5\frac{t^{11\alpha}}{\Gamma(1+11\alpha)}+e^x\left[\frac{t^{3\alpha}}{\Gamma(1+3\alpha)}\right.$$
$$-(1-c_1)\frac{t^{4\alpha}}{\Gamma(1+4\alpha)}-c_2\frac{t^{5\alpha}}{\Gamma(1+5\alpha)}-(2c_1c_6+c_2)\frac{t^{6\alpha}}{\Gamma(1+6\alpha)}+\frac{\left(1-c_1^2\right)c_3t^{7\alpha}}{\Gamma(1+7\alpha)}$$
$$\left.+2c_2c_7\frac{t^{8\alpha}}{\Gamma(1+8\alpha)}+2c_1c_2c_4\frac{t^{9\alpha}}{\Gamma(1+9\alpha)}-c_2^2c_5\frac{t^{11\alpha}}{\Gamma(1+11\alpha)}\right],$$

and so on.

where

$$c_1 = \frac{\Gamma(1+2\alpha)}{\left[\Gamma(1+\alpha)\right]^2},\ c_2 = \frac{\Gamma(1+4\alpha)}{\left[\Gamma(1+2\alpha)\right]^2},\ c_3 = \frac{\Gamma(1+6\alpha)}{\left[\Gamma(1+3\alpha)\right]^2},\ c_4 = \frac{\Gamma(1+8\alpha)}{\Gamma(1+3\alpha)\Gamma(1+5\alpha)},$$

$$c_5 = \frac{\Gamma(1+10\alpha)}{\left[\Gamma(1+5\alpha)\right]^2},\ c_6 = \frac{\Gamma(1+5\alpha)}{\Gamma(1+3\alpha)\Gamma(1+2\alpha)},\ c_7 = \frac{\Gamma(1+7\alpha)}{\Gamma(1+5\alpha)\Gamma(1+2\alpha)}.$$

Solution is

$$\left.\begin{aligned} u(x,t) &= \lim_{p\to 1}\sum_{n=0}^{\infty} p^n u_n(x,t), \\ v(x,t) &= \lim_{p\to 1}\sum_{n=0}^{\infty} p^n v_n(x,t). \end{aligned}\right\}$$

This solution converges [36] very fast.

9.4 NUMERICAL RESULTS AND DISCUSSION

Figures 9.1–9.4 depict the space–time graphs of exact and numerical solutions by HPTM at $\alpha = 1$, $0 \le t \le 1$, $-4 \le x \le 2$ for Example 9.1. Figures 9.5 and 9.6 show the comparison of solutions at $x = 0.01$ and $\alpha = 1$ at distinct t for Example 9.1, which confirms that HPTM is in full agreement with the exact solution. Figures 9.7–9.10

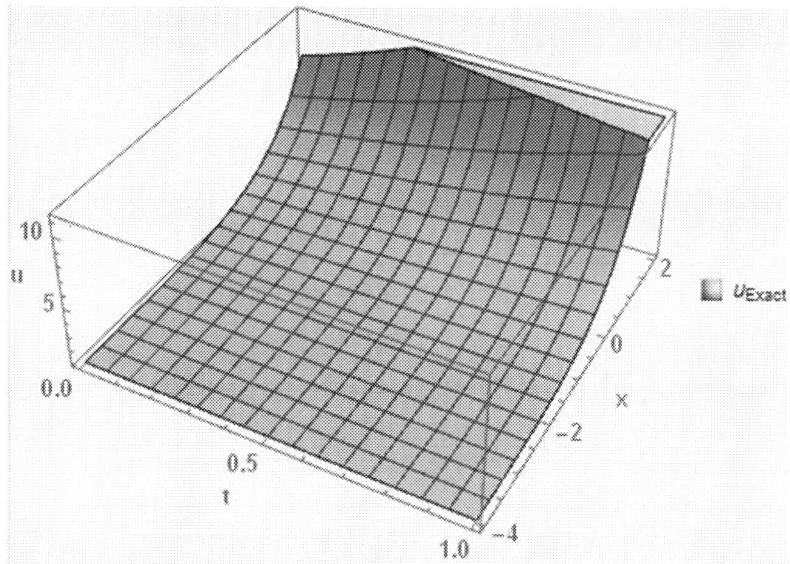

FIGURE 9.1 3D-plot of $u\ (x, t)$ at $0 \le t \le 1$, $-4 \le x \le 2$, and $\alpha = 1$.

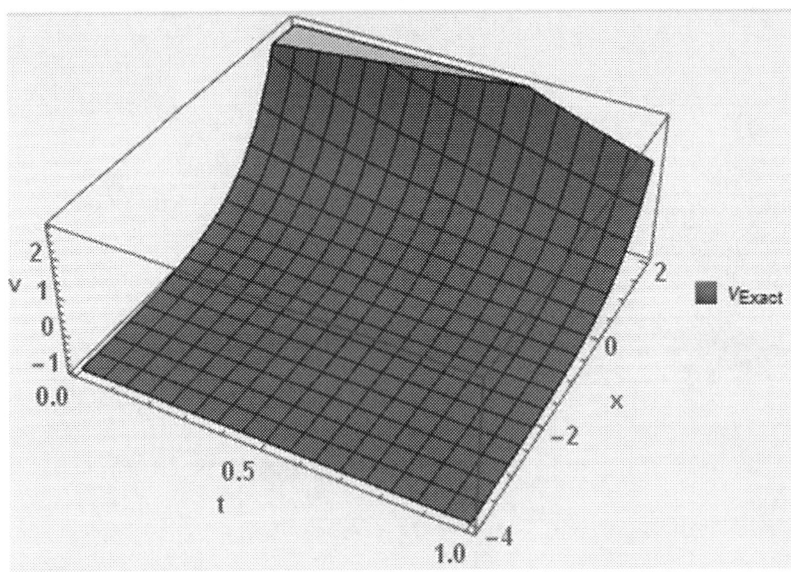

FIGURE 9.2 3D-plot of $v\ (x, t)$ at $0 \leq t \leq 1$, $-4 \leq x \leq 2$, and $\alpha = 1$.

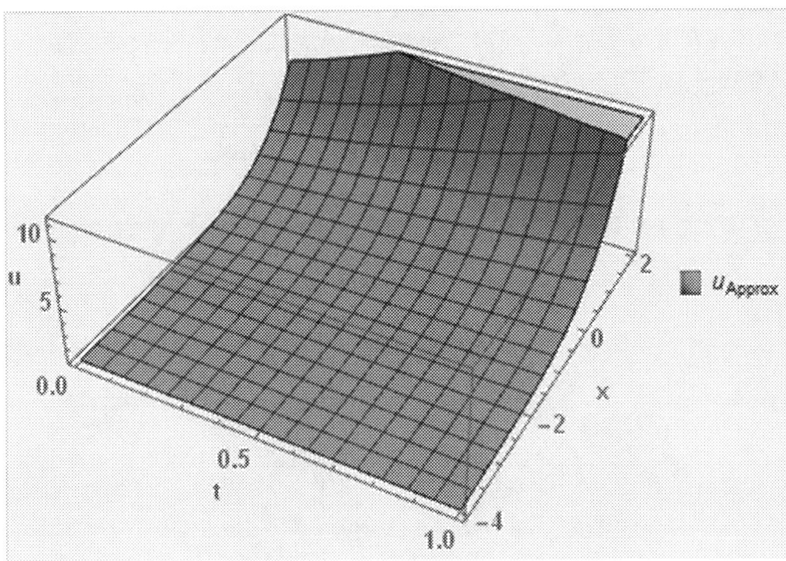

FIGURE 9.3 3D-plot of $u\ (x, t)$ by HPTM at $0 \leq t \leq 1$, $-4 \leq x \leq 2$, and $\alpha = 1$.

illustrate the space–time graphs of solutions at $\alpha = 1$, $0 \leq t \leq 1$, $-4 \leq x \leq 2$ for Example 9.2. Figures 9.11 and 9.12 show their comparison at $x = 0.01$, $\alpha = 1$ at diverse t for Example 9.2. It reconfirms the accuracy of HPTM. Graphs are plotted with Mathematica software package.

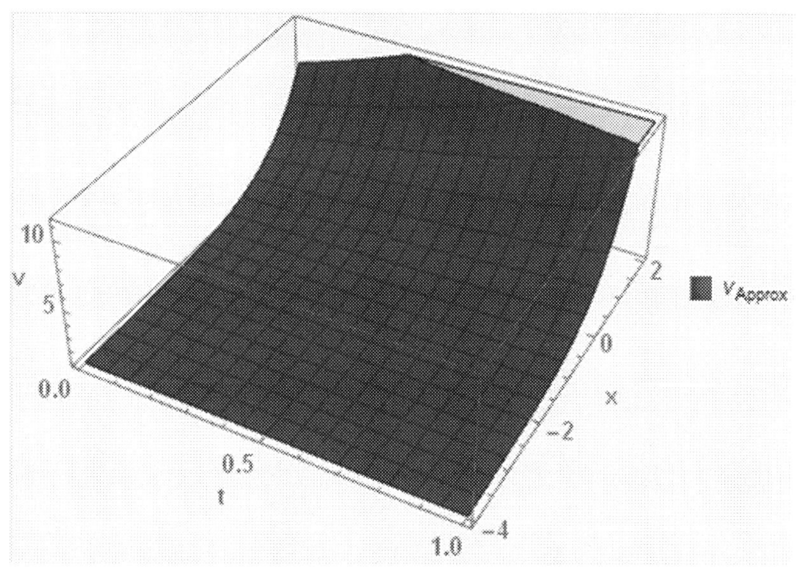

FIGURE 9.4 3D-plot of $v(x, t)$ by HPTM at $0 \le t \le 1$, $-4 \le x \le 2$, and $\alpha = 1$.

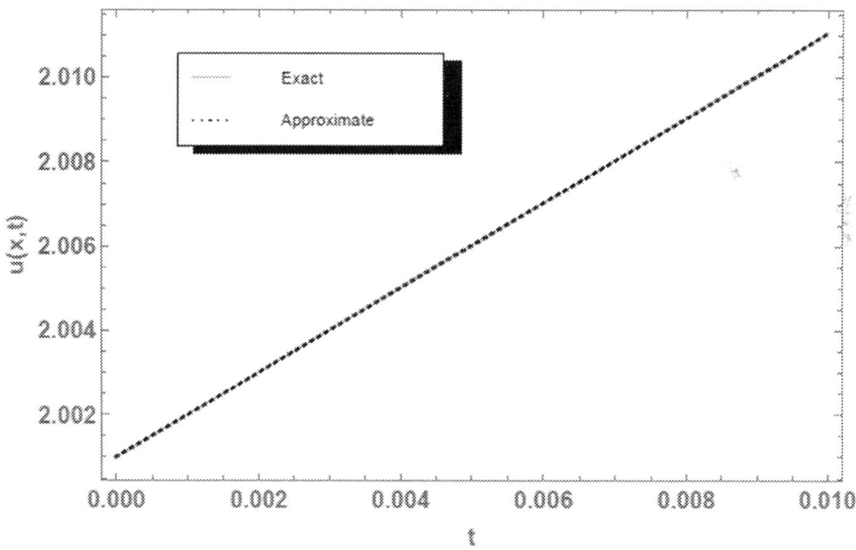

FIGURE 9.5 Comparison of solutions $u(x, t)$ at $x = 0.01$ for distinct t.

Table 9.1 represents the numerical solutions using Laplace Adomian decomposition method (LADM), Laplace variation iteration method (LVIM), and HPTM at $t = 0.05$ and 0.10;$-4 \le x \le 2$ and $\alpha = 0.25$, 0.50, and 0.75 for Example 9.1. Table 9.2 shows the comparison of these solutions at $t = 0.05$ and 0.10; $-4 \le x \le 2$; and for

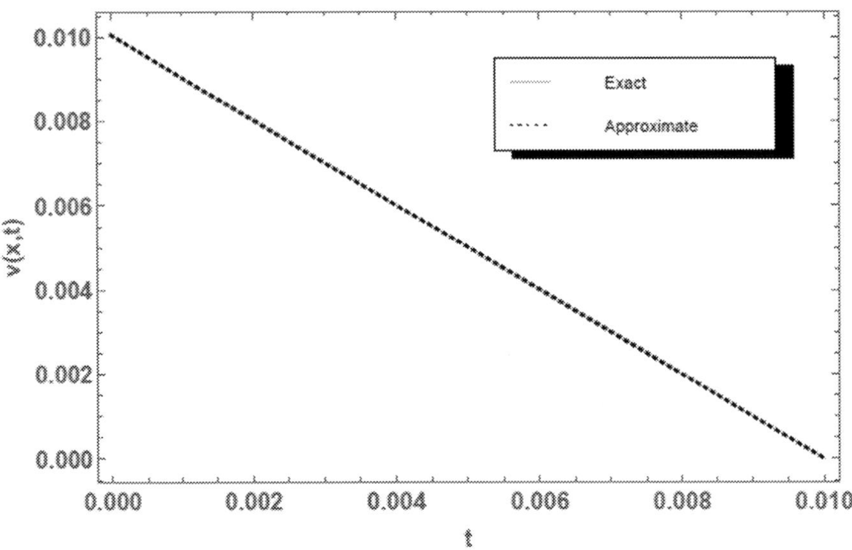

FIGURE 9.6 Comparison of solutions $v(x, t)$ at $x = 0.01$ for distinct t.

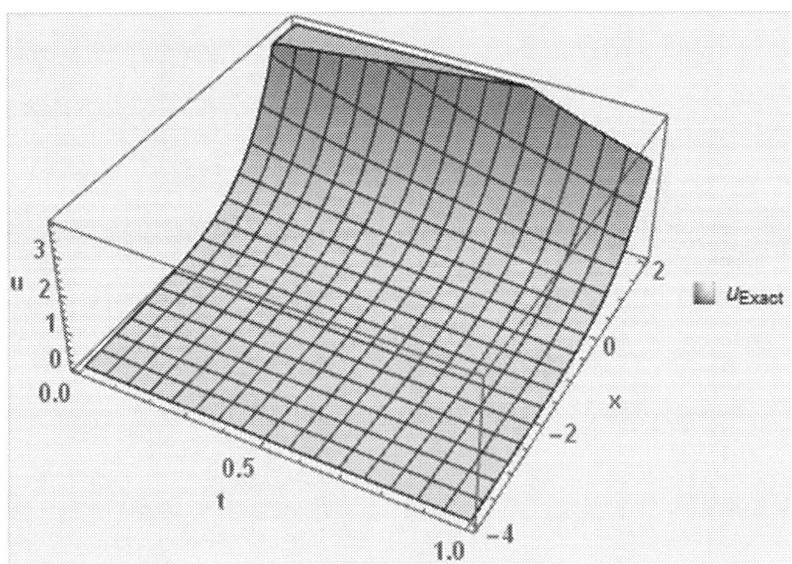

FIGURE 9.7 3D-plot of $u(x, t)$ at $0 \le t \le 1$, $-4 \le x \le 2$, and $\alpha = 1$.

$\alpha = 1$ for Example 9.1. Table 9.3 represents the numerical solutions using LADM, LVIM, and HPTM at $t = 0.05$ and 0.10; $-4 \le x \le 2$; and $\alpha = 0.25$, 0.50, and 0.75 for Example 9.2. Table 9.4 depicts the comparison of numerical solutions at $t = 0.05$ and 0.10; $-4 \le x \le 2$; and $\alpha = 1$ for Example 9.2. Tables 9.5 and 9.6 illustrate the solutions

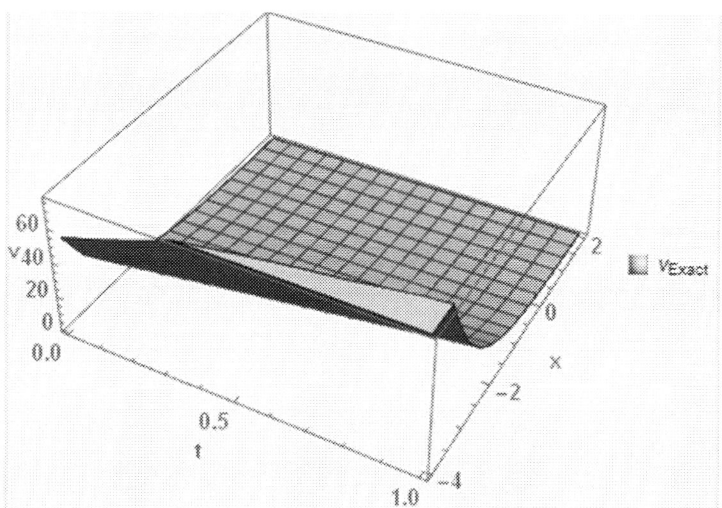

FIGURE 9.8 3D-plot of $v\ (x, t)$ at $0 \leq t \leq 1$, $-4 \leq x \leq 2$, and $\alpha = 1$.

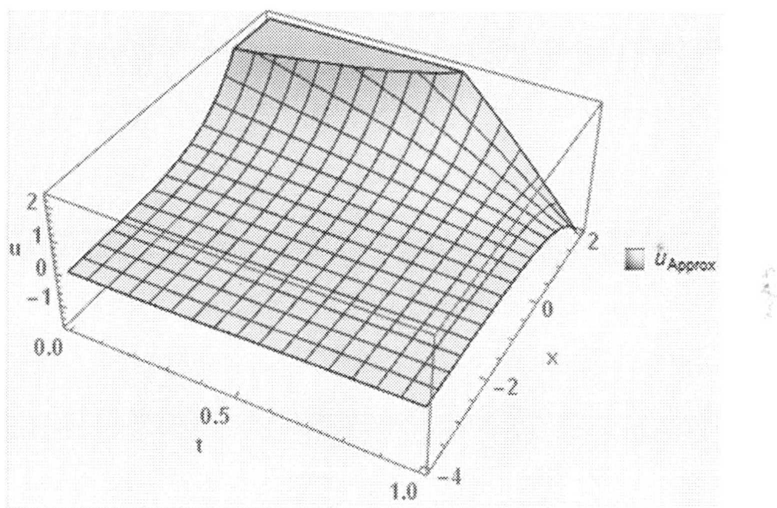

FIGURE 9.9 3D-plot of $u\ (x, t)$ by HPTM at $0 \leq t \leq 1$, $-4 \leq x \leq 2$, and $\alpha = 1$.

using LADM, LVIM, and HPTM at $t = 0.05$ and 0.10; $-1 \leq x \leq 1$; and $\alpha = 0.25$, 0.50, and 0.75 for Example 9.2. Table 9.7 shows the absolute error amid exact and numerical solutions at $t = 0.05$ and 0.10; $-1 \leq x \leq 1$ for standard case; and $\alpha = 1$. From the aforementioned tables, we see that the numerical results by HPTM are in complete agreement with LVIM [34] and LADM [34]. In Tables 9.2 and 9.7, clearly, absolute error is negligible. The numerical results direct that the HPTM works efficiently even if less-order iterations are used. Its efficiency can be enhanced with higher-order iterations.

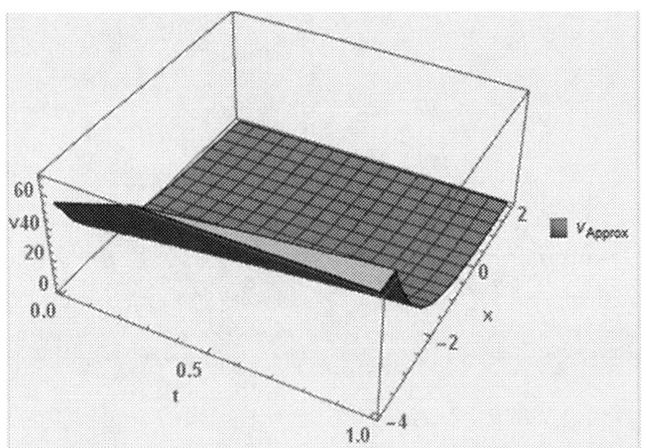

FIGURE 9.10 3D-plot of $v(x, t)$ by HPTM at $0 \leq t \leq 1$, $-4 \leq x \leq 2$, and $\alpha = 1$.

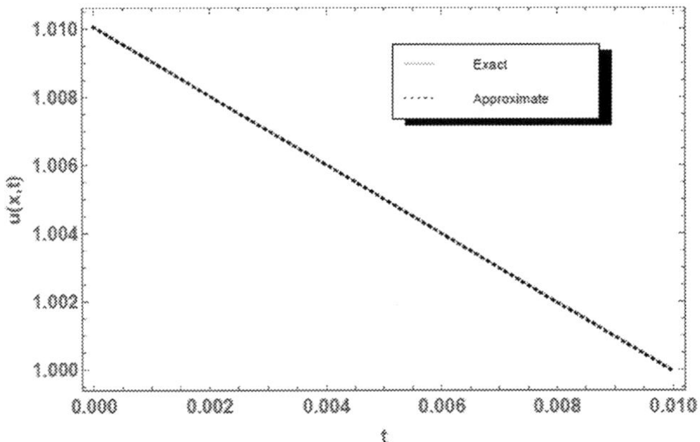

FIGURE 9.11 Comparison of $u(x, t)$ at $x = 0.01$, $\alpha = 1$ for distinct t.

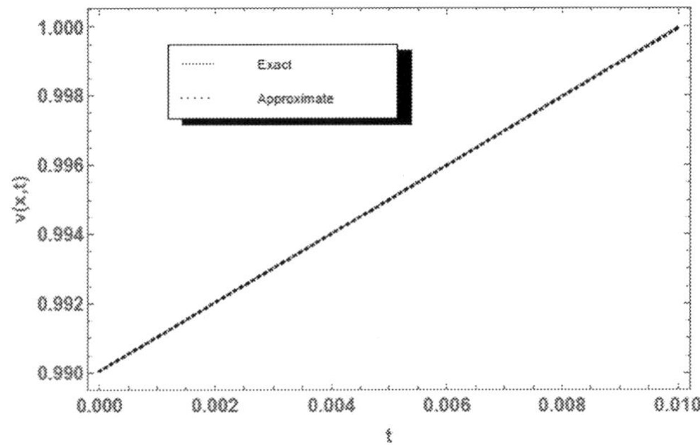

FIGURE 9.12 Comparison of $v(x, t)$ at $x = 0.01$, $\alpha = 1$ for distinct t.

TABLE 9.1
Comparison of Solutions of Eq. (9.10) by LADM, LVIM, and HPTM at $t = 0.05$, 0.10; $-4 \le x \le 2$; $\alpha = 0.25, 0.50, 0.75$.

		$\alpha = 0.25$	$\alpha = 0.50$	$\alpha = 0.75$
t	x	$u_{HPTM} = u_{LADM} = u_{LVIM}$	$u_{HPTM} = u_{LADM} = u_{LVIM}$	$u_{HPTM} = u_{LADM} = u_{LVIM}$
0.05	−4	1.0360514286	1.0240330988	1.0205857657
	−3	1.0979979434	1.0653287358	1.0559579128
	−2	1.2663860290	1.1775819155	1.1521093777
	−1	1.7241123020	1.4827176940	1.4134761575
	0	2.9683413123	2.3121627361	2.1239447256
	1	6.3505064216	4.566 8281215	4.0551985238
	2	15.5441843790	10.69564 40680	9.3048906298
0.10	−4	1.1273773921	1.0887978753	1.0637137293
	−3	1.3462476505	1.2413776510	1.1731918727
	−2	1.9411986966	1.6561324 827	1.4707843206
	−1	3.5584433140	2.7835530048	2.2797244639
	0	7.9545699696	5.8481997230	4.4786517558
	1	19.9044811732	14.1787732079	10.4559558553
	2	52.3877076498	36.8236197324	26.7039529724

TABLE 9.2
Comparison of Solutions of Eq. (9.10) by LADM, LVIM, and HPTM at $t = 0.05$, 0.10; $-4 \le x \le 2$; $\alpha = 1$.

t	x	u_{Exact}	$u_{HPTM} = u_{LADM} = u_{LVIM}$	$u_{Error} = \lvert u_{Exact} - u_{HPTM} \rvert$
0.05	−4	1.019254701775387	1.019254701775384	$2.8865798640e^{-15}$
	−3	1.052339705948432	1.052339705948425	$7.7715611723e^{-15}$
	−2	1.142274071586514	1.142274071586493	$2.1094237468e^{-14}$
	−1	1.386741023454501	1.38674102345444	$5.7287508071e^{-14}$
	0	2.051271096376024	2.051271096375860	$1.5631940187e^{-13}$
	1	3.857651118063164	3.857651118062740	$4.2410519541e^{-13}$
	2	8.767901106306770	8.767901106305619	$1.1510792319e^{-13}$
0.10	−4	1.049787068367864	1.049782924035073	$4.1443327905e^{-10}$
	−3	1.135335283236613	1.135324017772097	$1.1265464516e^{-6}$
	−2	.367879441171442	1.367848818463960	$3.0622707483e^{-5}$
	−1	2.000000000000000	1.999916758850712	$8.3241149288e^{-5}$
	0	3.718281828459045	3.718055555555555	$2.2627290349e^{-4}$
	1	8.389056098930650	8.388441025408820	$6.1507352183e^{-4}$
	2	21.08553692318767	21.08386498001012	$1.6719431776e^{-3}$

TABLE 9.3
Comparison of Solutions of Eq. (9.10) by LADM, LVIM, and HPTM at $t = 0.05$, 0.10; $-4 \le x \le 2$; $\alpha = 0.25, 0.50, 0.75$

		$\alpha = 0.25$	$\alpha = 0.50$	$\alpha = 0.75$
t	x	$v_{HPTM} = v_{LADM} = v_{LVIM}$	$v_{HPTM} = v_{LADM} = v_{LVIM}$	$v_{HPTM} = v_{LADM} = v_{LVIM}$
0.05	–4	–.9880379449	–.9855237049	–.9836456385
	–3	–.9674837631	–.9606493501	–.9555442365
	–2	–.9116117042	–.8930338435	–.8791567060
	–1	–.7597357019	–.7092358407	–.6715138700
	0	–.3468939244	–.2096210694	–.1070821220
	1	.7753263770	1.1484726845	1.4272024418
	2	3.8258374304	4.8401542572	5.5978202916
0.10	–4	–.9852248854	–.9911144679	–.9927214237
	–3	–.9598370746	–.9758466197	–.9802147784
	–2	–.8908258497	–.9343443054	–.9462181917
	–1	–.7032338912	–.8215293185	–.8538058878
	0	–.1933060793	–.5148663897	–.6026032014
	1	1.1928214256	.3187298772	.0802364961
	2	4.9607066343	2.5846794618	1.9363872380

TABLE 9.4
Comparison of Solutions of Eq. (9.10) by LADM, LVIM, and HPTM at $t = 0.05$, 0.10; $-4 \le x \le 2$; $\alpha = 1$

t	x	v_{Exact}	$v_{HPTM} = v_{LADM} = v_{LVIM}$	$v_{Error} = \lvert v_{Exact} - v_{HPTM} \rvert$
0.05	–4	–.9825776253605065	–.9825776253605036	2.8865798640e^{-15}
	–3	–.9526410756088590	–.9526410756088514	7.6605388699e^{-15}
	–2	–.8712650964121957	–.8712650964121749	2.0872192862e^{-14}
	–1	–.6500622508888447	–.6500622508887880	5.6732396558e^{-14}
	0	–.0487705754992860	–.0487705754991320	1.5398793352e^{-13}
	1	1.58570965931584600	1.5857096593162647	4.1877612489e^{-13}
	2	6.02868758058929300	6.0286875805904310	1.1386447341e^{-12}
0.10	–4	–.9932620530009145	–.9932588273534521	3.2256474624e^{-6}
	–3	–.9816843611112658	–.9816755928923834	8.7682188824e^{-6}
	–2	–.9502129316321360	–.9501890971420800	2.3834490061e^{-5}
	–1	–.8646647167633873	–.8645999279021774	6.4788861210e^{-5}
	0	–.6321205588285577	–.6319444444444444	1.7611438411e^{-4}
	1	.0000000000000000	.0004787285300654	4.7872853007e^{-4}
	2	1.71828182845904500	1.7195831475230867	1.3013190640e^{-3}

TABLE 9.5
Solutions u (x, t) of Eq. (9.17) by LADM, LVIM, and HPTM at $t = 0.05, 0.10; -1 \leq x \leq 1; \alpha = 0.25, 0.50, 0.75$

		$\alpha = 0.25$			$\alpha = 0.50$			$\alpha = 0.75$		
t	x	u_{LADM}	u_{LVIM}	u_{HPTM}	u_{LADM}	u_{LVIM}	u_{HPTM}	u_{LADM}	u_{LVIM}	u_{HPTM}
.05	−1	−.33618374	.03163789	.32829775	.32829775	.32829775	.27866738	.32829775	.32829883	.32596392
	−.5	−.26679276	.18647561	.54139518	.54139518	.54139518	.45475053	.54139518	.54139665	.53698843
	0	−.15238638	.44175984	.89273343	.89273343	.89273343	.74506256	.89273343	.89273555	.88490904
	.5	.03623785	.86265238	1.47199228	1.47199228	1.47199228	1.22370619	1.47199228	1.47199546	1.45853314
	1	.34722665	1.55658686	2.42702868	2.42702868	2.42702868	2.01285611	2.42702868	2.42703365	2.40427940
.10	−1	−1.43975842	−.28167840	.30379978	.30379978	.30379978	.24234793	.30379978	.30382684	.29845182
	−.5	−1.56505401	−.14388855	.50146699	.50146699	.50146699	.38953949	.50146699	.50150332	.49013587
	0	−1.77163150	.08328850	.82736511	.82736511	.82736511	.63221735	.82736511	.82741674	.80616946
	.5	−2.11222021	.45784015	1.36468028	1.36468028	1.36468028	1.03232550	1.36468028	1.36475713	1.32722075
	1	−2.67375606	1.07537143	2.25056323	2.25056323	2.25056323	1.69199231	2.25056323	2.25068167	2.18628910

TABLE 9.6
Comparison of solutions of Eq. (9.17) using LADM, LVIM, and HPTM at $t = 0.05, 0.10$; $-1 \le x \le 1$; $\alpha = 0.25, 0.50, 0.75$

		$\alpha = .25$			$\alpha = .50$			$\alpha = .75$		
t	x	v_{LADM}	v_{LVIM}	v_{HPTM}	v_{LADM}	v_{LVIM}	v_{HPTM}	v_{LADM}	v_{LVIM}	v_{HPTM}
.05	−1	4.51490076	5.11848384	4.42411221	3.56009092	3.56212327	3.44470891	3.05501302	3.05501696	3.03379533
	−.5	2.63113041	3.01970101	2.74800473	2.15640203	2.15762221	2.09437214	1.85287744	1.85287977	1.84039598
	0	1.48856594	1.74672487	1.73139415	1.30502168	1.30574925	1.27535149	1.12374536	1.12374670	1.11656268
	.5	.79556555	.97462581	1.11478867	.78863339	.78906218	.77859036	.68150439	.68150514	.67753559
	1	.37523957	.50632406	.74079853	.47542806	.47567562	.47728950	.41327169	.41327207	.41125220
.10	−1	3.89778488	5.58341855	4.70913170	4.01191701	4.02838986	3.79770710	3.31448167	3.31457074	3.25753092
	−.5	2.10060733	3.21653253	2.95887173	2.42414021	2.43428924	2.31818826	2.00993235	2.0099866	1.97729772
	0	1.01056404	1.78094359	1.89728540	1.46110490	1.46741834	1.42081473	1.21868319	1.21871481	1.20079702
	.5	.34941936	.91021489	1.25340074	.87699446	.88098150	.87653017	.73876631	.73878478	.72982555
	1	−.05158514	.38209123	.86286495	.52271357	.52528957	.54640490	.44768201	.44769250	.44416691

TABLE 9.7
Comparison of solutions of Eq. (9.17) by HPTM with solution (at $\alpha = 1$) for $t = 0.05, 0.10$, and $-1 \leq x \leq 1$

t	x	u_{Exact}	v_{Exact}	u_{HPTM}	v_{HPTM}	$u_{Error} = \lvert u_{Exact} - u_{HPTM} \rvert$	$v_{Error} = \lvert v_{Exact} - v_{HPTM} \rvert$
.05	−1	.34 993774 911115	2.857651118063164	.3495111020372	2.8543522183658	$4.266470739e^{-4}$	$3.2988996973e^{-3}$
	−.5	.576949810380487	1.733253017867395	.5762198852212	1.7312688236654	$72992515925e^{-4}$	$1.9841942019e^{-3}$
	0	.951229424500714	1.051271096376024	.9499994783112	1.0500843113655	$12299461894e^{-3}$	$1.1867850104e^{-3}$
	.5	1.568312185490169	.637628151621773	1.5662578439923	.6369250197342	$2.0543414978e^{-3}$	$7.0313188749e^{-4}$
	1	2.585709659315846	.386741023454501	2.5822961197377	.3863312420147	$3.4135395781e^{-3}$	$4.0978143978e^{-4}$
.10	−1	.332871083698080	3.004166023946433	.3312905903021	2.9913693618111	$1.5804933959e^{-3}$	$1.2796662135e^{-2}$
	−.5	.548811636094027	1.822118800390509	.5459983546653	1.8144929106841	$2.8132814287e^{-3}$	$7.6258897063e^{-3}$
	0	.904837418035960	1.105170918075648	.8999916127553	1.1006812603818	$4.8458052805e^{-3}$	$4.4896576938e^{-3}$
	.5	1.491824697641270	.670320046035639	1.4836278270528	.6677326092133	$8.1968705588e^{-3}$	$2.5874368222e^{-3}$
	1	2.459603111156950	.406569659740599	2.4458812679159	.4051359781985	$1.3721843241e^{-2}$	$1.4336815420e^{-4}$

9.5 CONCLUSION

In this paper, a hybrid and innovative scheme HPTM is efficaciously used to gain important solutions to the coupled systems of FDE useful in the modeling of epidemiological models. This scheme provides a solution in rapidly convergent series. It is important that HPTM is directly applied without linearization, special polynomials, or any restrictive norms. The results show that the implementation of this approach is easier and it is computationally very attractive. Hence, this scheme can be used further to solve new FDE of physical, biological, medical, and social importance.

REFERENCES

1. He JH. A Tutorial review of fractal space-time and fractional calculus. *Int. J. Theor. Phys.* 2014; 53 (11): 3698–3718. https://doi.org/10.1007/s10773-014-2123-8.
2. Wald R. Construction of solutions of gravitational, electromagnetic, or other perturbation equations from solutions of decoupled equations. *Phys. Rev. Lett.* 1978; 41(4): 203–206. https://doi.org/10.1103/PhysRevLett.41.203.
3. Sundnes J, Lines GT, Mardal KA, A Tveito. Multigrid block preconditioning for a coupled system of partial differential equations modeling the electrical activity in the heart. *Comput. Methods Biomech. Biomed. Engin.* 2002; 5 (6): 397–409. https://doi.org/10.1080/1025584021000025023.
4. Choi JS, Kumar D, Singh J, Swaroop R. Analytical techniques for system of time-fractional nonlinear differential equations. *J. Korean Math. Soc.* 2017; 54 (4): 1209–1229. https://doi.org/10.4134/JKMS.j160423.
5. Rafei M, Ganji DD, Daniali H. Solution of the epidemic model by homotopy perturbation method. *Appl. Math. Comput.* 2007; 187: 1056–1062. https://doi.org/10.1016/j.amc.2006.09.019.
6. Podlubny I. *Fractional Differential Equations.* Academic Press: New York, 1998.
7. Caputo M. *Elasticità e Dissipazione.* Zani-Chelli: Bologna, 1969.
8. Diethelm K. *The Analysis of Fractional Differential Equations* Springer-Verlag: Berlin; 2004.
9. Prakash A, Goyal M, Baskonus HM, Gupta S. A reliable hybrid numerical method for a time dependent vibration model of arbitrary order. *AIMS Mathematics* 2020; 5 (2): 979–1000. https://doi.org/10.3934/math.2020068.
10. Goyal M, Baskonus HM, Prakash M. An efficient technique for a time fractional model of Lassa hemorrhagic fever spreading in pregnant women. *Eur. Phys. J. Plus* 2019; 134 (10): 482. https://doi.org/10.1140/epjp/i2019-12854-0.
11. He JH. Homotopy perturbation technique. *Comput. Methods in Appl. Mech. Eng.* 1999; 178: 257–262. https://doi.org/10.1016/S0045-7825(99) 00018-3.
12. He JH. Homotopy perturbation method: A new nonlinear analytical technique. *Appl. Math. Comput.* 2003; 135: 73–79. https://doi.org/10.1016/S0096-3003(01)00312-5.
13. He JH. New interpretation of homotopy perturbation method. *Internat. J. Modern Phys. B.* 2006; 20: 2561–2568. https://doi.org/10.1142/S0217979206034819.
14. He JH. Homotopy perturbation method with an auxiliary term. *Abstr. Appl. Anal.* 2012; 2012: 857612. http://dx.doi.org/10.1155/2012/857612.
15. Mishra J. Unified fractional calculus results related with ¯H function. *J Int Acad Phys Sci.* 2016; 20 (3): 185–195.
16. Mishra J, Pandey N. An integral involving general class of polynomials with I-function. *Int J. Sci. Res. Pub.* 2013: 204

17. Mishra J. Fractional hyper-chaotic model with no equilibrium. *Chaos, Solitons Fract.* 2018; 116: 43–53.
18. Mishra J. A remark on fractional differential equation involving I-function. *Eur Phys. J. Plus* 2018 133 (2): 36.
19. Mishra J. Modified Chua chaotic attractor with differential operators with non-singular kernels. *Chaos, Solitons Fract.* 2019; 125: 64–72.
20. Mishra J. Analysis of the Fitzhugh Nagumo model with a new numerical scheme. *Discrete Cont Dyn Syst-S*, 2018; 13: 781.
21. Mishra J. Numerical analysis of a chaotic model with fractional differential operators: From Caputo to Atangana–Baleanu. *Methods Math Model: Fract. Diff. Eq.* 2019: 167
22. Ganji ZZ, Ganji DD, Jafari H, Rostamian M. Application of the homotopy perturbation method to coupled system of partial differential equations with time fractional derivatives. *Topol. Methods Nonlinear Anal.* 2008; 31: 341–348. https://projecteuclid.org/euclid.tmna/1463150278.
23. Ghori Q, Ahmed M, Siddiqui AM. Application of homotopy perturbation method to squeezing flow of a Newtonian fluid. *Int. J. Nonlinear Sci.* 2007; 8: 179–184. https://doi.org/10.1515/IJNSNS.2007.8.2.179.
24. Yildrim A. He's homotopy perturbation method for solving the space- and time-fractional telegraph equations. *Int. J. Comput. Math.* 2010; 87: 2998–3006. https://doi.org/10.1080/00207160902874653.
25. Prakash A, Veeresha P, Prakasha DG, Goyal M. A new efficient technique for solving fractional coupled Navier–Stokes equations using q-homotopy analysis transform method. *Pramana* 2019; 93 (1): 6. https://doi.org/10.1007/s12043-019-1763-x.
26. Prakash A, Veeresha P, Prakasha DG, Goyal M. A homotopy technique for a fractional order multi-dimensional telegraph equation via the Laplace transform. *Eur. Phys. J. Plus* 2019; 134: 19. https://doi.org/10.1140/epjp/i2019-12411-y.
27. Prakash A, Goyal M, Gupta S. Numerical simulation of space-fractional Helmholtz equation arising in seismic wave propagation, imaging and inversion. *Pramana* 2019; 93 (2): 28. https://doi.org/10.1007/s12043-019-1773-8.
28. Liu ZJ, Adamu MY, Suleiman E, He JH. Hybridization of homotopy perturbation method and Laplace transformation for the partial differential equations. *Therm. Sci.* 2017; 21 (4): 1843–1846. https://doi.org/10.2298/TSCI160715078L.
29. Goyal M, Prakash A, Gupta S. Numerical simulation for time-fractional nonlinear coupled dynamical model of romantic and interpersonal relationships. *Pramana* 2019; 92(5): 82. https://doi.org/10.1007/s12043-019-1746-y.
30. Prakash A, Goyal M, Gupta S. Fractional variational iteration method for solving time-fractional Newell-Whitehead-Segel equation. *Nonlinear Eng.* 2019; 8 (1): 164–171. https://doi.org/10.1515/nleng-2018-0001.
31. Prakash A, Goyal M, Gupta S. A reliable algorithm for fractional Bloch model arising in magnetic resonance imaging. *Pramana* 2019; 92 (2): 18. https://doi.org/10.1007/s12043-018-1683-1.
32. Yi M, Chen Y. Haar wavelet operational matrix method for solving fractional partial differential equations. *Comput. Model. Eng. Sci.* 2012; 88 (3): 229–244. https://doi.org/10.3970/cmes.2012.088.229.
33. Ertürk VS, Momani S. Solving systems of fractional differential equations using differential transform method. *J. Comput. Appl. Math.* 2008; 215 (1): 142–151. https://doi.org/10.1016/j.cam.2007.03.029.
34. Ahmed HF, Bahgat MSM, Zaki M. Numerical approaches to systems of fractional partial differential equations. *J. Egyptian Math. Soc.* 2017; 25 (2): 141–150. https://doi.org/10.1016/j.joems.2016.12.004.

35. Prakash A, Goyal M, Gupta S. q-homotopy analysis method for fractional Bloch model arising in nuclear magnetic resonance via the Laplace transform. *Indian J. Phys.* 2020; 94 (4): 507–520. http://doi.org/10.1007/s12648-019-01487-7.
36. Abbaoui K, Cherruault Y. New ideas for proving convergence of decomposition methods. *Comput. Math. Appl.* 1995; 29 (7): 103–108. https://doi.org/10.1016/0898-1221(95)00022-Q.
37. Biazar J, Ghazvini H. Convergence of the homotopy perturbation method for partial differential equations. *Nonlin Analy: Real World Appl.* 2009; 10 (5): 2633–2640. https://doi.org/10.1016/j.nonrwa.2008.07.002.

10 Balancing of Nitrogen Mass Cycle for Healthy Living Using Mathematical Model

Suresh Rasappan and Kala Raja Mohan
Vel Tech Rangarajan Dr. Sagunthala R&D
Institute of Science and Technology

CONTENTS

10.1 INTRODUCTION

Nitrogen is abundantly found in earth's atmosphere. It also finds its existence in other planets such as Venus, Mars, Pluto, and Titan. Nitrogen keeps on changing itself to several other forms with the help of nitrifying bacteria [8,4,11,13]. It gains its original form at the end of the cycle. In the earth, a period of 2–6 weeks is necessary for a complete nitrogen mass cycle [1,9,20].

The existence of nitrogen level varies from place to place [5,7,10]. In this part, the scientific properties of a framework wherein an underlying measure of nitrogen experiences the different procedure of nitrogen mass cycle are assessed [14,15,19].

10.2 MATHEMATICAL MODEL OF NITROGEN MASS CYCLE

Mathematical modeling is an art that links the theoretical information to numerical simulation. Ordinary differential equations play a vital role in confining the numerical model. The modeling of the nitrogen mass cycle is one of the paradigms in sustainable energy and biocycle research [24,26,30].

The following assumptions are made for this model.

Initially, the mass of the atmospheric nitrogen is bN.

x_1, x_2, x_3, x_4 are the amount of nitrogen, ammonia, nitrite, and nitrate, respectively (Figure 10.1).

In all the stages, the natural decay rate is considered as μ.

The dynamics of the nitrogen mass cycle is described by

$$\begin{aligned}
\frac{dx_1}{dt} &= bN - \alpha x_1 - \mu x_1 + \delta x_4 \\
\frac{dx_2}{dt} &= \alpha x_1 - \beta x_2 - \mu x_2 \\
\frac{dx_3}{dt} &= \beta x_2 - \gamma x_3 - \mu x_3 \\
\frac{dx_4}{dt} &= \gamma x_3 - \delta x_4 - \mu x_4.
\end{aligned} \tag{10.1}$$

The system (10.1) can be reframed as

$$\begin{aligned}
\frac{dx_1}{dt} &= bN - (\alpha + \mu)x_1 + \delta x_4 \\
\frac{dx_2}{dt} &= \alpha x_1 - (\beta + \mu)x_2
\end{aligned}$$

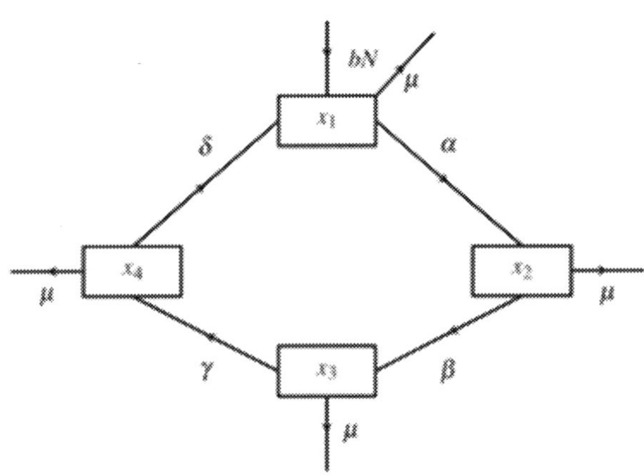

FIGURE 10.1 Flow diagram of nitrogen mass cycle.

$$\frac{dx_3}{dt} = \beta x_2 - (\gamma + \mu)x_3$$
$$\frac{dx_4}{dt} = \gamma x_3 - (\delta + \mu)x_4. \tag{10.2}$$

10.3 MATHEMATICAL PROPERTIES OF THE DETERMINISTIC MODEL

The deterministic model of the nitrogen mass cycle system is analyzed for its characteristics.

10.3.1 Boundedness of the Nitrogen Mass Cycle

The boundedness of the nitrogen mass cycle (10.2) is analyzed in this section.

Define a function

$$W = x_1 + x_2 + x_3 + x_4. \tag{10.3}$$

Differentiate (10.3) along with Eq. (10.2), which implies

$$\frac{dW}{dt} = \dot{x}_1 + \dot{x}_2 + \dot{x}_3 + \dot{x}_4 \tag{10.4}$$

From Eq. (10.4), the inequality holds that

$$\frac{dW}{dt} = bN - \mu W. \tag{10.5}$$

Then, (10.5) implies

$$\frac{dW}{dt} + \mu W = bN. \tag{10.6}$$

From the theorem of differential inequality in (10.6),

$$0 \leq W(x_1, x_2, x_3, x_4)$$
$$\leq \frac{bN}{\mu}(1 - e^{-\mu t}) + W(x_1(0), x_2(0), x_3(0), x_4(0))e^{-\mu t} \tag{10.7}$$

and for $t \to \infty$, then (10.7) implies

$$0 \leq W \leq \frac{bN}{\mu}.$$

Hence, the solution of (10.2) that lies in R_4^+ is confirmed in the area given by

$$A=\left\{(x_1,x_2,x_3,x_4)\in R_+^4 : W=\frac{bN}{\mu}+\epsilon \text{ for any } \epsilon>0\right\}.$$

10.3.2 Local Stability Analysis

The property of local asymptotic stability for several states of the described structure is investigated in this section.

The boundary equilibrium point E is given by

$$E=(\nabla_1,\nabla_2,\nabla_3,\nabla_4),$$

where

$$\nabla_1=\left(\frac{(\beta+\mu)(\mu+\delta)b_1N}{\alpha a(\mu+\delta)-\alpha b\delta},0,0,0\right),\quad \nabla_2=\left(0,\frac{(\delta+\mu)b_1N}{a(\mu+\delta)-b\delta},0,0\right),$$

$$\nabla_3=\left(0,0,\frac{(\mu+\delta)bb_1N}{\gamma\left(a(\mu+\delta)-b\delta\right)},0\right) \text{ and } \nabla_4=\left(0,0,0,\frac{bb_1N}{a(\delta+\mu)-b\delta}\right).$$

The equilibrium points at the interior are given by

$$E^*=\left(x_i^*, i=1,2,3,4\right),$$

where

$$x_1^*=\frac{bN+\delta x_4}{\alpha+\mu},\; x_2^*=\frac{\alpha x_1}{\beta+\mu},\; x_3^*=\frac{\beta x_2}{\gamma+\mu} \text{ and } x_4^*=\frac{\gamma x_3}{\delta+\mu}.$$

The Jacobian of the system (10.2) is

$$J=\begin{bmatrix} -\mu-\alpha & 0 & 0 & \delta \\ \alpha & -\mu-\beta & 0 & 0 \\ 0 & \beta & -\mu-\gamma & 0 \\ 0 & 0 & \gamma & -\mu-\delta \end{bmatrix}. \tag{10.8}$$

The characteristic equation of (10.8) is

$$\Delta_1\lambda^4+\Delta_2\lambda^3+\Delta_3\lambda^2+\Delta_4\lambda+\Delta_5=0,$$

where $\Delta_1=1$

$$\Delta_2=\alpha+\beta+\gamma+\delta+4$$

$$\Delta_3 = \alpha\beta + \alpha\gamma + \alpha\delta + \beta\gamma + 3\alpha\mu + \beta\delta + 3\beta\mu + \gamma\delta + 3\gamma\mu + 3\delta\mu + 6\mu^2$$

$$\Delta_4 = 3\alpha\mu^2 + 3\beta\mu^2 + 3\gamma\mu^2 + 3\delta\mu^2 + 4\mu^3 + \alpha\beta\gamma + \alpha\beta\delta + 2\alpha\beta\mu + \alpha\gamma\delta$$
$$+ 2\alpha\gamma\mu + \beta\gamma\delta + 2\alpha\delta\mu + 2\beta\gamma\mu + 2\beta\delta\mu + 2\gamma\delta\mu$$

$$\Delta_5 = \alpha\mu^3 + \beta\mu^3 + \gamma\mu^3 + \delta\mu^3 + \mu^4 + \alpha\beta\mu^2 + \alpha\gamma\mu^2 + \alpha\delta\mu^2 + \beta\gamma\mu^2$$
$$+ \beta\delta\mu^2 + \gamma\delta\mu^2 + \alpha\beta\gamma\mu + \alpha\beta\delta\mu + \alpha\gamma\delta\mu + \beta\gamma\delta\mu$$

Theorem 10.1

The interior equilibrium point E^* is locally asymptotically stable in the positive octant.

Proof:
By *divergence criterion* theorem, assume

$$\theta(x_1, x_2, x_3, x_4) = \frac{1}{x_1 x_2 x_3 x_4}, \tag{10.9}$$

where $\theta(x_i, i = 1,2,3,4) > 0$ if $x_i > 0, i = 1,2,3,4.$

Now consider

$$\begin{aligned} p_1 &= bN - (\alpha + \mu)x_1 + \delta x_4 \\ p_2 &= \alpha x_1 - (\beta + \mu)x_2 \\ p_3 &= \beta x_2 - (\gamma + \mu)x_3 \\ p_4 &= \gamma x_3 - (\delta + \mu)x_4. \end{aligned} \tag{10.10}$$

$$\text{Define } \nabla = \frac{\partial}{\partial x_1}(p_1\theta) + \frac{\partial}{\partial x_2}(p_2\theta) + \frac{\partial}{\partial x_3}(p_3\theta) + \frac{\partial}{\partial x_4}(p_4\theta). \tag{10.11}$$

Find ∇ in (10.11) along with the trajectories (10.9) and (10.10), which gives

$$\nabla = \frac{-bNx_2x_3x_4 - \delta x_2x_3x_4^2}{x_1^2x_2^2x_3^2x_4^2} - \frac{\alpha x_1^2x_3x_4}{x_1^2x_2^2x_3^2x_4^2} - \frac{\beta x_1x_2^2x_4 + \gamma x_3^2}{x_1^2x_2^2x_3^2x_4^2} - \frac{\gamma x_1x_2x_3^2}{x_1^2x_2^2x_3^2x_4^2},$$

which is less than zero.

From *Bendixson–Dulac criterion*, it is clear that the first octant does not contain any limit cycle.

Consequently, E^* is found to be locally asymptotically stable.

Theorem 10.2

There are no closed trajectories around the interior equilibrium point E^*.

Proof:

$$\text{Define } \Psi\left(x_i, i=1,2,3,4\right) = \frac{\partial p_1}{\partial x_1} + \frac{\partial p_2}{\partial x_2} + \frac{\partial p_3}{\partial x_3} + \frac{\partial p_4}{\partial x_4}. \tag{10.12}$$

Find Ψ along with the trajectories (10.12), which gives

$$\Psi = -\left(\alpha + \beta + \gamma + \delta\right) - 4\mu \neq 0. \tag{10.13}$$

Hence, applying *Bendixson's criteria theorem*, (10.13) shows that there are no closed trajectories around the point E^* [21].

Hence, the limit cycle does not exist encompassing E^*.

Therefore, the point E^* is evidential to be locally asymptotically stable [22,23,31] Nitrogen mass cycle occurs with the help of nitrogen-fixing, nitrifying, and denitrifying bacteria.

Theorem 10.3

The nitrogen mass cycle Eq. (10.2) has a nontrivial periodic solution.

Proof:

$$\text{Define } \Phi = \sum_{i=1}^{4} x_i \frac{dx_i}{dt}. \tag{10.14}$$

Find Φ (10.14) along the trajectories (10.2), which gives

$$\begin{aligned} \Phi &= bNx_1 + \alpha x_1 x_2 + \beta x_2 x_3 + \gamma x_3 x_4 + \delta x_1 x_4 \\ &\quad - \left[\left(\alpha+\mu\right)x_1^2 + \left(\beta+\mu\right)x_2^2 + \left(\gamma+\mu\right)x_3^2 + \left(\mu+\delta\right)x_4^2\right] \end{aligned} \tag{10.15}$$

From (10.15), observe that

$$bNx_1 + \alpha x_1 x_2 + \beta x_2 x_3 + \gamma x_3 x_4 + \delta x_1 x_4$$

is positive for

$$x_1^2 + x_2^2 + x_3^2 + x_4^2 > \frac{bN}{\left(\mu+\alpha\right)\left(\mu+\beta\right)\left(\mu+\gamma\right)\left(\mu+\delta\right)},$$

which implies $x_1^2 + x_2^2 + x_3^2 + x_4^2$ is increasing along any solution $x_i(t), i = 1,2,3,4$ of (10.2) and decreasing when

$$x_1^2 + x_2^2 + x_3^2 + x_4^2 > \frac{bN}{(\alpha+\mu)(\beta+\mu)(\gamma+\mu)(\delta+\mu)}.$$

Hence, by *Poincare–Bendixson* theorem, there exists at least one periodic solution $x_i(t), i = 1,2,3,4$ of (10.2) lying in this annulus.

Hence, the nitrogen mass cycle (10.2) has a nontrivial periodic solution.

10.3.3 Global Stability Analysis

Theorem 10.4

The global stability of interior equilibrium point exists if

$$\delta = \frac{-x_1\left(x_1 - x_1^*\right) - bN + (\mu+\alpha)x_1}{x_4};$$

$$\alpha = \frac{-x_2\left(x_2 - x_2^*\right) + (\beta+\mu)x_2}{x_1};$$

$$\beta = \frac{-x_3\left(x_3 - x_3^*\right) + (\gamma+\mu)x_3}{x_2};$$

$$\gamma = \frac{-x_4\left(x_4 - x_4^*\right) + (\delta+\mu)x_4}{x_3}.$$

Proof:
Consider the Lyapunov function

$$V(x_i) = \sum_{i=1}^{4} l_i \left[\left(x_i - x_i^*\right) - x_i^* \ln\left(\frac{x_i}{x_i^*}\right)\right]. \tag{10.16}$$

Differentiate (10.16) along with the solution of (10.1)

$$\dot{V} = l_1\left(x_1 - x_1^*\right)\frac{\dot{x}_1}{x_1} + \cdots + l_4\left(x_4 - x_4^*\right)\frac{\dot{x}_4}{x_4}. \tag{10.17}$$

Substitute (10.1) into (10.17), which implies that

$$\dot{V} = -\sum_{i=1}^{4} l_i\left(x_i - x_i^*\right)^2, \tag{10.18}$$

which is a negative definite function.

Applying *LaSalle's invariance principle* on (10.18), the global asymptotic stability of E^* is proved.

10.3.4 Global Stability Analysis of Nitrogen Mass Cycle by Pseudo-Back-Propagation

While studying the qualitative properties of the nitrogen mass cycle, it is necessary to analyze it global stability. Pseudo-back-propagation is one of the feedback techniques used to analyze global stability.

The system of nitrogen mass cycle can be rearranged as

$$\begin{aligned}
\frac{dx_4}{dt} &= \gamma x_3 - (\delta + \mu)x_4 \\
\frac{dx_3}{dt} &= \beta x_2 - (\gamma + \mu)x_3 \\
\frac{dx_2}{dt} &= \alpha x_1 - (\beta + \mu)x_2 \\
\frac{dx_1}{dt} &= bN - (\alpha + \mu)x_1 + \delta x_4.
\end{aligned} \tag{10.19}$$

Theorem 10.5

The system of nitrogen mass cycle (10.19) is globally asymptotically stable for all initial densities if

$$bN + \left[\alpha + \frac{\beta}{\alpha}\left(\frac{\beta^2 + \gamma^2}{\beta}\right)\right]\Delta_3 + \left[\delta - \frac{\gamma}{\alpha}\left(\frac{\beta^2 + \gamma^2}{\beta}\right) - \frac{\gamma}{\alpha}(\delta + \mu)\right]x_4$$

$$= (\alpha + \mu)\,p_3 + \Delta_2\left[\left(\frac{\gamma + \mu}{\alpha}\right)\left(\beta + \frac{\gamma^2}{\beta}\Delta_2 + \frac{\gamma^2}{\beta}\left(1 - \frac{\delta}{\beta}\right)\right)\right]$$

Proof:
Starting with the stability of nitrate, the back-propagation process proceeds

$$\dot{x}_4 = \gamma x_3 - (\delta + \mu)x_4,$$

where x_3 is regarded as pseudo-controller

$$x_3 = p_1(x_4).$$

Assume the Lyapunov function

$$V_1 = \frac{1}{2}x_4^2. \tag{10.20}$$

Equation (10.20) on differentiation gives

$$\dot{V}_1 = x_4\dot{x}_4 = x_4\left(\gamma p_1 - (\delta + \mu)x_4\right).$$

Assume the pseudo-controller $x_3 = p_1(x_4)$.

Then,

$$\dot{V}_1 = -(\delta + \mu)x_4^2. \tag{10.21}$$

Equation (10.21) is a negative definite function.

Hence, Eq. (10.19) is asymptotically stable.

The relation between x_3 and p_1 is defined by

$$\Delta_2 = x_3 - p_1 = x_3$$

$$\dot{\Delta}_2 = \dot{x}_3.$$

Consider that the (x_4, Δ_2) subcycle is given by

$$\dot{\Delta}_2 = \beta x_2 - (\gamma + \mu)x_3$$

$$\dot{x}_4 = \gamma\Delta_2 - (\delta + \mu)x_4$$

Let x_2 be a pseudo-controller in the subcycle (10.19).

Assume that when $x_2 = p_2$, the subcycle (10.19) is globally asymptotically stable.

Consider

$$V_2 = V_1 + \frac{1}{2}\Delta_2^2. \tag{10.22}$$

From Eq. (10.22), $\dot{V}_2 = -(\delta + \mu)x_4^2 + \Delta_2\left(\gamma x_4 + \beta x_2 - (\gamma + \mu)\Delta_2\right)$.

Assume the pseudo-controller $x_2 = p_2(x_4, \Delta_2)$.

On simplification

$$\dot{V}_2 = -(\delta + \mu)x_4^2 - (\gamma + \mu)\Delta_2^2,$$

which is found to be negative definite.

Hence, globally asymptotically stability of (10.19) is verified.

The relation between x_2 and p_2 is defined by

$$\Delta_3 = x_2 - p_2.$$

Now consider that the subcycle $\left(\dot{x}_4, \dot{\Delta}_2, \dot{\Delta}_3\right)$ is given by

$$\dot{x}_4 = \gamma\Delta_2 - (\delta + \mu)x_4$$

$$\dot{\Delta}_2 = \beta\Delta_3 - \gamma x_4 - (\gamma + \mu)\Delta_2$$

$$\dot{\Delta}_3 = \alpha x_1 + \frac{\gamma^2}{\beta}\Delta_2 - (\beta + \mu)\Delta_3 + \gamma\left(1 - \frac{\delta}{\beta}\right)x_4.$$

Consider the Lyapunov function

$$V_3 = \frac{1}{2}\left(x_4^2 + \Delta_2\dot{\Delta}_2 + \Delta_3\dot{\Delta}_3\right). \tag{10.23}$$

The derivative of Eq. (10.23) is

$$\dot{V}_3 = -(\delta + \mu)x_4^2 - (\gamma + \mu)\Delta_2^2 + \Delta_3$$

$$= \left(\beta\Delta_2 + \alpha x_1 + \frac{\gamma^2}{\beta}\Delta_2 - (\beta + \mu)\Delta_3 + \gamma\left(1 - \frac{\delta}{\beta}\right)x_4\right).$$

Consider the pseudo-controller $x_1 = p_3(x_4, \Delta_2, \Delta_3, \Delta_4)$.

On simplification, it follows that

$$\dot{V}_3 = -(\delta + \mu)x_4^2 - (\gamma + \mu)\Delta_2^2 - (\beta + \mu)\Delta_3^2,$$

which has a negative definite measure.

Thus, Eq. (10.19) is identified to be globally asymptotically stable. Now, the relation between x_1 and p_3 is given by

$$\Delta_4 = x_1 - p_3.$$

Consider that the subcycle $(x_4, \Delta_2, \Delta_3, \Delta_4)$ is given by

$$\dot{x}_4 = \gamma\Delta_2 - (\delta + \mu)x_4$$

$$\dot{\Delta}_2 = \beta\Delta_3 - \gamma x_4 - (\gamma + \mu)\Delta_2$$

$$\dot{\Delta}_3 = \alpha\Delta_4 - \beta\Delta_2 - (\beta + \mu)\Delta_3$$

$$\dot{\Delta}_4 = bN - (\alpha + \mu)x_1 + \delta x_4 - \frac{\gamma}{\alpha}x_4 - \frac{\gamma}{\alpha}\left(\beta + \frac{\gamma^2}{\beta}\right)x_4$$

$$-\frac{\gamma}{\alpha}(\delta + \mu)x_4 + \frac{\beta}{\alpha}\left(\beta + \frac{\gamma^2}{\beta}\right)\Delta_3 - \left(\frac{\gamma + \mu}{\alpha}\right)\left(\beta + \frac{\gamma^2}{\beta}\right)\Delta_2 + \frac{\gamma^2}{\beta}\left(1 - \frac{\delta}{\beta}\right)\Delta_2.$$

Now consider the Lyapunov function

$$V_4 = V_3 + \frac{1}{2}\Delta_4^2.$$

The derivative of V_4 is given by

$$\dot{V}_4 = -(\delta+\mu)x_4^2 - (\gamma+\mu)\Delta_2^2 - (\beta+\mu)\Delta_3^2 - (\alpha+\mu)\Delta_4^2,$$

which is definitely negative.

Hence, globally asymptotically stable condition of the nitrogen mass cycle (10.19) for all initial densities is verified.

10.4 NONDETERMINISTIC MATHEMATICAL MODEL OF NITROGEN MASS CYCLE

Nitrogen mass cycle occurs with the help of nitrogen-fixing, nitrifying, and denitrifying bacteria [2,6]. These bacteria involve themselves in the processes of nitrogen cycle to gain energy for their livelihood [27,29]. During the process of conversion of nitrogen into different forms, there is a possibility of the probabilistic behavior of the events [18,28]. The probabilistic behavior of the events of variables results in the formation of nondeterministic mathematical model of nitrogen mass cycle [3,12,16,17].

10.4.1 Description of the NonDeterministic Mathematical Model

Probabilistic events of the variables x_1, x_2, x_3, x_4 encompassing E^* are introduced to the mathematical model of nitrogen mass cycle. The proportionality of the distances from x_i^*, $i = 1,2,3,4$ to x_i, $i = 1,2,3,4$ is considered in this model.

The nondeterministic differential equation is given by

$$\begin{aligned}
dx_1 &= \left[bN - (\alpha+\mu)x_1 + \delta x_4\right]dt + \sigma_1\left(x_1 - x_1^*\right)d\omega t_1 \\
dx_2 &= \left[\alpha x_1 - (\beta+\mu)x_2\right]dt + \sigma_2\left(x_2 - x_2^*\right)d\omega t_2 \\
dx_3 &= \left[\beta x_2 - (\gamma+\mu)x_3\right]dt + \sigma_3\left(x_3 - x_3^*\right)d\omega t_3 \\
dx_4 &= \left[\gamma x_3 - (\delta+\mu)x_4\right]dt + \sigma_4\left(x_4 - x_4^*\right)d\omega t_4.
\end{aligned} \tag{10.24}$$

where $\sigma_1, \sigma_2, \sigma_3$, and σ_4 are the real constants, and $\omega t_1 = \omega_1 t$, $\omega t_2 = \omega_2 t$, $\omega t_3 = \omega_3 t$, $\omega t_4 = \omega_4 t$ are independent among themselves.

The asymptotic stability of the equilibrium is investigated. The dynamical behavior of the system of Eq. (10.1) is compared with asymptotic stability behavior of Eq. (10.24). The comparison results based on its robustness are furnished. Equation (10.24) is considered as the It ö nondeterministic differential system.

10.4.2 Nondeterministic Stability of the Positive Equilibrium

The change of variables approach is used to center the positive equilibrium of (10.24):

$$u_1 = x_1 - \overset{*}{x_1}, u_2 = x_2 - \overset{*}{x_2}, u_3 = x_3 - \overset{*}{x_3}, u_4 = x_4 - \overset{*}{x_4}.$$

The formation of linearized nondeterministic system encircling E^* is as follows:

$$du(t) = f(u(t))dt + g(u(t))d\omega t,$$

where $u(t) = \begin{bmatrix} u_1(t) & u_2(t) & u_3(t) & u_4(t) \end{bmatrix}^T$ and

$$f(u(t)) = \begin{bmatrix} -\alpha-\mu & 0 & 0 & \delta \\ \alpha & -\beta-\mu & 0 & 0 \\ 0 & \beta & -\gamma-\mu & 0 \\ 0 & 0 & \gamma & -\delta-\mu \end{bmatrix} u(t)$$

$$g(u(t)) = \begin{bmatrix} \sigma_1 u_1 & 0 & 0 & 0 \\ 0 & \sigma_2 u_2 & 0 & 0 \\ 0 & 0 & \sigma_3 u_3 & 0 \\ 0 & 0 & 0 & \sigma_4 u_4 \end{bmatrix}.$$

The positive equilibrium point with reference to the trivial solution $u(t) = 0$ is considered.

Let U be the set $U = (t \geq t_0)R^n, t_0 \in R^+$. Hence, $V \in C_2^0(U)$ is a double-time continuously differentiable function concerning u and a continuous function concerning t.

The It ö nondeterministic differential representation is

$$\mathcal{L}V(t,u) = \frac{\partial V(t,u)}{\partial t} + f^T(u)\frac{\partial V(t,u)}{\partial u} + \frac{1}{2}Tr\left[g^T(u)(\frac{\partial^2 V(t,u)}{\partial u^2}g(u)\right],$$

where $\frac{\partial V}{\partial u} = \text{Col}\left(\frac{\partial V}{\partial u_1}, \frac{\partial V}{\partial u_2}, \frac{\partial V}{\partial u_3}, \frac{\partial V}{\partial u_4}\right)$ and $\frac{\partial^2 V(t,u)}{\partial u^2} = \frac{\partial^2 V}{\partial u_j \partial u_i}, i, j = 1,2,3,4.$

Theorem 10.6

Suppose that there exists a function $V(t,u) \in C_2^0(U)$ satisfying the inequalities

$$K_1 \mid u \mid^p \leq V(t,u) \leq K_2 \mid u \mid^p$$

$$\mathcal{L}V(t,u) \leq -K_3 \mid u \mid^p, K_i 0, p0.$$

Then, the trivial solution of Eq. (10.24) is exponentially p-stable for $t \geq 0$.

If $p = 2$, the trivial solution of (10.24) ensures the global asymptotic stability in probability.

Theorem 10.7

The zero solution is asymptotically mean square stable when

$$\delta u_1 u_4 \omega_1 + \alpha u_1 u_2 \omega_2 + \beta u_2 u_3 \omega_3 + \gamma u_3 u_4 \omega_4 = 0.$$

Proof:

$$\text{Assume } V(u) = \frac{1}{2}\left[\omega_1 u_1^2 + \omega_2 u_2^2 + \omega_3 u_3^2 + \omega_4 u_4^2\right],$$

where ω_i are the constants, which are chosen to be real and positive.

The inequalities become true when $p = 2$.

The It ö process changes to

$$\mathcal{L}V(t,u) = \omega_1(-\alpha u_1 - \mu u_1 + \delta u_4)u_1 + \omega_2(\alpha u_1 - \beta u_2 \mu u_2)u_2$$
$$+ \omega_3(\beta u_2 - \gamma u_3 - \mu u_3)u_3 + \omega_4(\gamma u_3 - \delta u_4 - \mu u_4)u_4$$
$$+ \frac{1}{2}\text{trace}\left[g^T(u(t))\right]\left(\frac{\partial^2 V(t,u)}{\partial u^2} g(u(t))\right). \quad (10.25)$$

$$\text{Here, } \frac{\partial^2 V}{\partial u^2} = \omega_1 + \omega_2 + \omega_3 + \omega_4 \quad (10.26)$$

$$\frac{1}{2}\text{trace}\left[g^T(u(t))\right]\frac{\partial^2 V(t,u)}{\partial u^2} g(u(t))$$
$$= \frac{1}{2}\left[\omega_1 \sigma_1^2 u_1^2 + \omega_2 \sigma_2^2 u_2^2 + \omega_3 \sigma_3^2 u_3^2 + \omega_4 \sigma_4^2 u_4^2\right]. \quad (10.27)$$

Using Eqs. (10.25), (10.26), and (10.27)

$$\mathcal{L}V(t,u) = -\omega_1\left[\alpha + \mu - \frac{1}{2}\sigma_1^2\right]u_1^2 - \omega_2\left[\beta + \mu - \frac{1}{2}\sigma_2^2\right]u_2^2$$
$$- \omega_3\left[\gamma + \mu - \frac{1}{2}\sigma_3^2\right]u_3^2 - \omega_4\left[\delta + \mu - \frac{1}{2}\sigma_4^2\right]u_4^2,$$

which is negative definite, ensuring asymptotically mean square stability property.

10.5 NUMERICAL SIMULATION

With the help of MATLAB® ODE solver using Runge–Kutta method, the numerical simulation is carried out. The parameter values are evaluated between (0, 1). For the purpose of simulation, $b = 0.9$, $N = 10$, $\alpha = 0.95$, $\beta = 0.15$, $\gamma = 0.29$, $\delta = 0.234$, and $\mu = 0.1$ are considered.

Figure (10.2) portrays the stability of nitrogen mass cycle around the bounded equilibrium point.

Figure (10.3) describes the stability of nitrogen mass cycle for all initial density values.

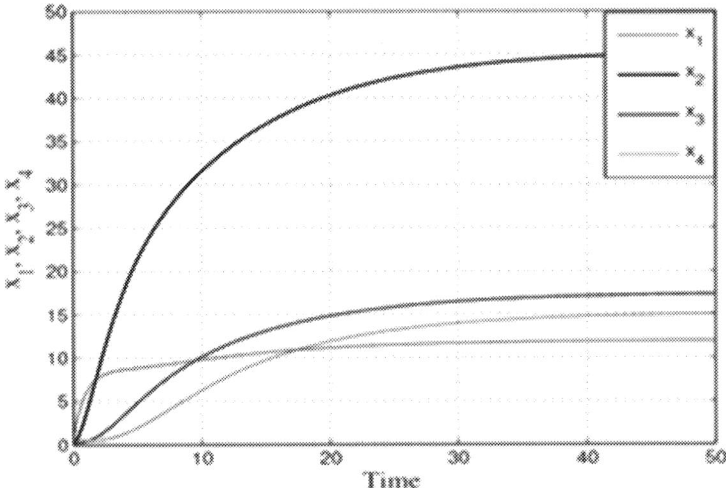

FIGURE 10.2 Stability of nitrogen mass cycle around the bounded equilibrium point.

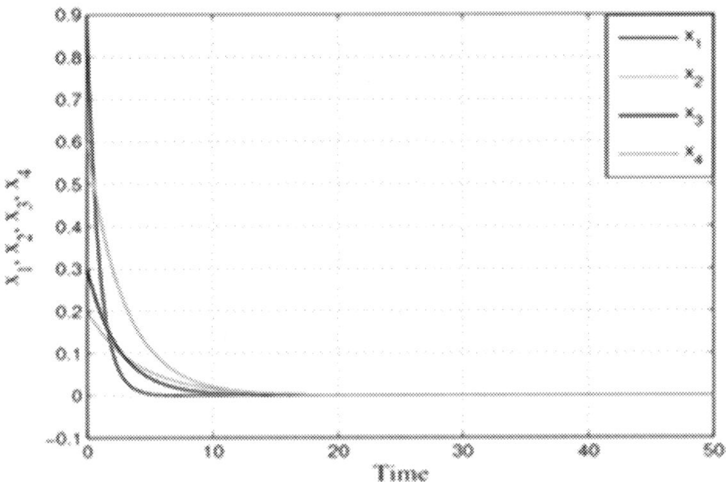

FIGURE 10.3 Stability of nitrogen mass cycle for all density values.

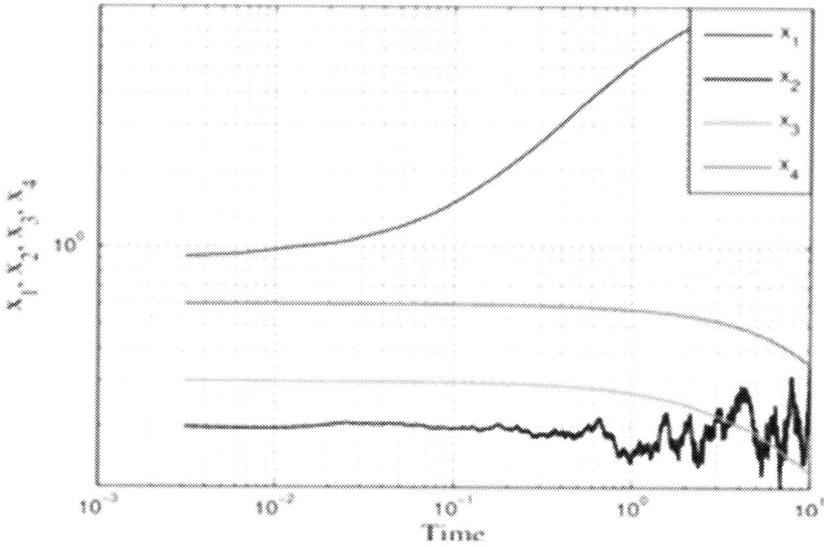

FIGURE 10.4 Stability of nitrogen mass cycle for nondeterministic model.

Figure (10.4) narrates the stability of nitrogen mass cycle for the nondeterministic model.

10.6 CONCLUSION

The mathematical model for nitrogen mass cycle has been framed. The boundary equilibrium point and interior equilibrium point have been identified. Local stability as well as global stability of the system has been analyzed. Asymptotic stability analysis using pseudo-back-propagation method has been carried out. In addition, the stability analysis of nondeterministic nitrogen mass cycle model has been presented. Numerical simulation to support the study is also given. This confirms the balancing of nitrogen content mathematically, which in turn helps to overcome many health issues.

REFERENCES

1. Cabello P., Roldan M.D. and Moreno-Vivian C., (2004), "Nitrate reduction and the nitrogen cycle in archaea", *Microbiology*, Vol. 150, pp. 3527–3546.
2. Cheng Q., (2008), "Perspectives in biological nitrogen fixation research", *Journal of Integrative Plant Biology*, Vol. 50, pp. 784–796.
3. Cameron D.R. and Kowalenko C.G., (1975), "Modelling nitrogen processes in soil: Mathematical development and relationship", *Canadian Journal of Soil Science*, Vol. 56, pp. 71–78.
4. Djaman K., Bado B.V. and Mel V.C., (2016), "Effect of nitrogen fertilizer on yield and nitrogen use efficiency of four aromatic rice varieties", *Emirates Journal of Food and Agriculture*, Vol. 28, pp. 126–135.

5. Dwivedi D., Steefel C.I., Arora B. and Bisht G., (2017), "Impact of intra-meander hyporheic flow on nitrogen cycling", *Procedia Earth and Planetary Science*, Vol. 17, pp. 404–407.
6. Espinosa R.M.M., Colet J.A., Richardson D.J. and Watmough N.J. (2011), "Enzymology and ecology of nitrogen cycle", *Biochemical Society Transactions*, Vol. 39, pp. 175–178.
7. Filoso S.L., Martenll L.A., Howarth R.W., Boyer E.W. and Dentener F. (2006), "Human activities changing the nitrogen cycle in Brazil", *Biogeochemistry*, Vol. 79, pp. 61–89.
8. Frak M., Kardel I. and Jankiewicz U., (2012), "Occurrence of nitrogen cycle bacteria in the Biebrza River", *Annals of Warsaw University of Life Sciences*, Vol. 44, pp. 55–62.
9. Galloway J.N., Dentener F.J., Capone D.G., Boyer E.W., Howarth R.W., Seitzinger S.P., Asner G.P., Cleveland C.C., Green P.A., Holland E.A., Karl D.M., Michaels A.F., Porter J.H., Townsend A.R. and Voro Smarty C.J. (2004), "Nitrogen cycles: Past, present, and future", *Biogeochemistry*, Vol. 70, pp. 153–126.
10. Gilliam F.S., Billmyer J.H., Walter C.A. and Peterjohn W.T., (2016), "Effects of excess nitrogen on biogeochemistry of a temperate hardwood forest: Evidence of nutrient redistribution by a forest understory species", *Atmospheric Environment*, Vol. 146, pp. 261–270.
11. Halbleib C.M. and Ludden P.W., (2000), "Regulation of biological nitrogen fixation", *Recent Advances in Nutritional Sciences*, Vol. 130, pp. 1081–1084.
12. Inomura K., Bragg J., Riemann L. and Follows M.J., (2018), "A quantitative model of nitrogen fixation in the presence of ammonium", *PLos One*, Vol. 13, pp. 1–16.
13. Khan I., Massod A. and Ahamed A., (2010), "Effect of nitrogen fixing bacteria on plant growth and yield of Brassica Juncea", *Journal of Phytology*, Vol. 2, pp. 25–27.
14. Kizilkaya R., (2009), "Nitrogen fixation capacity of Azotobacter spp. strains isolated from soils in different ecosystems and relationship between them and the microbiological properties of soils", *Journal of Environmental Biology*, Vol. 30, pp. 73–82.
15. McTigue N.D., Gardner W.S., Dunton K.H. and Hardison A.K., (2016), "Biotic and abiotic controls on co-occurring nitrogen cycling processes in shallow Arctic shelf sediments", *Nature Communications*, Vol. 7, pp. 1–11.
16. Mishra J., (2018), "Analysis of the Fitzhugh Nagumo model with a new numerical scheme", *Discrete and Continuous Dynamical Systems-S*, Vol. 13, pp. 781–795.
17. Mishra J., (2018), "Fractional hyper-chaotic model with no equilibrium," *Chaos, Solitons and Fractals*, Vol. 116, pp. 43–53.
18. Mishra J., (2019), "Modified Chua chaotic attractor with differential operators with non-singular kernels", *Chaos, Solitons and Fractals*, Vol. 125, pp. 64–72.
19. Mohammadi M., Sohrabi Y., Heidari G., Khalesro S. and Majidi M., (2012), "Effective factors on biological nitrogen fixation", *African Journal of Agricultural Research*, Vol. 7, pp. 1782–1788.
20. Moklyachuk L., Yatsuk I., Mokliachuk O. and Plasiuk L., (2016), "Mathematical modeling as a tool for determination of tendencies in changes of humus concentration in soil of arable lands", *Emirates Journal of Food and Agriculture*, Vol. 28, pp. 438–448.
21. Mukherjee D., (2003), "Stability analysis of a stochastic model for Prey-predator system with disease in the Prey", *Nonlinear Analysis: Modelling and Control*, Vol. 8, pp. 83–92.
22. Naresh K.J. and Suresh R., (2017), "Stability analysis of anopheles mosquito mathematical model with the unvarying controller", *International Journal of Pure and Applied Mathematics*, Vol. 116, pp. 431–444.
23. Naresh K.J. and Suresh R. (2017), "Stabilization and complexities of anopheles mosquito dynamics with stochastic perturbations", *International Journal of Applied Mathematics*, Vol. 47, pp. 307–311.

24. Onor I.O., Onor Junior G.I. and Kambhampati M.S., (2014), "Ecophysiological effects of nitrogen on Soyabean [Glycine max. (L.) Merr.]", *Open Journal of Soil Science*, Vol. 4, pp. 357–365.
25. Pinder R.W., Bettez N.D., Bonan G.B., Greaver T.L., Wieder W.R., Schlesinger W.H. and Davidson E.A., (2012), "Impacts of human alteration of the nitrogen cycle in the US on radiative forcing", *Biogeochemistry*, Vol. 114, pp. 25–40.
26. Proporato A., Odorico P.D., Laio F. and Rodriguez-Iturbe I. (2003), "Hydrolic controls on soil carbon and nitrogen cycles. I Modeling scheme", *Advances in Water Resources*, Vol. 26, pp. 45–58.
27. Roper M.M. and Gupta V.V.S.R., (2016), "Enhancing non-symbiotic N_2 fixation in agriculture", *The Open Agriculture Journal*, Vol. 10, pp. 7–27.
28. Sarkar R.R. (2004), "A stochastic model for autotroph-herbivore system with nutrient recycling", *Ecological Modelling*, Vol. 178, pp. 429–440.
29. Usta A., Yilmaz M. and Altun L., (2016), "Effect of Alder on nitrogen transport to surface waters and cation losses in natural ecosystems in Simsrili Watershed", *Journal of Bartin Faculty of Forestry*, Vol. 18, pp. 13–22.
30. Vitousek P.M., Aber J.D., Howarth R.W., Likens G.E., Matson P.A., Schindler D.W., Schlesinger W.H. and Tilman D.G., (1977), "Human alteration of the global nitrogen cycle: Sources and consequences", *Ecological Applications*, Vol. 7, pp. 737–750.
31. Watanabe M., Ortega E., Bergier I. and Silva J.S.V., (2012), "Nitrogen cycle and ecosystem services in the Brazilian La Plata Basin: Anthropogenic influence and climate change", *Brazilian Journal of Biology*, Vol. 72, pp. 691–708.

11 Neutralizing of Nitrogen when the Changes of Nitrogen Content Is Rapid

Suresh Rasappan and Kala Raja Mohan
Vel Tech Rangarajan Dr. Sagunthala R&D
Institute of Science and Technology

CONTENTS

11.1 INTRODUCTION

Nitrogen, a fundamental supplementary, is significant for the survival of each and every creature [18,25]. It is the essential part of proteins. Nitrogen is found boundlessly in the air. Still it can't be utilized straightforwardly by living beings. It has to be changed over into a couple of different structures to fit for consumption and usable by organisms [2,24]. It also finds its applications in several manufacturing industries [4,8,28].

The framework of the nitrogen mass cycle exponential growth model is accomplished [6,10,23]. The dynamical reaction of the model is determined using the point of interior equilibrium [1,3]. Lyapunov function is adopted to conduct the analysis of stability for the model [20]. In addition, qualitative properties of the model are also identified.

11.2 DESCRIPTION OF THE MATHEMATICAL MODEL

Nitrogen fixation, nitrification, and denitrification are the processes involved in nitrogen mass cycle [21,26]. Bacteria such as rhizobium, nitrobacter, pseudomonas, and paracoccus are involved in these processes. These bacteria convert nitrogen into several other forms, depending on the factors necessary during the processes [13,22]. The transition procedure may cause enormous growth at each state, which can be related to the exponential growth [11,16].

The formulation of the exponential growth mathematical model is expressed as follows:

$$\begin{aligned}\frac{dx_1}{dt} &= \omega_1 x_1^2 + \delta x_4 - \alpha x_1 \\ \frac{dx_2}{dt} &= \omega_2 x_2^2 + \alpha x_1 - \beta x_2 \\ \frac{dx_3}{dt} &= \omega_3 x_3^2 + \beta x_2 - \gamma x_3 \\ \frac{dx_4}{dt} &= \omega_4 x_4^2 + \gamma x_3 - \delta x_4,\end{aligned} \tag{11.1}$$

where x_1 is the total amount of nitrogen in the atmosphere, x_2 is the total amount of ammonia, x_3 is the total amount of nitrate, x_4 is the total reduction of nitrite, α is the conversion agent of ammonia, β is the conversion agent of nitrite from ammonification, γ is the conversion agent of nitrite from ammonification, and δ is the conversion agent of denitrification from nitrite to nitrogen gas. The inceptive amount at each state x_i, $i = 1,2,3,4$ is represented by ω_i, $i = 1,2,3,4$, respectively.

Figure (11.1) represents the flow diagram of nitrogen mass cycle with the exponential growth.

11.3 MATHEMATICAL PROPERTIES OF THE DETERMINISTIC MODEL

This section discusses the qualitative properties of the exponential growth model of nitrogen mass cycle.

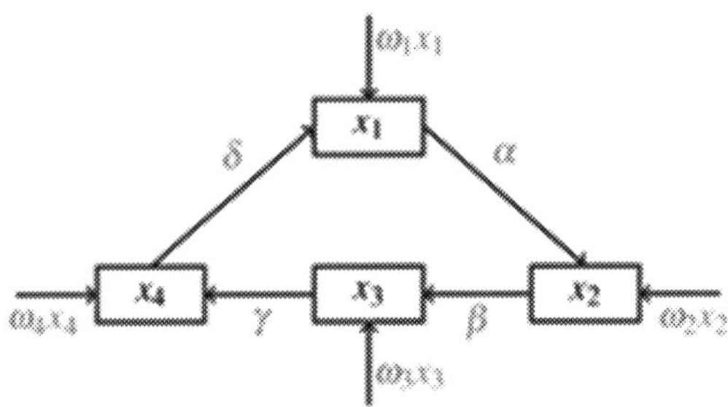

FIGURE 11.1 Flow diagram of nitrogen mass cycle with exponential growth.

11.3.1 Boundedness of the Nitrogen Mass Cycle with Exponential Growth

The boundedness of the solutions of Eq. (11.1) is discussed in this part.

$$\text{Define } W = x_1 + x_2 + x_3 + x_4. \tag{11.2}$$

Differentiate Eq. (11.2) along with Eq. (11.1)

$$\frac{dW}{dt} = w_1 x_1^2 + w_2 x_2^2 + w_3 x_3^2 + w_4 x_4^2.$$

For each N, the inequality holds that

$$\frac{dW}{dt} + NW = \left(N + w_1 x_1 - \eta\right)x_1 + \left(N + w_2 x_2 - \eta\right)x_2$$
$$+\left(N + w_3 x_3 - \eta\right)x_3 + \left(N + w_4 x_4 - \eta\right)x_4 + \eta W. \tag{11.3}$$

Assume $N = \max\{w_1, w_2, w_3, w_4\}$.

Then, Eq. (11.3) implies

$$\frac{dW}{dt} + NW \leq \eta W = L.$$

Consider the constant L such that

$$\frac{dW}{dt} + NW = L.$$

From the theorem of differential inequality

$$0 \leq W\left(x_1, x_2, x_3, x_4\right)$$
$$\leq \frac{L}{N}\left(1 - e^{-Nt}\right) + W\left(x_1(0), \ldots, x_4(0)\right)e^{-Nt}$$

and when $t \to \infty$,

$$0 \leq W \leq \frac{L}{N}.$$

Hence, all the solutions of Eq. (11.1) that commence in $\left\{R_+^4 \backslash 0\right\}$ are bounded in the region

$$A = \left\{\left(x_1, x_2, x_3, x_4\right) \in R_+^4 : W = \frac{L}{N} + \epsilon \text{ for any } \epsilon > 0\right\}.$$

11.3.2 Local Stability Analysis

The examination of the local asymptotic stability at different steady states of the system is done [9,15]. The boundary equilibrium point is $E_1(0,0,0,0)$ and the planar equilibrium point is

$$E_2\left(\frac{\alpha}{w_1},\frac{\beta}{w_2},\frac{\gamma}{w_3},\frac{\delta}{w_4}\right).$$

The point of interior equilibrium is given by

$$E^* = \left(x_1^*, x_2^*, x_3^*, x_4^*\right)$$

$$x_1^* = \frac{\alpha + \sqrt{\alpha^2 - 4\omega_1\delta x_4}}{2\omega_1}$$

$$x_2^* = \frac{\beta + \sqrt{\beta^2 - 4\omega_2\alpha x_1}}{2\omega_2}$$

$$x_3^* = \frac{\gamma + \sqrt{\gamma^2 - 4\omega_3\beta x_2}}{2\omega_3}$$

$$x_4^* = \frac{\delta + \sqrt{\delta^2 - 4\omega_4\gamma x_3}}{2\omega_4}.$$

The Jacobian matrix of the system of Eq. (11.1) is

$$J = \begin{bmatrix} -\alpha + 2w_1x_1 & 0 & 0 & \delta \\ \alpha & -\beta + 2w_2x_2 & 0 & 0 \\ 0 & \beta & -\gamma + 2w_3x_3 & 0 \\ 0 & 0 & \gamma & -\delta + 2w_4x_4 \end{bmatrix} \tag{11.4}$$

The eigen esteem of the Jacobian matrix at the equilibrium point E_1 is obtained by substituting into Eq. (11.4):

$$J(0,0,0,0) = \begin{bmatrix} -\alpha & 0 & 0 & \delta \\ \alpha & -\beta & 0 & 0 \\ 0 & \beta & -\gamma & 0 \\ 0 & 0 & \gamma & -\delta \end{bmatrix} \tag{11.5}$$

The characteristic equation of Eq. (11.5) is

$$\Delta_1\lambda^4 + \Delta_2\lambda^3 + \Delta_3\lambda^2 + \Delta_4\lambda + \Delta_5 = 0, \tag{11.6}$$

where

$$\Delta_1 = 1$$

$$\Delta_2 = \alpha + \gamma + \beta + \delta$$

$$\Delta_3 = \alpha(\gamma + \beta + \delta) + \beta(\gamma + \delta) + \gamma\delta.$$

$$\Delta_4 = \alpha\beta(\gamma + \delta) + (\alpha + \beta)\gamma\delta$$

$$\Delta_5 = 0$$

The discriminant of Eq. (11.6) is

$$\Delta = 18\Delta_1\Delta_2\Delta_3\Delta_4 - 4\Delta_2^3\Delta_4 + \Delta_2^2\Delta_3^2 - 4\Delta_1\Delta_3^2 - 27\Delta_1^2\Delta_4^2.$$

Here, one of the eigen values is zero. Hence, it is non-hyperbolic equilibrium point, which implies that the nitrogen mass cycle with exponential growth is unstable.

The nitrogen mass cycle with exponential growth has one zero eigen value, which implies that the equilibrium point $E_1(0,0,0,0)$ is the saddle-node equilibrium point.

11.3.3 Bifurcation

A bifurcation of the nitrogen cycle is a qualitative change in its dynamics produced by varying parameters [14,27]. Depending on the parameter values, there exist the following bifurcation types:

- One of the eigen values is that the characteristic Eq. (11.6) is zero and the discriminant equation is $\Delta > 0$, which has three distinct real roots. Hence, it has *cusp bifurcation.*
- One of the eigen values is that the characteristic Eq. (11.6) is zero and the discriminant equation is $\Delta = 0$, which has numerous real roots. Hence, it has *cusp bifurcation.*
- One of the eigen values is that the characteristic Eq. (11.6) is zero and the discriminant equation is $\Delta < 0$, in which one of the root is real and other two are nonreal complex conjugates. Hence, it has *fold-Hopf bifurcation.*

The eigen values of the Jacobian matrix at the equilibrium point E_2 are obtained by substituting into Eq. (11.4)

$$J\left(\frac{\alpha}{w_1}, \frac{\beta}{w_2}, \frac{\gamma}{w_3}, \frac{\delta}{w_4}\right) = \begin{bmatrix} \alpha & 0 & 0 & \delta \\ \alpha & \beta & 0 & 0 \\ 0 & \beta & \gamma & 0 \\ 0 & 0 & \gamma & \delta \end{bmatrix}. \tag{11.7}$$

The characteristic Eq. (11.7) is

$$\Delta_1\lambda^4 - \Delta_2\lambda^3 + \Delta_3\lambda^2 - \Delta_4\lambda + \Delta_5 = 0, \tag{11.8}$$

where

$$\Delta_1 = 1$$

$$\Delta_2 = \alpha + \beta + \gamma + \delta$$

$$\Delta_3 = \alpha(\beta + \gamma + \delta) + \beta(\gamma + \delta) + \gamma\delta.$$

$$\Delta_4 = \alpha\beta(\gamma + \delta) + (\alpha + \beta)\gamma\delta$$

$$\Delta_5 = 0$$

The discriminants of Eqs. (11.8) and (11.6) are the same.

Hence, the qualitative properties of the planner equilibrium point $E_2\left(\frac{\alpha}{w_1},\frac{\beta}{w_2},\frac{\gamma}{w_3},\frac{\delta}{w_4}\right)$ and $E_1(0,0,0,0)$ are the same.

Theorem 11.1

The interior equilibrium point E^* is locally asymptotically stable in the positive octant.

Proof:
By divergence criterion theorem, assume

$$\theta(x_1,\ldots,x_4) = \frac{1}{x_1x_2x_3x_4}, \tag{11.9}$$

where $\theta(x_1,\ldots,x_4) > 0$ if $x_1 > 0,\ldots,x_4 > 0$.

Now, consider

$$\begin{aligned} p_1 &= w_1x_1^2 + \delta x_4 - \alpha x_1 \\ p_2 &= w_2x_2^2 + \alpha x_1 - \beta x_2 \\ p_3 &= w_3x_3^2 + \beta x_2 - \gamma x_3 \\ p_4 &= w_4x_4^2 + \gamma x_3 - \delta x_4. \end{aligned} \tag{11.10}$$

Define $\Delta(x_1,x_2,x_3,x_4) = \frac{\partial}{\partial x_1}(p_1\theta) + \frac{\partial}{\partial x_2}(p_2\theta) + \frac{\partial}{\partial x_3}(p_3\theta) + \frac{\partial}{\partial x_4}(p_4\theta).$

Find Δ along with the trajectories of Eqs. (11.9) and (11.10), which gives

$$\Delta = \frac{w_1x_1 + w_2x_2 + w_3x_3 + w_4x_4}{x_1x_2x_3x_4} - \frac{\delta_2x_2x_4 + \alpha x_1^2}{x_1^2x_2^2x_3x_4} - \frac{\beta x_2x_4 + \gamma x_3^2}{x_1x_2x_3^2x_4^2},$$

which is less than zero.

Bendixson–Dulac criterion [7,12] confirms that there is limit cycle which does not exist in the first octant.

Thus, the interior equilibrium point E^* has locally asymptotically stable.

11.3.4 Global Stability Analysis

Theorem 11.2

Global stability of the interior equilibrium point exists when

$$\alpha = \frac{-x_2\left(x_2 - x_2^*\right) - \omega_2x_2^2 + \beta x_2}{x_1}; \beta = \frac{-x_3\left(x_3 - x_3^*\right) - \omega_3x_3^2 + \gamma x_3}{x_2};$$
$$\gamma = \frac{-x_4\left(x_4 - x_4^*\right) - \omega_4x_4^2 + \delta x_4}{x_3}; \delta = \frac{-x_1\left(x_1 - x_1^*\right) - \omega_1x_1^2 + \alpha x_1}{x_4}. \quad (11.11)$$

Proof:
Assume the Lyapunov function

$$V(x_i) = \sum_{i=1}^{4} l_i\left[\left(x_i - x_i^*\right) - x_i^*\ln\left(\frac{x_i}{x_i^*}\right)\right] \quad (11.12)$$

Differentiate Eq. (11.12) along with the solution of Eq. (11.1)

$$\dot{V}_i = l_1\left(x_1 - x_1^*\right)\frac{\dot{x}_1}{x_1} + \cdots + l_4\left(x_4 - x_4^*\right)\frac{\dot{x}_4}{x_4} \quad (11.13)$$

Substitute Eq. (11.11) into Eq. (11.13), which implies that

$$\dot{V} = -l_1\left(x_1 - x_1^*\right)^2 - \cdots - l_4\left(x_4 - x_4^*\right)^2,$$

which is negative definite.

Hence, by *LaSalle's invariance principle*, globally asymptotic stability of E^* is established [19].

11.3.5 Global Stability Analysis of Nitrogen Mass Cycle with Exponential Growth by Pseudo-Back-Propagation

For understanding the nature of the nitrogen cycle [5,17], the global stability of the cycle is needed to be determined. For this purpose, the backward pseudo-back-propagation is used. It is one way of feedback mechanism in this life cycle.

Consider the exponential growth mathematical model of the nitrogen mass cycle.

$$\begin{aligned}\frac{dx_4}{dt} &= \omega_4 x_4^2 + \gamma x_3 - \delta x_4 + u_1\\ \frac{dx_3}{dt} &= \omega_3 x_3^2 + \beta x_2 - \gamma x_3 + u_2\\ \frac{dx_2}{dt} &= \omega_2 x_2^2 + \alpha x_1 - \beta x_2 + u_3\\ \frac{dx_1}{dt} &= \omega_1 x_1^2 + \delta x_4 - \alpha x_1 + u_4,\end{aligned} \tag{11.14}$$

where u_1, u_2, u_3 and u_4 are the feedback controllers.

First consider the stability of nitrite

$$\dot{x}_4 = \omega_4 x_4^2 + \gamma x_3 - \delta x_4 + u_1,$$

where x_3 is observed as pseudo-controller.

$$\text{Define } V_1 = \frac{1}{2}x_4^2. \tag{11.15}$$

The derivative of V_1 from Eq. (11.15) is obtained as

$$\dot{V}_1 = x_4\dot{x}_4 = x_4\left(w_4 x_4^2 + \gamma x_3 - \delta x_4 + u_1\right).$$

Assume the pseudo-controller $x_3 = p_1(x_4)$.

Choose $p_1(x_4) = 0$ and $u_1 = -w_4 x_4^2$.

Then, $\dot{V}_1 = -\delta x_4^2$,

which is negative definite.

Equation (11.14) is globally asymptotically stable.

The relation between nitrite and nitrate is defined by

$$\Delta_2 = x_3 - p_1 = x_3.$$

Consider the (x_4, Δ_2) subcycle, given by

$$\begin{aligned}\dot{x}_4 &= \gamma\Delta_2 - \delta x_4\\ \dot{\Delta}_2 &= w_3\Delta_2^2 + \beta x_2 - \gamma\Delta_2 + u_2.\end{aligned} \tag{11.16}$$

Let x_2 be a pseudo-controller in the subcycle (11.16). Assume that when $x_2 = p_2$, the subcycle Eq. (11.16) is globally asymptotically stable.

The Lyapunov function is presumed as

$$V_2 = V_1(x_4) + \frac{1}{2}\Delta_2^2. \tag{11.17}$$

The derivative of V_2 from Eq. (11.17) is

$$\dot{V}_2 = -\delta x_4^2 + \Delta_2(\gamma x_4 + w_3\Delta_2^2 + \beta x_2 - \gamma\Delta_2 + u_2) \tag{11.18}$$

Assume the pseudo-controller $x_2 = p_2(x_4, \Delta_2)$ and choose

$$p_2(x_4, \Delta_2) = 0 \text{ and } u_2 = -\gamma x_4 - w_3. \tag{11.19}$$

Substituting Eq. (11.19) into Eq. (11.18), and simplifying, it follows that

$$\dot{V}_2 = -\delta x_4^2 - \gamma\Delta_2^2.$$

Thus, $\dot{V}_2$ is negative definite.

The existence of globally asymptotically stable property of Eq. (11.14) is evident.

The function $p_2(x_4, \Delta_2)$ is estimative.

The relation between x_2 and p_2 is

$$\Delta_3 = x_2 - p_2 = x_2.$$

Consider that $(x_4, \Delta_2, \Delta_3)$ subcycle is given by

$$\begin{aligned} \dot{x}_4 &= \gamma\Delta_2 - \delta x_4 \\ \dot{\Delta}_2 &= \beta\Delta_3 - \gamma\Delta_2 - \gamma x_4 \\ \dot{\Delta}_3 &= w_2\Delta_2^2 + \alpha x_1 - \beta\Delta_3 + u_3 \end{aligned}$$

Consider the Lyapunov function

$$V_3 = V_2(x_4, \Delta_2) + \frac{1}{2}\Delta_3^2 \tag{11.20}$$

The derivative of V_3 from Eq. (11.20) is

$$\dot{V}_2 = -\delta x_4^2 - \gamma\Delta_2^2 + \Delta_3(\beta\Delta_2 + w_2\Delta_2^2 + \alpha x_1 - \beta\Delta_3 + u_3). \tag{11.21}$$

Assume the pseudo-controller $x_1 = p_3(x_4, \Delta_2, \Delta_3)$ and choose

$$p_3(x_4, \Delta_2, \Delta_3) = 0 \quad \text{and} \quad u_3 = -\beta\Delta_2 - w_2\Delta_2^2. \tag{11.22}$$

Substituting Eq. (11.22) into Eq. (11.21), and simplifying, then it follows that

$$\dot{V}_3 = -\delta x_4^2 - \gamma\Delta_2^2 - \beta\Delta_3^2.$$

Thus, $\dot{V}_3$ is negative definite proving the globally asymptotic stability of Eq. (11.14).

The relation between x_1 and p_3 is $\Delta_4 = x_1 - p_3 = x_1$.

Consider that $(x_4, \Delta_2, \Delta_3, \Delta_4)$ subcycle is given by

$$\dot{x}_4 = \gamma\Delta_2 - \delta x_4$$

$$\dot{\Delta}_2 = \beta\Delta_3 - \gamma\Delta_2 - \gamma x_4$$

$$\dot{\Delta}_3 = \alpha\Delta_4 - \beta\Delta_3 - \beta\Delta_2$$

$$\dot{\Delta}_4 = w_1\Delta_4^2 + \delta x_4 - \alpha\Delta_4 + u_4.$$

Consider the Lyapunov function

$$V_4 = V_3(x_4, \Delta_2, \Delta_3) + \frac{1}{2}\Delta_4^2 \tag{11.23}$$

The derivative of V_4 from Eq. (11.23) is

$$\dot{V}_4 = -\delta x_4^2 - \gamma\Delta_2^2 - \beta\Delta_3^2 + \Delta_4\left(\alpha\Delta_3 + w_1\Delta_4^2 + \delta x_4 - \alpha\Delta_4 + u_4\right). \tag{11.24}$$

Now choose the back-propagation feedback control

$$u_4 = -\alpha\Delta_3 - w_1\Delta_4^2 - \delta. \tag{11.25}$$

Substituting Eq. (11.25) into Eq. (11.24), and simplifying, then it follows that

$$\dot{V}_4 = -\delta x_4^2 - \gamma\Delta_2^2 - \beta\Delta_3^2 - \alpha\Delta_4^2.$$

$\dot{V}_4$ is negative definite.

From *Lyapunov stability theory*, globally asymptotic stability of Eq. (11.14) for all initial densities for the nitrogen mass cycle with exponential growth is proved.

11.4 NUMERICAL SIMULATION

Runge–Kutta method is applied to bring out numerical simulation using MATLAB® ODE solver. Figure (11.2) portrays the stability of nitrogen mass cycle with exponential growth. Figure (11.3) describes the stability of the nitrogen mass cycle around the equilibrium point.

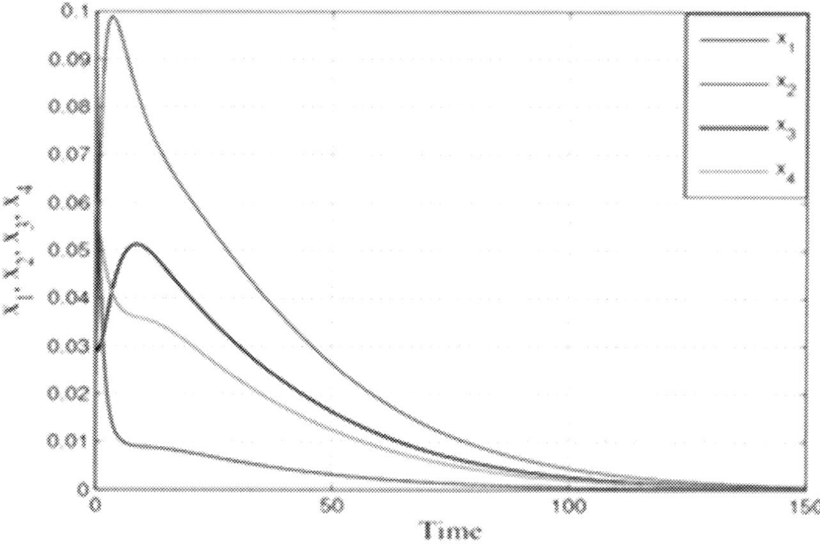

FIGURE 11.2 Stability of nitrogen mass cycle with exponential growth.

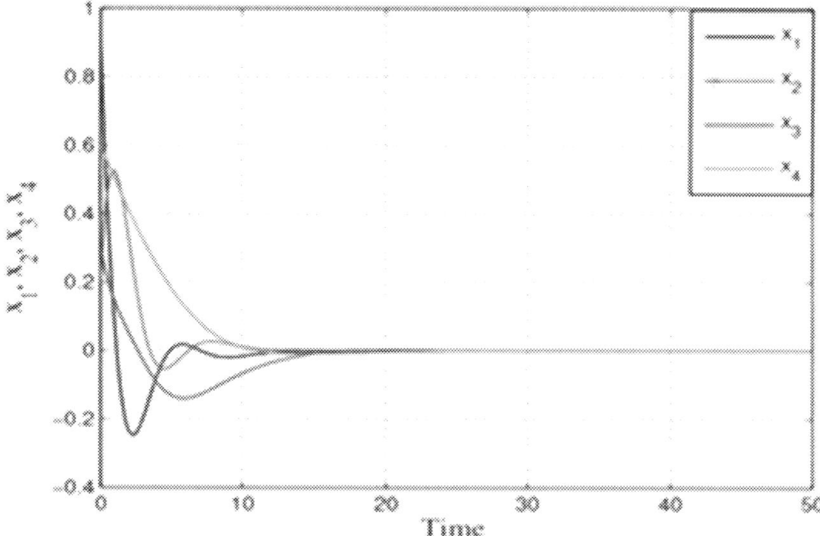

FIGURE 11.3 Stability of nitrogen mass cycle with exponential growth around the equilibrium point.

11.5 CONCLUSION

Description of the exponential growth nitrogen mass cycle mathematical model has been done. Various equilibrium points such as boundary, planner, and the interior equilibrium have been identified. The qualitative properties such as bifurcation, limit cycle, and local asymptotic stability are analyzed. The application of Lyapunov function investigates the global stability of the system. By back-propagation feedback method, the global stability is proved for all initial densities of the model. Numerical simulation in support of the analysis has also been presented. This concludes the neutralization of nitrogen content mathematically.

REFERENCES

1. Cabello P., Roldan M.D. and Moreno-Vivian C., (2004), "Nitrate reduction and the nitrogen cycle in Archaea", *Microbiology*, Vol. 150, pp. 3527–3546.
2. Cheng Q., (2008), "Perspectives in biological nitrogen fixation research", *Journal of Integrative Plant Biology*, Vol. 50, pp. 784–796.
3. Cameron D.R. and Kowalenko C.G., (1975), "Modelling nitrogen processes in soil: Mathematical development and relationship", *Canadian Journal of Soil Science*, Vol. 56, pp. 71–78.
4. Djaman K., Bado B.V. and Mel V.C., (2016), "Effect of nitrogen fertilizer on yield and nitrogen use efficiency of four aromatic rice varieties", *Emirates Journal of Food and Agriculture*, Vol. 28, pp. 126–135.
5. Dwivedi D., Steefel C.I., Arora B. and Bisht G., (2017), "Impact of intra-meander hyporheic flow on nitrogen cycling", *Procedia Earth and Planetary Science*, Vol. 17, pp. 404–407.
6. Espinosa R.M.M., Colet J.A., Richardson D.J. and Watmough N.J., (2011), "Enzymology and ecology of nitrogen cycle", *Biochemical Society Transactions*, Vol. 39, pp. 175–178.
7. Filoso S.L., Martenll L.A., Howarth R.W., Boyer E.W. and Dentener F., (2006), "Human activities changing the nitrogen cycle in Brazil", *Biogeochemistry*, Vol. 79, pp. 61–89.
8. Frak M., Kardel I. and Jankiewicz U., (2012), "Occurrence of nitrogen cycle bacteria in the Biebrza River", *Annals of Warsaw University of Life Sciences*, Vol. 44, pp. 55–62.
9. Galloway J.N., Dentener F.J., Capone D.G., Boyer E.W., Howarth R.W., Seitzinger S.P., Asner G.P., Cleveland C.C., Green P.A., Holland E.A., Karl D.M., Michaels A.F., Porter J.H., Townsend A.R. and Voro Smarty C.J., (2004), "Nitrogen cycles: Past, present, and future", *Biogeochemistry*, Vol. 70, pp. 153–126.
10. Gilliam F.S., Billmyer J.H., Walter C.A. and Peterjohn W.T., (2016), "Effects of excess nitrogen on biogeochemistry of a temperate hardwood forest: Evidence of nutrient redistribution by a forest understory species", *Atmospheric Environment*, Vol. 146, pp. 261–270.
11. Halbleib C.M. and Ludden P.W., (2000), "Regulation of biological nitrogen fixation", *Recent Advances in Nutritional Sciences*, Vol. 130, pp. 1081–1084.
12. Inomura K., Bragg J., Riemann L. and Follows M.J., (2018), "A quantitative model of nitrogen fixation in the presence of ammonium", *PLos One*, Vol. 13, pp. 1–16.
13. Khan I., Massod A. and Ahamed A., (2010), "Effect of nitrogen fixing bacteria on plant growth and yield of Brassica Juncea", *Journal of Phytology*, Vol. 2, pp. 25–27.
14. Kizilkaya R., (2009), "Nitrogen fixation capacity of Azotobacter spp. strains isolated from soils in different ecosystems and relationship between them and the microbiological properties of soils", *Journal of Environmental Biology*, Vol. 30, pp. 73–82.

15. McTigue N.D., Gardner W.S., Dunton K.H. and Hardison A.K., (2016), "Biotic and abiotic controls on co-occurring nitrogen cycling processes in shallow Arctic shelf sediments", *Nature Communications*, Vol. 7, pp. 1–11.
16. Mohammadi M., Sohrabi Y., Heidari G., Khalesro S. and Majidi M. (2012), "Effective factors on biological nitrogen fixation", *African Journal of Agricultural Research*, Vol. 7, pp. 1782–1788.
17. Moklyachuk L., Yatsuk I., Mokliachuk O. and Plasiuk L., (2016), "Mathematical modeling as a tool for determination of tendencies in changes of humus concentration in soil of arable lands", *Emirates Journal of Food and Agriculture*, Vol. 28, pp. 438–448.
18. Mukherjee D. (2003), "Stability analysis of a stochastic model for prey-predator system with disease in the Prey", *Nonlinear Analysis: Modelling and Control*, Vol. 8, pp. 83–92.
19. Naresh K.J. and Suresh R., (2017), "Stability analysis of anopheles mosquito mathematical model with the unvarying controller", *International Journal of Pure and Applied Mathematics*, Vol. 116, pp. 431–444.
20. Naresh K.J. and Suresh R. (2017), "Stabilization and complexities of anopheles mosquito dynamics with stochastic perturbations", *International Journal of Applied Mathematics*, Vol. 47, pp. 307–311.
21. Onor I.O., Onor Junior G.I. and Kambhampati M.S., (2014), "Ecophysiological effects of nitrogen on Soyabean [Glycine max. (L.) Merr.]", *Open Journal of Soil Science*, Vol. 4, pp. 357–365.
22. Pinder R.W., Bettez N.D., Bonan G.B., Greaver T.L., Wieder W.R., Schlesinger W.H. and Davidson E.A., (2012), "Impacts of human alteration of the nitrogen cycle in the US on radiative forcing", *Biogeochemistry*, Vol. 114, pp. 25–40.
23. Proporato A., Odorico P.D., Laio F., and Rodriguez-Iturbe I., (2003), "Hydrolic Controls on soil carbon and nitrogen cycles. I Modeling scheme", *Advances in Water Resources*, Vol. 26, pp. 45–58.
24. Roper M.M. and Gupta V.V.S.R., (2016), "Enhancing non-symbiotic N_2 fixation in agriculture", *The Open Agriculture Journal*, Vol. 10, pp. 7–27.
25. Sarkar R.R., (2004), "A stochastic model for autotroph-herbivore system with nutrient recycling", *Ecological Modelling*, Vol. 178, pp. 429–440.
26. Usta A., Yilmaz M. and Altun L., (2016), "Effect of Alder on nitrogen transport to surface waters and cation losses in natural ecosystems in Simsrili Watershed", *Journal of Bartin Faculty of Forestry*, Vol. 18, pp. 13–22.
27. Vitousek P.M., Aber J.D., Howarth R.W., Likens G.E., Matson P.A., Schindler D.W., Schlesinger W.H. and Tilman D.G., (1977), "Human alteration of the global nitrogen cycle: Sources and consequences", *Ecological Applications*, Vol. 7, pp. 737–750.
28. Watanabe M., Ortega E., Bergier I. and Silva J.S.V., (2012), "Nitrogen cycle and ecosystem services in the Brazilian La Plata Basin: Anthropogenic influence and climate change", *Brazilian Journal of Biology*, Vol. 72, pp. 691–708.

12 Application of Blockchain Technology in Hospital Information System

Deepa Elangovan
KPJ Healthcare University College

Chiau Soon Long
Quest International University Perak

Faizah Safina Bakrin and Ching Siang Tan
KPJ Healthcare University College

Khang Wen Goh
Quest International University Perak

Zahid Hussain
University of Canberra

Yaser Mohammed Al-Worafi
University of Science and Technology
University of Science and Technology of Fujairah

Kah Seng Lee and Yaman Walid Kassab
University of Cyberjaya

Long Chiau Ming
Universiti Brunei Darussalam

CONTENTS

12.1 INTRODUCTION

Nowadays, blockchain technology is increasing quickly and penetrating every aspect of information and communication technology [1]. According to Kamel Boulos, Wilson [2], and Engelhardt [3], blockchain technology is defined as decentralized, distributed ledger, immutable, and cryptographically secure technology, which consists of a series of transactions list with identical copies shared and maintained by many group or parties.

Blockchain is a sort of dispersed ledger of cryptographically chained blocks where value exchange transactions are consecutively aggregated. Each block in blockchain is immutably recorded across peer-to-peer (P2P) network, and each of them is chained to the former block by using cryptographic assurance system [4]. According to Hölbl and Kompara [1], blockchain can also be defined as a time-stamped chain of blocks which are linked together by using the cryptographic hashes, where each new block holds a reference to the previous block content, and where the new block are being appended to the end of the chain and the chain is growing constantly.

Benchoufi and Ravaud [5] define blockchain technology as a distributed ledger where a set of nodes report transaction, which is executed in a P2P network in the absence of a trusted mediator. The blockchain principle can be imposed by firm cryptography, which is unmanageable to compromise by a single individual.

Blockchain is a firm, decentralized, abiding, and distributed ledger technology, which is secured cryptographically and is applicable to transfer information without the trusted mediator [2].

Meanwhile, Hospital Information System (HIS) is a comprehensive, homogenized information system intended to organize the financial, administrative, and clinical sectors of the hospital [6]. The objective of the initiation of HIS was to improve the quality of public healthcare services particularly in enhancing patient satisfactions and management of record [7].

This review makes a comparison of the functions of blockchain technology and HIS in the real healthcare environment. An overall understanding on the usability of blockchain technology as well as HIS in healthcare is analyzed. This study will help other researchers and practitioners to gain better understanding of blockchain technology and HIS in the healthcare sector.

12.2 HOSPITAL INFORMATION SYSTEM

The current HIS, which is mainly on cloud-based system, has several issues that influence the efficiency and effectiveness of HIS in hospital [6]. The first issue in HIS, which leads to a necessity of a new technology in the healthcare sector, would be the fragmented health data where the health data in a current medical system are separated and difficult to share with others because of varying format and standard [8].

A lack of adequate confidentiality and security measures in the current medical system has resulted in numerous breach of data and data validity, leaving patient exposed to economical treats and possible social stigma [9]. Centralized data stores and authority providers in HIS are the attractive targets for cyberattack, and establishing a consistent view of the patient record across a data-sharing network is problematic [8]. Lack of patient centricity has become the real threat in the current HIS since patient must have a control over their own medical data to improve the quality of life.

HIS is a health management in the hospital which incorporates the system of computer and improves the service efficiency, which concentrates on patient, ancillary, admission, financial, and clinical applications [7]. According to Goldzweig and Towfigh [10], HIS improves the assembly, communication, and management of data, which is essential for an efficient hospital management.

According to Haux [11], HIS is a process of depository, recovery, distributed healthcare data, and knowledge for conveyance and resolution, which uses hardware and software of the computer system. HIS is an instance of health information systems which provide information and knowledge in healthcare environments. According to Ismail and Abdullah [12], HIS is a nonsegregated electronic systems that gather, keep, recover, and exhibit the patient's health data which are used within the hospitals.

According to Ismail and Abdullah [13], HIS consists of at least two of the following components:

1. Clinical Information System (CIS),
2. Financial Information System (FIS),
3. Laboratory Information System (LIS),
4. Nursing Information System (NIS),
5. Pharmacy Information System (PIS),
6. Picture Archiving Communication System (PACS),
7. Radiology Information System (RIS).

12.3 TYPE OF HOSPITAL INFORMATION SYSTEM

Three main categories of HIS are Total Hospital Information System (THIS) for hospital with 400 beds, Intermediate Hospital Information System (IHIS) for hospital with 200–400 beds, and Basic Hospital Information System (BHIS) for hospital with <200 beds. Due to having the complete HIS components, THIS hospitals are known as paperless hospitals. IHIS and BHIS retain manual and electronic systems that use the hybrid system, where only several forms of information system are adopted [13]. The difficulty and opportunity of the hospital system are defined by the terms such as basic, intermediate, and total that introduce HIS [14]. The implementation of different types of HIS depends on the number of beds and hospital size.

12.4 PURPOSE OF HOSPITAL INFORMATION SYSTEM

HIS was executed to produce nonsegregated care delivery system. It is a good automated filing system, has the ability to share information for a greater storage in public hospital, and uses information for research purposes and medical statistics [11]. The THIS system is used for the appropriate meeting arrangement of patient, admission, registration, discharge, and transfer, and also to manage the clinical information of patient. THIS is used to order job, drug, tests, and supplies, and also to enter and view the result of patient's tests [15].

THIS is also used in the hospital for easy approach to data, enhancing safety and quality of life of patient for the management of illness. Enhanced security and record management, including workflow through the reengineering of work process, is also one of the aims of HIS [15]. Reduction of patient's stay time and patient's discharge processing time are also the purposes of the HIS implementation [15]. The quality and effectiveness of healthcare are identified by the data, information, and knowledge obtained from the HIS [11].

12.5 ADVANTAGES OF HOSPITAL INFORMATION SYSTEM

HIS provides the required data to the management at the right time, form, and place, so that the decision to be made is effective and efficient particularly in a critical situation where only small margin of errors is allowed [16]. HIS improves the patient care by appraising data accurately, producing recommendations for management, and permitting a hospital to develop from retrospective view to contemporaneous view, which affect the quality and care of patient's life. HIS plays an essential role in designing, beginning, arranging, and managing the process of the subsystem of the hospital and providing a synergistic organization in the operation [6].

The HIS also improves the health status, which supports the continuity of care and focuses on wellness and empowerment of the individuals' and community responsibility for health [17]. Reduction in transcription error and duplication of information entries are the benefits of the implementation of HIS in hospital. HIS implementation also leads to improved monitoring of drug usage [7]. The HIS is known to enhance the efficacy and productivity of patient services in healthcare organization. From the HIS implementation, a good relationship between patients and healthcare providers is enhanced, which creates a greater mutual understanding, trust, and satisfaction of patients. This relation will enhance an effective communication, which leads to a more efficient chronic disease care, and higher quality of physical health and life [18].

12.6 DISADVANTAGES OF HOSPITAL INFORMATION SYSTEM

Although HIS offers various advantages, several disadvantages of HIS need to be highlighted.

12.7 FRAGMENTED HEALTH DATA (AGGREGATION)

Healthcare data consist of not just the patient's medical record but also other financial and visiting record, which included a patient's insurance and private information. Sending information across institutional boundaries such as between two hospitals needs a shared understanding of data structures and interpretation. If data can be shared efficiently and securely, then the interoperability matters are left abandoned, which will limit the usage of the data [8].

Health data contained in the current medical system in Malaysia are separated and hard to share with others due to different standards and formats. Due to privacy concern and fear of giving others competitive advantages, institutions are often reluctant to share data. The fragmented current healthcare data landscape is unsuitable to the immediate needs of modern users. Stakeholders are incentivized to retain their own records, which acts as a challenge to confirm the originality and validity of the record. This results in high server maintenance cost and high security cost.

A consensus of the supporting technical architecture and infrastructure posed as a major obstacle for sharing data. Either a centralized data source or the transmission of bulk data to other institutions is required for many attempts at data sharing, where both centralized data sources and transmission of bulk data initiate unique problems. The security risk footprint increased, and the centralized trust in a single authority is required by a centralization process, while the institutions are forced to yield operational control of the data by bulk data transmission [8].

12.8 INSUFFICIENT FINANCIAL SOURCES

The HIS applications in hospital lead to higher cost, including the cost for the initiation, maintenance, and training of this system. The majority of the public hospitals, including private hospitals in Malaysia, have finite financial sources which are insufficient to upgrade the HIS system, whereas the financial sources from the government are limited, thus hindering the full capacity of the HIS [12].

12.9 MAINTENANCE BY DIFFERENT DEPARTMENTS

Maintaining HIS and training new HIS user are the responsibility of the information technology (IT) department in hospital. But the IHIS hospital outsources the control to maintain the system, which leads to dissimilarity of the department jobs that cope for maintaining the HIS system worldwide.

12.10 CONFIDENTIALITY ISSUES

The security level of the HIS is not strong enough, where not only the authorized user can access the information but also the nurses and doctors can access the patient's record with or without the patient's approval. Financial, legal, and patient care issues may arise if the patient's information is not secured properly; thus, securing the electronic record is a difficult job [19].

12.11 ACCEPTANCE LEVEL IS LOW

The healthcare provider, especially an older people, has a difficulty in adapting to a new technology in healthcare sector, where they feel HIS is time-wasting and complex, thus favoring the manual method of data recording [12].

12.12 TECHNICAL AND INFRASTRUCTURE ISSUES

The application of HIS will become a problem when the hospitals have a limited number of laptops or computers. Other issues such as affinity, accessibility, and instability of network may also occur, which will lead to a failure of HIS. For instance, computers using the newer Window 10 system cannot be used to implement HIS that was started in early 2000s [16].

12.13 SYSTEM BREAKDOWN

When dealing with the patient, the breakdown of the system will lead to a delay in the patient's care management [17].

12.14 HISTORY OF BLOCKCHAIN TECHNOLOGY

In 2008, Satoshi Nakamoto's (pseudonym) whitepaper stated that blockchain was originally developed from the Bitcoin cryptocurrency. Bitcoin is an open, distributed public register reporting all the Bitcoin contracts in a confirmable and secure way with the absence of mediator [20]. In 2009, Nakamoto proposed the issue of ownership related to Bitcoin (digital currencies) and provided a new resolution using a blockchain [21]. In Bitcoin transaction, all data are combined in a block with time-stamped and a new block is added to the existing Bitcoin blockchain. Then, all the involved nodes are verified and the transaction is confirmed [22].

According to Kuo and Kim [23], the market for blockchain technology is forecasted to grow rapidly in the future. Bitcoin is a pioneer digital coin, which assures a decentralized transaction in the absence of middle authority. The individual identity is not revealed as the public keys are utilized. Lately, the research community realized the potential of blockchain as a decentralized technology utilized in the healthcare sector, apart from the financial application [24]. Blockchain is known as a new technological rebellion that is rarely seen in the history of technology. World Economic Forum report stated that, by the year 2025, 10% of global gross domestic output will be gathered or stored using blockchain technology [22].

12.15 FUNDAMENTAL PROPERTIES OF BLOCKCHAIN TECHNOLOGY

A blockchain is normally defined as blocks joined together by using cryptographic secure hash function, which forms a chain that is time-stamped [25]. The more widespread adoption of this technology was observed after the proposal of the Bitcoin cryptocurrency [1]. The most important features or functions of blockchain technology are decentralized, distributed ledger, immutable, private, and agent of trust instead of centralized, hidden, exclusive, and alterable architecture [3]. Through blockchain technology, confidential and verified transactions can be carried out directly among third parties by mass collaboration. Blockchain is known as decentralized technology because no single source of force can request for an original information [3].

Initially, the blocks are locked in a secure and rigid manner, where a new block is being appended to the end of the continuously growing blockchain. A new block will hold a reference known as hash value, which is dependent on the previous block. Hash is created by cryptographic one-way process, which is used to link the blocks in the blockchain causing immutability, invisibility, and compactness of the block [1].

Each block in the blockchain consists of a definite record of information, which includes when the data are created and the previous record with cryptographic signature. Therefore, each block has a specific signature or hash value, which is dependent on the content of the previous block and contains cryptographically created series of letter and number that specially recognize any digital entity. Modification of any record leads to a variation in signature or hash value and easily detectable break in the chain. Blockchain is said to be immutable since records cannot be eliminated, whereas only data or records can be added [3].

Blockchain is a distributed ledger where various parties maintains and shares a sequential list of identical copies. Distributed ledger is also known as distributed database, maintained by a consensus protocol in a P2P network which is run by nodes [26]. Each node consists of a private and public key. The public key encrypts the sent messages to a node and ensures the stability, nonreproducibility, and unchangeability of a blockchain. The mechanism where the private key decrypts the encrypted message and let the node to read and interpret the message is known as symmetric cryptography [1].

Before the transaction being broadcast to the network for later confirmation, each transaction should be endorsed by a specific private key, which leads to authentication and also provides the integrity of transaction [1]. Blockchain is private where a specific cryptographic key is needed to access the information in block. Blockchain acts as an agent of trust between two parties because of the decentralized and immutable nature of blockchain implementation combined with transparency of data, where everyone has the same data and well-designed blockchain, which assures the stakeholder can access data which are needed to analyze the transaction on the chain [3].

Miners are specific nodes that are ordered and packed into time-stamped blocks when transactions propagated in the network are considered valid. Based on the consensus protocol, the miner and type of data to be included in the block are chosen [1]. A node is described as peer when the node attaches and interacts with another node in the network. Nodes collect an up-to-date register, send an accurate agreement into the network, certify a newly covered blocks which approve the transaction, and build a new block in blockchain [1].

12.16 TYPES OF BLOCKCHAIN TECHNOLOGY

There are three different types of blockchain, which depend on the availability and management of data, and the action taken by the user. The three main types of blockchain are public permissionless, consortium (public permissioned), and private.

Entities can act as a reader or writer in a blockchain. A reader will read or inspect the content of record or verify the blockchain passively. In contrast, writers involve in the consensus protocol, who have the capability to expand the blockchain [27]. In public permissionless (public) blockchain, the data are accessible and visible to the public. In order to protect a participant's invisibility, some part of the blockchain may be encrypted. Anyone can join and act as a node or simple minor in blockchain without any approval [1]. In public blockchain, entities are free to participate as a writer or reader and in consensus process [27]. Bitcoin, Ethereum, and Litecoin are the examples of public blockchain [1].

Only a chosen category of nodes is enabled to involve in the distributed consensus, which leads to the consortium blockchain. Public-permissioned blockchain can be used within one industry like financial sector, where public uses are limited and are partially centralized. Consortium blockchain can also be used between many industries such as financial or governmental institutions, insurance companies, which is free for public use but still becomes moderately centralized trust [1]. If only chosen nodes are allowed to join the network means, it is known as a private blockchain. Private blockchain is a distributed and permissioned network, which performs the transaction by controlling the nodes, which is mainly used for private purposes [1]. Consensus decisions are taken either independently by an essential entity, or by a preselected group [27]. Examples of private blockchain would be Hyperledger Fabric and Ripple.

A differences between blockchain can be made based on the function. For example, Bitcoin is mostly used for tracking digital assets, while smart contract is used for running certain logic. The blockchain that utilizes tokens are Ripple, Bitcoin, and Ethereum, while others use Hyperledger Fabric [1].

12.17 THE NEED OF BLOCKCHAIN TECHNOLOGY AND ITS ADVANTAGES IN HEALTHCARE SECTOR

12.17.1 Patient Data Management

The valuable sources of healthcare intelligence are known as healthcare data. The blockchain technology in the healthcare sector leads to an efficient patient data management, where patients are able to exercise personal control over their own data. Blockchain technology lightens the physician burden by providing greater level of intelligibility and amenability organization, which acts as digital tools of time-saving and alluring patient for their own care [3].

Individual can retrieve the up-to-date information, statistics, and reviews about healthcare service provider due to decentralized features of blockchain technology in the healthcare sector, which create immutable and objective data resources [4]. Patient can freely share their private health records with medical providers without the worry of the information being stolen, with the help of patient-centered care, and with secure P2P network [3]. To improve the quality of healthcare services, data sharing is fundamental, which leads to a smarter healthcare system [8]. Interoperability has been defined by the Health Information and Management Systems Society as "the capability of various software application and information technology system to communicate, exchange data and use the information which have been exchanged" [28].

The speed to retrieve relevant information from different data sources and that to increase data quality and quantity for medical researchers are attributable to interoperability [4]. To store and maintain the patient's medical history, a blockchain technology in the healthcare sector is much needed. The patient's medical history record might not be available or not well maintained because of patient's disconnected hospital visits, which leads to discontinuity and unavailability of the previous record. Therefore, blockchain technology is used to efficiently maintain the record of patient's history for each visit to any hospital in a chronological order. Blockchain eliminates the need of the patient to do the same laboratory test as previously due to inaccessibility of disconnected data, which may decrease the cost and risk of doing the same test with high radiation repeatedly [24].

By the blockchain technology, the patient's medical data can be shared with the person residing in the other country and the patient would agree and have a control over their own data. At a foreign country, the medical history of the patient should be known in order to obtain a better treatment. For instance, the patient should be informed if they have any allergies to medication and their recent treatment. The service provider should assess the medical history of the patient securely [24].

Information sharing and storing using the blockchain technology lead to patient's satisfaction on the data security and privacy. The stakeholder involved is able to have a vision on the overall transaction and interaction. According to Kamel Boulos and Wilson [2], utilization of a blockchain-based ledger becomes more efficient when medical credentials and licensures are stored, shared, and verified. Blockchain technology will allow only authorized entities to access and change the data, and also allow the patients to gain access to medical data in a safer and secure way [24].

The usage of blockchain for electronic healthcare record (EHR/EMR) has become a healthcare application, which ensures the security of the patient information and makes sure that only valid and authorized entities are able to access the data. The usage of smart contract concept with no trusted third party enables trustless features in the system. The interoperability features are supported by blockchain-based healthcare system [24].

Medical chain is an application that uses the blockchain technology, which securely stores the health record. Medical chain develops a permissioned blockchain which shared across a network of trusted international healthcare institutions to help patients receive care internationally without a complicated collection and transfer of medical records. Medical chain enables an efficient decentralized sharing of data between stakeholders who will be able to trust that the patient data is private due to encryption and historically accurate due to the immutable nature of the blockchain [3]. Therefore, the blockchain could create a global interoperable ledger with a patient's complete medical information.

12.18 PAYMENTS AND REIMBURSEMENT

One of the aspects in healthcare that need to be eliminated and avoided is fraudulent billing and claims. Basically, health insurance is implemented to protect the individual assets from devastating cost of treatment. During the treatment period, the patient may pay a specific amount, but the remaining costs are submitted as claims to the healthcare companies. Through claim adjudication process, the providers are reimbursed, whereas the financial responsibilities for the payments are determined by the insurer and the amount to pay is made to the provider [29].

The insurer may decide to pay, deny the claim or reduce the amount paid to the provider. To prevent any fraudulent in the payments and reimbursement, a smart contract is implemented in blockchain technology. Smart contracts are often central and important aspect of the blockchain technology used in healthcare. The benefits of smart contract in the application of blockchain technology in healthcare are immutable and trustable to be operated using trusted information which is shared equally between the parties [3]. In short, blockchain could offer automated insurance payments and reimbursement procedures through smart contracts.

12.19 DRUG AND MEDICAL DEVICE TRACEABILITY

The blockchain technology is efficient in healthcare, and each step of transactional information process acts as a trusted tracking in transparent and immutable ways such as tracking and reconciling errors in patient's data, inspecting medical practitioner's credential, and justifying insurance claims. A company known as iSolve deployed end-to-end blockchain technology solution to track the medical distribution.

According to Engelhardt [3], counterfeit and fraudulent medication is alarming. The original detailed, trustworthy pharmaceutical information and the chronological documentation which records the sequence of custody, transfer, and control are built into the blockchain technology solution to prevent the fake and illegal medication. For instance, the illegal practices such as manufacturing of fake medicine and relabeling of expiration dates can be avoided when the local distributors and purchaser could audit their own supply through the blockchain technology [3]. Entering a complete accurate information into a system is essential, in order for the blockchain technology to be used as a trusted data storage and sharing [3].

One of the applications of blockchain technology in medical and device traceability in healthcare is tracing the prescription drug fraud. The fake drug prescription includes the replication of prescription, doctor-shopping, and altering number to change the prescription itself. Doctor-shopping is where the fraudsters collect many original prescriptions by visiting many doctors. The experts introduce the blockchain technology to address this problem where a monitoring program was installed to enhance the access and response time; to avoid any suspicious of buying patterns, the prescription data are scanned, which would alert the physicians and pharmacists [30].

The conditions when the feedback is incomplete between the physician (prescription writer) and pharmacists (prescription fillers) are known as open-ended loop. Blockchain technology is used to solve this open-ended loop. The first step of the blockchain technology solution for this prescription drug fraud is a machine-readable code, which is attached to the prescription written by the doctor that acts as a unique identifier. The block in the blockchain serves as a unique identifier, which includes the name, quantity of the drug, and the anonymized identity of the patient and a timestamp. After the pharmacists filled the prescription, the symbol was scanned. The fulfilling of the prescription is recorded and compared against the blockchain. From this process, the pharmacists will be acknowledged whether the prescription is desirable to be filled and the data is given to prove the accuracy [3].

According to the World Health Organization (WHO), the counterfeit medicines are alarming worldwide. Thirty percent of drugs are counterfeit in developing countries, and 10% of drugs are counterfeit worldwide [31]. This counterfeit drug problems are addressed by introducing blockchain technology, where the Counterfeit Medicine Projects were launched by the Hyperledger, a research network across industries. Each drug that is manufactured is marked with time-stamped, which leads to easier identification of when and where the drug was produced. This information from the blockchain technology was used to trace the manufacturing of fake medicine. An increase in safety in relation to drugs and a decrease in cost of health-related follow-up are ensured by the application of blockchain technology [22].

The information sourced for manufacturing, the production process, and the distribution of the manufactured items are tracked by using a pharmaceutical supply chain management. Delivering the fake medication has remarkable side effects on the patient. For the supply chain management technologies' complaints, the product recognition, tracking, validation, discovery and response to nonstandard drugs, alerting upon nonstandard drugs, and the capability to store pertinent information, including affirmation and product information, are the key necessity, where blockchain technology is compatible and relevant with each of the key requirements Kamel Boulos et al. [2].

The implementation of blockchain technology in healthcare enhanced the safety and security of medical devices and supplies by holding the unique device identifiers for each medical device, keeping track, and issuing firmware updates by using the smart contract. Through many responsive device recall and notices, the device tracking may enhance the safety and efficacy of the medical devices. To avoid the device loss, theft, or malicious interference, immutability is applied through the blockchain-based medical device tracking [2].

12.20 MEDICAL RESEARCH

For a better understanding of human health in the future, accumulation of medical data by people all around the world is essential and should be made easily accessible to researchers, which should be adhered with ethical standard and preservation of confidentiality through efficient anonymization and control of the information by the respective individual, including the capability to allow and repeal the access to information. The willingness of the people to contribute their own data for research may loss, if fail to address privacy and ownership concern [3].

Blockchain technology revolutionizes or has a greater effect on medical research and individual care due to tracking, sharing, and caring of data [5]. In blockchain technology, patient's medical record, genome and connectome files, and quantified self-data commons are aggregated in an encrypted pseudonymous form where the data can be analyzed although private. To make a new innovation, the medical data that is high in volume, variety, value, and accuracy is valuable for researchers for further improving public health [4].

The privacy and ownership concerns of the people's data for research are achieved by using the blockchain technology, where the first-level protection would be encryption and keyed access [3]. Blockchain is a suitable technology to increase access to medical data worldwide in a secure environment with patient's consent due to many reasons as follows:

1. Immutability of data, where the data is trusted to not change once those data are stored for research [3].
2. Transparent storage, where the participant involved in research would be clear of what data was and was not available [3].
3. Tested and tried smart contract, where the data owner could have the confidence that they own the data, and allow, repeal, and ingress those data in anonymized form to validate the research. The unchangeable smart contract provides confidence that the data will not be changed [3].

Thus, overall, blockchain increases access to medical data worldwide in a secure environment with the patient's consent.

12.21 REGULATORY PROCEDURE

Blockchain technology in the healthcare sector improves the chain of trust, which results in obtaining authorizations and certification. The blockchain technology leads to immutable audit data due to proof of work (PoW) of blockchain, which encounters difficulty in changing data or records. To hack the system, the assailant need to redo the PoW of the targeted block and all the blocks after this targeted block in blockchain, and then surpass all truthful sites. According to the blockchain and Bitcoin inventor, Satoshi Nakamoto, the possibility of an honest or normal node to discover the next block is greater than the possibility of an attacker to discover the next block. On a decentralized basis of cryptocurrency, blockchain technology was the first proposed protocol on the implementation of P2P timestamp sever [23].

Blockchain technology prevents the double-spend attacks and eliminates the necessity of trusted party due to the point-to-point communication, distributed consensus protocol, and PoW techniques. Smart contract is applied in blockchain technology without the necessity for trust between the parties while featuring some kinds of agreement to act or not to act [27]. Enhancement of the chain of trust by using the blockchain technology is achieved by using a hash function and the timestamp features. The block contains multiple transactions that have to be validated by this hash function. Each site of the block contains "mines," which solve a hard hashing issue. The site that satisfies the PoW first will verify the transaction and the confirmed block, which are considered as immutable and are then added at the last or end of the blockchain [23]. Blockchain thus could help to improve the "chain of trust," which results in obtaining authorizations and certification.

12.22 CLINICAL TRIALS

In the healthcare setting, clinical trial is a beneficial process which needs a suitable monitoring at each stage of trials, such as gathering of needed information, trial procedure, and monitoring and data management trials. So blockchain technology is an essential appliance to deal with every phase trial, where all phases are appropriately discovered and the information or data are managed and analyzed in the absence of resource wastes, which leads to a huge trust among the different entities [24].

Blockchain technology is time-stamped, thus leading to a consequential level of scared information for the full document progress in a clinical trial, which certainly automates the clinical trial through smart contract; ensures traceability; compacts control of data, safety, and shareable parameter; prevents posteriori reconstruction for patient or the clinical trial collaborator. Cryptographic validation for each transaction leads to data integrity, which ensures the reliability of data and limits the fake information. Blockchain technology is a not centralized protected tracking system for any data in clinical trials with a P2P network which allows data distribution in the research [32].

The consent and clinical trial protocol are packed into information structures which are kept on the blockchain before the clinical trials begin. Data structure accounts for robust proof, which avoids typical problems related to nondetectable clinical trial protocols [32]. The blockchain systems are best implemented to gather the agreement of participant for a clinical trial due to the unique characteristics of blockchain which is time-stamped, where a unique main document that holds a data structure is obtained. The original agreement and the protocol contract are hashed and formatted into cryptographic form, thus resulting in this master document to produce a secure and robust proof of the whole collected agreement [32]. Blockchain could be optimized to track and secure clinical trials' documents such as consent forms, reports, or results.

12.23 DISADVANTAGES OF BLOCKCHAIN TECHNOLOGY IN HEALTHCARE SECTOR

The disadvantages of blockchain technology in the healthcare sector that might occur are low quality or false information that is executed into the blockchain, which will lead to immutable and decentralized information. So this incorrect or low quality of information will exist in blockchain forever [3]. The loss of key which is used to access information is a real threat in blockchain technology. Key is defined as a unique series of digits and characters in blockchain technology. The data acquired become unrecoverable, and the ability to access to a lifetime health information in blockchain is lost if the key in blockchain technology is lost. In order to reconnect the user with their information in blockchain technology, a new solution should be sought [3].

Due to the distributed nature of blockchain, complexity and latency might occur. Normally, the blockchain-based agreement may need a longer time to complete the transaction, until every party updates their corresponding ledgers. This latency may lead to unreliability for participants involved in certain transactions and becomes an initial step for cybercriminals [27]. The issue that may arise from the applications of blockchain technology in the future might be the lack of awareness and adoption. A shallow understanding regarding the blockchain technology and its mechanism will lead to problems in the development of blockchain technology in the future, as blockchain works upon the number of participants adopting this blockchain technology [27].

Energy and cost are the problems faced in the blockchain technology, where the computational power is a treat. For example, in blockchain, the Bitcoin mining process would need a great level of power to calculate and validate the transaction, where this computational energy significantly becomes higher when the network spreads [27]. A greater operation and development costs are needed in the blockchain technology in the healthcare sector, which will be defined by the government as well as the healthcare sector [24].

12.24 CONCLUSION

Blockchain technology is still a fairly new technology that has not been widely implemented in the healthcare sector. Electronic medical record (EMR), biomedical research and education, remote patient monitoring, drug or pharmaceutical supply chain, health insurance claims, health data analytics, and other area were the commonly used cases of blockchain technology in healthcare. More research should be carried out regarding the implementation of blockchain technology in the real healthcare environment for a better understanding, characterization, and evaluation of blockchain technology in the healthcare. Additional focus should be given by researchers to carry out researches regarding safety on the implementation of blockchain technology in healthcare.

REFERENCES

1. Hölbl M, Kompara M, Kamišalić A, Nemec Zlatolas L. A systemic review of the use of blockchain in healthcare. *ResearchGate*. 2018. doi: 10.20944/preprints201809.0136.v1.
2. Kamel Boulos MN, Wilson JT, Clauson KA. Geospatial blockchain: Promises, challenges, and scenarios in health and healthcare. *International Journal of Health Geographics* 2018;17(1):25. doi: 10.1186/s12942-018-0144-x.
3. Engelhardt MA. Hitching healthcare to the chain: An introduction to blockchain technology in the healthcare sector. *Technology Innovation Management Review*. 2017;7(10):22–34.
4. Gökalp E, Gökalp MO, Çoban S, Eren PE. Analysing opportunities and challenges of integrated blockchain technologies in healthcare. *Information Systems: Research, Development, Applications, Education*. 2018:174–183. doi: 10.1007/978-3-030-00060-8-13.
5. Benchoufi M, Ravaud P. Blockchain technology for improving clinical research quality. *Trials*. 2017;18(1):335. doi: 10.1186/s13063-017-2035-z.
6. Ismail A, Jamil AT, Rahman AFA, Bakar JMA, Saad NM, Saadi H. The implementation of Hospital Information System (HIS) in tertiary hospitals in Malaysia: A qualitative study. *Malaysian Journal of Public Health Medicine*. 2010;10(2):16–24.
7. Ismail NI, Abdullah NH. An overview of Hospital Information System (HIS) implementation in Malaysia. *Proceeding of the 3rd International Conference on Business and Economic Research (3rd ICBER 2012)*, 2012, Bandung, Indonesia.
8. Peterson K, DR, Kanjamala P, Boles K. A blockchain-based approach to health information exchange networks. *Proceedings of the NIST Workshop Blockchain Healthcare*, 2016, vol. 1, pp. 1–10. [Online]. Available: https://www.healthit.gov/sites/default/files/12-55-blockchain-based-approach-final.pdf. [Accessed: July 08, 2018].
9. Yue X, Wang H, Jin D, Li M, Jiang W. Healthcare data gateways: Found healthcare intelligence on blockchain with novel privacy risk control. *Journal of Medical Systems*. 2016;40(10):218. doi: 10.1007/s10916-016-0574-6.
10. Goldzweig CL, Towfigh A, Maglione M, Shekelle PG. Costs and benefits of health information technology: New trends from the literature. *Health Affairs*. 2009;28(2):w282–w293.
11. Haux R. Health information systems: Past, present, future. *International Journal of Medical Informatics*. 2006;75(3–4):268–281. doi: 10.1016/j.ijmedinf.2005.08.002.
12. Ismail NI, Abdullah NH, Shamsuddin A. Adoption of Hospital Information System (HIS) in Malaysian Public Hospitals. *Social and Behavioral Sciences* 2015;172:336–343. doi: 10.1016/j.sbspro.2015.01.373.
13. Ismail NI, Abdullah NH, Shamsuddin A, Ariffin NAN. Implementation differences of Hospital Information System (HIS) in Malaysian Public Hospitals. *International Journal of Social Science and Humanity*. 2013;3(2):115.
14. Hassan R, Tajuddin MZM. Implementation of Total Hospital Information System (THIS) In Malaysian Public Hospitals: Challenges and future prospects. *International Journal of Business and Social Research (IJBSR)*. 2012;2(2):33–41.

15. Sulaiman H, Wickramasinghe N. Assimilating Healthcare Information Systems in a Malaysian Hospital. *Communications of the Association for Information Systems.* 2014;34(77):1290–1319.
16. Bakar NAA, ChePa N, Jasin NM. Challenges in the implementation of Hospital Information Systems in Malaysian Public Hospitals. *Proceedings of the 6th International Conference on Computing and Informatics, ICOCI 2017,* Kuala Lumpur, 2017, pp. 636–642.
17. Selvaraju Dr. S. Health Information Management: Malaysian Experience. Health Informatics Center, Ministry of Health, Malaysia. 2008.
18. Almunawar MN, Anshari M. Health Information Systems (HIS): Concept and Technology. 2012.
19. Littlejohns P, Wyatt JC, Garvican L. Evaluating computerised health information systems: Hard lessons still to be learnt. *BMJ* 2003;326(7394):860–863.
20. Yli-Huumo J, Ko D, Choi S, Park S, Smolander K. Where is current research on blockchain technology? A systematic review. *PLoS One.* 2016;11(10):e0163477. doi: 10.1371/journal.pone.0163477.
21. Nakamoto S. Bitcoin: A peer-to-peer electronic cash system. *ResearchGate.* 2009; 1–9. Available: https://bitcoin.org/en/bitcoin-paper [Accessed August 13, 2020]
22. Mettler M., Blockchain technology in healthcare: The revolution starts here. e-Health Networking, Applications and Services (Healthcom), *2016 IEEE 18th International Conference on e-Health Networking, Applications and Services (Healthcom),* Munich, Germany, 2016: IEEE.
23. Kuo TT, Kim HE, Ohno-Machado L. Blockchain distributed ledger technologies for biomedical and health care applications. *Journal of the American Medical Informatics Association.* 2017;24(6):1211–1220. doi: 10.1093/jamia/ocx068.
24. Kumar T, Ramani V, Ahmad I, Braeken A, Harjula E, Ylianttila M. Blockchain utilization in healthcare: Key requirements and challenges. *IEEE 20th International Conference on e-Health Networking, Applications and Services (Healthcom),* Ostrava, Czech Republic, 2018, pp. 1–7.
25. Meng W, Tischhauser EW, Wang Q, Wang Y, Han J. When intrusion detection meets blockchain technology: A review. *IEEE Access.* 2018;6:10179–10188.
26. Brogan J, Baskaran I, Ramachandran N. Authenticating health activity data using distributed ledger technologies. *Computational and Structural Biotechnology Journal.* 2018;16:257–266. doi: 10.1016/j.csbj.2018.06.004.
27. Meng W, Tischhauser EW, Wang Q, Wang Y, Han J. Research challenges and opportunities in security and privacy of blockchain technologies. *When Intrusion Detection Meets Blockchain Technology: A Review,* IEEE Access. 2017;6:10179–10187. doi: 10.1109/ACCESS.2018.2799854.
28. Gordon WJ, Catalini C. Blockchain technology for healthcare: Facilitating the transition to patient-driven interoperability. *Computational and Structural Biotechnology Journal.* 2018;16:224–230. doi: 10.1016/j.csbj.2018.06.003.
29. Parekh N, Sadanand P, Jain S. Blockchain technology. *Journal of Emerging Technology and Innovative Research (JETIR).* 2017;4(0):177–179.
30. McDonald DC, Carlson KE. Estimating the prevalence of opioid diversion by "doctor shoppers" in the United States. *PLoS One.* 2013;8(7):e69241. doi: 10.1371/journal.pone.0069241.
31. World Health Organisation. Growing threat from counterfeit medicines. *Bulletin of the World Health Organization.* 2010;88(4):247–248. doi: 10.2471/BLT.10.020410.
32. Benchoufi M, Porcher R, Ravaud P. Blockchain protocols in clinical trials: Transparency and traceability of consent. *F1000Research.* 2017;6:66. doi: 10.12688/f1000research.10531.5.

13 Complexity Analysis of Pathogenesis of Coronavirus Epidemiological Spread in the China Region

Rashmi Bhardwaj and Aashima Bangia
GGS Indraprastha University

Jyoti Mishra
Gyan Ganga Institute of Technology and Sciences

CONTENTS

13.1 INTRODUCTION

These CoVs are a family of enormous viruses infecting wide-ranging animal species. Predominant infections in association with these genomic viruses have respiratory and gastroenteric syndromes, along with occurrences of hepatic and neurological disorders. Human-infecting CoVs acknowledged in the 1960s (including the prototype viruses HCoV-OC43 and HCoV-229E) were accountable for up to 30% of the respiratory tract disorders (Figure 13.1).

Basically, these are encased nonsegmented positive-sense RNA viruses and have its place in the clan of *Coronaviridae* majorly disseminated among other living mammalia. In general, human-affecting coronavirus contagions are believed to be mild, which are the two *Betacoronaviruses*: severe acute respiratory syndrome coronavirus (SARS-CoV) and Middle-East respiratory syndrome coronavirus (MERS-CoV). In December 2019, succession of pneumonia cases from an unknown cause appeared across the city of Wuhan, China, and clinical tests detected them as viral pneumonia. Deep-sequencing analyzers depicted that the lower respiratory tract samples indicated a novel coronavirus coined as 2019 novel coronavirus (2019-nCoV). As far as approximately 800 confirmed cases taking down the healthcare workers of this genomic virus are existent in Wuhan. The exported population have been found in other provinces of China, Thailand, Japan, South Korea, and the United States as of now.

Coronaviruses (CoVs) are majorly identified in the respiratory tract in addition to the gastrointestinal tract taints that can be hereditarily categorized as four key genera: Alphacoronavirus, Betacoronavirus, Gammacoronavirus, and Deltacoronavirus. The first two genera predominantly infect mammalia, and the

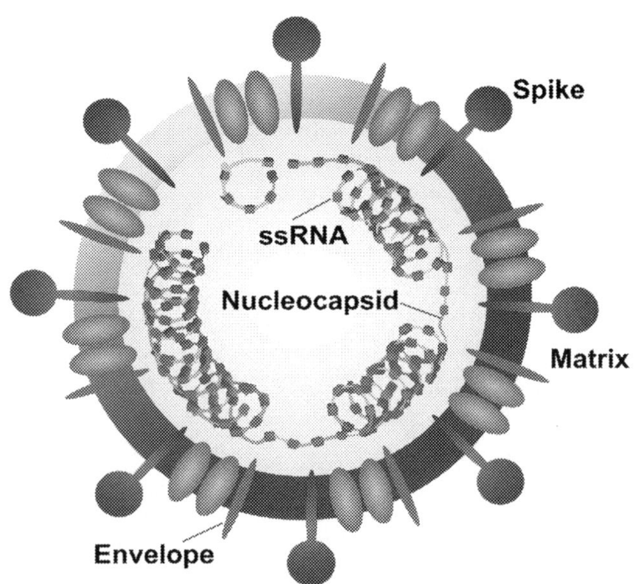

FIGURE 13.1 Detailed diagram of coronavirus affecting the host RNAs.

last two primarily infect birds. Various kinds of human-destroying CoVs have been previously identified. These comprise HCoV-NL63 and HCoV229E, which may belong to *Alphacoronavirus* genus, and HCoV-OC43, HCoVHKU1, SARS-CoV, and MERS-CoV belonging to *Betacoronavirus* genus. Coronaviruses did not attract worldwide responsiveness until the 2003 SARS pandemic followed by the year 2012 MERS and, most recently, the 2019 nCoV outbreaks. SARS-CoV and MERS-CoV are considered highly pathogenic. Also, it is very probable that both SARS-CoV and MERS-CoV were communicated from bats to palm civets and dromedary camels and further transferred ultimately to humans.

Novel coronavirus (CoV)-"*2019 novel coronavirus*"/"*2019-nCoV*" by the World Health Organization (WHO) is accountable for the recent pneumonia outbreak that started in early December 2019 in Wuhan City, Hubei Province, China. This outbreak is connected to the large seafood and animal market, and investigations are ongoing to determine the origins of the infection. Till today, thousands of human infections have been confirmed in China along with many exported cases across the globe (Figure 13.2).

Bhardwaj [1–2] discussed environmental concerns involving pollution through wavelet, non-linearity analysis etc. Bhardwaj and Bangia [3] contributed towards hybridized fuzzified PID controller for nonlinear control surfaces for improving

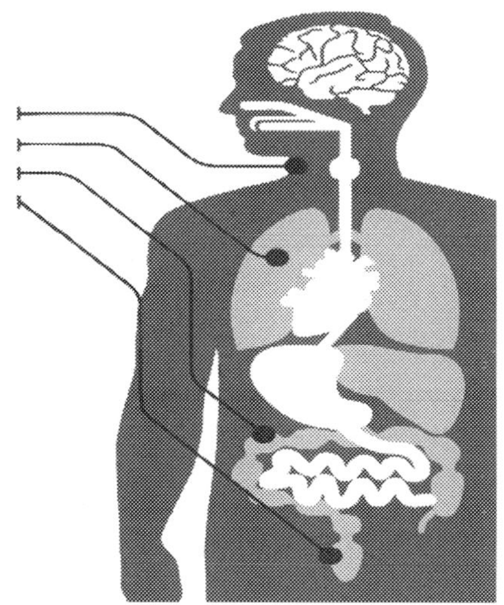

FIGURE 13.2 Coronavirus, its symptoms, and its transmission as instructed by CDCP/USA Today/WHO.

the designing of electric battery-driven vehicles. Dynamism of atmospheric pollutants was also studied [4]. Various methodologies for nonlinear analysis of the real-life case studies that involve meditating body [5] were discussed. Soft computing analysis for the prediction of demonetization scenario for the stock markets' volatility was carried out [6]. Complexity of HIV dynamics via statistical simulations was studied [7]. Implications of demonetization via artificial neural fuzzy model were described for National Stock Exchange [8]. The forensics of malwares in IoT-based wireless transmissions were inferred via dynamic study [9, 10]. Bangia et al. [8] studied water quality through AI. Bhardwaj & Bangia [11–12] discussed intelligent ML and hybrid soft-computing techniques. Bhardwaj & Datta [13] explored consensus algo. Bhardwaj et al [14–24] studied various real life applications and analyzed case studies through artificial intelligence and nonlinear time series processes. Box and Cox [25] gave an analysis on various necessary transformations. Chen et al. [26] modeled a mathematical prototype for the transfer on the basis of phases for the novel coronavirus. Rosenblum [27] studied the corona theorem for countably many functionalities. Fuhrmann [28] analyzed the applications of corona theorem to spectral problems. WHO [29] released a detailed report on the MERS-CoV updates. Wu et al. [30] described in detail minute molecules that target the severe acute respiratory infections, leading to coronavirus. Zhou et al. [31] discussed glycopeptide antibiotics that would inhibit cathepsin L into late endosome/lysosome, which blocks the entry of Ebola, MERS-CoV, and SARS-CoV. WHO [32] released an extensive report on novel coronavirus spreading through China. Mishra [33] discussed fractional hyper-randomized modeling.

None of the authors have studied the mathematical model design for the spread of coronavirus. In this paper, the impact of time on the amount of uninfected cells, virus particles in the blood cells, and infected cells is studied. Also, state dynamics through time progression, and phase and Lyapunov characteristics have been analyzed. Data-fitted regression procedures with various other outcomes for the prediction towards the behavior of those genomic RNAs are observed.

13.2 CASE STUDY: PATHOGENESIS OF CORONAVIRUS EPIDEMIOLOGICAL SPREAD IN THE CHINA REGION

The occurrences of unpredictable pneumonia syndromes, stated as the severe acute respiratory syndromes (SARS), can be believed to be extended through markets of Wuhan, China, fast spreading to the distant parts of the world. Brutality of this novel infection carries a high mortality rate ranging from 3% to 6% through a recent WHO report, suggesting this rate can go as high as 43%–55% due to senior citizens above the age of 60 years. SARS-CoV has been originated in bats, and these infections were transmitted directly to humans from market civets and dromedary camels, respectively. The spread of virus is supposed to be analogical to the susceptibles–infecteds–recovered (SIR) biological viral spread model. SIR system is complex because of the nonlinear nature of the interactions that govern the real-world systems. The systems belong to some physical or biological sciences which have to be analyzed using mathematical theories whose validation is then drawn from computer simulations. Under various disciplines, the content of the system exists from study in which the understanding of the system is derived. After this, mathematical equations are developed, which approximate the logic or rule that governs the system behavior. The simulations of the system flow in the form of phase portraits, time series, and recurrence plots finally help in determining the complexity of system evolution.

Let x_1, x_2, and x_3 be the number of susceptible people (susceptibles), asymptomatically and symptomatically infected people (infecteds), and the recovered and removed people (removed ones). Then, we consider various factors on which these variables majorly depend. The model so developed is as follows:

$$\dot{x}_1 = n_b - m_d x_1 - t_r x_1 x_2$$

$$\dot{x}_2 = (1-\delta_p)\omega_p x_1 - (\gamma_p + m_d) x_2 x_3 + t_r x_1 x_2,$$

$$\dot{x}_3 = (\gamma_p + m_d) x_2 x_3 - m_d x_3 + \varepsilon_v x_2$$

where

$\dot{x}_1$: rate of change of susceptibles
$\dot{x}_2$: rate of change of infecteds
$\dot{x}_3$: rate of change of removed ones
n_b: birth rate of people
m_d: death rate of people
t_r: transmission rate from x_1 to x_2
δ_p: proportion of infection rate of people
$1/\omega_p$: incubation period of people
$1/\gamma_p$: infectious period of people
$1/\varepsilon_v$: lifetime of virus.

The model is designed to study the dynamics of coronavirus spread in the citizens of China transmitted due to various mammals of the food chain. This virus having single-stranded RNAs has become communicable disease with no or very less symptoms such as cold, cough, and fever. It is proving to be a pandemic for the mankind as stated by WHO.

13.3 PHASE, TIME PROGRESSION, AND LYAPUNOV CHARACTERISTICS EXPONENT ANALYSIS FOR THE PREDICTION OF ITS SPREAD

13.3.1 Phase Analysis

Inside a phase space, all degrees of freedom (DOF), i.e., parameters of that particular system, are exemplified as the alignment of multidimensional space where one-dimensional system is referred as the phase line; two-dimensional system exists as the phase trajectory. At every probable state of the modeled system, there gets evolved an arrangement of values of system's considerations wherein the fixed point is simulated for that multidimensional structure. System's evolved phase with respect to time traces the pathway through the phase space trajectory for that particular system via n-dimensional space.

Phase space is generally symbolized through cotangent bundle:

$$T * M := \{(q, p) : q \in M, p \in T_q^* M\},$$

which may emanate as the canonical symplectic form $\omega := dp \wedge dq$.

Hamilton's equation of motion describes the movement mapping: $t \mapsto (q(t), p(t))$ of the structure in phase structure as the function of time in the form of a Hamiltonian, $H : M \rightarrow R$, via Hamilton's equation of motion:

$$\frac{d}{dt} q(t) = \frac{\partial H}{\partial p}(q(t), p(t));$$

$$\frac{d}{dt} p(t) = -\frac{\partial H}{\partial q}(q(t), p(t)).$$

Unvaryingly, in the form of the symplectic version, ω as:

$$\dot{x}(t) = \nabla_{\omega} H(x(t)),$$

where $x(t) := (q(t), p(t))$ is the trail of setup in the phase-state space and $\nabla\omega$ is the symplectic gradient; hence, $\omega(\cdot, \nabla_{\omega} H) = dH$.

13.3.2 Lyapunov Characteristic Exponent (LCE)

Lyapunov exponents are equivalent to the real values of eigenvalues at the calculated fixed points.

- If Lyapunov exponents tend towards values less than zero, it means adjoining initial conditions (ICs) all converge on each other and the initial delta errors diminish with respect to time domain.
- If any of the Lyapunov exponents is positive, then infinitesimally nearby ICs depart exponentially fast, i.e., delta differences in the initial conditions evolve with respect to the considered time.
- If trajectories diverge with time, then this state is sensitivity dependent on ICs), which is coined as chaos.

13.3.3 Algorithm for the Computation of LCE

1. Let $x_n = f_n(x_0)$; $y_n = f_n(y_0)$.
2. Then, the nth iterations of orbits of x_0 and y_0 under f are: $|x_0 - y_0| \ll 1$; $|x_n - y_n| \ll 1$.
3. Then, separation at nth iteration for one-dimensional system can be given by:

$$|x_n - y_n| \approx \left(\prod_{t=0}^{n-1} |f'(t)| |x_0 - y_0| \right),$$

where $|x_0 - y_0| \ll 1, |x_n - y_n| \ll 1$, and $x_n = f_n(x_0), y_n = f_n(y_0)$ are, respectively, the nth iterations of orbits of x_0 and y_0 under f.

4. Then, exponential separation rate $\log|f'(x)|$ of two proximate ICs, averaged over the complete trail is represented through:

$$\lambda(x_0) = \lim_{n\to\infty} \frac{1}{n} \log\left(\prod_{t=0}^{n-1} |f'(x_t)|\right) \quad \text{Where} \prod_{t-0}^{n-1} |f'(x_t)| \approx e^{\lambda(x_0)n} \text{ for } n \gg 1.$$

5. $\lambda(x_0)$ is defined as the LCE of orbit of x_0.
6. Quantitatively, two trajectories in phase space with the initial separation δx_0 diverge $|\delta \times (t)| \approx e^{\lambda t} |\delta \times (0)|$, where $\lambda > 0$ is the Lyapunov exponent.
7. Let $\lambda_1, \lambda_2, \ldots, \lambda_n$ be the eigen values of the linearized equation $\frac{du}{dt} = A\left(u^*\right)$ such that $m_i(t) = e^{\lambda t}$ and $\tilde{\lambda i} = \lim_{t\to\infty} \frac{1}{t} \ln\left|e^{\lambda_i t}\right| = \text{Re}\left[\lambda_i\right]$.

13.3.4 Attractors

The behavior of the attractor as per the value of the Lyapunov exponents is defined as:

- Equilibrium: $0 > \lambda_1 \geq \lambda_2 \ldots \geq \lambda_n$;
- Periodic limit cycle: $\lambda_1 = 0, 0 > \lambda_2 \geq \lambda_3 \ldots \geq \lambda_n$;
- k-periodic cycle: $\lambda_1 = \lambda_2 = \ldots \lambda_k = 0,\ 0 > \lambda_{K+1} \geq \lambda_{K+2} \ldots \geq \lambda_{K+n}$;
- Strange chaotic: $\lambda_1 > 0, \sum_{i=1}^{n} \lambda_i < 0$;
- Hyper-chaotic: $\lambda_1 > 0,\ \lambda_2 > 0,\ \sum_{i=1}^{n} \lambda_i < 0$.

13.3.5 Nonlinear Regression

Nonlinear regression is used to model complex phenomena, which cannot be handled by the linear model.

13.3.6 Box–Cox Time Transformation Measure

The Box–Cox transformation is principally useful family of transformations.

Theorem 13.1

Suppose a sample of n response values $t_1, t_2, \ldots, t_n$. Let δ be a value such that $t + \delta > 0$. Compute the set of f_i(t)'s with respect to t_i's as:

$$f(t) = \left\{ \begin{array}{ll} \left((t+\delta)^\omega - 1\right)/\left(\omega B_c^{\omega-1}\right), & \omega \neq 0 \\ B_c \ln(t+\delta), & \omega = 0 \end{array} \right\}$$

where B_C: geometric mean of $(t+\delta) = \sqrt[n]{\prod_{i=1}^{n}(t+\delta)}$

t: response variable, ω: transformation parameter.

Natural log is applied in the case of $\omega = 0$ instead of the aforementioned formula. It helps to define the measure of normality of the resulting transformation. It is meant to moderate non-normal dependent variable into normal contour. The measure computes the correlation coefficient of normal probability. Correlation is simulated for the variables of probability plot and a scale of linearity of probability plot. Vertical axis encapsulates the correlation coefficient of normal probability, and horizontal axis stands for the values of ω. This stationarity test is applied towards the positive and negative values.

13.3.7 Autocorrelation and Partial Correlation Function (ACF & PACF)

The statistics under consideration are said to have an autocorrelation whenever the response variables x_i's at time domain and t_i are determined to be correlated through the values X_{i+d}'s at time domain t_{i+d}, where d refers to the time increment that lies in the upcoming events. It can be observed that within the long memory process, autocorrelation deteriorates over time, resulting in the power-law trend which is written as:

$$p(k) = Ck^{-\alpha},$$

where C is the constant and $p(k)$ is the autocorrelation function having lag "k." Generalizing, consider a set of responses as $X_1, X_2, \ldots, X_n$ at time $t_1, t_2, \ldots, t_n$ along with k-lag autocorrelation function is represented via:

$$r_k = \frac{\sum_{i=1}^{n-k}(X_i - \bar{X})(X_{i+k} - \bar{X})}{\sum_{i=1}^{n}(X_i - \bar{X})^2},$$

where $\bar{X} = \dfrac{X_1 + X_2 + \ldots.. + X_n}{n}$.

The interpretations should be uniformly sampled. Unlike cross-correlation, ACF results in a correlation coefficient signifying the degree of resemblance of two response variables at time, t_i and t_{i+k}. ACF is used to identify nonrandomness in data and propagate appropriate time regressiveness when data has no chaos. Whenever ACF is applied for locating apt time successive regression, there "k"-lag ACF gets charted.

13.3.8 Augmented Dickey–Fuller Stationarity Test (ADF)

Basically, a unit root test is used to check stationarity as these unit roots can cause unpredictable results in the autoregressive models of time-series analysis. Time series is different in comparison with the predictive modeling. As in modeling, the assumptions exist that summary statistics of observations are consistent. In context with time series, these expectations are referred as time domain, which is stationary.

Time series is taken to be stationary when it does not contain trend or seasonal effects. So these summary statistics computed on time domain is said to be consistent over time. Thus, statistical modeling considers that stationarity in the series makes it effective. In particular, it concludes how strongly a time series is defined by a trend.

13.4 RESULTS AND DISCUSSION

With the equations of the corona spread model, phase-space diagram, time series, and Lyapunov characteristics have been simulated. Also, with the daily data of number of cases of virus-infected people in China ranging from January 22, 2020, to March 10, 2020, the statistical analyses, including nonlinear regressive (NR) series, autoregressive (ACF), partial autoregressive (PACF), mixture modeling, and classification and regression tree (CART), have been carried out. Figure 13.3a–c depict the LCEs of all three variables for initial phase, during the phase, and then after 60 days of spread. Similarly, Figure 13.4a–c show the phase diagram of the three variables in the three phases. Figure 13.5 shows the time series of all three variables, i.e., susceptibles, infected, and removed ones. Also, Figure 13.6 shows the relative time series of all three variables depicting the statistics of recovered or removed ones. Figure 13.7a–d shows nonlinear regression that is simulated for the data fitting and prediction of virus-infected cases with residuals of active and validated data. Figure 13.8a shows daily increase in the number of cases in China. Figure 13.8b shows the linearly fitted regression model of the daily data. Figure 13.8c denotes the plot wit Box–Cox transformation of data, and Figure 13.8d shows

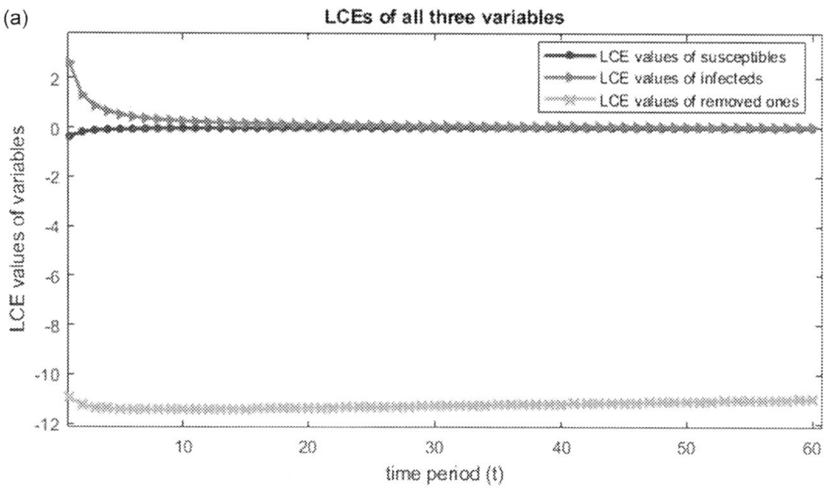

FIGURE 13.3 (a) LCE values of all variables during 50 days of spread.

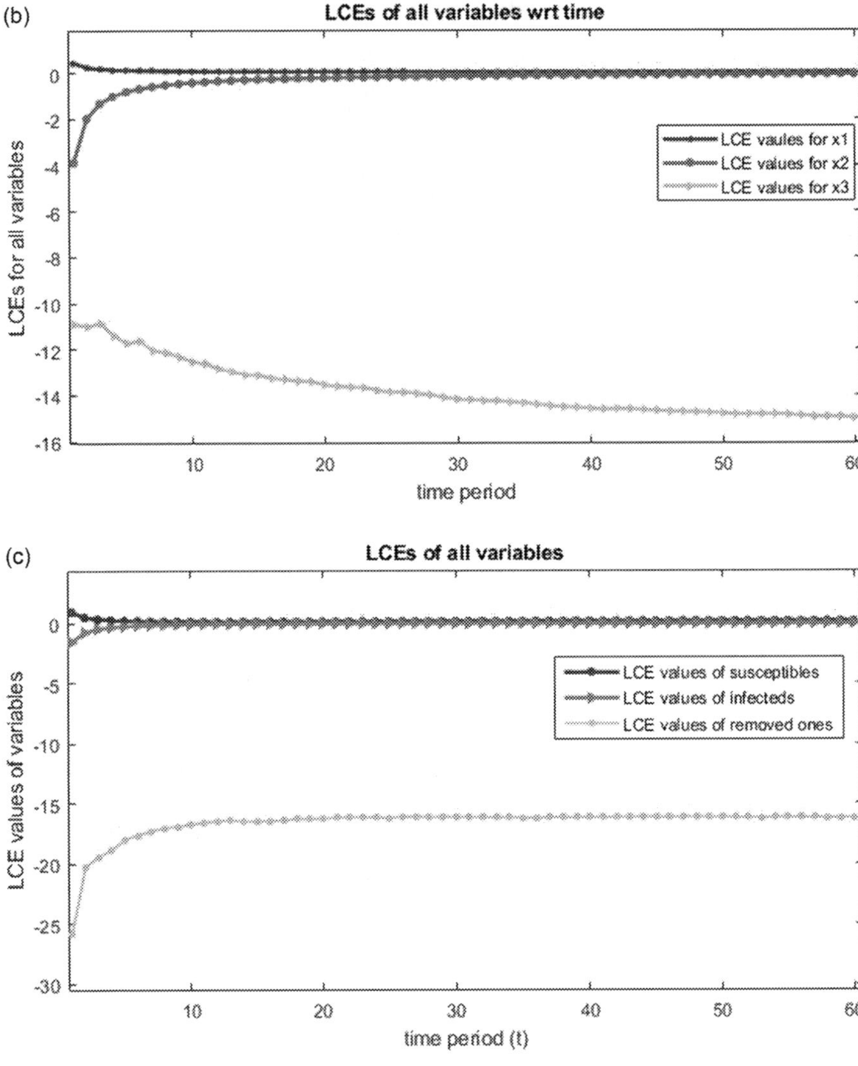

FIGURE 13.3 (*Continued*) (b) LCE values of all variables during the initial days of spread, and (c) LCE values of all variables after 60 days of spread.

the polynomial regression of the daily data. Figure 13.9a–c depicts the autocorrelation, partial autocorrelation, and cross-correlation of the infected cases with respect to time. Figure 13.10 shows the Maximum a posteriori (MAP) classification obtained from Gaussian mixture modeling. Figure 13.11 shows the fitted plot of outcomes with the mixture modeling. Figure 13.12 denotes the cumulative distribution functionality via mixture modeling. Figure 13.13 shows the CART of the statistics referring to the nodes at which the linearity is lost.

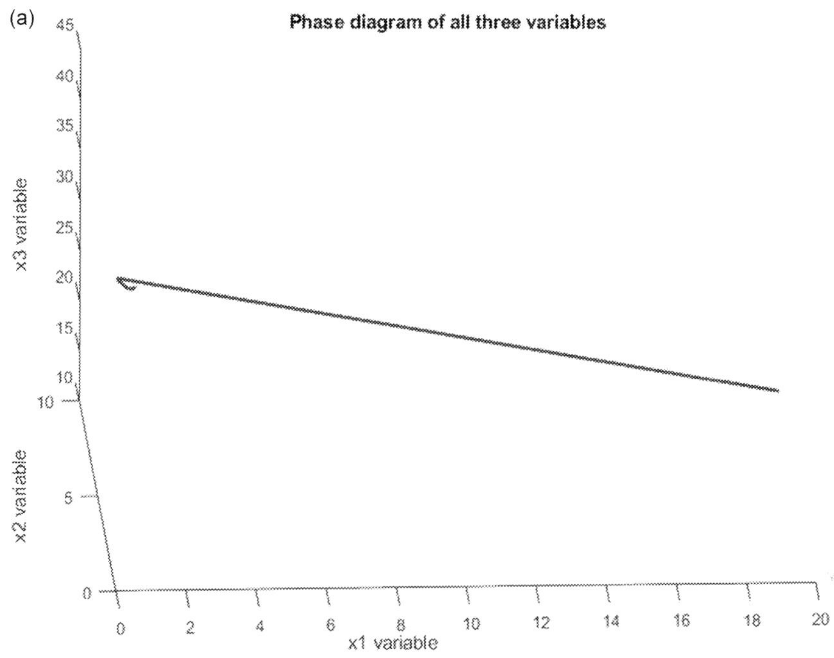

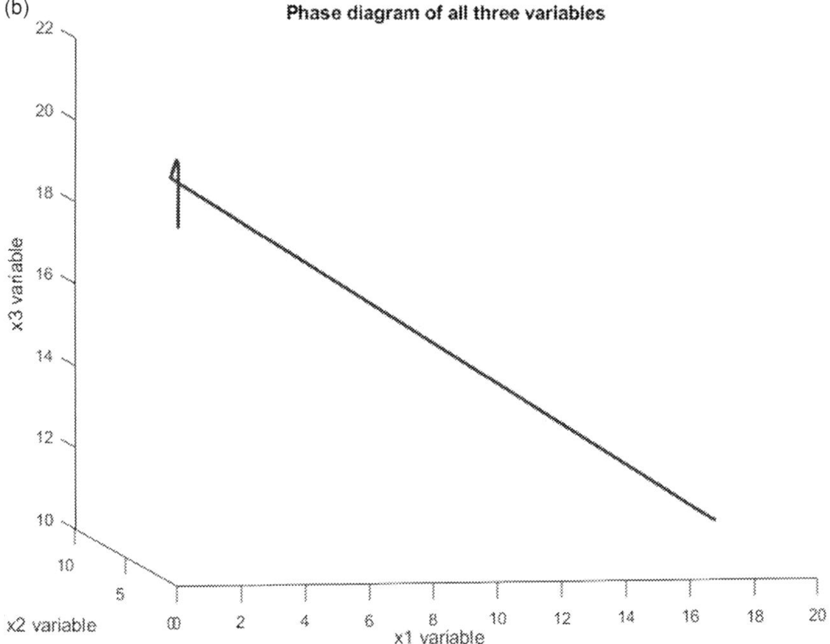

FIGURE 13.4 (a) Phase diagram of three variables during the initial 30 days of spread of corona in China, (b) phase diagram of three variables during the 50 days of spread of corona in China.

(*Continued*)

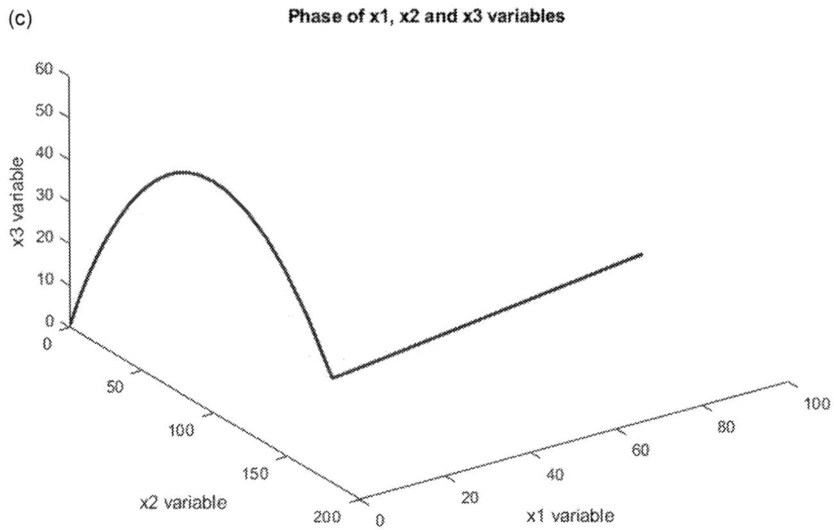

FIGURE 13.4 (CONTINUED) (c) phase diagram of three variables during the 60 days of spread of corona in China.

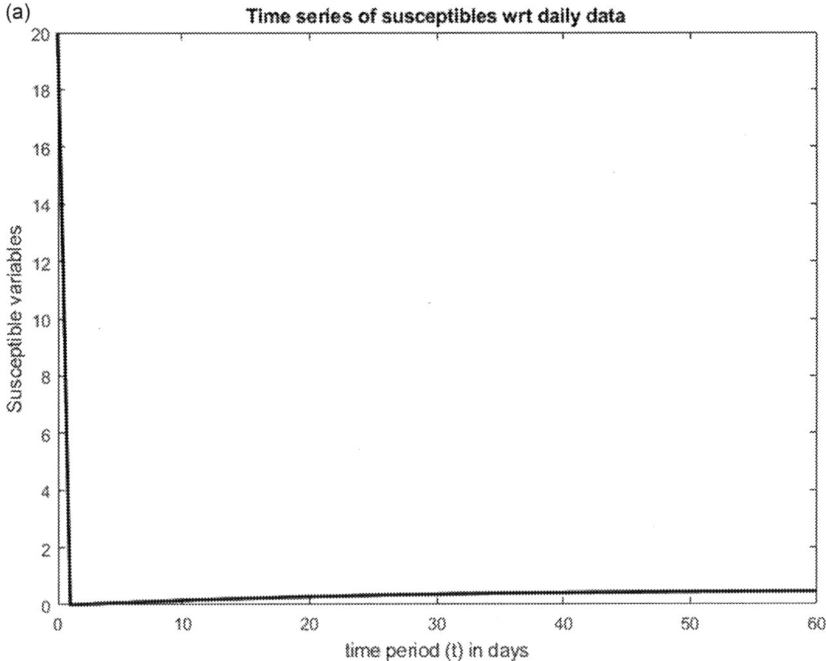

FIGURE 13.5 (a) Time series of susceptibles' population.

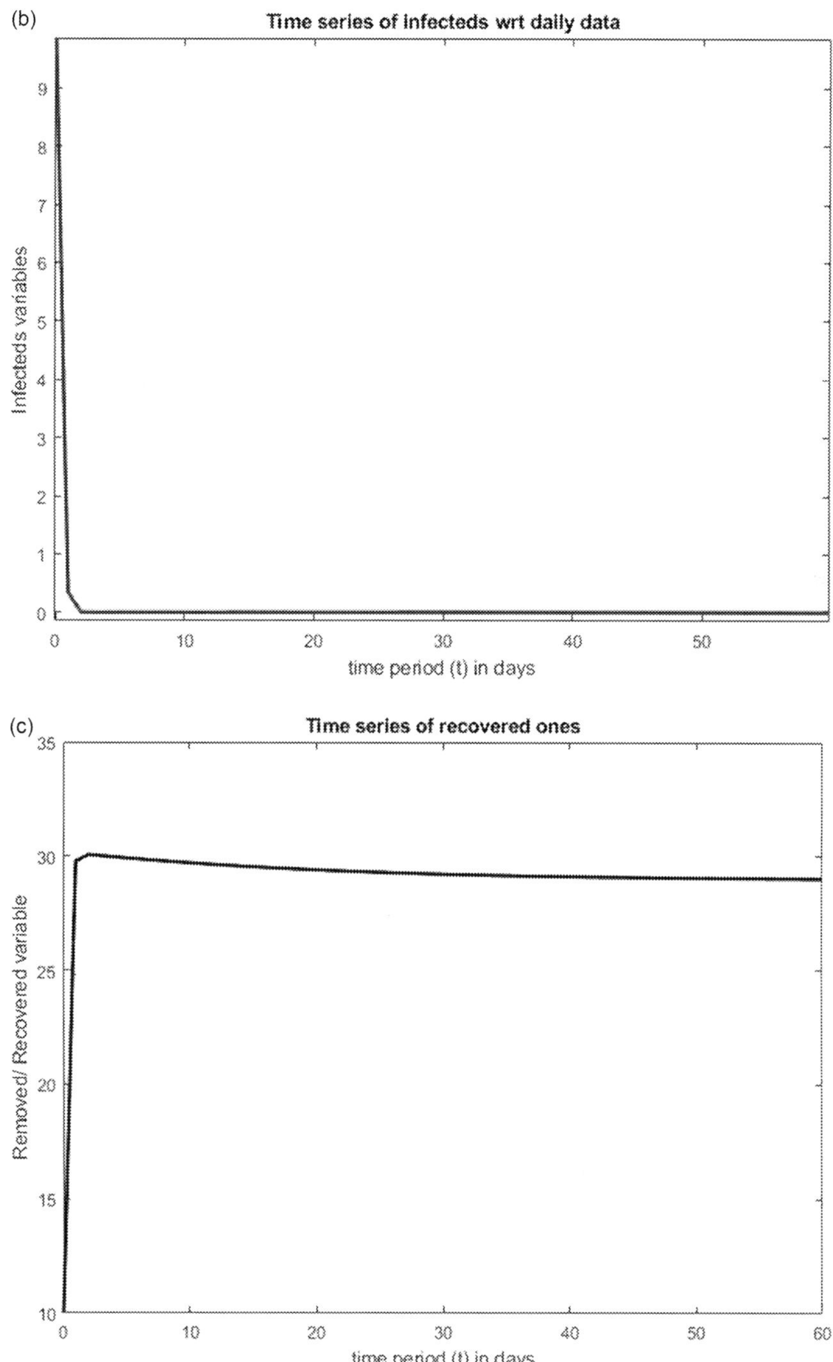

FIGURE 13.5 (CONTINUED) (b) time series of infected ones' population, and (c) time series of removed ones' population.

(Continued)

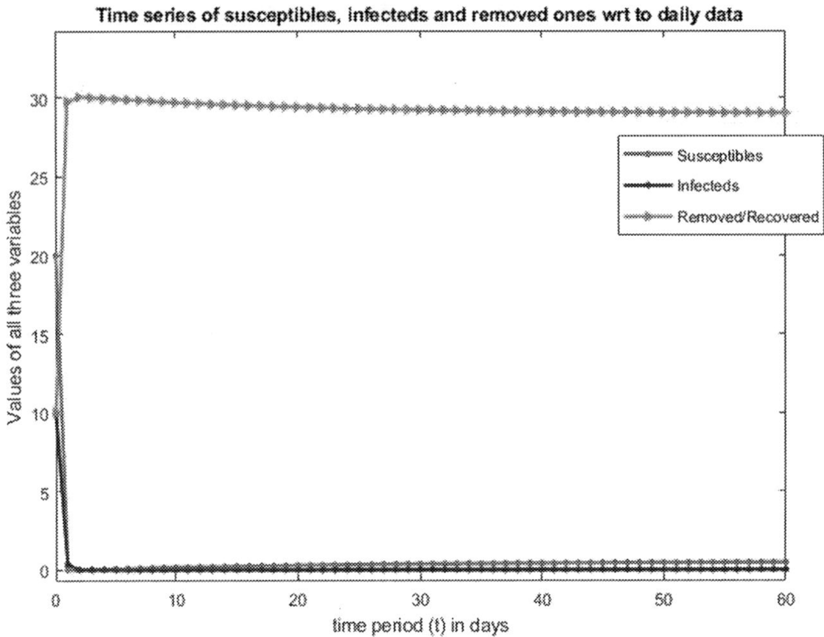

FIGURE 13.6 Relative time series of all three variables for the 60-day period considered.

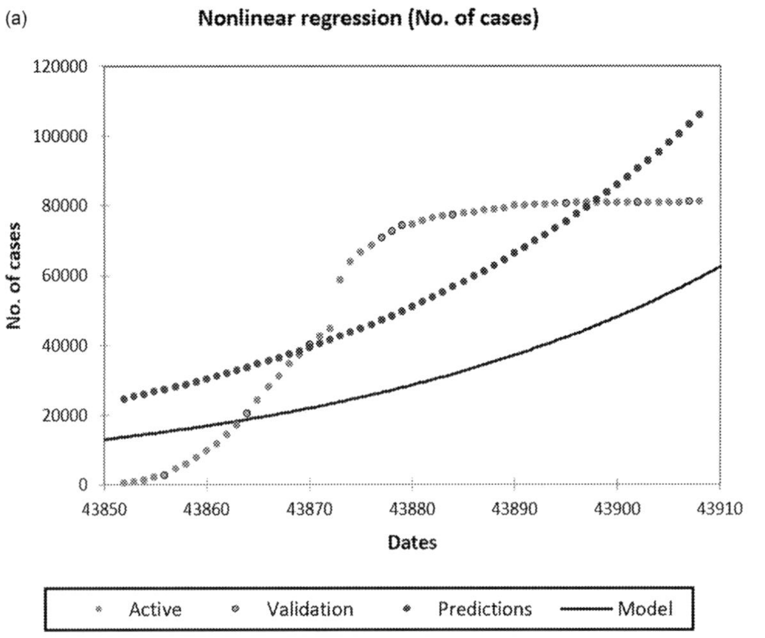

FIGURE 13.7 (a) Nonlinear regression of the daily basis data of corona cases in China.

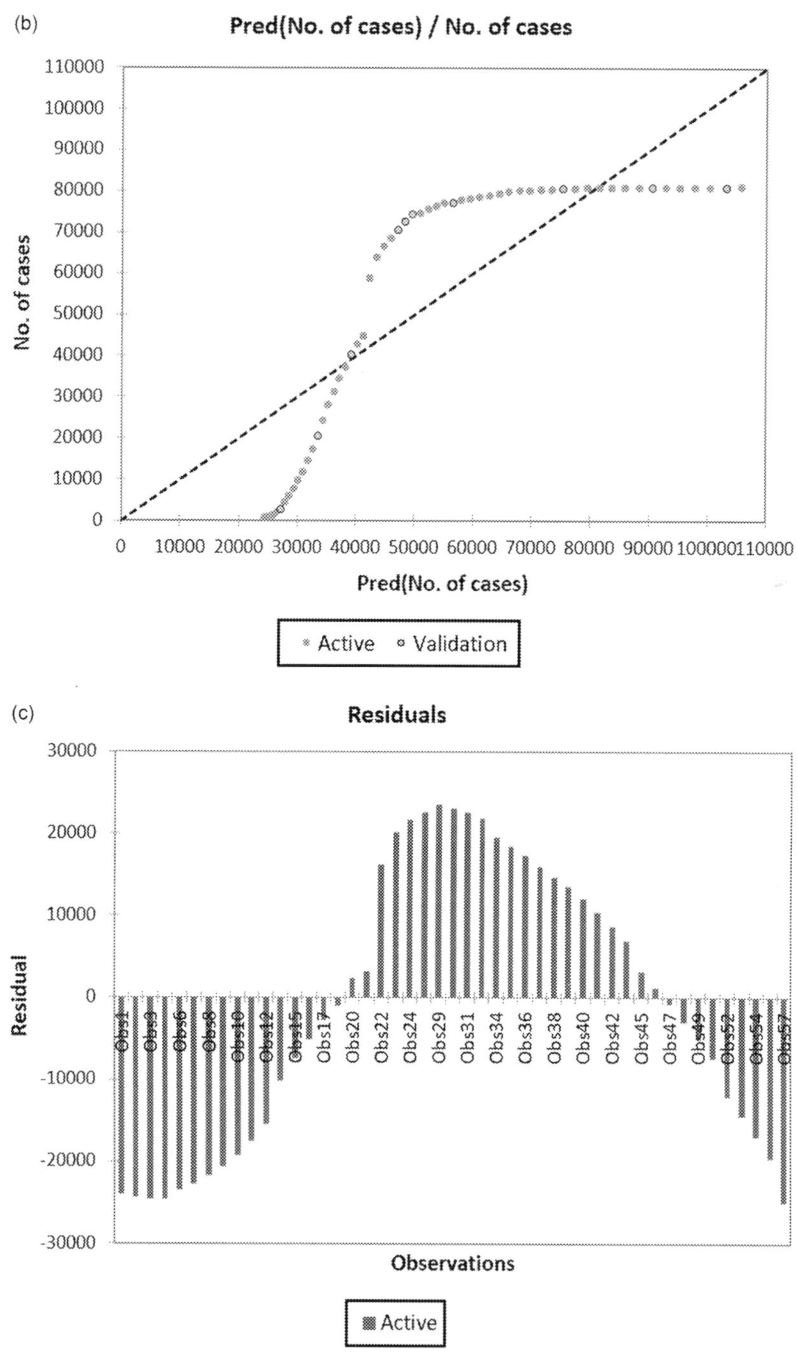

FIGURE 13.7 (CONTINUED) (b) prediction of cases vs. actual number of cases showing active and validation data, and (c) residuals of active data of nonlinear regression.

(Continued)

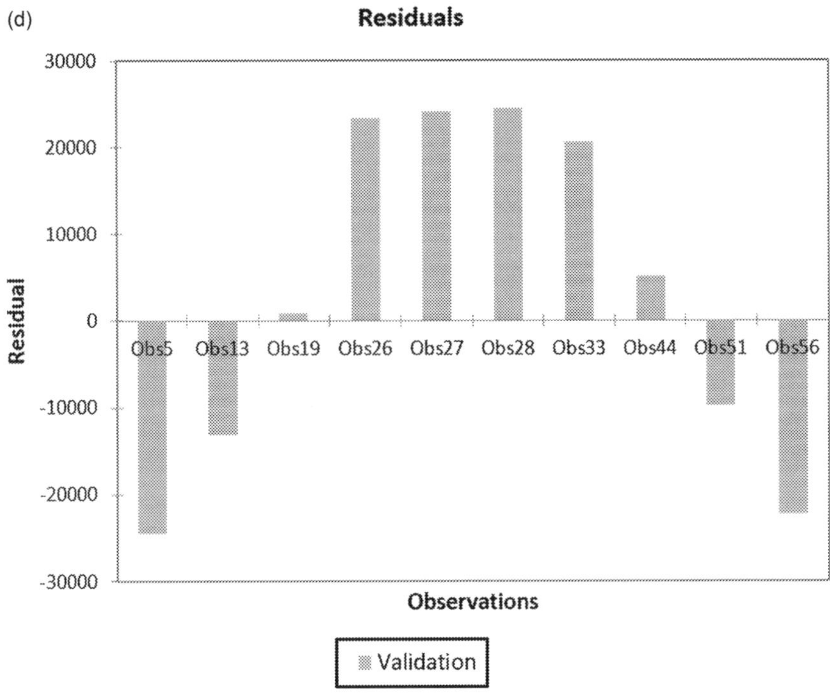

FIGURE 13.7 (CONTINUED) (d) residuals of validation data of nonlinear regression.

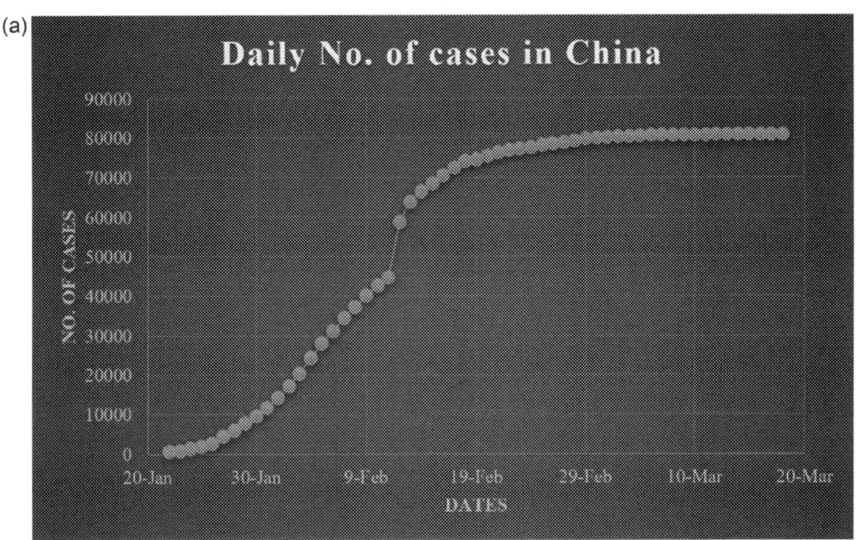

FIGURE 13.8 (a) Daily increase in cases recorded in China from January 22, 2020, to March 20, 2020.

(Continued)

(b)
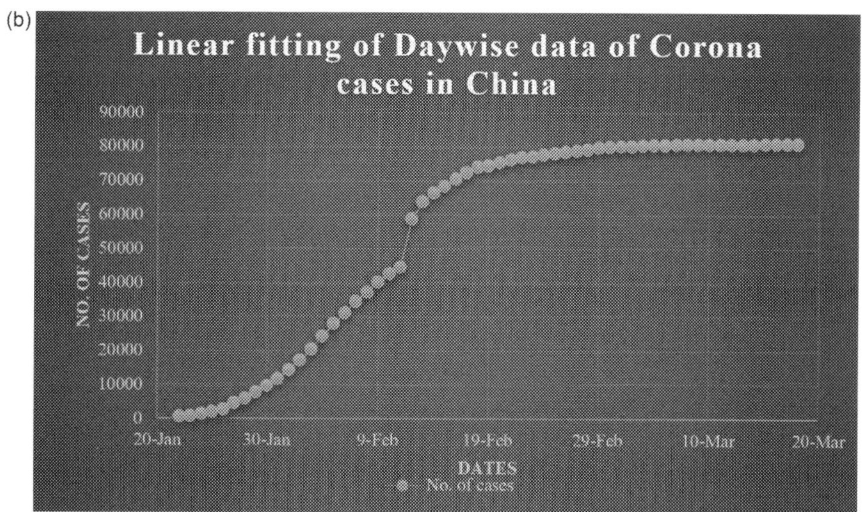

(c)
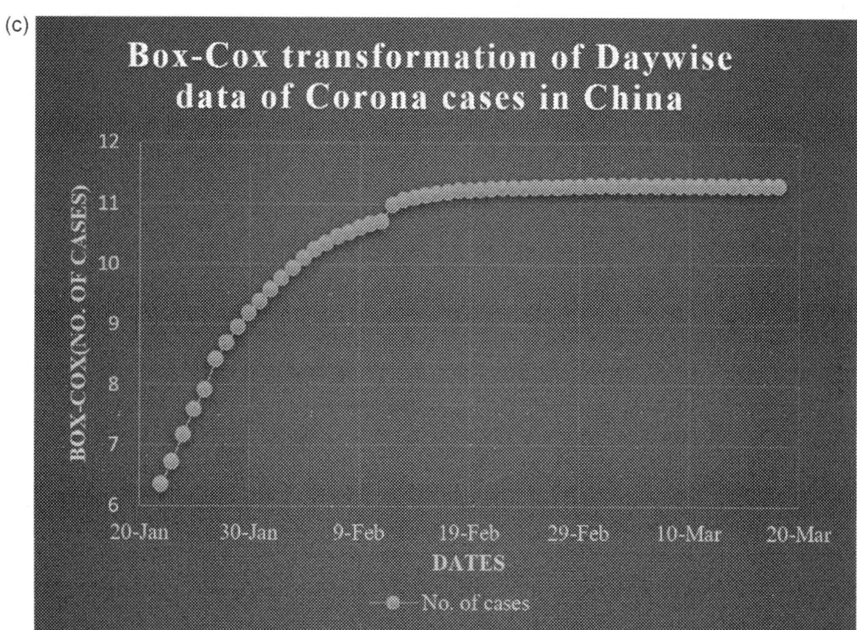

FIGURE 13.8 (CONTINUED) (b) linear fitting of daily data of corona cases in China, (c) Box–Cox transformation of daily data of corona cases in China.

(*Continued*)

(d)

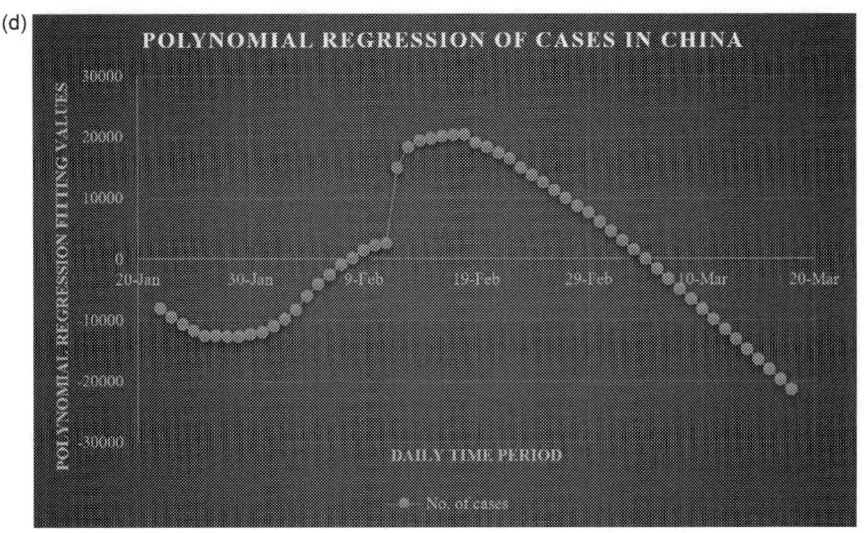

FIGURE 13.8 (CONTINUED) (d) polynomial regression of degree 6 for daily data of number of cases in China.

(a)

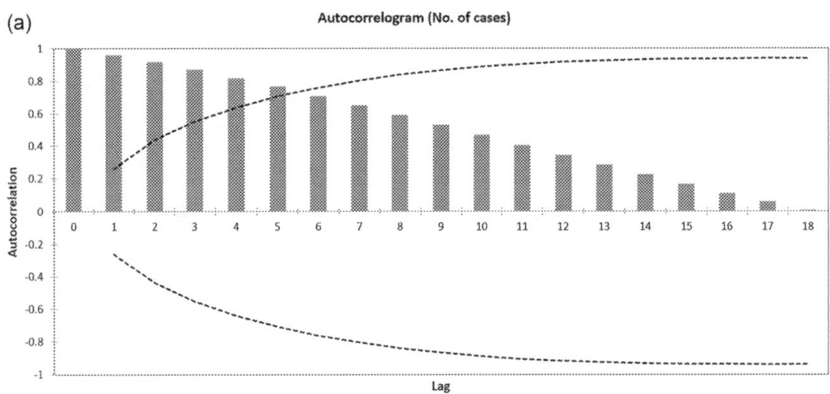

FIGURE 13.9 (a) Autocorrelation (ACF) of everyday number of cases.

(Continued)

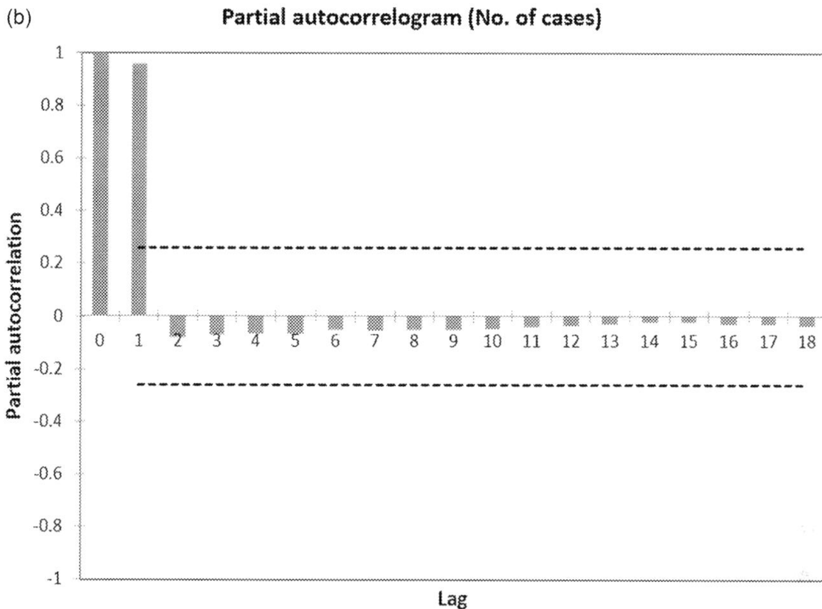

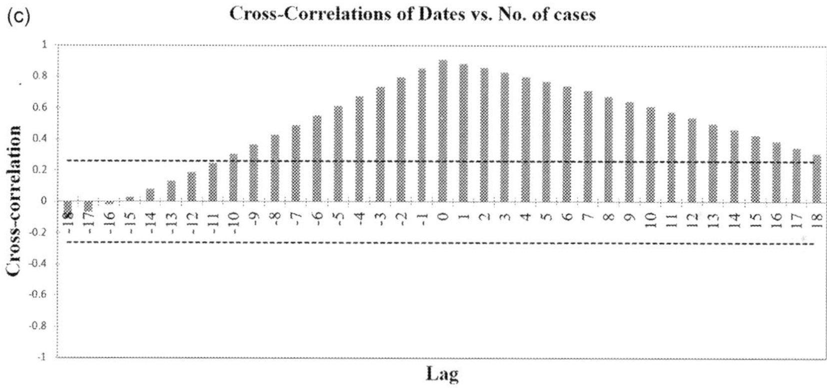

FIGURE 13.9 (CONTINUED) (b) partial autocorrelation (PACF) of everyday number of cases, and (c) cross-correlation of everyday number of cases.

(*Continued*)

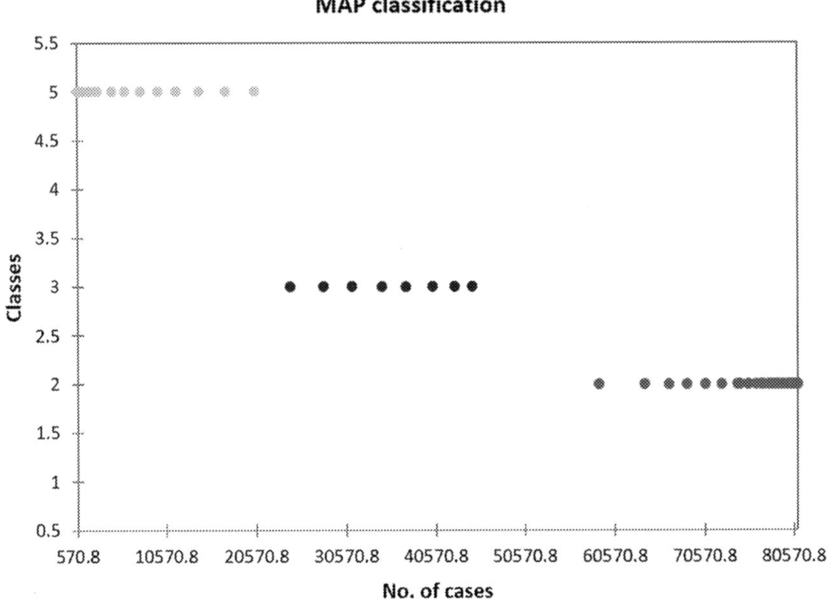

FIGURE 13.10 MAP classification categorizes into classes based on the spread of the corona.

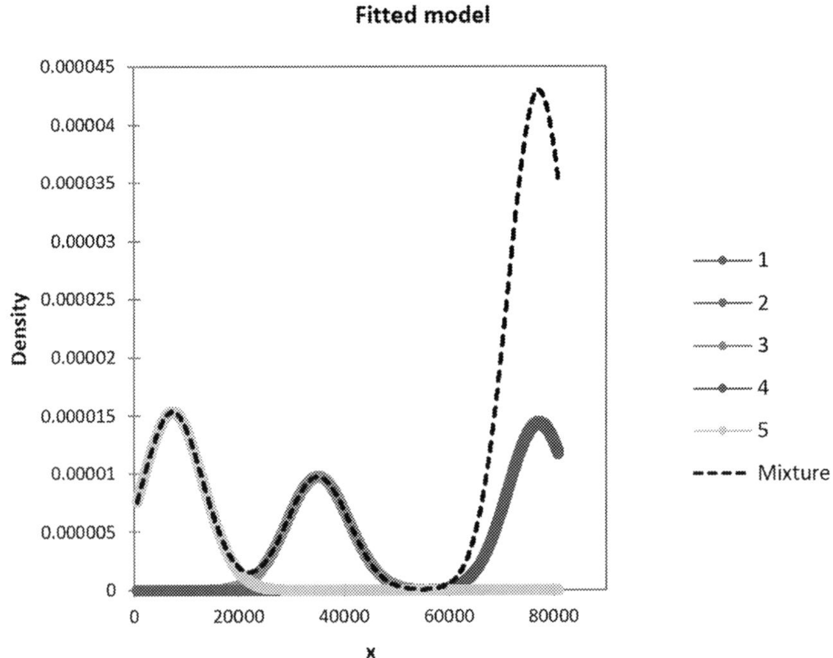

FIGURE 13.11 Fitting through mixture modeling.

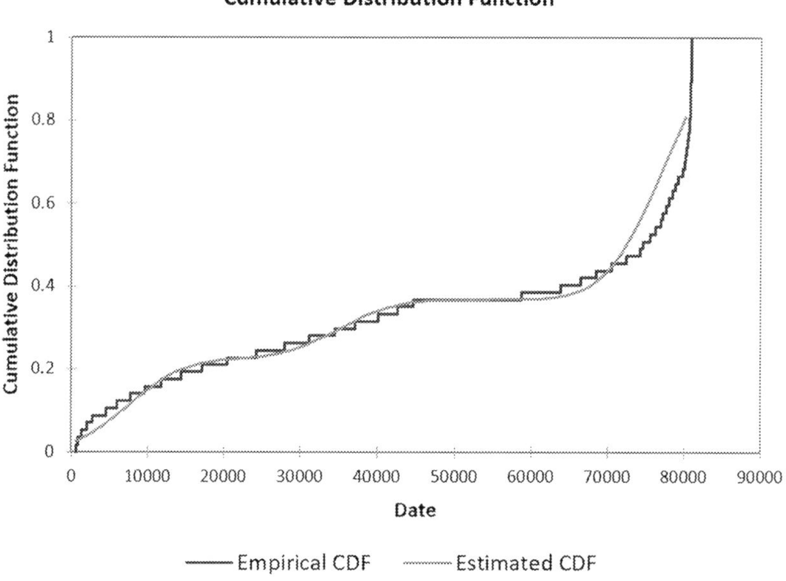

FIGURE 13.12 Cumulative distribution function via mixture modeling.

13.5 CONCLUSION

It is the need of the hour to model the factors of corona transmission in order to minimize its spread and the extent to which it can be harmful. Since China is the first country to record and report such cases, it is in a way the breeding place of this epidemic. Thus, it is necessary to understand the scenario at its core. Preventive measures should be followed at its best so that the virus does not communicate to more people and stops its breeding further. In this study, it can be observed that the theoretical model so designed has been evaluated along with the daily basis data of corona cases being recorded in China. And it can be concluded that the model through its time series, phase trajectories, and characteristic exponents agrees with the data analyzed through various techniques and regressors such as nonlinear regressors, Box–Cox transformations, polynomial regressors, autocorrelation, partial autocorrelation, regression trees, MAP classification, and mixture modeling. These techniques essentially predicted a trend comparable to that of the SIR analogous model.

The coronaviruses already identified may be only the beginning of the destruction followed by possibly more novel and austere zoonotic occurrences waiting to be unfolded. No antiviral treatment for coronavirus infection has been proven to be effective. It is concluded that for China, the spread follows Gaussian fitting with peak on February 13, 2020. Also, it is observed that on March 10, 2020, the death rate reduced to 4.89% and the survival rate goes to 95.11%. Now, almost it is following the log-normal behavior with long tail. Accordingly, the cases have been reduced remarkably, and the state is almost on the verge of recovering citizens. Also, nearly no or very few new cases are being recorded to this date.

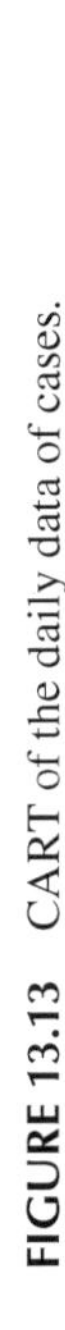

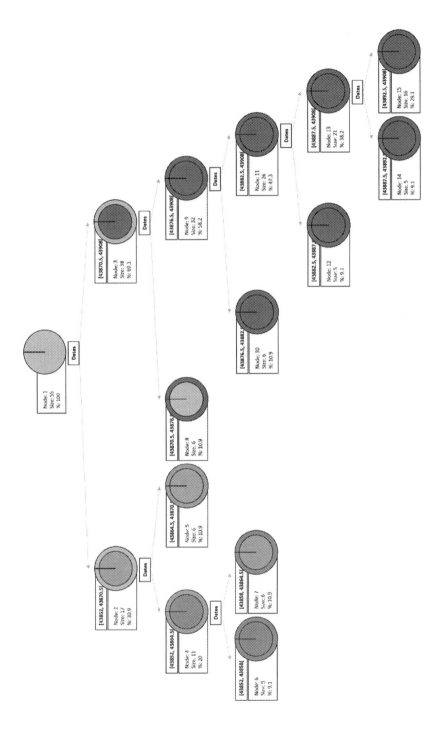

FIGURE 13.13 CART of the daily data of cases.

ACKNOWLEDGMENTS

Guru Gobind Singh Indraprastha University provided financial support and research facilities to the first and second author of this chapter. The third author is thankful to Gyan Ganga Institute of Science and Technology. There is no conflict of interest among the authors.

REFERENCES

1. Bhardwaj, R. (2016). Wavelets and Fractal Methods with environmental applications. *Mathematical Models, Methods and Applications* Singapore: Springer Science+Business Media, 173–195. ISBN: 978-981-287-971-4.
2. Bhardwaj, R. (2019). *Nonlinear Time Series Analysis of Environment Pollutants. Mathematical Modeling on Real World Problems: Interdisciplinary Studies in Applied Mathematics*, New York: NOVA Publisher, 71–102.
3. Bhardwaj, R. & Bangia, A. (2019). Hybrid fuzzified-PID controller for non-linear control surfaces for DC motor to improve the efficiency of electric battery driven vehicles. *International Journal of Recent Technology and Engineering™ (IJRTE)*. 8(3), 2561–2568.
4. Bhardwaj, R. & Bangia, A. (2019). Dynamic indicator for the prediction of atmospheric pollutants. *Asian Journal of Water, Environment and Pollution*. 16(4), 39–50.
5. Bhardwaj, R. & Bangia, A. (2016). Complexity dynamics of meditating body. *Indian Journal of Industrial and Applied Mathematics*. 7(2), 106–116.
6. Bhardwaj, R. & Bangia, A. (2019). Stock market trend analysis during demonetization using soft-computing techniques. *2018 International Conference on Computing, Power and Communication Technologies (GUCON-2018). IEEE Xplore Digital Library*, 696–701.
7. Bhardwaj, R. & Bangia, A. (2018). Statistical time series analysis of dynamics of HIV. *JNANABHA*. Special Issue 48, 22–27.
8. Bhardwaj, R. & Bangia, A. (2019). Neuro-fuzzy analysis of demonetization on NSE. In Bansal, J.C., Das, K.N., Nagar, A., Deep, K., & Ojha, A.K (Eds). *Advances in Intelligent Systems and Computing (AISC) (ISSN:*2194-5357) *Springer Proceedings: Soft Computing for Problem Solving (SocPros*-2017), 853–861.
9. Bhardwaj, R. & Bangia, A. (2020). Dynamical forensic inference for malware in iot-based wireless transmissions. In K. Sharma, M. Makino, G. Shrivastava, & B. Agarwal (Eds.) *Forensic Investigations and Risk Management in Mobile and Wireless Communications*, IGI Global, 51–79.
10. Bangia, A., Bhardwaj, R., & Jayakumar, K.V. (2020). River water quality estimation through Artificial Intelligence conjuncted with Wavelet Decomposition. 979. *Numerical Optimization in Engineering and Sciences*, Springer, 107–123.
11. Bhardwaj, R. & Bangia, A. (2020). Assessment of Stock prices variation using Intelligent Machine Learning Techniques for the prediction of BSE. In J. Kacprzyk, D. Dutta, & B. Mahanty (Eds.) *Advances in Intelligent Systems and Computing (AISC) Volume 979. Numerical Optimization in Engineering and Sciences*, 107–123.
12. Bhardwaj, R. & Bangia, A. (2020). Data Driven Estimation of Novel COVID-19 Transmission Risks through Hybrid Soft-Computing Techniques. *Chaos, Soliton and Fractals*. doi:10.1016/j.chaos.2020.110152
13. Bhardwaj, R. & Datta, D. (2020). Consensus Algorithm. 71, Studies in Big Data, 91–107, ISBN: 978-3-030-38676-4.

14. Bhardwaj, S., Ashish, K., & Deepak, G. (2020). Water quality evaluation using soft computing method. In D. Gupta, A. Khanna, S. Bhattacharyya, A. E. Hassanien, S. Anand, A. Jaiswal (Eds.) *Advances in Intelligent Systems and Computing* Vol. 1166. Springer.
15. Sharma, S. K., Bhardwaj, S., Alowaidi, M., & Bhardwaj, R. (2020). Nonlinear Time series analysis of Pathogenesis of COVID-19 Epidemiology Spread in Saudi Arabia *Computers, Materials and Continua*. In press.
16. Bhardwaj, R., Datta, D., Bhardwaj, R., Bhardwaj, S., Sharma, S. K., Al Shehri, M. (2020). An apparatus and method with IOT to detect and control temperature change simulation case. Indian Patent Application No. 202011021339.
17. Bhardwaj, R. & Datta, D. (2020). Development of epidemiological modeling RD_COVID-19 of coronavirus infectious disease and its numerical simulation. In P. Agarwal, J. J. Nieto, D. F. M. Torres (Eds.) *Mathematical Modelling and Analysis of Infectious Disease Problems (COVID-19)*, Springer.
18. Bhardwaj, R. & Datta, D. (2020). Development of a recommender system health Mudra using blockchain for prevention of diabetes. In S. Mohanty, A. A. Elengar, S. Jain, P. Gupta, & J. Chhatterjee (Eds.) *Recommender System with Machine Learning and Artificial Intelligence: Particle Tools and Applications in Medical and Agricultural Domains*. Wiley & Sons, USA.
19. Datta, D. & Bhardwaj, R. (2020). Fuzziness-randomness modeling of plasma disruption in first wall of fusion reactor using type I fuzzy random set. In *An Introduction to Fuzzy Sets*, Nova publisher Inc., ISBN: 978-1-53618-012-1
20. Bhardwaj, R. & Datta, D. (2020). Optimization techniques. *Revista INGLOMAYOR Ingeniena Global Mayor*. 18(A), 54–82. ISSN 0719 7578, INGLOMAYOR, Chile.
21. Bhardwaj, R. & Pruthi, D. (2020). Evolutionary techniques for optimizing air quality model. *Procedia Computer Science*. 167C, 1872–1879. ISSN: 1877-0509, Elsevier.
22. Bhardwaj, R. & Pruthi, D. (2020). Development of model for sustainable nitrogen dioxide prediction using neuronal networks. *International Journal of Environmental Science and Technology*. 17, 2783–2792. doi:10.1007/s13762-019-02620-z. ISSN: 1735-1472 (Print) 1735-2630, Springer.
23. Bhardwaj, R. & Duhoon, V. (2020). Auto-regressive integrated moving-averages model for daily rainfall forecasting. *International Journal of Scientific and Technology Research*. 9(2), 793–797. ISSN: 2277-8616.
24. Bhardwaj, R. & Das, S. (2020). Synchronization of two three-species food chain system with Beddington- DeAngelis functional response using active controllers based on the Lyapunov function. *Italian Journal of Pure and Applied Mathematics*. 44. 00–00.
25. Box, G. E. P. & Cox, D. R. (1964). An analysis of transformations. *Journal of the Royal Statistical Society, Series B*. 26, 211–252.
26. Chen T.-M., Rui J., Wang Q.-P., Zhao Z.-Y., Cui J-A, Y. L. (2020).A mathematical model for simulating the phase-based transmissibility of a novel coronavirus. *Infectious Diseases of Poverty*. 9(24), 1–8.
27. Rosenblum, M. A. (1980). Corona theorem for countably many functions. *Integral equations Operator Theory*. 3, 125–137.
28. Fuhrmann, P. A. (1968). On the Corona theorem and its application to spectral problems in Hilbert space. *Transactions of the American Mathematical Society*. 132(1), 55–66.
29. WHO. (2013). Middle East respiratory syndrome coronavirus (MERS-CoV) - update: 2 December 2013 http://www.who.int/csr/don/2013_12_02/en/.
30. Wu C.-Y., Jan J.-T., Ma S.-H., Kuo C.-J., Juan H.-F., Cheng Y.-S.E., et al. (2004). Small molecules targeting severe acute respiratory syndrome human coronavirus. *Proceedings of National Academy of Sciences*. 101, 10012–10017.

31. Zhou N., Pan T., Zhang J., Li Q., Zhang X., Bai C., et al. (2016). Glycopeptide antibiotics potently inhibit cathepsin L in the late endosome/lysosome and block the entry of ebola virus, Middle East Respiratory Syndrome Coronavirus (MERS-CoV), and Severe Acute Respiratory Syndrome Coronavirus (SARS-CoV). *Journal of Biological Chemistry*. 291, 9218–9232.
32. WHO. (2020). Novel Coronavirus – China. 12 January 2020. http://www.who.int/csr/don/12-january-2020-novel-coronavirus-china/en/.
33. Mishra J. (2018). Fractional hyper-chaotic model with no equilibrium. *Chaos, Solitons & Fractals*. 116, 43–53.

14 A Mathematical Fractional Model to Study the Hepatitis B Virus Infection

Ritu Agarwal and Kritika
Malaviya National Institute of Technology

S. D. Purohit
Rajasthan Technical University

Jyoti Mishra
Gyan Ganga Institute of Technology and Sciences

CONTENTS

14.1 INTRODUCTION

The hepatitis B virus (HBV) is one of the most important viruses causing hepatitis in humans. HBV leads to critical health problems and is the leading cause of death around the world. Due to the hepatitis B, nearly two billion individuals have been infected, and 7,80,000 individuals died worldwide. The contagious hepatitis B infection is the main cause of the primary liver cancer. Through many ways, this infection can be transmitted from one person to another; it involves skin cut or contact of mucosa with the blood or with the fluid of an infected person's body, and also through the semen and saliva, by sharing the syringes. The open sores on the body of infected person can also transmit the infection from one person to another.

The area of mathematical epidemiology is helpful in understanding the rapid transmission of the infection via the good comprehensive models and also helpful in evaluating the effectiveness of control mechanism through proper way.

The dynamic aspects of HBV infection can be addressed by the use of mathematical modeling. Among the efforts of modeling, the set of mathematical equations plays a key role in explaining the dynamics of the host cell, virus, and probably the immune system.

HBV is the type of the Hepadnaviridae material in a short (3kb) partially double-stranded circular DNA molecule, which is transformed into covalently closed circular DNA (cccDNA). Certain copies of cccDNA are found in the nucleus of the infected cell and are involved in the production of mRNDA. By the process of viral polymerase, the full-length unspliced mRNA pregenome get transcribed into DNA for the formation of new virions [13].

HBV chiefly infects the hepatocytes entirely. The HBV can also infect the other cell types by replicating at both intra- and extra-hepatical sites. But infection on the cell types other than the hepatocytes is not well documented. The basic model incorporates uninfected cells (i.e., target cells: hepatocytes), infected cells, and the virus. The formation of target cells occurs at a constant rate. These cells are created by the differentiation of progenitor cells or by a direct procreation from the mature hepatocytes. The literature behind the modeling of the dynamics of hepatitis B is very rich: [22,23,26,30]. A number of models have been used by various authors to explain the dynamics of the contagious infections such as HIV, hepatitis B, and vector-borne diseases (see, e.g., [6,14,16,33]).

Following HBV infection model of hepatocytes of curable infected cells, which was considered by Vargas–De–León [31] (see also [18,24]), we have

$$\frac{dX}{dt} = a - bX - cXZ + dY. \tag{14.1}$$

$$\frac{dY}{dt} = cXZ - (v + d)Y, \tag{14.2}$$

$$\frac{dZ}{dt} = sY - gZ, \tag{14.3}$$

where X, Y, and Z denote the density of the hepatocytes; the density of the uninfected one and the infected cells; and the density of virions, respectively. Let a be the production rate of the susceptible cells and are dying at the rate bX. At cXZ, these cells (susceptible) are getting infected; here, c denotes the rate of infection "cure." The infected hepatocytes are dying at the rate vY, the free virions from the infected hepatocytes are produced at the rate sY, and the viral particles are cleared at the rate of gZ. The term d appeared in Eq. (14.1) corresponding to the rate at which uninfected hepatocytes are created through 'cure'.

The mathematical operations such as differentiation and integration are generalized to arbitrary order under the area of mathematics: fractional calculus. Recently, this area is applied at a large scale in various fields [2,3,11,12,15,17,27,32]. Differential equations are frequently used in science and engineering, and that's why the concept of fractional calculus (introduction of arbitrary order in them) attracted many applied researchers [1,4,9,10]. The fractional calculus is capable to perform integration and differentiation of fractional order. To understand the dynamics of biological systems using mathematical models, the integer-ordered differential equations are valuable. But many biological systems have memory or after effects. The models involving fractional-ordered differential equations are advantageous over the integer-ordered models as these classical models ignore such effects. The nonlocal property of fractional operators makes them more beneficial over the use of integer-ordered differential operators; i.e., the next state of system depends not only upon current but also upon all its previous states. Due to this nonlocality property of fractional operators, the nonlinear fractional model considers the full memory effects and explains the problem accurately.

Salman and Yousef [28] considered the fractional model for modeling the phenomenon and obtained the numerical solution using predictor-corrector approach for the numerical solution of fractional differential equations.

In this study, we examine the fractional form of the above-mentioned HBV infection model. The Caputo fractional derivative that has been used to fractionalize the model is beneficial because it provides the freedom to use the classical initial condition for the respective initial value problem without facing problem during solvability. Due to the nonlocality of the Caputo derivative, the nonlinear model considers the full memory effects and explains the problem accurately.

In this work, we employed the FHATM (fractional homotopy analysis transform method). This method is a result of the modification in the homotopy analysis method (HAM) (given by Liao [19,20,21] by merging it with the Laplace transform technique. This modification provides the simplified procedure to reach the solution as compared to the standard methods. Unlike HAM, in FHATM there is no requirement of assuming an auxiliary linear operator.

Definition 14.1

Let the real-valued function g and its ith-order derivatives ($i = 1,2,3,\ldots,n$) be continuous in the interval $(0,\infty)$. Then, the fractional derivative of order η given by Caputo is [7, Eq. 5, p. 30] defined as

$$ {}^{C}D_0^{\eta}\left(g(t)\right) = \begin{cases} \dfrac{1}{\Gamma(n-\eta)} \displaystyle\int_0^t \dfrac{g^{(n)}(\tau)}{(t-\tau)^{\eta+1-n}}\, d\tau, & \text{if } n-1<\eta\le n, \quad n\in\mathbb{N}, \\ \dfrac{d^n}{dt^n} g(t), & \text{if } \eta = n \end{cases} \tag{14.4} $$

Definition 14.2

The Laplace transform of piecewise continuous function $g(t)$ of exponential order $\alpha > 0$ with respect to variable t is defined as (see, e.g., [29])

$$ L\left[g(t);p\right] = \bar{f}(p) = \int_0^\infty e^{-pt} g(t)\,dt, \quad \Re(p) > \alpha, t \ge 0. \tag{14.5} $$

The inverse Laplace transform of function $\overline{g}(p)$ with respect to $t \geq 0$ is given by

$$L^{-1}\left[\overline{g}(p);t\right] = f(t) = \frac{1}{2\pi i}\int_{\gamma-i\infty}^{\gamma+i\infty} e^{pt}\overline{g}(p)dp, \tag{14.6}$$

where γ is the fixed real number.

Definition 14.3

The Laplace transform of the Caputo fractional derivative ${}^{C}D_0^{\eta}\left(g(t)\right)$ is given by Caputo [8] as:

$$L\left[{}^{C}D_0^{\rho}g(t)\right] = k^{\rho}L\left[g(t)\right] - \sum_{i=0}^{n-1} k^{\rho-i-1}g^{i}(0), \quad n-1 < \rho \leq n. \tag{14.7}$$

14.2 THE FRACTIONAL MODEL

Here, we are considering the fractionalized HBV infection model [28] as follows:

$$D_t^{\rho}X(t) = a - bX - cXZ + dy, \tag{14.8}$$

$$D_t^{\rho}Y(t) = cXZ - (v+d)Y, \tag{14.9}$$

$$D_t^{\rho}Z(t) = sY - gZ, \tag{14.10}$$

where $\rho \in (0,1)$ and the initial conditions for the system are as follows [31] (see also, [28]): $X(0) = 1.73 \times 10^{8}$ cells $(\text{mL})^{-1}$, $Y(0) = 0$, and $Z(0) = 400$ copies $(\text{mL})^{-1}$.

The relation of the fractional-ordered differential equations with the systems with memory is the reason behind considering the fractional model. Here, this reason is applicable to the immune system, which creates the memory T cells and B cells which are from their experiences capable to fight any threat; these cells have the proficiency to identify and fight the same threat later. On the contrary, there is no information regarding the memory of neither the hepatocytes nor the free virions in integer-ordered models.

14.3 SOLUTION OF THE HBV INFECTION MODEL

For finding the solution of the system of Eqs. (14.8)–(14.10), taking the Laplace transform, and using the initial conditions thereupon

$$k^{\rho}L\left[X(t);k\right] - k^{\rho-1}X_0 + L\left[-a + bX + cXZ - dY\right] = 0, \tag{14.11}$$

$$k^{\rho}L\left[Y(t);k\right] - k^{\rho-1}Y_0 + L\left[-cXZ + (v+d)Y\right] = 0, \tag{14.12}$$

$$k^{\rho}L[Z(t);k]-k^{\rho-1}Z_0+L[-sY+gZ]=0. \tag{14.13}$$

For homotopy analysis, the nonlinear operators are defined as follows:

$$N_1\left[f_1(t,q),f_2(t,q),f_3(t,q)\right]=L\left[f_1(t),q\right]-\frac{X_0}{k}+\frac{1}{k^{\rho}}L\left[-a+bf_1+cf_1f_2-df_2\right], \tag{14.14}$$

$$N_2\left[f_1(t,q),f_2(t,q),f_3(t,q)\right]=L\left[f_2(t),q\right]-\frac{Y_0}{k}+\frac{1}{k^{\rho}}L\left[-cf_1f_2+(v+d)f_2\right], \tag{14.15}$$

$$N_3\left[f_1(t,q),f_2(t,q),f_3(t,q)\right]=L\left[f_3(t),q\right]-\frac{Z_0}{k}+\frac{1}{k^{\rho}}L\left[-sf_2+gf_3\right], \tag{14.16}$$

where $q\in[0,1]$ denotes the embedding parameter and $f_1(t,q)$, $f_2(t,q)$, and $f_3(t,q)$ are the real-valued functions. Now, developing the homotopy as

$$(1-q)L\left[f_i(t,q)-I_i(t)\right]=hqN_i\left[f_1,f_2,f_3\right],\quad i=1,2,3. \tag{14.17}$$

Here, L stands for the Laplace transform operator; $h\neq 0$ is for auxiliary parameter; and $I_i(t)$ represents the initial estimates of $X(t)$, $Y(t)$, and $Z(t)$, respectively, corresponding to $i=1,2,3$. If the embedding parameter $q=0$, it gives

$$f_1(t;0)=I_1(t), \tag{14.18}$$

$$f_2(t;0)=I_2(t), \tag{14.19}$$

$$f_3(t;0)=I_3(t), \tag{14.20}$$

and for $q=1$, we have

$$f_1(t;1)=X(t), \tag{14.21}$$

$$f_2(t;1)=Y(t), \tag{14.22}$$

$$f_3(t;1)=Z(t). \tag{14.23}$$

As a result, as the values of q vary from 0 to 1, the values of $f_i(t,q)$ $(i=1,2,3)$ vary from the initial guess $I_i(t)$ $(i=1,2,3)$ to the solutions of $X(t)$, $Y(t)$, and $Z(t)$, respectively. On writing the function $f_i(t;q)$, $(i=1,2,3)$ in the form of series using Taylor's theorem around q, we have

$$X(t;q)=X_0(t)+\sum_{m=1}^{\infty}X_m(t)q^m,\quad \text{where } X_m=\frac{1}{m}\frac{\partial^m f_1(t;q)\,|_{q=0}}{\partial q^m}, \tag{14.24}$$

$$Y(t;q)=Y_0(t)+\sum_{m=1}^{\infty}Y_m(t)q^m,\quad \text{where } Y_m=\frac{1}{m}\frac{\partial^m f_2(t;q)\,|_{q=0}}{\partial q^m}, \tag{14.25}$$

and

$$Z(t;q) = Z_0(t) + \sum_{m=1}^{\infty} Z_m(t)q^m, \quad \text{where } Z_m = \frac{1}{m}\frac{\partial^m f_3(t;q)|_{q=0}}{\partial q^m}. \tag{14.26}$$

Now on making the use of initial approximation $I_1(t) = X_0(t)$, $I_2 = Y_0(t)$, and $I_3(t) = Z_0(t)$, selecting the parameter h appropriately, we reach at the result given below as the series (14.24) converges at $q = 1$,

$$X(t) = X_0(t) + \sum_{m=1}^{\infty} X_m(t), \tag{14.27}$$

$$Y(t) = X_0(t) + \sum_{m=1}^{\infty} Y_m(t), \tag{14.28}$$

$$Z(t) = X_0(t) + \sum_{m=1}^{\infty} Z_m(t). \tag{14.29}$$

Now, defining the *m*th-order deformation equations,

$$L\left[X_m - \phi_m X_{m-1}\right] = h_1 R_{1,m}\left(X_{m-1}(t), Y_{m-1}(t), Z_{m-1}(t)\right), \tag{14.30}$$

$$L\left[Y_m - \phi_m Y_{m-1}\right] = h_2 R_{2,m}\left(X_{m-1}(t), Y_{m-1}(t), Z_{m-1}(t)\right), \tag{14.31}$$

$$L\left[Z_m - \phi_m Z_{m-1}\right] = h_3 R_{3,m}\left(X_{m-1}(t), Y_{m-1}(t), Z_{m-1}(t)\right), \tag{14.32}$$

where

$$R_{1,m}\left(X_{m-1}(t), Y_{m-1}(t), Z_{m-1}(t)\right) = L\left[X_{m-1}(t)\right] - \left(1 - \phi_m\right)\frac{X_0}{k} + \frac{1}{k^{\rho}} L\left[-a + bX_{m-1} + cA_{m-1} - dY_{m-1}\right], \tag{14.33}$$

$$R_{2,m}\left(X_{m-1}(t), Y_{m-1}(t), Z_{m-1}(t)\right) = L\left[Y_{m-1}(t)\right] - \left(1 - \phi_m\right)\frac{Y_0}{k} + \frac{1}{k^{\rho}} L\left[-cA_{m-1} + (\upsilon + \mathrm{d})Y_{m-1}\right], \tag{14.34}$$

$$R_{3,m}\left(X_{m-1}(t), Y_{m-1}(t), Z_{m-1}(t)\right) = L\left[Z_{m-1}(t)\right] - \left(1 - \phi_m\right)\frac{Z_0}{k} + \frac{1}{k^{\rho}} L\left[-8Y_{m-1} + gZ_{m-1}\right], \tag{14.35}$$

respectively. On taking the Laplace inverse transformation,

$$X_m = \phi_m X_{m-1} + h_1 L^{-1} R_{1.m}\left(X_{m-1}(t), Y_{m-1}(t), Z_{m-1}(t)\right), \tag{14.36}$$

$$Y_m = \phi_m Y_{m-1} + h_2 L^{-1} R_{2.m}\left(X_{m-1}(t), Y_{m-1}(t), Z_{m-1}(t)\right), \tag{14.37}$$

$$Z_m = \phi_m Z_{m-1} + h_3 L^{-1} R_{3.m}\left(X_{m-1}(t), Y_{m-1}(t), Z_{m-1}(t)\right). \tag{14.38}$$

By substituting (14.33), (14.34), and (14.35) in above equations, we have

$$X_m = \phi_m X_{m-1} + h_1 L^{-1}\left[L[X_{m-1}] - (1-\phi_m)\frac{X_0}{k} + \frac{1}{k^\rho} L[-a + bX_{m-1} + cA_{m-1} - dY_{m-1}]\right], \tag{14.39}$$

$$Y_m = \phi_m Y_{m-1} + h_2 L^{-1}\left[L[Y_{m-1}] - (1-\phi_m)\frac{Y_0}{k} + \frac{1}{k^\rho} L[-cA_{m-1} + (v+d)Y_{m-1}]\right], \tag{14.40}$$

$$Z_m = \phi_m Z_{m-1} + h_3 L^{-1}\left[L[Z_{m-1}] - (1-\phi_m)\frac{Z_0}{k} + \frac{1}{k^\rho} L[-sY_{m-1} + gZ_{m-1}]\right], \tag{14.41}$$

where

$$\phi_m = \begin{cases} 0, & \text{if } m \le 1, \\ 1, & \text{if } m > 1 \end{cases} \tag{14.42}$$

A_m is the homotopy polynomial and is expressed as

$$A_m = \frac{1}{\Gamma(m+1)} \frac{d^m}{dq^m}\left[\sum_{j=0}^{m} q^j X_j(t) \sum_{j=0}^{m} q^j Z_j(t)\right]_{q=0}. \tag{14.43}$$

For ease, we will take $h_1 = h_2 = h_3 = h$ and take the initial approximations as $X_0 = 1.73 \times 10^8$ cells$(\text{mL})^{-1}$, $Y_0 = 0$, and $Z_0 = 400$ copies$(\text{mL})^{-1}$.

By using the recursive schemes (14.39), (14.40), and (14.41), we will obtain the following components of series solution:

$$\begin{aligned} X_1(t) &= \phi_1 X_0 + h_1 X_0 - (1-\phi_1) X_0 + L^{-1}\left[\frac{1}{k^\rho}\left(\frac{-a}{k} + \frac{bX_0}{k} + \frac{cA_0}{k} - \frac{dY_0}{k}\right)\right] \\ &= h_1 X_0 - X_0 + \frac{-at^\rho}{\Gamma(\rho+1)} + \frac{-bX_0 t^\rho}{\Gamma(\rho+1)} + \frac{cX_0 Z_0 t^\rho}{\Gamma(\rho+1)} - \frac{dY_0 t^\rho}{\Gamma(\rho+1)} \\ &= -2X_0 + (-a + bX_0 + cA_0)\frac{t^\rho}{\Gamma(\rho+1)}, \end{aligned} \tag{14.44}$$

$$Y_1(t)=\phi_1 Y_0+h_2 Y_0-\left(1-\phi_1\right)\frac{Y_0}{k}+L^{-1}\left[\frac{1}{k^{\rho}}\left(\frac{-cA_0}{k}+\frac{(\upsilon+\mathrm{d})Y_0}{k}\right)\right]$$

$$=\phi_1 Y_0+h_2 Y_0-\left(1-\phi_1\right)Y_0-\frac{cA_0 t^{\rho}}{\Gamma(\rho+1)}+\frac{(\upsilon+\mathrm{d})Y_0 t^{\rho}}{\Gamma(\rho+1)}$$

$$=\frac{-cA_0 t^{\rho}}{\Gamma\rho+1}, \tag{14.45}$$

$$Z_1(t)=\phi_1 Z_0+h_3 Z_0-\left(1-\phi_1\right)Z_0+L^{-1}\left[\frac{1}{k^{\rho}}\left(\frac{-sY_0}{k}+\frac{gZ_0}{k}\right)\right]$$

$$=\phi_1 Z_0+h_3 Z_0-\left(1-\phi_1\right)Z_0-\frac{sY_0 t^{\rho}}{\Gamma\rho+1}+\frac{gZ_0 t^{\rho}}{\Gamma(\rho+1)}$$

$$=-2Z0+\frac{t^{\rho}gZ_0}{\Gamma(\rho+1)}, \tag{14.46}$$

$$X_2=\phi_2 X_1+h_1 X_1-\left(1-\phi_2\right)\frac{X_0}{k}+L^{-1}\left[\frac{1}{k^{\rho}}L\left[-a+bX_1+cA_1-dY_1\right]\right]$$

$$=L^{-1}\left[\frac{1}{k^{\rho}}\left[\frac{-a}{k}+b\left(\frac{-2X_0}{k}+\frac{1}{k^{\rho+1}}(-a+bX_0+cA_0)\right)\right.\right.$$

$$\left.\left.+c\left(X_0\left(\frac{-2Z_0}{k}+\frac{gZ_0}{k^{\rho+1}}\right)+Z_0\left(\frac{-2X_0}{k}+\frac{1}{k^{\rho+1}}(-a+bX_0+cA_0)\right)\right)+\frac{dcA_0}{k^{\rho}+1}\right]\right]$$

$$=(-a-2bX_0-2Z_0X_0c-2X_0Z_0)\frac{1}{k^{\rho+1}}$$

$$+\left(-a+bX_0+cA_0+\left(cg-a+bX_0+cA_0+dcA_0\right)Z_0\right)\frac{1}{k^{2\rho+1}}$$

$$=(-a-2bX_0+2cZ_0X_0+2X_0Z_0)\frac{t^{\rho}}{\Gamma\rho+1}+\left(-a+bX_0+cA_0+cgZ_0\right.$$

$$\left.-aZ_0+bX_0Z_0+cZ_0A_0+dcA_0\right)\frac{t^2\rho}{\Gamma(2\rho+1)}, \tag{14.47}$$

$$Y_2 = \phi_2 Y_1 + h_2 Y_1 - (1-\phi_2)Y_0 + L^{-1}\left[\frac{1}{k^\rho} L[-cA_1 + (\upsilon + d)Y_1]\right]$$

$$= L^{-1}\left[\frac{1}{k^\rho}\left(-cX_0\left(\frac{-2Z_0}{k} - \frac{gZ_0}{k^{\rho+1}}\right) - cZ_0\left(\frac{-2X_0}{k} + (-a + bX_0 + cA_0)\frac{1}{k^{\rho+1}}\right)\right.\right.$$

$$\left.\left. - (\upsilon + d)\frac{cA_0}{k^{\rho+1}}\right)\right]$$

$$= L^{-1}\left[4cX_0Z_0\frac{1}{k\rho+1} + \left(acZ_0 + cgX_0Z_0 - bcX_0Z_0 + c^2A_0Z_0 - cA_0(\upsilon + d)\right)\frac{1}{k^{2p+1}}\right]$$

$$= 4cX_0Z_0\frac{t\rho}{\Gamma(\rho+1)} + \left(acZ_0 + cgX_0Z_0 - bcX_0Z_0 + c^2A_0Z_0 - cA_0(\upsilon + d)\right)\frac{t^2\rho}{\Gamma(2\rho+1)}, \tag{14.48}$$

$$Z_2 = \phi_2 Z_1 + h_3 Z_1 - (1-\phi_2)\frac{Z_0}{k} + L^{-1}\left[\frac{1}{k^\rho} L[-sY_1 + gZ_1]\right]$$

$$= L^{-1}\left[\frac{1}{k^\rho}\left(\frac{csA_0}{k^{\rho+1}} + g\left(\frac{-2Z_0}{k^{\rho+1}}\right)\right)\right]$$

$$= -2gZ_0\frac{t^\rho}{\Gamma(\rho+1)} + \left(g^2Z_0 + csA_0\right)\frac{t^\rho}{\Gamma(2\rho+1)}. \tag{14.49}$$

Thus, we obtain the values of X, Y, Z as follows:

$$X = X_0 + X_1 + X_2$$

$$= 1.73 \times 10^8 - 2X_0 + (-a + bX_0 + cA_0)\frac{t^\rho}{\Gamma(\rho+1)}$$

$$+ (-a - 2bX_0 + 2cZ_0X_0 + 2X_0Z_0)\frac{t^\rho}{\Gamma\rho+1}$$

$$+ (-a + bX_0 + cA_0 + cgZ_0 - aZ_0 + bX_0Z_0 + cZ_0A_0 + dcA_0)\frac{t^{2\rho}}{\Gamma(2\rho+1)}, \tag{14.50}$$

$$Y = Y_0 + Y_1 + Y_2$$

$$= \frac{-cA_0 t^{\rho}}{\Gamma(\rho+1)} + 4cX_0 Z_0 \frac{t^{\rho}}{\Gamma(\rho+1)}$$

$$+\left(acZ_0 + cgX_0 Z_0 - bcX_0 Z_0 + c^2 A_0 Z_0 - cA_0(\upsilon + d)\right)\frac{t^{2\rho}}{\Gamma(2\rho+1)}, \quad (14.51)$$

$$Z = Z_0 + Z_1 + Z_2$$

$$= 400 + 2Z_0 + \frac{t^{\rho} g Z_0}{\Gamma(\rho+1)} - 2gZ_0 \frac{t^{\rho}}{\Gamma(\rho+1)} + \left(g^2 Z_0 + csA_0\right)\frac{t^{2\rho}}{\Gamma(2\rho+1)}. \quad (14.52)$$

14.4 CONVERGENCE ANALYSIS

Theorem 14.1

The solution obtained for the HBV infection model (14.8)–(14.10) by the use of FHATM is unique, wherever $0 < \omega_i < 1$, $i = 1,2,3$, where $\omega_1 = 1 + hT_1(b + cG)$, $\omega_2 = 1 + h + hT(v + d)$, and $\omega_3 = 1 + h + hgT_3$.

Proof: The solution of the fractional HBV infection model is

$$X = \sum_{n=0}^{\infty} X_n(t), \quad (14.53)$$

$$Y = \sum_{n=0}^{\infty} Y_n(t), \quad (14.54)$$

$$Z = \sum_{n=0}^{\infty} Z_n(t), \quad (14.55)$$

$$X_m = (\phi_m + h)X_{m-1}(t) - h(1-\phi_m)L^{-1}\left[\frac{X_0}{k} + \frac{h}{k^{\rho}}L\left[-a + bX_{m-1} + cA_{m-1} - dY_{m-1}\right]\right], \quad (14.56)$$

$$Y_m = (\phi_m + h)Y_{m-1}(t) - h(1-\phi_m)L^{-1}\left[\frac{Y_0}{k} + \frac{h}{k^{\rho}}L\left[-cA_{m-1} + (v+d)Y_{m-1}\right]\right], \quad (14.57)$$

$$Z_m = (\phi_m + h)Y_{m-1}(t) - h(1-\phi_m)L^{-1}\left[\frac{Z_0}{k} + \frac{h}{k^{\rho}}L\left[-sY_{m-1} + gZ_{m-1}\right]\right]. \quad (14.58)$$

Assuming two sets of solutions X, Y, Z and X^*, Y^*, Z^* for the above system of equations such that $|X| \le E$, $|Y| \le F$, and $|Z| \le G$,

$$\left|X - X^*\right| = \left|(1+h)\left(X - X^*\right) + hL^{-1}\left[\frac{1}{k^\rho}L\left[b\left(X - X^*\right) + cZ\left(X - X^*\right)\right]\right]\right|, \quad (14.59)$$

$$\left|Y - Y^*\right| = \left|(1+h)\left(Y - Y^*\right) + hL^{-1}\left[\frac{1}{k^\rho}L\left[(\upsilon + d)\left(Y - Y^*\right)\right]\right]\right|, \quad (14.60)$$

$$\left|Z - Z^*\right| = \left|(1+h)\left(X - X^*\right) + hL^{-1}\left[\frac{1}{k^\rho}L\left[g\left(Z - Z^*\right)\right]\right]\right|. \quad (14.61)$$

By applying the convolution theorem for the Laplace transform, we have

$$\left|X - X^*\right| \le (1+h)\left|X - X^*\right| + h\int_0^t \left(b\left|X - X^*\right| + cZ\left|X - X^*\right|\right)\frac{(t-\varsigma)^\rho}{\Gamma(1+\rho)}d\varsigma, \quad (14.62)$$

$$\left|Y - Y^*\right| \le (1+h)\left|Y - Y^*\right| + h\int_0^t (\upsilon + d)\left|Y - Y^*\right|\frac{(t-\varsigma)^\rho}{\Gamma(1+\rho)}d\varsigma, \quad (14.63)$$

$$\left|Z - Z^*\right| \le (1+h)\left|Z - Z^*\right| + h\int_0^t g\left|Z - Z^*\right|\frac{(t-\varsigma)^\rho}{\Gamma(1+\rho)}d\varsigma. \quad (14.64)$$

and hence,

$$\left|X - X^*\right| \le (1+h)\left|X - X^*\right| + h\int_0^t \left(b\left|X - X^*\right| + cG\left|X - X^*\right|\right)\frac{(t-\varsigma)^\rho}{\Gamma(1+\rho)}d\varsigma, \quad (14.65)$$

$$\left|Y - Y^*\right| \le (1+h)\left|Y - Y^*\right| + h\int_0^t (\upsilon + d)\left|Y - Y^*\right|\frac{(t-\varsigma)^\rho}{\Gamma(1+\rho)}d\varsigma \quad (14.66)$$

$$\left|Z - Z^*\right| \le (1+h)\left|Z - Z^*\right| + h\int_0^t g\left|Z - Z^*\right|\frac{(t-\rho)^\rho}{\Gamma(1+\rho)}d\varsigma. \quad (14.67)$$

Now, applying the mean value theorem,

$$\begin{aligned}
\left|X - X^*\right| &\le \left(1 + h + hb + hcGT_1\right)\left|X - X^*\right| \le \left|X - X^*\right|\omega_1, \\
\left|Y - Y^*\right| &\le (1+h)\left|Y - Y^*\right| + h(\upsilon + d)\left|Y - Y^*\right|T_2 \\
&\le \left(1 + h + hT_2(\upsilon + d)\right)\left|Y - Y^*\right| \le \left|Y - Y^*\right|\omega_2, \quad (14.68)
\end{aligned}$$

$$\begin{aligned}\left|Z-Z^*\right| &\le (1+h)\left|Z-Z^*\right|+hg\left|Z-Z^*\right|T_3\\ &\le \left(1+h+hgT_3\right)\left|Z-Z^*\right|\\ &\le \left|Z-Z^*\right|\omega_3.\end{aligned} \tag{14.69}$$

It gives $(1-\omega_1)\left|X-X^*\right|\le 0$, $(1-\omega_2)\left|Y-Y^*\right|\le 0$ and $(1-\omega_3)\left|Z-Z^*\right|\le 0$. Since $0<\omega_1,\omega_2,\omega_3<1$, $\left|X-X^*\right|=0$, $\left|Y-Y^*\right|=0$, and $\left|Z-Z^*\right|=0$, which implies $X=X^*$, $Y=Y^*$, and $Z=Z^*$. Therefore, the solution is unique.

Theorem 14.2

Let us suppose that K, L, and M be the Banach spaces, and $\mathbb{F}_1$, $\mathbb{F}_2$, and $\mathbb{F}_3$ be the nonlinear mappings $\mathbb{F}_1: K\to K$, $\mathbb{F}_2: L\to L$, and $\mathbb{F}_3: M\to M$. Also assume $\left\|\mathbb{F}_1(X)-\mathbb{F}_1(X')\right\|\le\omega_1\left\|X-X'\right\|$, $\forall\ X,X'\in K$, $\left\|\mathbb{F}_2(Y)-\mathbb{F}_2(Y')\right\|\le\omega_2\left\|Y-Y'\right\|$, $\forall$ $Y,Y'\in L$ and $\left\|\mathbb{F}_3(Z)-\mathbb{F}_3(Z')\right\|\le\omega_1\left\|Z-Z'\right\|$, $\forall\ Z,Z'\in M$. Then, each mapping $\mathbb{F}_1$, $\mathbb{F}_2$, and $\mathbb{F}_3$ has a fixed point as per the Banach's fixed point theorem [5,25]. The sequences corresponding to the solution obtained by the FHATM with $X_0\in K$, $Y_0\in L$, and $Z_0\in M$ chosen arbitrarily will converge to the fixed points of $\mathbb{F}_1$, $\mathbb{F}_2$, and $\mathbb{F}_3$, respectively,

$$\left\|X_m-X_n\right\|\le\frac{\omega_1^n}{1-\omega_1}\left\|X_1-X_0\right\|\forall X_m,X_n\in K, \tag{14.70}$$

$$\left\|Y_m-Y_n\right\|\le\frac{\omega_2^n}{1-\omega_2}\left\|Y_1-Y_0\right\|\forall Y_m,Y_n\in L, \tag{14.71}$$

$$\left\|Z_m-Z_n\right\|\le\frac{\omega_3^n}{1-\omega_3}\left\|Z_1-Z_0\right\|\forall Z_m,Z_n\in M. \tag{14.72}$$

Proof: Let us consider $\left(C_1[J_1],\|.\|\right)$, $\left(C_2[J_2],\|.\|\right)$, and $\left(C_3[J_3],\|.\|\right)$ of all continuous functions of J_1, J_2, and J_3 with the norm $\left\|g_1(t)\right\|=\max_{t\in J_1}\left|g_1(t)\right|$, $\left\|g_2(t)\right\|=\max_{t\in J_2}\left|g_2(t)\right|$, and $\left\|g_3(t)\right\|=\max_{t\in J_3}\left|g_3(t)\right|$, respectively. Now, we will show that X_n, Y_n, and Z_n are the Cauchy sequences in the aforesaid Banach spaces.

$$\left\|X_m-X_n\right\|=\max_{t\in J_1}\left|X_m-X_n\right|, \tag{14.73}$$

$$\left\|Y_m-Y_n\right\|=\max_{t\in J_2}\left|Y_m-Y_n\right|, \tag{14.74}$$

$$\left\|Z_m-Z_n\right\|=\max_{t\in J_3}\left|Z_m-Z_n\right|, \tag{14.75}$$

$$\|X_m - X_n\| = \max_{t\in J_1}\Big|(1+h)(X_{m-1}-X_{n-1}) + hL^{-1}$$

$$\times\left[\frac{1}{k^{\rho}}L\left[b(X_{m-1}-X_{n-1})+cZ(X_{m-1}-X_{n-1})\right]\right]\Big|$$

$$\le \max_{t\in J_1}\Big[(1+h)|X_{m-1}-X_{n-1}| + hL^{-1}$$

$$\times\left[\frac{1}{k^{\rho}}L\left[\left|b(X_{m-1}-X_{n-1})\right|+\left|cZ(X_{m-1}-X_{n-1})\right|\right]\right]\Big], \tag{14.76}$$

$$\|Y_m - Y_n\| = \max_{t\in J_2}\left|(1+h)(Y_{m-1}-Y_{n-1}) + hL^{-1}\left[\frac{1}{k^{\rho}}L\left[(v+d)(Y_{m-1}-Y_{n-1})\right]\right]\right|$$

$$\le \max_{t\in J_2}\left[(1+h)|Y_{m-1}-Y_{n-1}| + hL^{-1}\left[\frac{1}{k^{\rho}}L\left[\left|(v+d)(Y_{m-1}-Y_{n-1})\right|\right]\right]\right], \tag{14.77}$$

$$\|Z_m - Z_n\| = \max_{t\in J_3}\left|(1+h)(Z_{m-1}-Z_{n-1}) + hL^{-1}\left[\frac{1}{k^{\rho}}L\left[g(Z_{m-1}-Z_{n-1})\right]\right]\right|,$$

$$\le \max_{t\in J_3}\left[(1+h)|Z_{m-1}-Z_{n-1}| + hL^{-1}\left[\frac{1}{k^{\rho}}L\left[\left|g(Z_{m-1}-Z_{n-1})\right|\right]\right]\right]. \tag{14.78}$$

Now applying the convolution theorem, we have

$$\|X_m - X_n\| \le \overset{\max}{t\in J_1}\Big[(1+h)|X_{m-1}-X_{n-1}|$$

$$+h\int_0^t\left(\left|b(X_{m-1}-X_{n-1})\right|+\left|cZ(X_{m-1}-X_{n-1})\right|\right)\frac{(t-\varsigma)^{\rho}}{\Gamma(1+\rho)}d\varsigma\Big]$$

$$\le \overset{\max}{t\in J_1}\Big[(n+h)|X_{m-1}-X_{n-1}| + h\int_0^t b|X_{m-1}-X_{n-1}|$$

$$+cG|X_{m-1}-X_{n-1}|\frac{(t-\varsigma)^{\rho}}{\Gamma(1+\rho)}d\varsigma\Big]. \tag{14.79}$$

Next by using the integral mean value theorem, we have

$$\|X_m - X_n\| \le \overset{\max}{t\in J_1}\left[(1+h)|X_{m-1}-X_{n-1}| + h\left(\left|b(X_{m-1}-X_{n-1})\right|+\left|cZ(X_{m-1}-X_{n-1})\right|\right)T_1\right]$$

$$\le \overset{\max}{t\in J_1}\left[(1+h)|X_{m-1}-X_{n-1}| + h\left(\left|b(X_{m-1}-X_{n-1})\right|+\left|cZ(X_{m-1}-X_{n-1})\right|\right)T_1\right]$$

$$\le \omega_1\|X_{m-1}-X_{n-1}\|, \tag{14.80}$$

$$\|Y_m - Y_n\| \le \max_{t\in J_2}\left[(1+h)|Y_{m-1} - Y_{n-1}| + h\left(|(v+d)(Y_{m-1} - Y_{n-1})|\right)T_2\right]$$

$$\le \omega_2\|Y_{m-1} - Y_{n-1}\|, \tag{14.81}$$

$$\|Z_m - Z_n\| \le \max_{t\in J_3}\left[(1+h)|Z_{m-1} - Z_{n-1}| + h\left(|g(Z_m - Z_n)|\right)T_3\right]$$

$$\le \omega_3\|Z_{m-1} - Z_{n-1}\|. \tag{14.82}$$

Taking $m = n+1$ gives

$$\|X_{n+1} - X_n\| \le \omega_1\|X_n - X_{n-1}\| \le \omega_1^2\|X_{n-1} - X_{n-2}\| \le \ldots \le \omega_1^n\|X_1 - X_0\|. \tag{14.83}$$

Now, on using the triangle inequality,

$$\|X_m - X_n\| \le \|X_{n+1} - X_n\| + \|X_{n+2} - X_{n+1}\| + \cdots + \|X_m - X_{m-1}\|, \tag{14.84}$$

we obtain

$$\|X_m - X_n\| \le \left[\omega_1^n + \omega_1^{n+1} + \cdots + \omega_1^{m-1}\right]\|X_1 - X_0\| \tag{14.85}$$

$$\le \omega_1^n\left[1 + \omega_1 + \omega_1^2 + \cdots + \omega_1^{m-n-1}\right]\|X_1 - X_0\| \tag{14.86}$$

$$\le \omega_1^n\left[\frac{1-\omega_1^{m-n-1}}{1-\omega_1}\right]\|X_1 - X_0\|. \tag{14.87}$$

Similarly,

$$\|Y_m - Y_n\| \le \omega_2^n\left[\frac{1-\omega_2^{m-n-1}}{1-\omega_2}\right]\|Y_1 - Y_0\| \tag{14.88}$$

$$\|Z_m - Z_n\| \le \omega_3^n\left[\frac{1-\omega_3^{m-n-1}}{1-\omega_3}\right]\|Z_1 - Z_0\|. \tag{14.89}$$

Since, $0 < \omega_i < 1$, so $1 - \omega_i^{m-n-1} < 1$ for $i = 1,2,3$. Hence,

$$\|X_m - X_n\| \le \frac{\rho^n}{1-\rho}\|X_1 - X_0\|, \tag{14.90}$$

$$\|Y_m - Y_n\| \le \frac{\omega_2^n}{1-\omega_2}\|Y_1 - Y_0\|, \tag{14.91}$$

$$\|Z_m - Z_n\| \le \frac{\omega_3^n}{1-\omega_3}\|Z_1 - Z_0\|. \tag{14.92}$$

Thus $\|X_1 - X_0\| < \infty$, $\|Y_1 - Y_0\| < \infty$, and $\|Z_1 - Z_0\| < \infty$ so as $m \to \infty$, then $\|X_m - X_n\| \to 0$, $\|Y_m - Y_n\| \to 0$, and $\|Z_m - Z_n\| \to 0$. Hence, the sequences X_n, Y_n, and Z_n are Cauchy sequences in $C_1[J_1]$, $C_2[J_2]$, and $C_3[J_3]$, respectively. So these sequences are convergent.

14.5 RESULT

In this section, the numerical simulation has been done. The values of the parameters considered here are as follows [28]: $a = 5 \times 10^5$ cells (mL d)$^{-1}$, $b = 0.003d^{-1}$, $c = 4 \times 10^{-10}$ mL (copies d)$^{-1}$, $d = 0.502\,\text{d}^{-1}$, $s = 6.24\,\text{d}^{-1}$, $g = 0.65\text{d}^{-1}$, $v = 0.1\text{d}^{-1}$, $X(0) = 1.73 \times 10^8$ cells (mL)$^{-1}$, $Y(0) = 0$, and $Z(0) = 400$ copies (mL)$^{-1}$.

The graphs here show the dynamical behavior of the susceptible, HBV-infected, and the virions. From Figure 14.1, for the integral value of ρ, we observe that the susceptible cells are raising with time and spreading of the infection is rising. But for the fractional order model, it can be revealed that the rise in susceptible units is at slower rate as compared to the integral one, and hence, on behalf of this observation and making use of fractional model to study the dynamics of infection, the susceptible cells can be prevented from getting infected at earlier step. As the susceptible cells are increasing with time, the infected ones are also increasing, as shown in Figure 14.2. For the comparison between the integral and the fractional order modeling, from Figure 14.3 we can observe that the virions replicate much faster for the integral values than for the fractional one, thus resulting in the proliferation of disease.

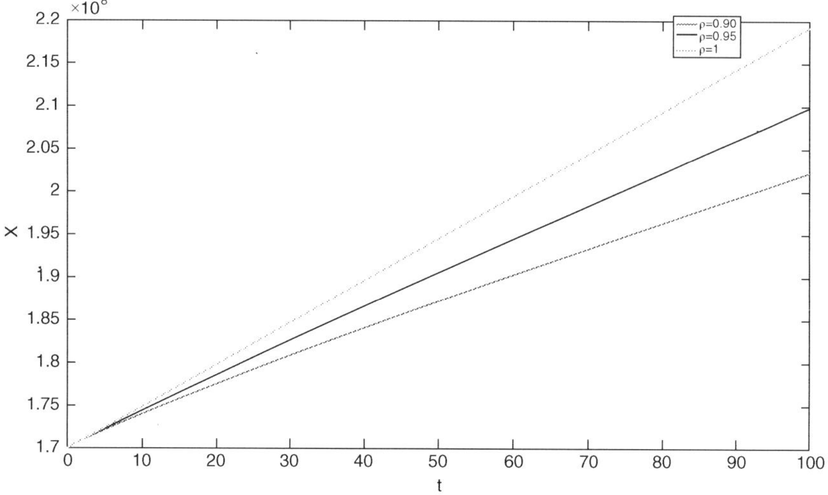

FIGURE 14.1 Graph between $X(t)$ and t for values of ρ.

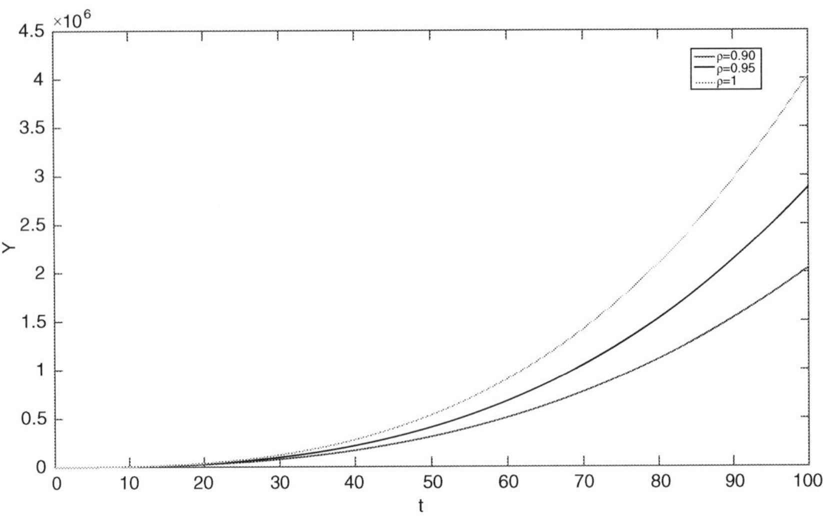

FIGURE 14.2 Graph between $Y(t)$ and t for values of ρ.

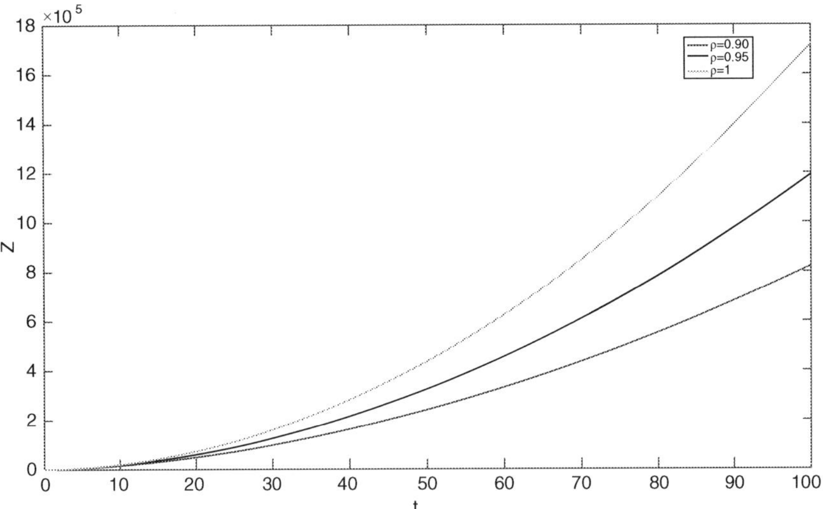

FIGURE 14.3 Graph between $Z(t)$ and t for values of ρ.

Hence, the Caputo fractional derivative involves the power-law memory kernel that is useful in describing the memory effects in the spreading of disease. Let the population ignore the disease history until $t = 100$ days; i.e., by considering the memoryless integral model system, the threshold point is at higher side—large population gets infected and measured at $t = 100$ days, but the inclusion of memory effects (consideration of the fractional model system) slows the growth of disease as with the knowledge of history of spreading of disease, the fraction of susceptible individuals can be protected against the infection. Thus, the fractional model is more advantageous over the integral model, as from the information obtained from the study of fractional model, the treatment can be initiated at very early stage with small damage of cells and lesser replication of virions.

14.6 CONCLUSION

The fractional calculus is extensively used to generalize the models and helps in elaborating them more precisely. The reason behind this generalization is that modeling a phenomenon makes some simplifications, which on consideration in the model decrease the rate of variation in it. The fractional model shows accuracy, improves the results, and has the potential to explain the computational dynamics problem here; hence, it is useful in investigating the hepatitis B disease.

REFERENCES

1. Agarwal, R., Jain, S., and Agarwal, R. P. (2018). Mathematical modeling and analysis of dynamics of cytosolic calcium ion in astrocytes using fractional calculus. *Journal of Fractional Calculus and Applications*, 9(2):1–12.
2. Agarwal, R., Purohit, S. D. and Kritika (2019). A mathematical fractional model with non-singular Kernel for thrombin receptor activation in calcium signalling. *Mathematical Methods in the Applied Sciences*, 42:1–12.
3. Agarwal, R., Yadav, M. P., Agarwal, R. P., and Goyal, R. (2019). Analytic solution of fractional advection dispersion equation with decay for contaminant transport in porous media. *Matematicki Vesnik*, 71(1):5–15.
4. Ahmed, E., El-Sayed, A., and El-Saka, H. A. (2006). On some routhhurwitz conditions for fractional order differential equations and their applications in Lorenz, Rossler, Chua and Chen systems. *Physics Letters A*, 358(1):1–4.
5. Argyros, I. K. (2008). *Convergence and Applications of Newton-Type Iterations*. Springer Science & Business Media: Berlin.
6. Blayneh, K. W., Gumel, A. B., Lenhart, S., and Clayton, T. (2010). Back- ward bifurcation and optimal control in transmission dynamics of west nile virus. *Bulletin of Mathematical Biology*, 72(4):1006–1028.
7. Caputo, M. (1967). Linear models of dissipation whose Q is almost frequency independent—II. *Geophysical Journal International*, 13(5):529–539.
8. Caputo, M. (1969). *Elasticity and Anelastic Dissipation. Zanichelli*: Bologna.
9. Demirci, E., Unal, A., and Ozalp, N. (2011). A fractional order SEIR model with density dependent death rate. *Hacettepe Journal of Mathematics and Statistics*, 40(2):287–295.
10. Ding, Y. and Ye, H. (2009). A fractional-order differential equation model of HIV infection of CD4+ T-cells. *Mathematical and Computer Modelling*, 50(3-4):386–392.
11. Djordjevic, V., Jaric, J., Fabry, B., Fredberg, J., and Stamenovic, D. (2003). Fractional derivatives embody essential features of cell rheological behavior. *Annals of Biomedical Engineering*, 31(6):692–699.
12. El-Sayed, A., El-Mesiry, A., and El-Saka, H. (2007). On the fractional-order logistic equation. *Applied Mathematics Letters*, 20(7):817–823.
13. Fields, B. N., Knipe, D. M., Howley, P. M., Everiss, K. D., and Kung, H.-J. (1996). *Fundamental Virology*. Lippincott-Raven: Philadelphia, PA.

14. Gupta, E., Bajpai, M., Sharma, P., Shah, A., and Sarin, S. (2013). Unsafe injection practices: A potential weapon for the outbreak of blood borne viruses in the community. *Annals of Medical and Health Sciences Research*, 3(2):177.
15. Jesus, I., Tenreiro Machado, J., and Boaventure, C. (2008). Fractional electrical impedances in botanical elements. *Journal of Vibration and Control*, 14(9-10):1389–1402.
16. Kabir, K. A., Kuga, K., and Tanimoto, J. (2019). Analysis of SIR epidemic model with information spreading of awareness. *Chaos, Solitons & Fractals*, 119:118–125.
17. Kilbas, A., Srivastava, H., and Trujillo, J. (2006). *Theory and Applications of Fractional Differential Equations*, vol. 204. Elsevier Science Limited: Amsterdam.
18. Lewin, S., Ribeiro, R., Walters, T., Lau, G., Bowden, S., Locarnini, S., and Perelson, A. (2001). Analysis of hepatitis b viral load decline under potent therapy: Complex decay profiles observed. *Hepatology*, 34:1012–1020.
19. Liao, S. (1997). Homotopy analysis method: A new analytical technique for nonlinear problems. *Communications in Nonlinear Science and Numerical Simulation*, 2(2):95–100.
20. Liao, S. (2004). On the homotopy analysis method for nonlinear problems. *Applied Mathematics and Computation*, 147(2):499–513.
21. Liao, S. J. (1992). The proposed homotopy analysis technique for the solution of nonlinear problems. PhD thesis, Shanghai Jiao Tong University Shanghai.
22. Liu, S., Wang, S., and Wang, L. (2011). Global dynamics of delay epidemic models with nonlinear incidence rate and relapse. *Nonlinear Analysis: Real World Applications*, 12(1):119–127.
23. Liu, W.-M., Hethcote, H. W., and Levin, S. A. (1987). Dynamical behavior of epidemiological models with nonlinear incidence rates. *Journal of Mathematical Biology*, 25(4):359–380.
24. Nowak, M.A., Bonhoeffer, S., Hill, A. M., Boehme, R., Thomas, H.C., and McDade, H. (1996). Viral dynamics in hepatitis b virus infection. *Proceedings of the National Academy of Sciences of the United States of America*, 93:4398–4402.
25. Magrenan, A. A. (2014). A new tool to study real dynamics: The convergence plane. *Applied Mathematics and Computation*, 248:215–224.
26. Mann, J. and Roberts, M. (2011). Modelling the epidemiology of hepatitis B in New Zealand. *Journal of Theoretical Biology*, 269(1):266–272.
27. Podlubny, I. (1998). *Fractional Differential Equations: An Introduction to Fractional Derivatives, Fractional Differential Equations, to Methods of Their Solution and some of Their Applications*. Elsevier: Amsterdam.
28. Salman, S. M. and Yousef, A. M. (2017). On a fractional-order model for HBV infection with cure of infected cells. *Journal of the Egyptian Mathematical Society*, 25(4):445–451.
29. Sneddon, I. N. (1951). *Fourier Transforms*. McGraw-Hill Book Co, NewYork.
30. Thornley, S., Bullen, C., and Roberts, M. (2008). Hepatitis B in a high prevalence New Zealand population: A mathematical model applied to infection control policy. *Journal of Theoretical Biology*, 254(3):599–603.
31. Vargas-De-Leon, C. (2012). Stability analysis of a model for HBV infection with cure of infected cells and intracellular delay. *Applied Mathematics and Computation*, 219(1):389–398.
32. Ye, H. and Ding, Y. (2009). Nonlinear dynamics and chaos in a fractional-order HIV model. *Mathematical Problems in* Engineering, 2009, Article ID 378614.
33. Zhao, S., Xu, Z., and Lu, Y. (2000). A mathematical model of hepatitis B virus transmission and its application for vaccination strategy in china. *International Journal of Epidemiology*, 29(4):744–752.

15 Nonlinear Dynamics of SARS-CoV2 Virus

India and Its Government Policy

Aditya Mani Mishra
VBS Purvanchal University
Rajasthan Technical University

Ritu Agarwal
Malaviya National Institute of Technology

Sunil Dutt Purohit and Kamlesh Jangid
Rajasthan Technical University

CONTENTS

15.1 INTRODUCTION

Recently, an outbreak of severe acute respiratory syndrome coronavirus declared by the World Health Organization is a Public Health Emergency of International Concern (PHEIC). More than 39,000 people have died from COVID-19, as the disease is officially known, while over 800,000 infections have been confirmed in at least 178 countries and territories in which four countries have reported human-to-human transmission [1]. More than 172,000 people have recovered from the corona-virus. About 665231 cases have been reported worldwide, leading to 30900 deaths as of March 30, 2020 [2]. First case of this virus has been reported in Wuhan, a city of Hubei Province of China, in December 2019 [3]. Since then, SARS-CoV-2 (severe acute respiratory syndrome coronavirus 2) incited an enormous scope scourge stating in China and spread most of the countries of the world [3].

Infected countries are working to set up countermeasure to stem conceivable decimating impacts of SARS-CoV-2. WHO facilitates data streams and asks scientific community to slow down the risk of this disease. Since the vaccine to counter the virus is not available till now, the remedial systems to manage the contamination are just strong, and counteraction planned for decreasing transmission in the network is our best weapon. Understanding the early transmission elements of the contamination and assessing the viability of control measures are essential for evaluating the potential for continued transmission to happen in new regions. We try to model the infection of SARS-CoV-2 transmission in the context of India and find a conclusion with reference to government measure.

15.2 BRIEF REVIEW

Since the vaccination of the disease is not available presently, as a mathematician we can best model the disease mathematically and search the best possible ways to control the virus. Mathematical modeling of the infectious disease has an old history. In the years 1348–1350, Black Death spread Mediterranean and Europe resulting approximately 50–100 million causalities. Another catastrophic epidemic was smallpox, killed approximately 35 million people. In early twentieth century, influenza affected around 20 million of world population. We can name others like Bombay plague (1905–1906), SARS (2003), H1N1 (2009), and MARS (2012) [4].

Modeling of disease was first reported by John Graunt in 1663 who statistically analyzed the public health in his book. After a century, Bernoulli studied smallpox mathematically. Louis Pasteur in the mid-nineteenth century modeled rabies and anthrax, and created vaccines for these diseases too. After that, Sir Ronald Ross developed the mathematical epidemiology branch by analyzing malaria [5]. In 1927, Kermack and Mackendrick [6] published a deterministic epidemic model, which is widely used as a first deterministic model in the analysis of epidemiology known as susceptible, infected, and recovered model (SIR model). The model is simple but effective to conclude the results for epidemic, especially airborne childhood diseases with lifelong immunity upon recovery, such as measles, mumps, rubella, and pertussis [7].

Recently, COVID-19 transmission disease has been studied stochastically in [8] and concluded that COVID-19 transmission most likely declined in Wuhan during late January 2020, concurring with the presentation of movement control measures. As more cases show up in global areas with comparative transmission potential to Wuhan before these control measures, it is likely that numerous chains of transmission will neglect to set up at first, yet may prompt new flare-ups in the long run. A statistical analysis is also given in [9]. Similarly in [10], the authors have modeled the disease deterministically using the fractional derivatives.

SIR model is deterministic and based on compartmental model. For our purpose of study, we take SEIR model (susceptible–pre-infectious–infected–recovered model) (which is applied successfully for measles), and then, we modified it by introducing another group, called death group.

15.3 SEIR MODEL

Before modeling the coronavirus, we should take the basic model that is based on measles. This model is known as SEIR model (susceptible–pre-infectious–infected–recovered model) (Figure 15.1).

In this model, society is broken into four compartments: People belonging to the susceptible group have not yet contacted to the disease. Once people get infected, they get moved to the pre-infectious group. (The letter E, instead of P, is traditionally used to denote this group.) At this stage, the infected person might not have symptoms and does not infect other people. After some time passes, the infected person himself becomes infectious. Anyone who contacts infectious person will also have the probability to get infected. And after some more time, he recovers and gains immunity, so he cannot become infected again (underlying assumption). The time that passes before an infectious person recovers, depends on each person, but we can use the average rate and apply it to everyone and symboled as r. Similarly, we use the letter f to denote the rate at which an infected person becomes infectious. Notice that f and r depend on the disease. There is also a rate at which a susceptible person becomes infected. This rate depends on the infectiousness of the disease as well as how many people are infectious at that time. Since the number of infectious people, $I(t)$, changes over time, this rate, which we shall call the force of infection, and label with λ, also changes over time. λ is expressed as β times $I(t)$, where β is the per-capita rate at which two people come into an effective contact. Based on the model, we can write out four differential equations that show how each compartment changes over time.

$$\frac{dS(t)}{dt} = -\lambda S(t) \tag{15.1}$$

$$\frac{dE(t)}{dt} = \lambda S(t) - fE(t) \tag{15.2}$$

$$\frac{dI(t)}{dt} = fE(t) - rI(t) \tag{15.3}$$

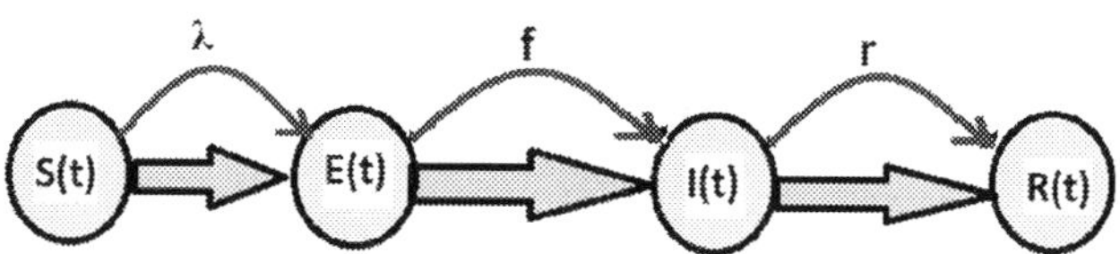

FIGURE 15.1 Semantic diagram of SEIRD model.

$$\frac{dR(t)}{dt} = rI(t). \tag{15.4}$$

Again suppose the total population is N. We observe

$$N = S(t) + E(t) + I(t) + R(t). \tag{15.5}$$

Notice that the susceptible group is always shrinking at a rate of λ. This will shrink over time as there are fewer people to infect. The shrink in S is precisely added into the E-group, and with the rate of f, the people in the E-group get moved into the infectious I-group. Similarly, we can make sense of how the infectious and the recovered groups change over time.

There are some important assumptions that we need to remember when dealing with this model:

- First, the model is deterministic. This means that, eventually, everyone gets the disease and goes through recovery.
- Second, it is short term. For COVID-19 disease, this incubation period is 14 days. The population is constant, and there are no births or deaths. In particular, there are no deaths caused by the disease.
- Third, it assumes a random mixing. In other words, we are assuming that people behave like air particles and can be expected to contact people far away.

These are strong assumptions that allow us to have an easier job during modeling the disease, but at the same time, it can lead to unrealistic predictions. The key parameter to portray an outbreak is the transmission rate, which is a proportion of the average number of people tainted by an individual conveying the infection. A general utilized measurement of transmission is the basic reproductive number R_0, which is the transmission rate given that the population has no invulnerability from past exposures or immunization, nor any intentional intercession in illness transmission [12]. When $R_0 > 1$, the number of infections grows and spreads in the population, but not for $R_0 < 1$. R_0 is commonly displayed from information on the spread of the disease, with the supposition that the dynamic of spread is in progress in the population and that the measure along these lines parameterizes the degree of intra-community spread (Figure 15.2).

15.4 MODIFICATION OF SEIR MODEL: SEIRD MODEL

We shall modify SIR model to fit the current situation for the coronavirus. The death rate is estimated to be at 2.49% as of March 30, 2020 [13], in India, so we shall add that to the model as the death rate, d. We shall let it flow from the infectious population to the D-group, which represents everyone who suffered the disease-related death (Figure 15.3).

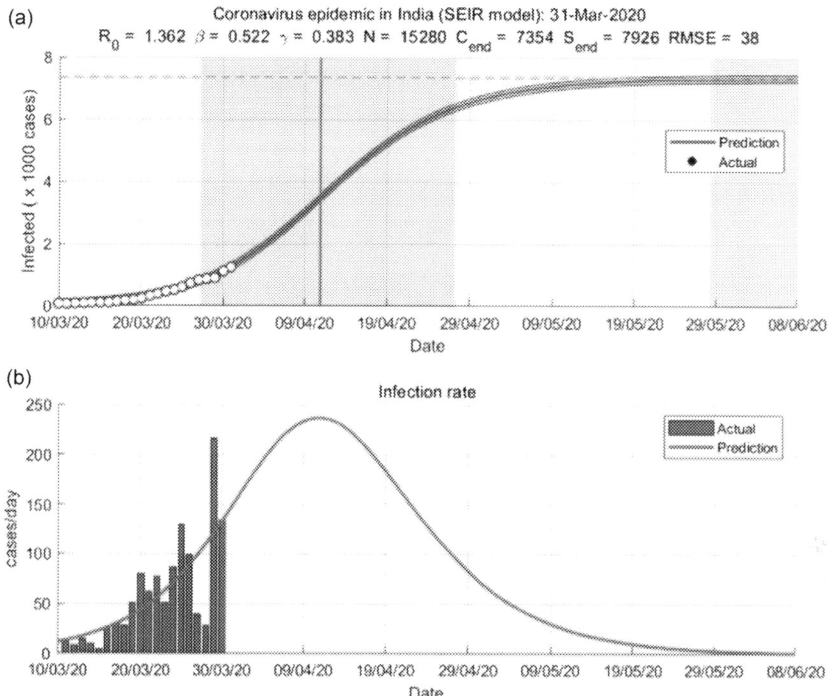

FIGURE 15.2 Estimation of parameter using SIR model. Data [11] has been taken from March 03, 2020, to March 30, 2020.

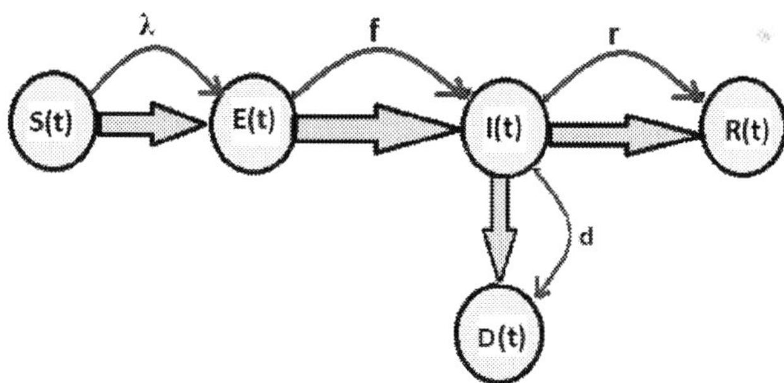

FIGURE 15.3 Semantic diagram of SEIRD model.

The rest of the equations are the same as before. Hence, we remodel Eqs. (15.1)–(15.4) as

$$\frac{dS(t)}{dt} = -\lambda S(t) \tag{15.6}$$

$$\frac{dE(t)}{dt} = \lambda S(t) - fE(t) \tag{15.7}$$

$$\frac{dI(t)}{dt} = fE(t) - rI(t) \tag{15.8}$$

$$\frac{dR(t)}{dt} = (r - d)I(t) \tag{15.9}$$

$$\frac{dD(t)}{dt} = dI(t). \tag{15.10}$$

Force of infection $\lambda(t)$ is related to $I(t)$ from the equation:

$$\lambda(t) = \beta I(t). \tag{15.11}$$

The model is nonlinear, which can be analyzed numerically.

15.5 ANALYTICAL STUDY

We have studied the model especially in the context of India. Currently, there are 15,24,266 persons scanned at airport, in which there are 867 total cases in which 87 total recovered, including migrated to other country, 755 active cases and 25 deaths [11] on March 30, 2020. Although these 755 cases are likely to be quarantined or in the hospital, we will use these numbers as our initial values (Figures 15.4–15.9).

We have taken the initial number of infectious people at three on March 03, 2020 [1], and died or recovered from the disease to be zero. The estimation of parameter provides the basic reproduction number $R_0 = 1.362$ and the force of infection $\beta = 0.522$ with 95% of confidence bounds. The rate of death is estimated to be $d = 0.0249$ with 95% confidence interval $(0.02193, 0.02791)$, and the rate of recovery is estimated to be $\gamma = 0.383$ (the study is based on the data [4,13,14] and MATLAB®).

We can make some modifications in our model. For example, after everyone starts being more cautious, or if the government subsidizes hand sanitizer, we can expect the effective contact rate to decrease. For example, we can show what the effects of reducing the effective contact rate will have on the simulation. In Figure 15.9, we can see that after implementing these measures, curve of infected people decreased significantly and also the span of disease reduced. The infectious curve can be beneficial for hospitals, as they are less likely to be overcrowded when the outbreak peaks. These measures reduced the contact rate, i.e., force of infection. We can name it as "policy effect" for simplicity. Note that the total deaths have decreased, but the recovery curve have stayed the same, because the policy took an effect after the peak. The effects are even more pronounced if the policy effect takes place before one week. The curve explains that the policy effect takes place twenty days after the outbreak; the effects are so great that the effective period of COVID-19 reduces significantly. The slowed effects of the outbreak would be great for any country, so for India, as there is more time to develop a vaccine.

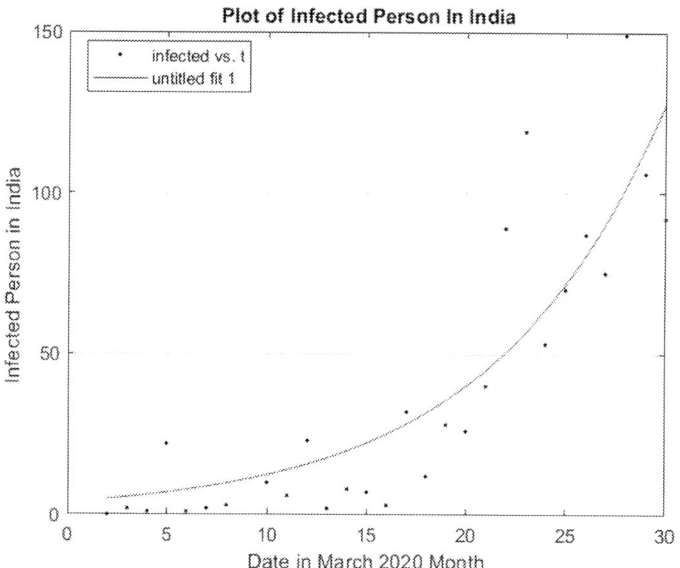

FIGURE 15.4 Plot of infected data in India from March 03, 2020, to March 30, 2020 (Source Refs. [13,14]); estimation gives $a = 3.904(-0.01556, 7.823), b = 0.1163(0.07854, 0.1541)$ (with 95% confidence bounds): goodness of fit: SSE : $1.173e + 04$, R-square: 0.7639, adjusted R-square: 0.7548, RMSE: 21.24.

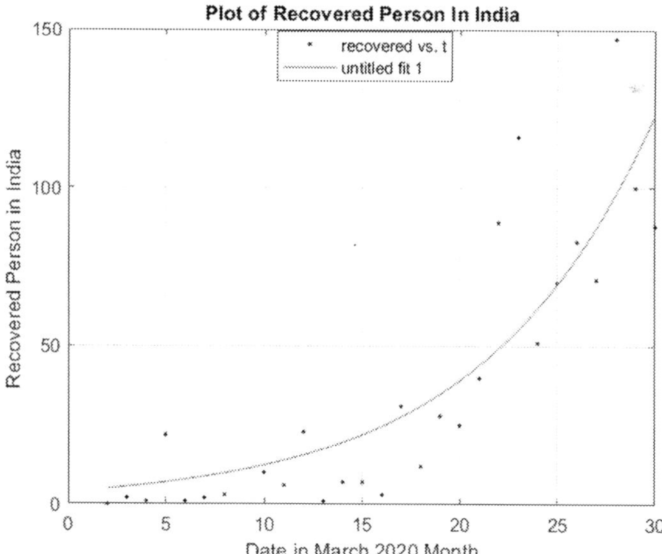

FIGURE 15.5 Plot of recovered data in India from March 03, 2020, to March 30, 2020 (Source Refs. [13,14]); estimation gives $a = 3.938(-0.0962, 7.973), b = 0.1148(0.07614, 0.1534)$ (with 95% confidence bounds): goodness of fit: SSE: 1.176e + 04, R-square: 0.7498, adjusted R-square: 0.7402, RMSE: 21.27.

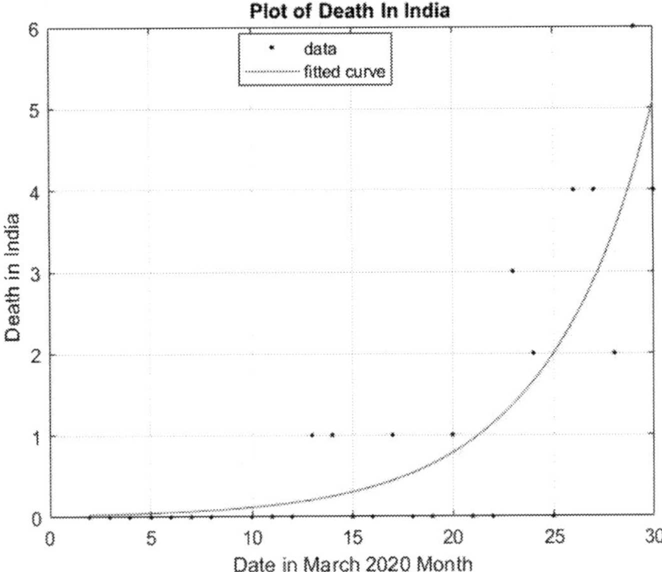

FIGURE 15.6 Plot of death in India from March 03, 2020, to March 30, 2020 (Source Refs. [13,14]); estimation gives $a = 10.28(7.649, 12.92), b = 0.157(0.1477, 0.1663)$ (with 95% confidence bounds): goodness of fit: SSE: 2.545e + 04, R-square: 0.9909, adjusted R-square: 0.9905, RMSE: 31.29.

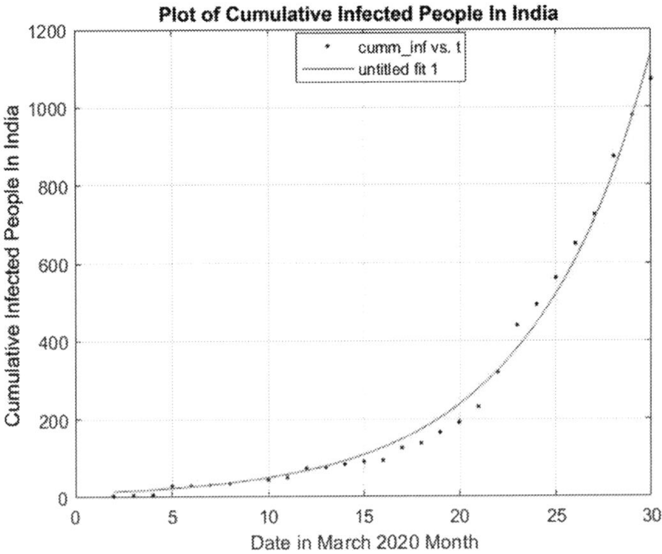

FIGURE 15.7 Plot of cumulative infection in India from March 03, 2020, to March 30, 2020 (Source Refs. [13,14]); estimation gives $a = 10.28(7.649, 12.92), b = 0.157(0.1477, 0.1663)$ (with 95% confidence bounds): goodness of fit: SSE: 2.545e+04, R-square: 0.9909, adjusted R-square: 0.9905, RMSE: 31.29.

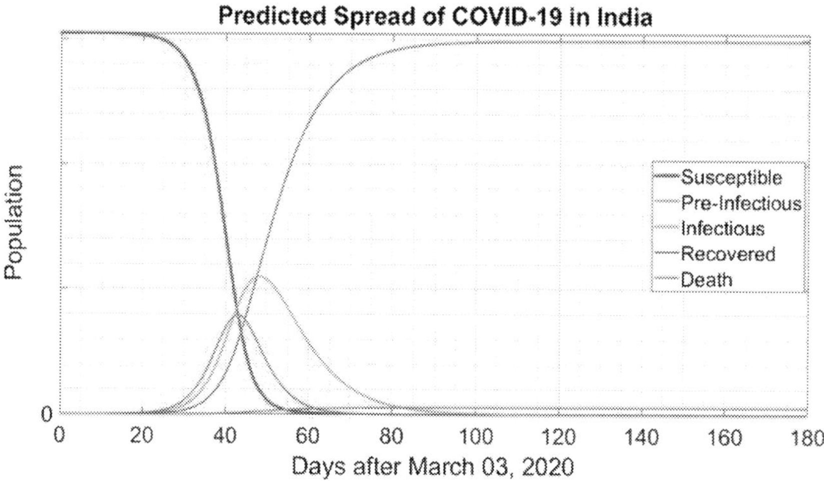

FIGURE 15.8 SEIRD model plot [13,14].

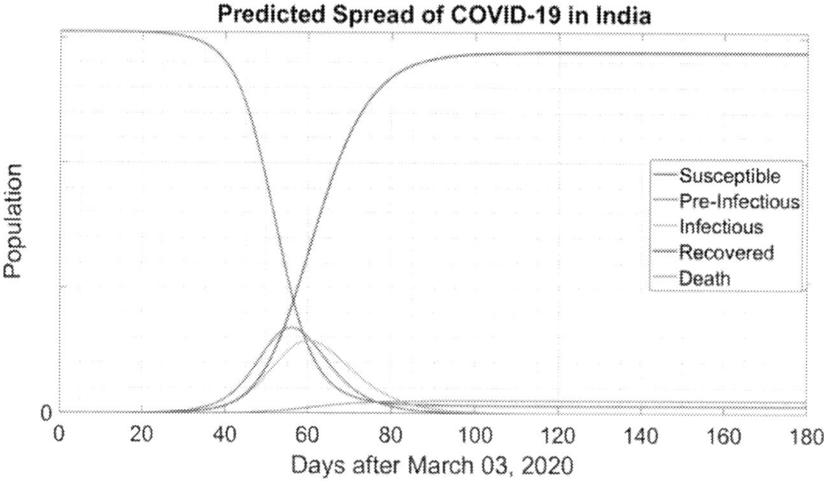

FIGURE 15.9 SEIRD model after taking government measures [13,14].

15.6 HOW CAN WE MAKE THE MODEL BETTER?

In this section, we discuss the limitation of model and further scope of improvement. We have included the death rate as 2.49%, but this death rate varies depending on your age. Perhaps one way we can improve our model is to stratify the population of the country we are modeling into different age groups. For that, we shall need a contact matrix instead of a single contact parameter. This is called the WAIFW (Who Acquires Infection From Whom) matrix. The WAIFW matrix is more realistic because schoolchildren tend to contact each other a lot, while the elderly do not contact people as much. Because of that only, Government of India issued an order to close the school and colleges as a first measure to combat the coronavirus. We can also try and model populations at the large country-level scale instead of picking some more areas which are more prone to disease. In case of India, the areas affected by coronaviruses are cities such as Delhi, Mumbai, Pune, and those areas in which individuals have been directly exposed to infections. We can see the effects of closing airports and seaports.

15.7 CONCLUSION

We studied the early policy effects can save significant lives in India. Though this may not be a realistic, it can still give us a sense of how large the benefits of locking down cities, or banning travel, can have. These estimates could help governments make important decisions at the international level.

ACKNOWLEDGMENT

We thank the Indian Institute of Technology Indore for providing resources for preparing the manuscript.

REFERENCES

1. https://www.who.int/docs/default-source/coronaviruse/situation-reports/20200326-sitrep-66-covid-19.pdf?sfvrsn=81b94e61_2.
2. https://economictimes.indiatimes.com/news/international/world-news/chinas-corona-virus-death-toll-touches-3042-confirmed-cases-rise-to-80552/ articleshow/74505414.cms?from=mdr.
3. Batista, Milan. "Estimation of the final size of the coronavirus epidemic by the SIR model." (2020).
4. Martcheva, Maia. *An Introduction to Mathematical Epidemiology*, Vol. 61. New York: Springer, 2015.
5. Ross, Sir Ronald. "Medicine and Hygiene, July, 1911, IV, 8, pp. 233–237. The author proposes a new explanation to account for the very limited extent to which yellow fever has spread in the Old World as compared with the New." *Bulletin* 1: 161 (1911).
6. Kermack, William Ogilvy, and Anderson G. McKendrick. "A contribution to the mathematical theory of epidemics." *Proceedings of the Royal Society of London. Series A, Containing Papers of a Mathematical and Physical Character* 115(772): 700–721 (1927).
7. Choisy, Marc, JeanFrancois Guegan, and P. Rohani. "Mathematical modeling of infectious diseases dynamics." In: Tibayrenc Michel (Ed.) *Encyclopedia of* Infectious Diseases: Modern Methodologies. Hoboken: Wiley, pp. 379–404 (2007).
8. Kucharski, Adam J., et al. "Early dynamics of transmission and control of COVID-19: A mathematical modelling study." *The Lancet Infectious Diseases* 20(5): 553–558 (2020).
9. Rabajante, Jomar F. "Insights from early mathematical models of 2019-nCoV acute respiratory disease (COVID-19) dynamics." arXiv preprint arXiv:2002.05296 (2020).
10. Khan, Muhammad Altaf, and Abdon Atangana. "Modeling the dynamics of novel coronavirus (2019-nCov) with fractional derivative." *Alexandria Engineering Journal* 59(4): 2379–2389 (2020).
11. https://www.worldometers.info/coronavirus/ data is taken on 31 March 2020.

12. Zhao, Shi, et al. "The basic reproduction number of novel coronavirus (2019-nCoV) estimation based on exponential growth in the early outbreak in China from 2019 to 2020: A reply to Dhungana." *International Journal of Infectious Diseases* 94: 148–150 (2020).
13. https://www.mohfw.gov.in/ data is taken on 31 March 2020.
14. https://www.ecdc.europa.eu/en/publications-data/download-todays-data- geographic-distribution-covid-19-cases-worldwide data is taken on 31 March 2020.

16 Ethical and Professional Issues in Epidemiology

Manoj Dubey
IES-IPS Indore

Ramakant Bhardwaj
Amity University Kolkata

Jyoti Mishra
Gyan Ganga Institute of Technology and Sciences

CONTENTS

16.1 INTRODUCTION

Epidemiology is concerned with studying the various reasons for the sudden increase in human outbreaks. In the case of epidemics, due to the prevalence of infectious and fatal diseases, and the sudden increase in the spread area, the number of diseases increases dramatically, so that a type of medical emergency situation arises. This scripture not only discusses the outbreak of pandemonium, but also deals with the study of the collective effects of any disease or disorder on the public in both ordinary and pandemic conditions. Under this science, apart from parasitic bacterial infectious diseases, noncommunicable diseases prevailing in the public and the condition of the body are also discussed. For this reason, the field of epidemiology has become more extensive. Therefore, we can say that the knowledge of the correlation between different causes and conditions that determine the frequency and distribution of any infectious, functional disorder, or disease within the medical field is epidemiological.

For the successful infection [1–5] of pathogenic parasitic bacteria, it is necessary that when bacteria enter the body of a human who is pathogenic, they can produce a toxin through the growth of cells of that human body. The origin of the disease depends on the destructive power of the parasitic microbes and the strength of the disease-resistant power of the host. If the destructive power is weaker than the resistance of humans, then disease does not occur. But when the destructive power is more powerful than the resistance of humans, the progress of infection work keeps on increasing and the besetting person becomes diseased. It is clear from this that the intensity of the disease is not only the measure of destructive power but also the measure of the lack of human resistance. The disease is actually the result of the failure of the resistance power of the aggressor against the destructive power of the aggressor. This can be understood from the following formula:

$$\text{Pathogenesis} = \frac{\text{Number of bacteria} \times \text{Destruction power of bacteria}}{\text{Human body resistance}}.$$

The pandemic spread at a particular place depends on whether the destructive power of the germ, its number, aggression, fertility, and the ability to make livelihood are more or less than the disease resistance of the population living at that place. The effect of these two opposing forces on the collective public is the result of the spread of diseases. Many facts can be obtained from studying the effects of disease infection in the laboratory and the effect of artificial infection on immunized animals. There are two types of epidemics: short term and long term. Short-term epidemics usually occur every year due to the increase in bacteria when the climate changes. Chronic epidemics occur when the intensity of infection or the number of diseased individuals increases. For these diseases, resistance cannot be produced inside the patient; they continue to occur frequently and persist.

The destructive power of a germ depends on its number, aggressiveness, fertility, and ability to make livelihood, and the immunity of the population depends on collective immunity as a result of the resistance power of each individual. As a result of the impact of these two opposing forces on the collective public, there is the possibility of spread, prevalence, distribution, recurrence, etc. of the particular disease. Many facts have been obtained from studying the effect of artificial infection on disease ejaculation and immune-suppressed groups in the laboratory. The population can be divided into the following special classes on the basis of immunity and the pathogenesis that depends on the mutual ratio of bacteria to destructiveness. There is no distinct boundary between these classes, but each class is hierarchically differentiated from each other, with a gradual distinction between them [6–10].

- *Uninfected, immunized population*: In this class, those humans are counted, in which infection of particular disease has not been found, and if ever it happens, due to being immune from that disease, the destruction power of bacteria will fail and these humans will remain from disease. Immunity is relative. For this reason, if the infection is very acute, then there is no possibility of complete protection from the disease. If there are a large number of animals of this class in the public, then the effects of mass immunity do not take the form of infection epidemic, but some sparsely immunized humans may be affected by occasional, or even, gastrointestinal infections. But if the number of immunized persons is relatively small, the velocity of the disease increases in the same proportion. The possibility of infection is dependent on the hygiene of the environment, public health, the prevalence of disease, unhealthy living conditions of the people, etc. For example, in soldiers immunized by an anticold vaccine, who live in a hygienic environment, Sheetla's disease is not particularly acute [11–17].
- *Unsanctioned, disease-prone population*: In this class, there are humans who have not been able to get the infection of a particular disease, but due to not being immune to the disease, they cannot remain free from the effects of the disease if the infection occurs. The level of collective immunity of the public falls due to the excess of this class of humans, and the outbreak of the disease becomes epidemic after infection. If the number of susceptible people in the public is not limited or accidentally increases, the disease spreads horribly. For first-class humans whose immunity has decreased, diseases can be transmitted by the affected class humans. The movement of human beings from place to place in fairs, festivals, and pilgrimages or flood, famine, war, commerce, behavior, industrialization, etc. increases the number of disease-prone persons, and there is a possibility of taking the contagious form of the disease. In order to remove the possibility of an epidemic, the number of diseased people is as limited as possible by immunization vaccines. Humans of this class are like fuel to provoke the disease and prove objectionable to the public. Almost all children are diseased. Disease resistant power decreases with malnutrition, fatigue, indigestion, anemia, anxiety, contaminated air, dampness, populism, insomnia, extreme cold or heat chronic disease, etc. Artificial immunity is temporary and decreases after some time.
- *Symptomatic infected population*: Humans of this class are immune and do not get sick on their own due to the insulating power. Parasitic bacteria thrive in their body when infection occurs, but they are not able to produce disease. These humans, despite being healthy themselves, are also curative and spread disease among disease-prone individuals. In some diseases, humans of this class continue to spread pathogens for many years. The number of pathogens increases as the epidemic spreads. If there is no shortage of disease-causing creatures and some remain constant, then the pathogens continue to manifest the disease by causing infections in them. Thus, disease takes a permanent form in that place [18–22]. Due to the healthy host of

parasitic bacteria, vector humans of this class remain a haven for infection. Diphtheria, enteric fever, circulatory meningitis, etc. are mostly caused by the pathogens. In the absence of symptoms, it is difficult to detect, and treatment attempts to remove their vasculature often fail.

- *Untargeted infected population*: The real symptoms of the disease do not appear in this class of human beings, but include mild madness or some malaise. Unhealthy symptoms with unexplained symptoms can be a form of many diseases. In the absence of indicative symptoms, the disease is not diagnosed, and in the absence of pathogenic pain, these untargeted or missing patients continue in their daily work and spread the disease to the uninfected persons. They are also pathogens and keep spreading the disease.
- *Atypical cases*: The symptoms of the disease are not obvious in this class of humans, but due to some symbolic symptoms, the suspicion of the disease arises. There may be some difficulties in diagnosis. The reason for this is that the symptoms are atypical. The form of the disease is underdeveloped or indigestible, and the patients are pathogens.
- *Typical cases*: In this class of people, the symptoms of the disease are obvious and typical, but the disease is not particularly troublesome or difficult. Such patients are usually pathogens, but if they are bedridden, disease prevalence is limited due to noncontact with humans other than family members. If they are diagnosed soon and admitted to the hospital, then it helps to prevent the spread of the disease. Even after the disease has calmed down, these humans continue to be pathogens for some time.
- *Difficult patient*: Due to the intensity of this class of patients, they themselves become bedridden, and in spite of having a disease, they are not able to spread the disease specifically. If the disease is chronic like tuberculosis, then the disease continues in the family or in nearby people. Death patients are also curative, but only for nearby people.

16.2 IMMUNIZATION

For more than 200 years, vaccines have remained a weapon for humans to fight against the disease. Smallpox and polio were eliminated by the worldwide vaccination campaign. Today, childhood immunization has resulted in a significant reduction in morbidity and death from infectious diseases in the most developed countries. But even today, there are diseases for which researchers have not been able to produce effective vaccines, such as HIV/AIDS and malaria. The first vaccine (smallpox vaccine) was produced from a live debilitated virus. "Vulnerability" refers to the weakening of a virus to the extent that it can stimulate immune response, but does not make the human host unwell. In most vaccines used today, live debilitated viruses are used, e.g., measles and rubella vaccines. Other vaccines use inactive forms of dead viruses, bacterial fractions, or bacterial toxins. Dead viruses, bacterial fractions, and inactivated toxins do not cause ill-health, but can catalyze immune responses that protect against the future infections. Different varieties of vaccines such as live recombinant vaccines and DNA vaccines are currently being manufactured using new techniques. Live recombinant vaccines use bacterial strains. DNA vaccines involve DNA coding for a particular antigen that has already been injected into the muscle.

16.3 EPIDEMIOLOGICAL SURVEILLANCE

It is a continuous process of research, evaluation, and control of public health. It includes the collection of epidemiological data for analysis and interpretation, and dissemination of information. In addition, it serves as a basis for designing the short- and long-term strategies to combat infectious diseases. Through epidemiological surveillance, it is possible to identify and prevent cases of dangerous diseases or events, such as epidemics, outbreaks of infectious diseases, cases of poisoning by pesticides, and other poisons. The objectives of epidemiological surveillance are divided into two types.

Individual	**Collective**
Health hazards detection	Diagnosis of health condition
Identifying human groups susceptible to risk	Timely detection of new risks
Optimizing population for schemes	Prioritize and plan the necessary preventive actions

16.4 IMPORTANCE OF EPIDEMIOLOGICAL SURVEILLANCE

Epidemiological surveillance is important to detect serious public health problems. It serves to design the short- and long-term strategies aimed at combating infectious diseases. It is also important in case of situations or events that endanger the lives of a certain population. Surveillance through epidemiological surveillance systems is generally carried out by governments at all levels (national, regional, and local). These include the evaluation of individual cases and collective cases. It seeks to identify systematic, timely, and reliable compilation of data on the causes and populations of epidemics or cases. The analysis and interpretation of these data will be the main input that the government will use to make decisions about situations, e.g., gastrointestinal, skin-related events, viral hepatitis, sexually transmitted diseases. The EVS serves as the basis for formulating public policies for the prevention and protection of the health of the population.

16.5 ETHICAL ISSUES IN EPIDEMIOLOGY

The contribution of vaccines to the eradication of the epidemic is notable. Despite this, vaccination has long been the subject of various ethical controversies. Typically, it includes major ethical debates related to the regulation, development, and use of vaccines. Let us discuss this issue point-by-point.

16.5.1 Mandates and Objections

The reason for the ethical debate and objection to the mandate of vaccination is that religious, philosophical, or health beliefs of some individuals or community are contrary to the concept of vaccination. For example, in an effort to protect large numbers of people from diseases, public health vaccine regulations may infringe on personal autonomy and independence. If people do not accept the existing medical or safety evidence, or if their ideological beliefs do not support vaccination, then the individual wants to exercise his or her right to himself/herself or his/her children, which causes the real problems.

Many local establishments waived the need for vaccination due to personal, philosophical, and religious belief; however, scientific and medical research studies have found that people who miss vaccination based on their religious or philosophical beliefs are at greater risk of spreading the infection, which increases their risk as well as their communities.

16.5.2 Vaccine Research and Testing

Research and testing of vaccines is also involved in ethics discussions, which mainly include topics such as vaccine development and study area, population, and testing location. Prior to obtaining a license, vaccines undergo several years of research and are required to meet the strict safety and efficacy standards. Vaccine development and research includes many specialists in scientific and social disciplines (e.g., public health, epidemiology, immunology, and statistics), and pharmaceutical companies. The priorities and objectives of these partners may be conflicting, which promotes many ethical discussions. Sometimes researchers are skeptical about who should be included in a vaccine test. An appropriate test of the effectiveness of a vaccine requires a clinical trial, which includes a control group. However, it can also be a difficult decision not to provide the appropriate protective options while vaccinating against potentially serious, incurable, or fatal infections. For example, TB vaccine researchers had much difficulty moving through ethical control group procedures. The current TB vaccine, called Bacillus Calumet–Guerin (BCG), has not always been found to be effective in protecting against TB and is vulnerable. Infections occurred in people with weak immune systems. Researchers often discuss whether controlling participants for such vaccines is safe and ethical. In addition, understanding the safety and efficacy of a vaccine in different populations is also extremely important.

When testing vaccines in developing countries, it is important to understand the many ethical concerns associated with this. Such concerns include engaging local communities in the research design process meaningfully as well as the following issues to consider:

- How the tests can be monitored by the local review panel.
- How the investigation and treatment can be provided for the presence of the disease during testing.

Several challenges arose during the testing of the malaria vaccine. As such, it was not understood how to opt out of the trial of the participants, what were the side effects of the vaccine, and what is the difference between a research study and treatment. Ethics is an important element in the research and development of HIV vaccines, as HIV vaccines have many unique ethical challenges. For example, if discriminatory behavior is adopted with participants involved in a vaccine trial, AIDS may be considered a stigma due to psychological problems. In addition, researchers also need to ensure that HIV-positive participants can be protected from the stigma of the disease and how they can be treated. And researchers have to consider that if the participant does not understand the test well, they may feel that they are safe from the virus and thus put themselves at risk. The complexities of these issues reinforce ethical analysis at the borderline of HIV vaccine research.

16.5.3 Informed Consent

Informed consent is a process by which a participant's risks, benefits, and expectations are disclosed, so that they can make informed decisions about what to participate. Broadly speaking, it requires competence, autonomy, and consent, as well as a full understanding of the risks, burdens, and benefits. Specifically, informed consent consists of three major components:

- *Information*: There should be sufficient disclosure of information about risks, burdens, and benefits, which enables the patient or subject to be informed.
- *Understanding*: A person should have the ability to understand based on the information or ability that he has been asked to make a reasonable choice.
- *Voluntarism*: There should be a voluntary decision or compromise on behalf of a competent person. Where informed consent has been obtained, it should be clearly documented.

Research participants have the right to refuse to participate in a study. Informed consent should be given freely on indictments without any external pressure and without any thought. There should always be an assessment of what incentives, if any, may be offered to potential respondents to participate in the study. While some reimbursements like travel costs may be reasonable, participants may not have to pay to participate. For some epidemiological studies, particularly case–control studies and historical cohort studies, nondisclosure of the study's full objectives may be permissible, as the study's full disclosure may bias the hypothesis investigation.

One of the most basic ethical principles of medicine and epidemiology is the moral obligation not to harm the participants, whether physical or psychological. Although the risk in an epidemiological investigation is usually minimal, most people who take part do not derive any personal benefit. Epidemiologists often use personal data, so confidentiality should be respected. Data should only be stored with personal identifiers if absolutely necessary and identifiable information should never be stored on computers outside research establishments. Files with personal identifiers (names, security numbers, addresses, telephone numbers, etc.) should be stored in closed cabinets.

Legal experts and advocates for patient rights believe that obtaining specific consent for testing is ethical and appropriate, so that parents and relatives can be better informed about vaccines and ask questions when needed to find suitable time. However, some believe that the written consent process may cause unnecessary fear and anxiety about the vaccine process.

16.5.4 Learning Issues

Vaccine-related ethical debates focus on evidence that access to vaccination depends to a large extent on social, economic, ethnic, and minority status. The latent question in these debates is whether all lives are of equal importance. The lack of vaccination has come from some vaccine manufacturers and suppliers. There are several factors that inhibit vaccine production, including accountability, expenditure, and the country's political and religious environment. In an ethical sense, the health sector will be positively affected by the increase in the number of vaccine producers. When there is a shortage of vaccines, medical providers decide which ones to protect and which ones are left to fight against diseases. Health inequality underscores the moral dilemma. Developing countries face the risk of disability and fatal infections, while most people in the developed world are not even aware of diseases like tuberculosis. In many developing countries, there is a complete lack of infrastructure to promote a widespread vaccination in places affected by diseases, and many contestants face health and social priorities such as poverty and violence. Public health and medical authorities have to make difficult decisions about which health needs to be dealt with and how to include vaccination in scarcity services. With the global spread of the coronavirus (COVID-19) epidemic, anomalies of every society, whether it is a developed society or a developing society, are coming to the fore. By the time this paper was written, globally, the number of infected people has crossed 6,78,969 and the number of deaths has increased to 31,771. The coronavirus is telling us things that we don't normally want to accept. This is forcing them to recognize inequalities that exist in rich countries. Like the United States, where there have been more than 1,23,781 cases of infection and more than 2,229 deaths, experts are advising people to avoid going into the crowd. Stay home and contact a doctor if you are ill. But a significant problem with that advice is that many low-income people cannot follow it. Low-income jobs (such as cooks, nurses, grocery store workers, midwives) cannot be turned away, and most low-income jobs do not pay for sick days. Most such people either do not have insurance or have a small amount. For many such people, it may be impossible to collect and keep food due to financial constraints. What about India? More than 987 cases of coronavirus have already been confirmed here, and 25 people have died due to this virus. To avoid any symptoms similar to corona, all people have to work from home, and avoid congestion. By way of stated, we have to wash our hands repeatedly with soap and use sanitizer (Figure 16.1).

There are so many people who can work from home and follow these tips! It has a large population of people working in the unorganized sector. There are millions

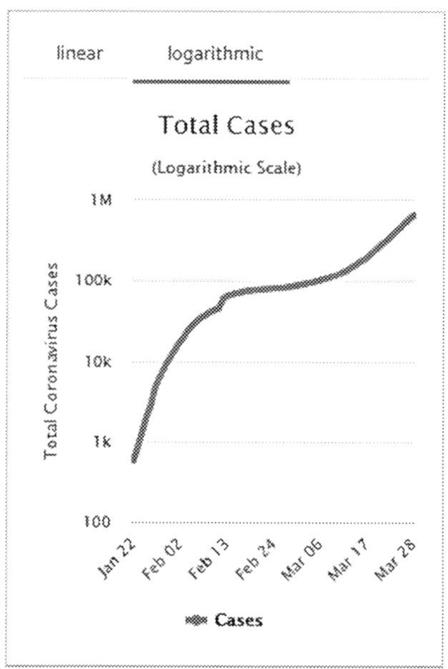

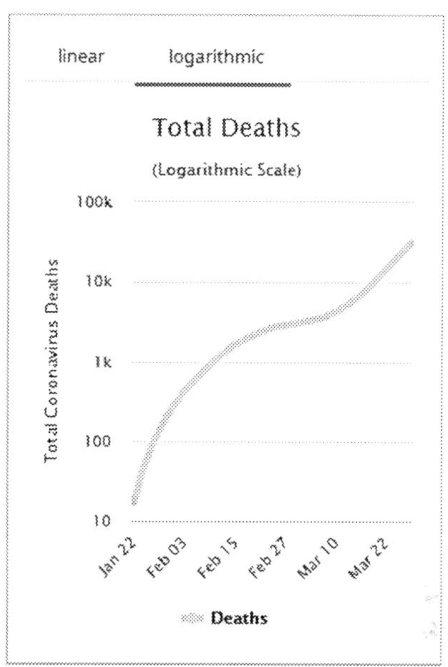

FIGURE 16.1 Logarithmic scale. Courtesy-https://www.worldometers.info/coronavirus/.

of daily salaried people who are not paid leave during sickness. If the coronavirus infection occurs locally, the poorer segment of the population, e.g., drivers, domestic maids, migrant laborers, and others, will certainly be at a huge disadvantage, because they are not so educated about the new coronavirus. Be sufficiently aware. Then, the virus is not tested for everyone; in that case, they will not be investigated and their access to essential medical facilities is difficult and impossible. There are neither sufficient ICUs nor ventilators in government hospitals, so poor people may be more affected by this. In India, the community-wide spread of the virus has not yet been confirmed. When infection occurs at the community level, the disease spreads in such a way that the source of the infection is not known. Infections can spread from one person to another at the workplace or while shopping, and also spread to people who may not know they are infected. Public health experts point out that not many people have been tested in India so far, and with it, education, lack of awareness of healthy conduct, poverty, and weak health system in many parts of the country will be a challenge in the coming days. Frequently washing hands with soap may be a good remedy, which one can do with the standard health precautions. India's position is also challenging in this case. Even though almost all households in India (97% of households according to recent surveys) have a wash basin, only rich and more educated families in urban areas use soap to wash hands. There is a huge disparity between rich and poor families. Only two out of ten poor families use soap to wash their hands, while nine out of ten rich families use soap. Caste and class play an equal role in deepening inequalities. In India, soap is the least used for washing hands among the families of scheduled castes and tribes.

A crisis can also become an opportunity. People know how the spread of plague in Surat became a reason for improvement in local administration in the year 1994. In the mid- and late 1990s, two cities of Surat underwent radical changes in garbage collection and cleaning of roads, hygiene standards were implemented in food establishments (hotels), and paved roads and toilets were provided in slums. As a result of such other changes, a dirty, flood-affected and sick city "Surat" became one of the cleanest cities in the country. Despite a sharp increase in population, cases of mosquito-borne parasitic diseases like malaria have steadily decreased in Surat. Covid-19 is a new virus. No vaccine has been made so far. But basic sanitation and prevention measures matter a lot. Kerala has been highly praised in India for its handling of the cases of Covid-19. But not every Indian state is Kerala, which has a literate population and excellent health system. If the virus spreads to the general population, prevention measures will be very important, as the health system is not as strong as in most states.

16.5.5 Conflicts of Interest

A conflict of interest is a situation in which a researcher has, or appears to have, a private or personal interest, e.g., a financial investment, sufficient to influence the objective exercise of their professional judgments. Researchers must disclose actual, apparent, or potential conflicts of interest to their colleagues, to the ethics committee, and subsequently to a journal publishing their work. All sponsorship of research should also be publicly acknowledged. Research results should be published in an appropriate journal without undue delay. As a general rule, research findings should be subject to independent peer review prior to publication or submission to the media. The nonpublication of research with "negative" findings is also seen as unethical.

16.5.6 Scientific Malpractice

Large-scale study provides ample opportunities to manipulate data. Scientific malpractice is defined as deviations that have not come without the objective of good faith and fairness. External pressures to obtain funding for research and publication are strong risk factors for scientific malpractice. Recent examples suggest that there is seriously no system for protecting data in epidemiological research.

16.5.7 Professional Issues in Epidemiology

The timely detection, investigation, control, and prevention of epidemic outbreaks and major long-term public health problems require a well-trained and competent epidemiological workforce as a key component of the national public health infrastructure. Epidemiology is an integral part of public health practice. Its purpose is to provide a foundation for preventing disease and to promote the health of the population through studies of the distribution of health-related states or events, including the study of determinants affecting such states. Professional epidemiological methods, defined as the application of epidemiological methods to public health practice, apply a combination of analytical methods and orient epidemiology to problem-solving in public health. Major areas of professional epidemiology include epidemiological assessment of public health data, health status and trend analysis, public health monitoring, and health program impact assessment. These areas are closely linked to essential public health function and services.

Epidemiology is the study of how often diseases occur in different groups of people and why. The information obtained from such research is used to prevent the management of individual patients as well as to evaluate strategies for management. In theory, research should be conducted with a desire to establish the truth. However, in the real world, other factors often interfere with this aspiration and may result in a conflict of interest. Research should be funded, and eventually published, while researchers seek to boost their reputation and career. The public has the right to learn about threats to health and be equipped to make evidence-based choices related to treatment and prevention. This is not possible without research. Doing no research is often unacceptable from an ethical standpoint, where there is clinical uncertainty. The quality of research should also be optimized. Poor research can lead to wrong decisions, which can have a profound negative impact on the patient's health.

Epidemiology is an integral part of public health practice. Its purpose is to provide a basis for preventing disease and to promote the health of populations through the study of the distribution of health-related states or events, including the study of determinants affecting such states. Professional epidemiological methods, defined as the application of epidemiological methods to public health practice, apply a combination of analytical methods and orient the epidemiology to problem-solving in public health. Major areas of professional epidemiology include epidemiological assessment of public health data, health status and trend analysis, public health monitoring, and health program impact assessment. These areas are closely linked to essential public health function and services. As the size of public health workers decreases, their responsibilities increase, so the time has come to ensure that young students learn their role in public health by learning epidemiology. Epidemiology should be designed for public health in such a way that students get more employment; only then, it is likely that more of them will consider a career in public health. Knowledge of epidemiology will undoubtedly prepare students to be scientifically literate individual and public health decision-makers, who will be able to appreciate epidemiological evidence and provide support for public health initiatives. Epidemiology should be seen as an essential core curriculum for public health professionals because its importance exceeds commercial value. Even if there were no issues with public health staff issues, there would be a valid scientific literacy argument for converting epidemiological education into a grade 4–12 curriculum. Are biology, chemistry, and physics taught for business purposes only? This chapter is beginning to explore (which is necessary, but not sufficient) what kind of professional development experiences are needed to prepare school and high school teachers to teach epidemiology. What is the need for knowledge of epidemiology to effectively teach epidemiology? How can teachers prepare to teach this science?

To begin planning for professional development teaching, epidemiology teachers will be required from grades 6th to 12th, we can take some epidemiological textbooks from our shelves, and with special focus on the content tables, the subjects start making a list. We can recall some of our favorite epistemological lessons experienced as students. We can start creating a collection of assigned readings such as textbooks, a combination of documents related to historical epidemiology, some dramatic case studies, and some current newspaper columns. For a change of pace, we can think of a video and a guest speaker. In short, we can begin this professional development plan by thinking about what we will do as providers of professional development. We can plan this professional development as a series of activities, during which we cover taxonomy of epidemiological disciplines that are sure to engage budding teachers of epidemiology. Upon completion of the course, individuals should acquire the following specialized knowledge and skills on the concepts of epidemiology and methods of public health problems:

- Understand the place of epidemiology in public health, particularly how epidemiology is used to identify causes of disease, identify populations at high risk for disease, develop preventive methods, and evaluate public health strategies.
- Calculate and explain the basic epidemiological measures of disease frequency, identify sources of data to measure health outcomes, and identify the key aspects of measurement problems.
- Identify the specific features of fundamental study designs, including randomized trial, cohort and case–control studies, birth cohort and ecological studies, and pre–post and quasi-experimental studies. Enable students to describe the strengths and limitations of various study designs in epidemiology and science, and the major sources of confusion and bias.
- Be able to interpret and draw conclusions from the results of epidemiological studies.

16.5.8 Epidemiological Ethics Roots

Ethical barriers must be balanced against opportunities to expand knowledge and improve patient care. In the past 50 years, several guidelines aimed at improving the ethical standards of epidemiological research have been published. Epidemiology and ethics of public health have emerged from several sources: including following:

- *Bioethics*: It is an area of inquiry that originally traced problems in medical and biomedical research, and now includes clinical practice, regulatory policy, research practice, and cultural and social concerns.
- *Philosophical ethics*: Theories and concepts drawn from the writings of Aristotle, Kant, Mill, Rawls, etc.
- Experiences of public health practitioners around ethical problems facing professionals, such as advocacy, conflict of interest, and scientific malpractice.

The Declaration of Helsinki was developed by the World Medical Association as a set of ethical principles that "provide guidance to physicians and other participants in medical research involving human subjects." Although not legally binding, it has been widely accepted as a cornerstone. Human Research Ethics, the first version, was adopted in 1964 and has since been revised seven times, most recently in 2013. It includes the following principles:

- Protecting research subjects;
- Informed consent;
- Reducing risk;
- Adherence to approved research plan.

REFERENCES

1. Broadbent A. *Philosophy of Epidemiology*. Springer, Berlin, Germany, 2013, 238 p.
2. Buck C. *The Challenge of Epidemiology: Issues and Selected Readings*. Pan American Health Org, Washington, D.C., 1988, 1046 p.
3. Coughlin SS, Beauchamp TL, Weed DL. *Ethics and Epidemiology*. Oxford University Press, Oxford, 2009, 328 p.
4. Friis RH, Sellers TA. *Epidemiology for Public Health Practice*. Jones & Bartlett Learning, Burlington, MA, 2004, 650 p.
5. Koepsell TD, Wagner EH, Cheadle AC, Patrick DL, Martin DC, Diehr PH, Perrin EB, Kristal AR, Allan-Andrilla CH, Dey LJ. Selected methodological issues in evaluating community-based health promotion and disease prevention programs. *Annu Rev Public Health*. 1992;13:31–57.
6. Krieger N. *Epidemiology and the People's Health: Theory and Context*. 1st edn. Oxford University Press, Oxford, USA, 2013. 400 p.
7. Kuller LH. Epidemiology is the study of "epidemics" and their prevention. *Am J Epidemiol*. 1991;134(10):1051–6.
8. MacMahon B, Pugh TF. *Epidemiology: Principles and Methods*. Little, Brown, Boston, MA, 1970, 408 p.
9. Merrill RM. *Introduction to Epidemiology*, 6th edn. Jones & Bartlett Learning, Burlington, MA, 2013, 434 p.
10. Mishra J. A remark on fractional differential equation involving I-function. *Eur Phys J Plus*. 2018;133(2):36.
11. Mishra J. Analysis of the Fitzhugh Nagumo model with a new numerical scheme. *Discrete Continuous Dyn Syst-S*. 2018;13(3):781.
12. Mishra J. Fractional hyper-chaotic model with no equilibrium. *Chaos, Solitons Fractals*. 2018;116:43–53.
13. Mishra J. Modified Chua chaotic attractor with differential operators with non-singular kernels. *Chaos, Solitons Fractals*. 2019;125:64–72.
14. Mishra J. Numerical Analysis of a Chaotic Model with Fractional Differential Operators: From Caputo to Atangana–BaleanuMethods of Mathematical Modelling: Fractional Differential Equations, 2019, 167.
15. Mishra J. Unified fractional calculus results related with H function. *J Int Acad Phys Sci*. 2016;20(3):185–195.
16. Mishra J, Pandey N. An integral involving general class of polynomials with I-function. *Int J Sci Res Publ*. 2013;204:121–127.
17. Moon G, Gould M. *Epidemiology: An Introduction*. McGraw-Hill Education, London, UK, 2000, 210 p.
18. Paul JR. President's address clinical epidemiology. *J Clin Invest*. 1938;17(5):539–541.
19. Rollins D.M. Epidemiology and Public Health, 1986.

20. Schwartz S, Susser E, Susser M. A future for epidemiology? *Annu Rev Public Health.* 1999;20:15–33.
21. Webb P, Bain C. *Essential Epidemiology: An Introduction for Students and Health Professionals*, 2nd edn. Cambridge University Press, Cambridge, UK; New York, 2010, 466 p.
22. Weed D, McKeown R. Ethics in epidemiology and public health I. Technical terms. *J Epidemiol Community Health.* 2001;55:855–857.

17 Cloud Virtual Image Security for Medical Data Processing

Shiv Kumar Tiwari, Deepak Singh Rajput, Saurabh Sharma, and Subhrendu Guha Neogi
Amity School of Engineering and Technology

Ashish Mishra
Gyan Ganga Institute of Technology and Sciences

CONTENTS

17.1 INTRODUCTION

Cloud technology is based on a virtualization model that can provide on-demand services remotely with virtualization balance, user-oriented architecture, and web application infrastructure [1,2]. There are several security concerns related to virtualization, databases, resource management, and load balancing as well as centralized data centers. There are several protocols that help us make use of the tools that cloud computing provides. The National Institute of Standards and Technology (NIST) concept lists five main cloud computing features: on-demand self-service, comprehensive network access, resource pooling, rapid elasticity or extension, and calculated operation. This classification approach lists three "business models" to provide cloud services (software, platform, and infrastructure) and four "deployment models" (private, public, group, and hybrid) [3]. The following are the various security issues facing different layers [4]:

i. *Application-level issues:* Services used by the authorized users unable to be used by them.
ii. *Network-level issues:* Attack in opposition to a single network from multiple computer or network systems.
iii. *Data storage-level issues:* It identifies the issues to the cloud data storage, such as theft and unavailability of cloud data.
iv. *Access control-level and authentication issues:* The unauthorized user access to network data due to confidentiality of traffic retention and consequent failure.

This classified approach consisting of three "service models" (software, platform, and the central role of virtualization in the cloud) is measured as it helps boost the efficiency of the low-cost IT industry and its applications. Virtualization provides a specific digital computer with a way of constructing a physical machine as if it were two or more computers, where a similar basic architecture is available [4]. The cloud offers three types of resources: It can run a set of virtual machine (VM) files, and it can potentially be permanent storage device for storing user's data. VM images are separate entities with special properties in the cloud. VM images need a high degree of credibility as they decide in the early stages of VM's operation, including their security states. In other words, the security and integrity of such images form the basis of the cloud's overall security. Worms such as viruses and VM images are an effective way of defining all threats to security. Nuwa leverage is called a series of technologies developed by IBM, called Mirage, which is used for an efficient offline introspection and manipulation of a wide set of VM files, granting multiple simultaneous VM permissions to cloud administrators. Eventually, VMs will encrypt anything to repair [5].

17.2 VIRTUALIZATION IN CLOUD COMPUTING

Virtualization is the foundation of distributed computing; distributed computing makes virtualization progressively productive with the assistance of proficient and it gives answers for significant difficulties in the field of information security and security insurance. There are several causes of virtualization: productive utilization of resources, expanded security, transportability, issue-free testing, simple administration, expanded adaptability, advantages of issue disengagement, and quick organization virtualization.

- For consolidating nearby and organization assets information stockpiling virtualization;
- For grouping physical capacities of devices into the single unit;
- For arriving at the significant level of accessibility or improving accessibility utilizing virtualization;
- Improving execution utilizing virtualization;
- Using virtualization utilizing stripping and storing;
- Capacity improvement.

17.2.1 Importance of Virtualization

Virtualization is necessary for the maintenance of resources in cloud computing environment, as it makes it easier. Cloud computing virtualization increases security as it protects the integrity of both guest VMs and components in the cloud. Virtualized machines from the cloud component can also be scaled up or down on request, or can provide reliability. Resource sharing, heavy use of shared resources, and quick provisioning are just some of the reasons that managed service provider VA has to offer. Most resource vulnerabilities have very little risk or thread to displace computing resources in a computerized environment. For example, if there are inherent vulnerabilities in a service and the service is moved to a nonvirtual server, then the service is still vulnerable to exploitation. Using virtualization, however, may help to reduce the effects of such misuse. But virtualization can also provide additional vectors for attacks and thus increase the likelihood of successful attacks (Figure 17.1).

Many of the virtualization aspects provide safety benefits as well as drawbacks. Cloud in the cloud user rewards many of the big problems involved in researching the usage and computing of public cloud IT resources [6].

- Increased security flaws;
- Low operating governance control;
- Limits between portability of cloud providers;
- Multi-sectored regulatory and legal issues.

Protected images of the shared VM in cloud computing security specifications define security problems in cloud computing in different layers. Virtualization image 13 is defined as the outline design (conceptual) used for the cloud computing (Figure 17.2).

17.2.2 Kerberos

The FTP is designed for Kerberos as a client/server authentication protocol that uses the database to access the network of the user server. The three servers include the Kerberos Ear authentication server (AS), ticket granting server (TGS), and a main server (data) providing user services. Kerberos is a tool used particularly in cloud-based servers for large-scale, critical network interrelationships. This moved to a modern client–cloud server partnership. Many Kerberos intercommunication cloud networks have other protocols for network authentication, which enable you to gather together [4,7].

Today more than ever safe contact is a critical thing. More and more companies tend to run their company using network infrastructure. The crucial element in maintaining connectivity over the distributed network is authentication—the method of assuring someone else's identity. Kerberos, the network protocol, is widely used

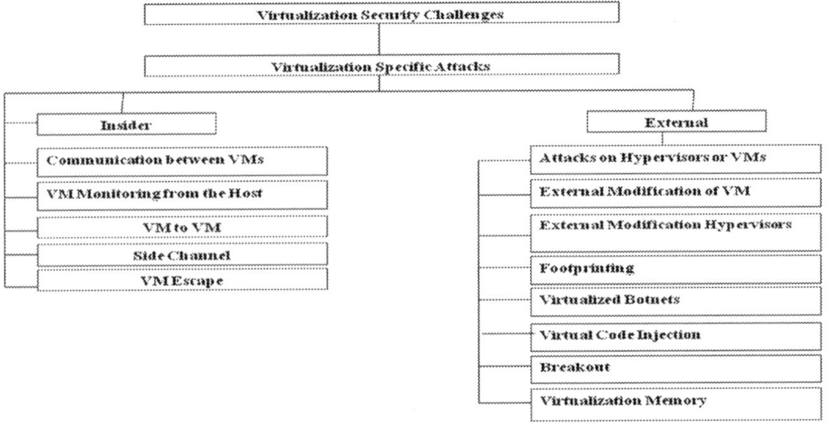

FIGURE 17.1 Challenges of cloud computing virtualization.

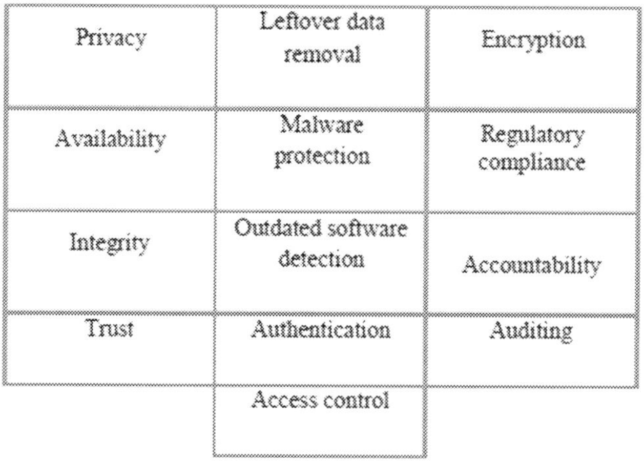

Privacy	Leftover data removal	Encryption
Availability	Malware protection	Regulatory compliance
Integrity	Outdated software detection	Accountability
Trust	Authentication	Auditing
	Access control	

FIGURE 17.2 Security issues [6].

to address the part of authentication, and acts as a vital building block to ensure a secure networked environment. While running other protocols, this protocol will run between two contact parties. Massachusetts Institute of Technology (MIT) created this authentication service. It includes the operation of symmetric key encryption and a KDC (key distribution center). On an unprotected network, it verifies the identities of communication parties. It is done without the host OS, which has to rely on authentication mechanism, without the host's protection being needed and without building trust on the host address.

17.2.2.1 How Does Kerberos Authentication Works

This area is given to clarify the procedure stream of the Kerberos authentication convention. To reduce convention complication, here we show five stages, which are as follows:

i. Simple mutual authentication based on symmetric key cryptography;
ii. KDC;
iii. Session key transmission in session tickets;
iv. Ticket granting tickets (TGTs);
v. Ticket service exchange grant.

17.2.2.1.1 Simple Mutual Authentication Based on Symmetric Key Cryptography

Consider an example: Alice (or client) wants to initiate a secure server communication. All ends agreed to use protocol Kerberos to ensure their identities. Let's get into the workings of the authentication process. As stated earlier, the Kerberos provides symmetric key cryptography, where both encryption and decryption are used with a single key.

The main concept involved here is, before processing the request, the server needs to confirm that Alice's message is originally from Alice. To ensure this, they decided to use the shared secret key information to confirm another end party's identity. Alice sends the server a message, as shown in Figure 17.3.

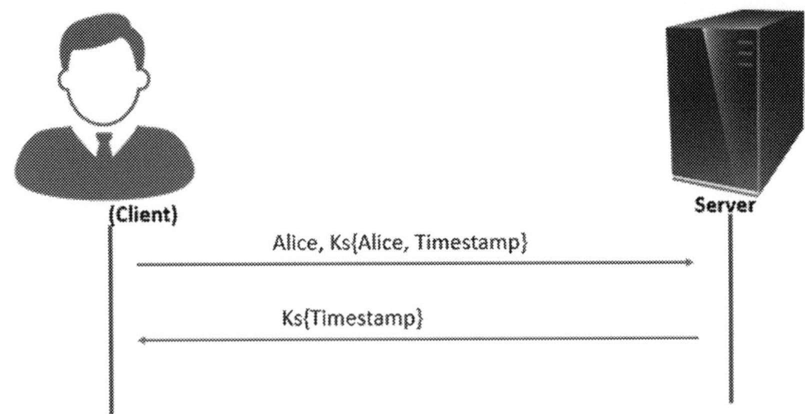

FIGURE 17.3 Simple mutual authentications based on symmetric key cryptography [8].

17.2.2.1.2 Key Distribution Centre

Since both parties use the same secret key, we need to understand how the client and the server exchanged the secret key here. Adding to that, the client may want to communicate with several servers, so each server requires one key. Similarly, the server may communicate with several clients, and each client requires one key. Consequently, several keys on many systems are hard to store as well as secure. A trusted intermediary, KDC is a solution to this issue from Kerberos protocol.

KDC maintains a database which includes in its empire account details of entire security principals. One of the account information in the database is the cryptographic key, which is kept confidential between the KDC and the security principal. The key is usually referred to as a long-term key used in communication between the KDC and the director of security (Figure 17.4).

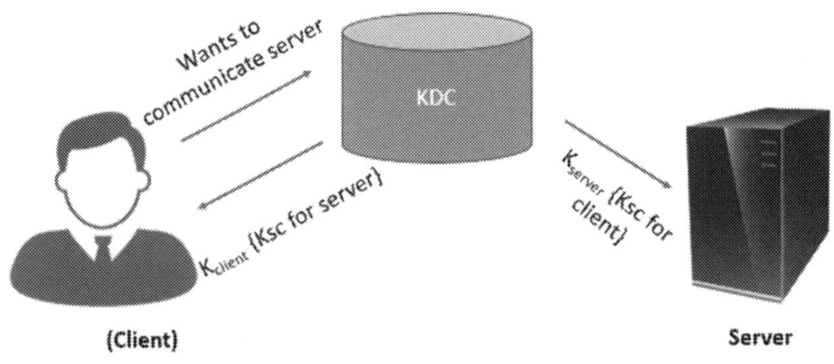

FIGURE 17.4 Key distribution center [8].

17.2.2.1.3 Session Key Transmission in Session Tickets

If a client is given the session key, it can send a message to the server either immediately or after some time at any time. In that scenario, we have two important things to think about. If a client decides to communicate later, the server will need to store the session key it received from KDC as a result of request from the client. And it needs that this client's key is remembered to manage if he asks for service. In addition, due to network traffic, the server may not receive the session key from KDC until the client message is reached. Consequently, the server need to suspend its reaction while waiting to receive a KDC key. That is where the session ticket idea comes in (Figure 17.5).

KDC sends all copies of the session key to the client itself, as a response to the client's inquiry. Here, the database copy is encrypted with the long-term key of the database, while the session key of the server is transmitted in a data structure known as the session ticket. Now the client is responsible for handling the ticket to the session before it hits the server. No harm will happen here if the KDC's message gets to the wrong hand. Because the session key copy of the client can only be obtained by someone who is aware of the secret key of the client and the session key copy of the server can only be obtained by someone who is aware of the secret key of the server. The client sends the message in this mutual authentication process, which includes the authenticator encrypted with the session key received from the KDC and the session ticket. Now together, the session ticket and authenticator become identity credentials for the client. Using the hidden key (server long-term key), which is exchanged between the server and KDC, it decrypts the session ticket when the server receives a message. Then, the session key for authenticator decryption is removed. After verifying the timestamp of the application, it will respond with the time stamp encrypted as mentioned in phase 1.

17.2.2.1.4 Ticket Granting Tickets

We need to find yet another important transmission in this protocol. The session key of the client sent to the client from KDC is encrypted using the long-term key of a client. Here, the long-term key of the client is taken from the credentials used by the one-way hash function to log on Alice's workstation. On the other hand, KDC extracts from the database its copy of Alice's long-term key from its account record. Here, the long-term key is replaced with the session key to improve security.

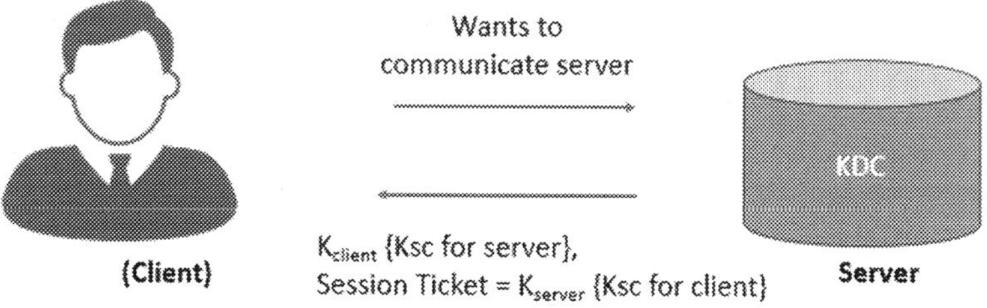

FIGURE 17.5 Session key transmission in session tickets [8].

It sends a request to the KDC for a session key that needs use between the client and the KDC after creating a long-term key when Alice logs on at his workstation. After receiving the request from Alice, the KDC gathers the long-term key from Alice's database and responds with a session ticket called a ticket granting ticket (TGT) to the client. The TGT contains the session key which the KDC must use when interacting with Alice. The answer message along with TGT also contains the client's session key to use in communication with the KDC. The TGT is encrypted with KDC's long-term session key, and the client's long-term key is encrypted. After receiving the KDC update, the client extracts the session key by decrypting it with its own long-term key copy. After the session key is retrieved, the client discards the long-term key and continues using the session key for more communication with the KDC. Bear in mind that after the TGT ends, the session must terminate.

17.2.2.1.5 Ticket Granting Service Exchange

Now, to request a service from the server, the client must get the KDC session ticket. Here, the client sends a message, including TGT and authenticator, which is encrypted using the shared session key between client and KDC. The KDC's function is broken down into two sections to operate around the cross-domain.

1. *Authentication service*: It issues TGTs.
2. *Ticketgranting service*: It issues session tickets.

When the KDC receives the message from the client, the ticketgranting service will validate the request and deliver the ticket and session key as described in phase 3.

Figure 17.6 illustrates the entire authentication process occurring in the Kerberos protocol as a summary of the workflow above.

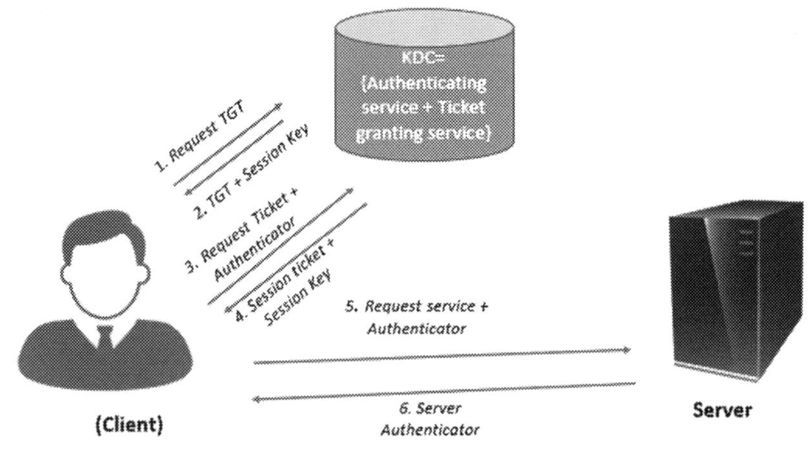

FIGURE 17.6 Ticket granting service exchange [8].

Using authentication methods, which do not reveal the credentials of the identity, is required. The Kerberos protocol is well adapted in the cloud computing environment.

17.2.2.2 Challenges in Kerberos

i. It does not provide end-to-end security; here, we are applying PKI (public key infrastructure) to achieve this process.
ii. In order to do this, Kerberos does not have layered entry; rather, we apply Lightweight Directory Access Protocol (LDAP) and PKI.

Every OpenStack is operated using private and public cloud, and virtual tools. OpenStack devices are platforms that compile the main cloudcomputing, networking, storage, identity, and image services. The components themselves are Open Stack, access management rules, logging systems, security monitoring software, and much more. It should come as no surprise that security issues are equally diverse, and a lot of leaders will need their in-depth analysis. Further [6], the authors balance external reference efforts to explore OpenStack security issues, and provide sufficient context for understanding and providing their handling. Personalization requires the special layers, which can increase the burden of security management and entail additional security checks. When a protection is breached, multiple systems combined on a single physical device may trigger a significant impact:

i. To identify security challenges in VM.
ii. To identify major threats while accessing cloud VM.
iii. What are the major data losses in virtual environment?

17.3 DATA CENTER TECHNOLOGY

In contrast to dispersed geological, shared IT assets use expensive power sharing to bundle IT resources grouping with each other as they consider better availability of higher efficiency, and IT forces work. These are places of interest to promote common ideas for the server farm [9]. Since there are currently server farms, IT systems, in particular IT properties, combine network programming, e.g., servers, databases, system management, and media broadcast gadgets, and use them sparingly. Server farms are typically creative and parts-integrated [10]:

- Virtualization;
- Network;
- Modularity and standardization;
- Storage;
- Remote management and operation;
- High availability;
- Computing hardware;
- Security-aware design, operation.

17.3.1 Virtualization Technology

Virtualization is the conversion of a physical IT asset into a virtual IT asset [11]. Most types of IT asset virtualization can occur, including:

- *Servers*: A server can be preoccupied into a virtual server.
- *Storage*: A physical storage component can be a virtual storage system, or a virtual circle.
- *Network*: Physical switches, e.g., virtual LANs (VLANs), should be concerned with legal network textures.
- *Power*: Physical uninterrupted power supply and strength dissemination units can be concerned with what is generally referred to as virtual UPSs.
- The first step is the development of a virtual server through virtualization programming, and the identification of physical IT assets leads to the creation of a working environment. Virtual servers use their own operating system for tourists, which is autonomous of the work framework in which they are developed.
- Visitors have unconscious working structures and virtual server applications that also exist in virtual IT assets that both run the virtualization programming process and run as if they were running on a separate physical server. This efficiency has excellent reliability, making it possible to operate on physical infrastructure projects as virtual systems are a fundamental aspect of virtualization.
- Virtualization programming is called a physical server display host or physical host, and secret computers are opened by virtualization. Virtualization programming utility envelopes are the system advantages specifically defined with the on-board VM and were generally not standard operating frameworks [12].

17.3.1.1 Hardware Independence

Virtualization is a method of transformation that defines novel IT methods to replicate and repeat institutional programming. Through the independence function, the virtual server goes to a large degree without any stretch in other virtualizations, resulting in anomaly leading to several computer programming problems being resolved. Digital cloning and monitor copying of IT properties is therefore much easier than physical ones.

17.3.1.2 Server Consolidation

Separate virtual server virtualization allows programs to coordinate tasks while all were in the same virtualization. A physical server creates powerful multiple virtual servers that share innovation in virtualization. This method is known as server association and is used for routinely employed tools, adjusting load, and organizing accessible IT assets.

17.3.1.3 Resource Replication

The data server is built as data plate images which are simulated parallel to the content of the hard cycle [10]. These virtual plate images are available to host working frameworks, which means, e.g., direct document activity, copying moves and glue, replicating virtual servers, migrating with and being back. This simplicity of control and replication is one of virtualization innovation's most important highlights, because it is powerful:

- Creation of institutionalized VM images usually organized to integrate virtual equipment capacities, visitor working frameworks, and extra application programming, for pre-bundling in virtual circle images on the side of prompt sending.
- Increased deftness in moving and sending new cases of a VM by having the option to quickly scale out the ability to quickly deliver production VMs depicting host-based memory and hard-plate virtual server diagrams recording conditions. (Administrators cannot restore a lot of preview status in front of a VM.)
- To congratulate the IT consolidation and restoration systems manufacturer for doing business, such as restoring numerous instances of original IT properties and applications.

17.3.1.4 Operating System-Based Virtualization

Establishing system-based virtualization programming is the first one's working framework, known as the host operating framework. When the host is working and system devices gadgets can provide considerable support, working framework virtualization software can change similarity problems, and virtualization programming can be inaccessible to the application driver [13]. Virtualization programming can make the VM accessible to five system connectors even if the virtualized operating framework is typically unused; typically, five system connectors are recorded.

17.3.1.5 Hardware-Based Virtualization

This option consists of devices that legally circumvent the virtualization program of the host operating framework, which explicitly represents the installation of a virtualization working framework. For the most part, allowing VMs to host business systems makes interacting with devices increasingly productive without the need for movement between Go-device-based virtualization. Such a hypervisor is generally found in programming tools for virtualization. One problem for virtualization dependent on hardware is the commonality of system components.

17.3.1.6 Virtualization Management

IT virtualized assets; the board is also retained, depending on the devices virtual infrastructure manager (VIM), which details all virtual IT asset virtualization and a corporate governance module, which is often called a dedicated PC for a sudden increase in the demand of controllers. Typically, VIM originates from the executive system instrument [14].

17.4 VIRTUAL MACHINE IMAGES

Virtualization is a major advancement associated with distributed computing. Virtualization is achieved through the use of computing in the real cloud. Cloud computing, interactive VMs are built to provide the real structure to the ultimate customer or developer at a remote location. A virtual machine, or VM, essentially uses any computing device or computer for product which performs physical (real) machine projects as system guidelines.

VM image is a format designed to create new cases. You can create pictures or drawings, or you can select pictures to clear your own picture of current affairs. The expert in those steps is constantly making list sketches, making sure they get started with the appropriate patch, and making plans for every product with great default settings. Pictures may have simple working systems, or may have started programming them, such as databases, application servers, or may have gone to different applications. In order to retrieve the information for the most part, a certain series of images was made acquainted with the events; information implanted with IP locations or host names was introduced for the swap information to get implanted. Image correction was transformed into a large and increasingly specialized field (Figure 17.7).

17.4.1 System Virtual Machine

The VM architecture organizes a membership framework that facilitates device task execution. This is a replica of an existing framework, a build mechanism, and a step-back justification for running the software that is not open to use the actual gear or for different VM activities, with all tools facilitating increasingly effective use. The need to use and cost of practicality are understood, or both (virtualization of hardware, as a route to a distributed state of processing).

17.4.2 Process Virtual Machine (Language Virtual Machine)

This type of VM is planned to perform a single task, which infers that it supports a philosophy of the lone. These VMs are almost fit for one or more of the additional programming lingos, and are manufactured with the inspiration driving that gives

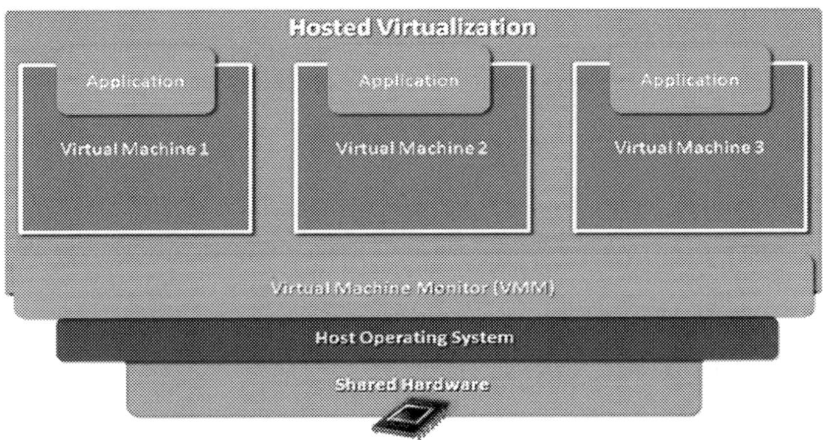

FIGURE 17.7 VM's architecture [15].

venture movement ability and versatility (among different things). A typical imperative for a VM is that the item that runs inside is bound by the VM's benefits and reflections—it cannot area or breaks its virtual environment.

17.5 VIRTUAL MACHINE SWITCH AND SESSION MANAGEMENT

17.5.1 Virtual Machine Switch

A virtual switch is a product program that lets you talk to another VM. The approach helps VMs to coordinate to avoid one of the main virtualization problems without physical hosts transferring VMs for approach and they were explicitly reconfigured. As for VMs to travel physically in a favorable manner, repetitive and potentially open device security breaks down if not handled properly, extinguishing their virtualization operation right by performing specific property functions on the server will lead to progress on these concerns. Which is why improvement can be beneficial in virtual switches? Because virtual switches are fetched so they can be used for the safety of the great VM's profile, the security setting (VM) for its device and VM is passed to the entire network. VMs are put together to further build each other. You should expect the edges of each VM to shift secluded in the virtual switch on the off-chance that you rely on estimation to make custom layer 2 switches work. It is generally not the case that VMs are isolated on a contingent server collect, equivalent VM virtualization process.

17.5.2 Session Management

Within the VM, the program is executed; within the case, it is an intuitive program, a login meeting of a handle, or a virtual showcase meeting such as Virtual Network Computing (VNC) is returned to the client. A large part of the current development, a modest customer structure, includes most private enterprise performance areas, a complete customerserver (including an attractive customer terminal and servers), and a meeting of officials, especially display areas [4] excluding a particular order thin terminal. Another commendable feature of the current building is a meeting of a slender executive's customer group, a union target setting table, which also contains a display area in which display divisions are organized to store details about customer results individually. A customer terminal information is sent to the customer terminal by the customer; terminal information is sent to the dependent execution area from customer information; and small customer association information was configured to look through the terminal board's goal setting and through the unique execution section of the thin customer terminal-specific execution section information gone for the design based on experience to interface.

17.6 MEDICAL DATA IN CLOUD COMPUTING

Cloud computing is a growing modern invention that will influence our lives a great deal. This innovation is used to process assets and offices anywhere, whenever possible. The social insurance industry is growing steadily, and the potential for future data-based medical service models. Enterprises can benefit from overseeing the multifaceted nature of change and innovation in the cloud. It promises innovation can help foster communication, joint efforts, and coordination among various providers of medicinal services. The cloud human resources industry is in a position to have more dollar opportunities. It is fast adaptable and versatile, and can provide framework and application for lovers. Cloud repositories can include international, safe, schedule, and documented electronic health records (EHRs), data frameworks for research facilities, data frameworks for drugstores, and help with diagnostic photographs. Patients provide a record of forward thinking to a large degree, and may have a clearer understanding of ongoing cooperation between different providers of human services. Due to the difference over the absence of issues such as gages, and directions, human service largely concerns cloud selection, which disrupts suppliers' confidentiality and trust [16–19].

17.6.1 e-Health Cloud Benefits

Cloud computing has numerous advantages:

i. *Improved understanding consideration*

Improved mutual appreciation as a result of the patient's nonstop collaboration with different stakeholders in the health services. Learning information for experts to interpret and evaluate is available anytime and wherever.

ii. *Cost investment funds*

There is no good justification to purchase the costly programming and equipment. Reserve funds consider the immediate cost of purchasing equipment and programming for buildings, as well as the expense of support and repair.

iii. *Energy investment funds*

The vitality bill will be cut in view of the fact that there is no requirement on the premises for server farms; there is no requirement for expensive cooling accordingly.

iv. *Robust debacle recuperation*

If a crisis should occur, virtually all cloud specialist organizations offer an excess framework and administrations.

v. *Research*

The cloud is a focal knowledge storehouse that can be used to assist national health study, monitor illnesses, and test pestilences.

vi. *Solving the shortage of assets*

Telemedicine may be used by the specialists in remote regions to perform interviews.

vii. *Rapid arrangement*

Frameworks for the programming and equipment can be used very easily.

viii. *Data accessibility*

Data is open to all participants in community care such as doctors, hospitals, clinics, and insurance providers.

17.6.2 e-Health Cloud Limitations

There are many obstacles in the cloud:

- **i.** Reliance on the Internet Availability and Standard Administration Association can be mild, unintentional, or intrusion. This would have a huge effect on the consumer experience.
- **ii.** *Interoperability*: It is the basis for achieving effective communication, alignment, and cooperation between the relevant standards for various human service providers [5].
- **iii.** *Security and rescue*: Transparent and mutual misfortune gave details intended to steal the situation.
- **iv.** *Laws and guidelines*: The widespread implementation of distributed computing includes regulations, guidelines, and ethical and legal structures.
- **v.** *Limited control and adaptability*: It banned command information capture as a result of centralization. Cloud applications are no regular special, and leasing custom programming can be difficult to do.
- **vi.** *Risk for attacks*: Cloud security prefers for different types of attacks.

17.6.3 Ownership and Privacy of Healthcare Information

The proprietor is in general described as the data producer. Data proprietorship building is vital for insurance against unapproved access or abuse of clinical data from patients. Responsibility for data can be maintained through a combination of encryption and watermarking systems that result in ensuring that data on human services cannot be distributed, accessed, or discharged without the mutual understanding of all substances involved in the proprietorship/forming of data on medicinal services. Patients may require or prohibit the sharing of their data with other social security professionals. Patient can offer rights to clients dependent on a job or property held by the particular client to impart specific social security information to that client in order to actualize the understanding of information sharing in a human services context.

17.6.4 Authenticity

Credibility broadly alludes to the honesty of starting points, assignments, duties, and goals. It guarantees reputation of the item which mentions access. Data validation can pose unusual problems, similar to man-in-the-center attacks, and is frequently moderated with a mix of usernames and passwords. Most cryptographic conventions explicitly include some type of endpoint confirmation in forestalling man-in-the-center assaults. All the human resources data provided by suppliers and the characteristics of shoppers should be reviewed at each entrance in a medicinal services system.

17.6.5 Non-Repudiation

Threats of repudiation worry about consumers who reject their label in the wake of having details about wellness. In the situation of human care, e.g., neither the patients nor the experts can dispute their markreality in the midst of abuse of knowledge about well-being. Human services cloud systems can use computerized marks and encryption to create genuineness and non-renouncement in much the same way as electronic commerce.

17.6.6 Audit

The evaluation guarantees a security commitment that the management of the human services is done well. Later, consumer tracking requests for medical care, e.g., using the system to hold every portal and information adjustment. Health Insurance Portability and Accountability Act (HIPAA) and Health Information Technology for Economic and Clinical Health (HITECH) are both considered responsible for their activities during the handling of secure data when the association needs patients within the clients' social insurance suppliers. There are various ways to analyze and control data; for example, there is a profile for incorporating the healthcare company audit trail and a customer record that the patient encountered that includes enough data to respond to the address: Some patients for record: who will receive this customer? Was the customer responsible for verifying this disappointment? Such activity is not incoherent in a social context. They will, by guaranteeing the unlawful disclosure of documents, eliminate and relieve from insider risks. Evaluation will assist the programmer's efforts to determine whether an open human service has broken into the cloud by assisting the frame and system order head to determine possible vulnerabilities [20].

17.6.7 Access Control

Access control is a device used to monitor a patient's general health data, which restricts access to the actual elements as they were. Usually, every expert advantage approves the access control system, and is silent on the right or accepted by an outsider. A few proposals to resolve the security and access control issues have been suggested. Job-based access control and property-based access control are the most famous models for social insurance application mists.

17.6.8 Cloud-Specific Security Aspects for e-Health Systems: The Case of VM Image Management

When e-health technologies are being implemented with emerging technological paradigms like cloud computing, they offers opportunities for increasing performance in the management of EHRs and helps to minimize costs. Those opportunities, however, introduce new safety risks that cannot be ignored. Based on our experience with installing part of the Swedish EHR management system in an infrastructure cloud, we offer an overview of the main criteria that need to be taken into account when transitioning e-health systems into the cloud. In the case of e-health systems, while migrating systems to a cloud infrastructure, certain security risks characteristic of cloud computing become particularly acute [8].

17.6.8.1 VM Images as an Attack Vector

VM image management is a fairly new feature of IS management, which is present in IaaS clouds. Long-term operation of an IaaS cloud leads to the possible build-up of ancillary data generated by the cloud platform, such as logs, data backups, configuration backups, and VM image snapshots. While all of the above data streams that contain confidential information may require organized and systematic management, future VM images can be mounted at an arbitrary point and become a part of the IaaS cloud. The security approach obscures the edge security of conveying VMs, taking into account the use of various earnestness wellsprings, perhaps allowing the production of yearly VM images, unpublished work frameworks, and application-level programming. Then again, the structure of different renditions of work frameworks and applications could itself produce new weaknesses in different segments. It is generally valid for IaaS clouds, and encourages complex e-health applications architecture. In the case where a component is configured to be deployed on a separate virtual server (such as a front end), backend database server will not occur, deploying them with unnecessary open ports in the required VM instance unnecessarily increasing the attack surface happened.

17.6.8.2 The Way Forward

Based on the security criteria outlined in collaboration with field, as well as existing best practices and our experience with migrating "Melior" to a completely virtualized IaaS cloud, map the process below to the steps taken during the migration of "Melior" to an IaaS deployment [8]. Protection of VM images is one of the cornerstones in maintaining data confidentiality for implementations of e-health systems and other security-critical applications.

The principles for the creation of tailored VM images for e-health applications are as follows.

17.6.8.2.1 Reliable Source Operating System

Trusted operating system source (OS): The OS images are obtained from a valid source via a protected, authenticated channel, as an initial bootstrap step. The validity of the images obtained is checked by comparing their hashes with the respective hashes that the OS provider provides.

17.6.8.2.2 Thorough Hardening of the OS According to Organization Policies and Best Practices

The default OS configuration is changed to, e.g., reduce the application attack surface, install the e-health application target, allow mandatory access policies, and remove redundant remote access channels.

17.6.8.2.3 Traceability of the VM Image Creation Process

The method of generating the VM image is automated and made traceable, in accordance with the relevant security policies. Deviation from the established process of creating VM images is considered a security vulnerability and leads to the discard of the created VM image. This meant having the operating system directly from the vendor's authenticated website using Secure Socket Layer (SSL) while migrating "Melior" to an IaaS server. Additionally, the integrity of the VM image was corroborated with the one the vendor presented. Next, the images were hardened by removing unnecessary components, and only the ports required by "Melior" components were left open. For each type of frontend and backend database server segments, an image was customized. The key and depiction of the sort permit gave traceability VM pictures.

17.6.8.3 Management of Virtual Image

Cloud platform provides useful information that most generates during the dynamic use of VM images. In this particular case, "dynamic use" begins from the second when the Richard picture was created and went to satisfy the variety, which is coupled with all the examples of destruction that the record creates. In this stage, the accompanying standards are relevant. Concentrated, discovered, and repeated VM image fixes and controls image variants and testing of similarities. Both securities patching and the resulting versioning of VM images are managed centrally to prevent inconsistencies in versions. Considering the combined complexity of the operating system and e-health system software stacks, and the frequent changes in the combination of VM image and e-health application versions, automated security checks are needed when patches are applied to any part of the deployment. VM instances can also benefit from additional security services during normal activity. One example is to launch on verified platforms to make sure the VM instance runs on an uncompromised host. Another such example is the provision of high-quality entropy pool for instance, which is essential for the confidentiality and integrity of the e-health data processed by the instance for cryptographic operations performed on the VM picture.

17.6.8.4 Image Archival and Destruction

Depending on the related organizational policies, the life cycle of a VM picture ends with its (optional) archiving and eventual destruction. If VM image archival is applicable, the archived images used in e-health deployments are encrypted and assigned the highest classification of information sensitivity available in the e-health application context. The reason for this is that such images are likely to contain aggregated traces of information from medical records which adversaries may misuse. Destruction of VM images is the final step in the management of VM images. The authors provide taxonomy of secure data deletion techniques analyzed by the target adversary, level of integration, granularity, scope, and efficiency, in a survey of secure data deletion techniques. Deletion of VM image files takes the above aspects into consideration when choosing the approach to deletion. However, e-health system users are ensuring the cloud provider has followed the appropriate deletion approach in the context of IaaS clouds. VM images with other components of the frontend and backend components often require careful deletion when discarded, as they may include session keys and cached data which may affect the confidentiality and integrity of the EHR.

17.7 LITERATURE REVIEW

The new technology can offer on-demand resources remotely with the help of new emerging expertise called cloud computing. Cloud computing helps users to work in an environment where physical infrastructure is not required at user end. There are many protocols that help to access the resources provided by the cloud computing. Cloud computing is an Internet-based technology, which is charged on the basis of usage and demand, and is offered to users for resources and services on them. The exact location where the user is to be configured is hidden from all cloud servers. Research is to improve performance while reducing costs.

Ristenpart et al. [1] stated there are several approaches to interpretive risk. First, the cloud provider could be confused with both the internal structure of its services and the placement policy in order to ascertain that its target-based co-residence was used to measure load, cash usage as a V for M to try to put on the same physical machine. Cryptographic keys to solve problems of that type will be stolen when their functions are isolated through machine virtualization like third parties within the cloud computing service.

Liu et al. [21] have developed a security measure to appropriate the framework for author mother. The purpose of this framework is to determine the status of user-level applications in the guest VM that runs for a period of time. Based on the theory, it combines measurement with VM monitoring. Future research should focus on the measurement framework applicable to many physical machines.

Revar and Bhavsar [22] proposed the method of SSO (single sign-on),the use of single disposable, sign-on for the purpose of single analysis to extend the utility of measuring value in terms of effectiveness and efficiency of using by doing. It allows them to authenticate to access applications only on SSO server device authentication mechanisms in cloud environments, resulting in once multi-searched user by moving cloud computing and storage.

Suen et al. [23] proposed binary compressed difference, which is used to implement VM instances based on volume and mapped volume to achieve optimal storage efficiency. A volume-based storage device is used in Ref. [23] in a quantitative comparison, which has already not performed xdelta3 achieving maximum storage efficiency.

Singh et al. [5] proposed the patch to provide virtual image security Nuwa simulation-based offline display cloud, but not in guaranteed to detect all malware in the image.

Srivastava et al. [2] proposed reliable snapshot as a new cloud service for customers and a system design for reliable snapshot generation on VMs, including compromising the route which malicious admin or root VM client traces attack VMs, can be cleared. The snapshot process begins.

Caliskan et al. [24] explained the use of detection systems and intrusion prevention systems in virtualization. The system is different from each other and simplicity brings manageable issues, but if this operation is not done with careful planning, many dangerous situations can occur.

Pandey and Srivastava [9] proposed hashing techniques to provide integrity and hashing, and virtual images are also used to secure encryption to generate keys. This is achieved by a key component that is safely generated, and the available resource configuration information itself is to deliver the other key components of the host platform. Future plans are to implement the cloud framework, and its performance and enhancement.

Zeb et al. [13] have attempted to defeat the restraints of existing hardware support. In particular, they proposed a safe and efficient protocol that is designed to take privacy, integrity, access control, and non-abandonment basic security services. Destination domain includes a domain account on the cloud VM. Proposed protocol enabled local authentication and authorization of the client as possible. Authenticated digital domain signed through ticket exchange using FIPS-196 on sources in destination domain authority servers. Generate asymmetric session key at both ends of encrypted AES (Advanced Encryption Standard) during the transition of data to the VM using ECDH, and access and data integrity using SHA-256.

Varadharajan and Tupakula [14] proposed the intrusion detection technology, which runs on virtual systems to provide reliable technology for intrusive policy-based distributed applications. It provides integrated security control and integrated security structure. A physical server security structure of their virtual interaction between applications and VMs on secure access control policy. Intrusion detection is an efficient way to explore dynamic attack and working with them to provide a reliable verification technology. It is used for decision evaluation engine (DEE), entity validation (EV), information capture and logging (ICL), and taint analysis (TA), which showed how to use integrated security structure for VM life cycle including dynamic allocation of resources for various physical servers.

Modi and Acha [10], different vulnerabilities and related assaults with which administrations are talked about with the secrecy, trustworthiness and accessibility of virtualization layer-cloud assets can influence. Execution of firewall goals issues may not be an adequate answer for virtualization security in particular. Directions extend the existing arrangements and examine new achievable arrangements that can send frameworks that identify and convey half-and-half interruptions on numerous endpoints cloud.

Hussein et al. [4] have proposed a security framework to protect the virtualization layer in a VM in a cloud environment. In order to maintain the VM image, it is necessary to protect cloud computing. Multi-tenancy VMs have a virtualization layer that affects the VM and an introduction to the VMM problem. The initial state of infected VMs can affect the security of VM images. Therefore, to protect VM images, create a structure that is capable of securing the repository VM image.

Hou et al. [25] formulated a new RA model to identify the risk of VM for certain time. In addition, heuristic has proposed a novel algorithm for physical separation between risky and secure VMs. Simulation results are better guaranteed than RVNE ODCN protection with an average coverage ratio of 0.75. In particular, we have been recognized by a larger number of secure servers than the benchmark RVNE. The solution algorithm is very close to the upper limit that appears optimality algorithm, which has been obtained by us. In the future work, we will focus on designing an offline heuristic new algorithm, which can affect the high optical bandwidth between VMs with two weaknesses. In addition, we prefer to estimate accurately the new application scenarios to the RA model which will confirm the Internet of Things.

Islam and Rahman [12] proposed cloud Kerberos security structure using new security encryption and decryption mechanisms, and VM images storage. They proposed enhanced VM images through Kerberos with encryption and decryption processes security framework. In the proposed model, VM security image has a wide variety of attacks, like theft and session hijacking. Image storage in a VM is vulnerable to unauthorized access to various types of attacks, such as data breach, malware wire attack, configuration, image snapshots, and malicious code installation. The proposed security model can be used by large IT organizations to ensure protection against such encryption, breach, disk encryption, security protection, and integration with OpenStack Kerberos to improve the guarantee. Some overhead is when Kerberos integrates with disk encryption, but is paid better [12] by improving the overall security of the cloud system.

Kumar and Rathore [11] proposed the point result of the security vulnerabilities assault surfaces alongside the essential presentation of distributed computing and virtualization. The virtualization is the cloud insurance that it is fundamental to have the haze of security down to the current ideal level. Highlighted attributes related to photographs of VM furthermost show region of research. Numerous frameworks are expending VM depiction downpours. Another conceivable territory of security for setting plate pictures in the cloud through the Cloud Swift help is additionally a region of conceivable research.

Janjua and Ali [26] evaluated the security ESMV2M system is evaluated between security performance and AVISP, which successfully identify security threats in SAFE. In results. In the future, evaluation criteria should be proposed by which we can analyze the effectiveness of the counter-remedy.

Borisaniya and Patel [27] proposed the method to monitor user activity and to find out the proposal based on a Virtual Machine Introspection (VMI) Malicious Security Framework. They said that to discuss the main advantages of the proposed framework, it was decided to implement other methods of VMI based on material knowledge. They are capable of analyzing the proposed design of VMs hosted on the cloud. A multi-threaded server component can be extended in real time. Experimental results suggest that the structure IaaS clouds perform well with a set of measures and parameters suitable for the cloud environment. They stated that the monitoring and their characteristics of a VM call classification process systems, which are described in our modified model vector space. The proposed framework is specifically designed to host multiple VMs on multiple physical machines to run multiple processes.

Pal et al. [28] discussed the performance of resource allocation algorithms that use CPU's best time and memory. They showed how to handle live migration of VMs in cloud computing across the cloud. They suggested a possible resource migration algorithm for optimal use of unused memory VMs. In the memory process occupied setting off, VM and persistent memory probably occupied for VM memory and total memory demand. Their proposed algorithm has not been published as part of the stack flow VM architecture of cloud storage memory measurement work for efficient use of memory.

Bhardwaj and Krishna [15], who have made progress, propose the identity-modified bus technology and modify the content of VM memory pages and only the dirty content. The proposed technique intensive workload virtualization environment configured on QEMU-KVM was revealed by VM. Results show that in a copy of the former VM immigration plan to read intensive workloads, we provide technology that is 69.4%, 73.5% possible, and reduce downtime for network traffic by 74.21%. Similarly, for write-intensive workloads, downtime, migration time, and network traffic decreased by 47.23%, 49.16%, and 51.37%, respectively. Thus, the proposed technique has increased effective performance of VM migration planning.

Jyoti et al. [29] proposed a Trusted Computing Framework Security for VM's life cycle. The use of the proposed framework achieved the integrity and security of VMs, a reliable technology life cycle. In particular, the concept of life cycle VM presents the breakdown of the various operating conditions of VM. Then, to develop a computational-based security framework that is reliable, trusted relationship platforms can be the way to extend VM modules and protect the security of trusted VM life cycles and reliability. Theoretical analysis suggests that a comprehensive security framework for TV can be supplied by M, which is offered in all states. In addition, it is possible to use the results of events to offer a higher level of security structures and some sophisticated schemes.

Al Hamid et al. [30] focus on the classification of medicinal services patient's mixed media information in the cloud by proposing an outsider one-round checked key on bilinear mixing cryptography. They can make a gathering key among the individuals to pass on securely.

Galletta et al. [31] present a structure developed that is claimed to address the patient's data security and insurance. They showed that system relies upon two programming portions, namely, the anonymizer and splitter. First accumulates anonymized clinical data, while the subsequent tangles and stores data in different dispersed stockpiling providers, simply endorsed clinical overseers can get to data over the cloud. They present a logical examination that uses appealing resonation imaging (MRI) data to assess the introduction of the realized structure.

Alexander and Sathyalakshmi [32] propose a protection-mindful framework and anonymization strategies for information distributing on cloud for PHRs. The proposed framework utilizes k-secrecy and AES.

Mishra [33] gives an extensive answer for safe access to protection touchy EHR information (1) through a cryptographic job-based strategy to convey meeting keys utilizing Kerberos convention, (2) area and biometrics-based confirmation technique to approve the clients, and (3) a wavelet-based steganographic method to install EHR information safely in distributed storage. This chapter likewise shows the versatility of the proposed answer for man-in-the center and replay assaults. Be that as it may, they didn't examine the versatility of the methodology and its flexibility to other critical security dangers, including honesty and accessibility of the information just as the computational overhead.

Researchers had proposed several methodologies to secure virtual images to their convenience over the cloud environment. An effort is being made by various researchers to develop better security cloud platform, but there is still room for improving the existing framework. Virtual cloud machine design requires a secure authentication, authorization, and virtual image access accounting. Virtual images require the processing of sensitive data, so it is important to protect the attacker's running virtual images by improving the key management algorithm. Virtual image protection is one of the biggest challenges in cloud computing, although a lot of work has not been done in this respect and there are many limitations in the computing of different metrics used by different researchers to design cloud computing. Using Kerberos KDC, the latest cluster design will allow the Kerberos security by providing additional enhancement with support.

17.8 OBJECTIVE

The following are the objectives:

i. Designing cloud security framework through Kerberos;
ii. Propose a framework for the modification of key manager algorithm to provide better quality of service.
iii. Comparing the proposed framework with the existing cloud virtual image securing mechanism.
iv. Implementation of the proposed model for improving the performance of cloud VM images.

17.8.1 Implementation Model

To achieve the objectives, the research will be carried out in the following phases:

Phase I:

Activity 1: Creating virtual cloud images for access server, AS, computation server, and storage server for topology formation.

Activity 2: Creating topology for virtual cloud images with Kerberos.

Activity 3: Providing authentication, authorization, and accounting for virtual cloud images with Kerberos.

Phase II:

A security framework needed to be designed for

i. The identification of security mechanism in Kerberos;
ii. The identification of suitable parameters for the modification of security features.

Phase III:

Managing VM images with key management algorithm

i. VM operations to obtain VM states using introspection;
ii. Validating VM states according to intrusion detection policies for privilege escalation and checking the unauthorized access to resources.

Phase IV:

Implementing Kerberos for managing sessions, and handling authentication and communication between client and server using different techniques

i. Use of virtual image management module for the identification of pattern in Kerberos;
ii. Kerberos building blocks of authentication and authorization;
iii. Design and modification of algorithms for hardware security module (HSM) and key management module (KMM).

Phase V:

i. Using OpenStack to provide massive scale shared resources for cloud computing platform;
ii. Cloud infrastructure services with the help of open nebula.

17.8.2 Terminologies Used in Implementation Model

17.8.2.1 Access Server

An access server is a networked computer, which is set up and responsible for granting rights to other computers which are not members of its network.

17.8.2.2 Authentication Server

An AS is a system that provides authentication services to other network systems. An AS is a form of network server validating and authenticating remote users or IT nodes that are linked to an application or service. It ensures that access to the server, application, storage, or any other IT resources behind the authentication system is only given to authorized and authenticated nodes.

17.8.2.3 Storage Server

A storage server is a type of server used for storing, accessing, securing, and managing digital data, files, and services. It is a built-in server which is used to store and access small to large quantities of data over a shared network or over the Internet. A file server could also be called a storage server.

17.8.2.4 Kerberos Security Mechanism

Kerberos is one of the well-known safety protocols that will provide authentication. It can be utilized in both distributed architecture and centralized architecture. But it is primarily designed for the distributed environment. The user initially creates the password that is called the long-term secret key. Every customer should have TGT, commonly known as an auto AS ticket. This TGT can be used in multiple servers as this TGT is used for client verification. Once the TGT is received from the server, the client requests that the ticket be granted for service. User-assigned entities are stored in the database. If we want to access some info, the servicegrant ticket will always demand the password. Set of authentication system, ticket granting system, and database is the main distribution center. The AS is responsible for granting tickets to all users, and TGS is responsible for granting users tickets to the facility.

For authentication Kerberos, you can follow these steps:

1. On the workstation, initial logging is performed; the message is sent to the AS for ticket grant request.
2. The AS check is performed, which checks the user's data being fed. If data input is correct, TGT is assigned to the user and a user session decides to use key as well. The user can communicate to the server in that time of session.
3. Also included in the ticket is the same copy of the session key which AU issues to the client. Both client and server both keep the key.

Using the user's password, both TGT and session key are then encrypted. Since the encryption is performed using the user key, there is no likelihood of any security vulnerability. The TGT and the session key are both encrypted using the user's password. As encryption is using the user key, there is no possibility that any can access it. The key is known use it.

17.8.2.5 Hardware Security Module (HSM)

HSMs are hardware-resistant tools that improve encryption activities by generating keys, encrypting and decrypting data, and producing and verifying digital signatures. HSM is a special "trusted" network computer that performs a number of cryptographic operations: key control, key exchange, encryption, etc. Enterprises buy HSMs to secure transactions, identities, and applications, as HSMs excel in protecting cryptographic keys and delivering encryption, decryption, authentication, and digital signature services for a wide range of applications. Cryptographic keys are designed to secure HSMs. Large-sized banks or corporate offices often simultaneously operate a variety of HSMs. These keys are controlled and updated by key management systems according to internal security policies and external standards. A centralized key management design brings the advantage of streamlining key management and providing the best overview of the keys across multiple systems.

17.8.2.6 OpenStack

OpenStack is an open-source cloud computing platform that enables businesses to control large computing, storage, and networking pools within a data center. OpenStack is a set of software tools for the construction and management of public and private cloud computing platforms. Backed by some of the largest software development and hosting firms, as well as thousands of individual group leaders, many believe that the future of cloud computing is OpenStack.

OpenStack clients tried to convey VMs and other instances that handle various tasks to control a cloud environment. It enables horizontal scaling, which means that tasks that benefit from running simultaneously can easily support more or less users on the fly by simply spinning up more instances. For example, a mobile application that needs to communicate with a remote server might break the work of interacting with each user into several instances, all interacting with each other but scaling quickly and efficiently as the application absorbs more users. OpenStack provides on-demand networks, IP addresses, firewalls, and routers. Basic features for these are built-in, and OpenStack can also be introduced from network equipment vendors such as Juniper, Cisco, and Nokia with telco-oriented SDN deals. For a company, the ability to identify networks through an application programming interface (API) allows for fast-paced automation of infrastructure and cloud-style operations.

Thus, OpenStack contains a diagram for programming characterized capacity, together with both square ("circle") and article stockpiling frameworks. There are programming tasks of these that are worked in to OpenStack, in spite of the fact that it is additionally doable to incorporate with outsider contributions from EMC, Pure-Storage, NetApp, and others.

17.8.2.6.1 Components of OpenStack

OpenStack consists of loads of different moving pieces. Because of the open design of OpenStack, anyone can add additional components to help it meet their needs. But the OpenStack community has collaboratively defined nine key components which form part of OpenStack's "heart," which are distributed as part of any OpenStack framework and officially maintained by the OpenStack community.

- *Nova*

  Nova is the primary motor behind OpenStack computing. It is used to deploy and manage large numbers of VMs and other instances for the processing of computing tasks.
- *Swift*

  Swift is an object and filestorage system. Instead of the traditional idea of referring files by their location on a disk drive, developers can instead refer to a unique file or piece of information identifier and let OpenStack decide where to store this information. This makes it easy to scale as developers are not concerned about the capability behind the applications on a single system. It also allows the system to worry about how best to ensure data are back up in case of machine or system and network connection failure.
- *Cinder*

  Cinder is a component of block storage, which is more analogous to the traditional notion that a computer can access specific locations on a disk drive. For situations where data access speed is the most, the more conventional method of accessing files may be useful.
- *Neutron*

  Neutron provides the OpenStack networking capability. This helps ensure that every aspect of an OpenStack deployment is able to interact easily and efficiently with one another.
- *Horizon*

  Horizon is OpenStack's Dashboard. It is OpenStack-only graphical interface, and for users who want to try OpenStack, this could be the first feature they actually "see." Developers can access any of OpenStack's components individually via an API, but the dashboard provides system administrators with an overview of what's going on in the cloud and handles it as required.
- *Keystone*

  Keystone provides OpenStack identification services. It is essentially a central list of all OpenStack cloud users, mapped against all of the cloud services they have permission to use. It provides multiple access methods, which means developers can easily map their current user access methods against Keystone.
- *Glance*

  Glance is provided by OpenStack with image services. "Images" refers in this case to images (or virtual copies) of hard disks. Glance makes the use of such images as models when installing new instances of a VM.

- *Ceilometer*

  Ceilometer provides telemetry services, enabling the cloud to provide billing services to individual cloud users. It also maintains a verifiable count of the system usage by each user for each of the different components of an OpenStack cloud. Think metering and monitoring of use.
- *Heat*

  Heat is OpenStack's orchestration feature, which allows developers to store a cloud application's specifications in a file that specifies what resources are needed for that application. It helps manage the infrastructure required for running a cloud service in this way.

17.8.2.7 Virtual Image Management

An image on a VM is a template for creating new instances. Virtual Image Management System (VIMS) is a high-performance cluster file system that converts data stores to VIMS format and connects the data stores to CNAs. This system allows multiple VMs to access an interconnected storage pool to make resource usage more efficient. VIMS is the basis for virtualizing multiple storage servers, delivering various services such as live storage replication, dynamic scheduling of storage resources, and high availability.

There are three CNA servers deployed, each running 2 VMs. The VMs regard their virtual disks as targets for local SCSI, but in fact they are volume files for VIMS. VM files of each CNA server are stored in the specified subdirectories of the VIMS. The VIMS locks its virtual disks when a VM is running, to ensure shared reading and exclusive writing. Every CNA server can link to all of the VIMS room. A VIMS volume is a cluster volume, which provides the distributed lock management to balance access and enables CNA servers to share the clustered storage pool.

1. Hierarchical directories;
2. Applying to VMs in a cluster;
3. Distributed lock management and logical volume management;
4. Scaling among multiple storage disks and dynamic data store expansion;
5. Quick restoration on clustered file systems with logs;
6. Independent encapsulation for VM files and VIMS uses abstract processing to simplify storage infrastructure architecture. It allows multiple virtualization servers to access storage devices concurrently to provide efficient storage pooling.

17.9 CONCLUSION

Cloud computing security is one of the fundamental issues that prevents the innovation and research from being quickly received. Unmistakably, the security subjects have assumed the most important job in obstructing the acknowledgment of distributed computing. Issues such as board personality and virtual cloud access control, Internet-based access, confirmation and approval, and cybercrime are significant concerns in the distributed computing medicinal services. Cloud information is handled over VM numbers. The security risks and solution relevant to the deployment of an EHR system in an IaaS cloud presented in this research cover only a fraction of the challenges facing a large-scale migration of public e-health systems using VMs to IaaS clouds. In case electronic medical records are prepared in the web, the protection of VM is increasingly important at that level. For various types of handling prerequisites, exchanging over VMs is essential; it will result in a variety of security that should be settled.

REFERENCES

1. Thomas Ristenpart, Earn Tromer, Hovav Shacham, and Stefan Savage, "Hey, you, get off of my cloud: Exploring information leakage in third-party compute clouds", *16th ACM Conference on Computer and Communications Security Nov 2009*, 199–212.
2. Abhinav Srivastava, Himanshu Raj, Jonathon Giffin, and Paul England "*Trusted VM Snapshots in Untrusted Cloud Infrastructures*", Springer-Verlag, Berlin Heidelberg, 2012.
3. Peter Melland Tim Grance, "The NIST definition of cloud computing", Version 15, October 7, 2009, http://csrc.nist.gov/groups/SNS/cloud-computing/cloud-def-v15.doc.
4. Raid Khalid Hussein, Ahmed Alenezi, Gary B. Wills, and Robert J. Walters "A framework to secure the virtual machine image in cloud computing", *International Conference on Smart Cloud*, New York, IEEE 2016.
5. Gundeep Singh, Prashant Kumar Singh, Krishan Kant Kandewal, and Seema Khanna "Cloud Security: Analysis and Risk Management of VM Images", *International Conference on Information and Automation Shenyang*, Shenyang, China, IEEE 2012.
6. B. Dhivya, S.P.S. Ibrahim, and R. Kirubakaran, "Hybrid cryptographic access control for cloud based electronic health records systems", *International Journal of Scientific Research in Computer Science, Engineering and Information Technology*, 2(2), 2017, 9.
7. Ashish Mishra et al., 'An authentication mechanism based on client-server architecture for accessing cloud computing', *International Journal of Emerging Technology and Advanced Engineering,* 2(7), 2012, 95–99.
8. Michalas Antonis, Paladi Nicolae, and Gehrmann Christian, "Security aspects of e-health systems migration to the cloud" Researchgate, October 2014.
9. Anjali Pandey and Shashank Srivastava, "An approach for virtual machine image security", *International Conference on Signal Propagation and Computer Technology*, 2014, 616–623.
10. Chirag N. Modi and Kamatchi Acha, *"Virtualization Layer Security Challenges and Intrusion Detection/Prevention Systems in Cloud Computing: A Comprehensive Review*", Springer, Berlin Heidelberg, 2016.
11. Vimlesh Kumar and Rajkumar Singh Rathore, "Security issues with virtualization in cloud computing", *International Conference on Advances in Computing, Communication Control and Networking (ICACCCN2018)*, Greater Noida (UP), India, IEEE 2018.
12. S.M. Neamul Islam and Md. Mahbubur Rahman, "Securing virtual machine images of cloud by encryption through Kerberos", *2017 2nd International Conference for Convergence in Technology (I2CT)*, Mumbai, India, IEEE, 2017.
13. Tayyaba Zeb, Abdul Ghafoor, Awais Shibli, and Muhammad Yousaf, "*A Secure Architecture for Inter-Cloud Virtual Machine Migration*", Institute for Computer Sciences, Social Informatics and Telecommunications Engineering, Springer, Berlin Heidelberg, 2015.
14. Vijay Varadharajan and UdayaTupakula "On the design and implementation of an integrated security architecture for cloud with improved resilience", *IEEE Transactions on Cloud Computing*, 99, 2015, 1–1.

15. Aditya Bhardwaj and C. Rama Krishna, "Improving the performance of pre-copy virtual machine migration technique", *International Conference on Communication, Computing and Networking,* Springer, Berlin Heidelberg, 2019.
16. Xiang Li, Xingshu Chen, and Wei Wang, "Cloud virtual machine lifecycle security framework based on trusted computing", *Tsinghua Science and Technology*, 24(5), IEEE 2019, 520–534.
17. Ashish Mishra et al., 'Secure cloud storage architecture for digital medical record in cloud environment using blockchain', *Elsevier SSRN International Conference on Intelligent Communication and Computation Research*, April 1, 2020. doi: 10.2139/ssrn.3565922.
18. Ashish Mishra et al., 'An enhanced DDoS TCP flood attack defence system in a cloud computing', *Elsevier SSRN International Conference on Intelligent Communication and Computation Research*, 2020. doi:10.2139/ssrn.3565916.
19. Ashish Mishra et al., 'A review on DDOS attack, TCP flood attack in cloud environment', *Elsevier SSRN International Conference on Intelligent Communication and Computation Research*, Available at https://ssrn.com/abstract=3565043, March 31, 2020.
20. Integrating the Healthcare Enterprise, IHE Profiles, Oak Brook, IL, USA, 2017.
21. Q. Liu, C. Weng, M. Li, andY. Luo, "Anin-VM measuring framework for increasing virtual machine security in clouds", *IEEE Security & Privacy, 8(6), Nov 2012*, 56–62.
22. Ashish G. Revar and Madhuri D. Bhavsar, "Securing user authentication using single sign-on in cloud computing", *2011 Nirma University International Conference, 2011*, 1–4.
23. Chun-Hui Suen, Markus Kirchberg, and Bu Sung Lee, "Efficient migration of virtual machines between public and private cloud", *Third IEEE International Conference on Cloud Computing Technology and Science*, Athens, Greece, IEEE 2011.
24. M. Caliskan, M. Ozsiginan, and E. Kugu, "Benefits of the virtualization technologies with intrusion detection and prevention systems", *International Conference on Application of Information and Communication Technologies*, 2013. doi:10.1109/IEEE 2013.
25. Weigang Hou, Zhaolong Ning, Lei Guo, Zhikui Chen, and Mohammad S. Obaidat, "Novel framework of risk-aware virtual network embedding in optical data center networks", *IEEE Systems Journal*, 12(3), Sept. 2018, 2473–2482.
26. Kanwal Janjua and Waris Ali, "Enhanced secure mechanism for VM migration in clouds", *International Conference on Frontiers of Information Technology (FIT)*, Islamabad, Pakistan, IEEE 2018.
27. Bhavesh Borisaniya and Dhiren Patel, *"Towards Virtual Machine Introspection Based Security Framework for Cloud"*, Indian Academy of Sciences, Bengaluru, 2019.
28. Souvik Pal, Raghvendra Kumar and Le Hoang Son, *"Novel Probabilistic Resource Migration Algorithm for Cross-Cloud Live Migration of Virtual Machines in Public Cloud"*, Springer, Berlin Heidelberg, 2019.
29. Jyoti Mishra. Analysis of the Fitzhugh Nagumo model with a new numerical scheme. *Discrete and Continuous Dynamical Systems-S*, 13(3), 2018, 781.
30. H.A. Al Hamid, S.M.M. Rahman, M.S. Hossain, A. Almogren, and A. Alamri, "A security model for preserving the privacy of medical big data in a healthcare cloud using a fog computing facility with pairing-based cryptography," *IEEE Access*, 5, 2017, 22313–22328.
31. A. Galletta, L. Bonanno, A. Celesti, S. Marino, P. Bramanti,and M.Villari, "An approach to share MRI data over the cloud preserving patients' privacy," in *Proceedings of the 2017 IEEE Symposium on Computers and Communications (ISCC 2017)*, pp. 94–99, Heraklion, Greece, July 2017.
32. E. Alexander and Sathyalakshmi, "Privacy-aware set-valued data publishing on cloud for personal healthcare records," in *Artificial Intelligence and Evolutionary Computations in Engineering Systems*, pp. 323–334, Springer, Berlin, Germany, 2017.
33. Jyoti Mishra. Fractional hyper-chaotic model with no equilibrium, *Chaos, Solitons & Fractals* 116, 2018, 43–53.

18 Medical Data Security Using Blockchain and Machine Learning in Cloud Computing

Deepak Singh Rajput, Saurabh Sharma, Shiv Kumar Tiwari, and A. K. Upadhyay
Amity School of Engineering and Technology

Ashish Mishra
Gyan Ganga Institute of Technology and Sciences

CONTENTS

18.1 INTRODUCTION

18.1.1 Cloud Computing

Cloud computing is a way to access data and storage resources without clear control and active resource management. Therefore, purchasing, maintaining, and updating systems may be a massive investment of time, and resources in today's world computing and storage demands are very diverse. Cloud computing provides various facilities and amenities over the Internet, such as databases, servers, storage, and applications. Rather than storing data on a hard disk in local storage, cloud computing allows us to store and save data on a centralized database.

Cloud computing provides the computer we use has Internet access [1–9]; it will also have access to the data. Cloud computing essentially outsources computer programs a bit. Such computer programs are managed by an outside party and are in the cloud. Because of this, users have no worries about storage and power, and can be at ease when it comes to their data (Figure 18.1).

Modern systems have always hated the required hardware and software because these systems were complicated and costly to run. We need to keep such large-scale applications installed, and configure, run, and check to protect and manage them. Cloud computing helps reduce the problems around the handling of one's own data by raising the software and hardware management. We could pay for remote storage of the data to vendors, who would charge us only on the basis of need, making it easy to scale up or down. Cloud computing is so called because in the cloud, which is nothing but virtual space, the knowledge and data that we are trying to access are discovered away. Cloud users can store files, documents, and programs on remote servers, and use the Internet to access these documents. This works in such a way that the user does not need to be in a specific position to get access to the data and can retrieve the data from anywhere. Cloud computing decreases the amount of data involved in the processing, and allows the machine to do all the work [10–13].

18.1.1.1 Service Models

Since hearing about the transition of cloud computing, we are now researching the business models that are categorized in terms of end-user abstraction [14]:

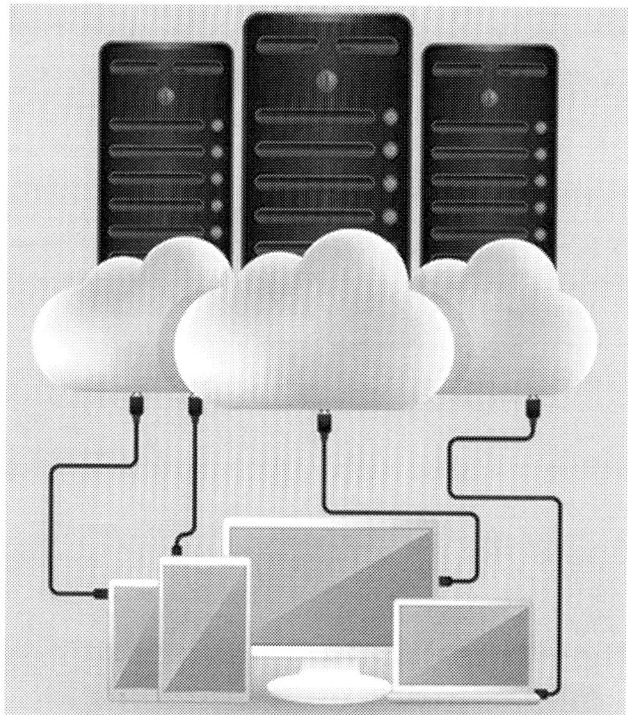

FIGURE 18.1 Cloud computing.

i. *SaaS (software as a service)*:

The cloud service provider (CSP) meets all the computing and storage needs in the SaaS-based model, and the user needs only to upload and access data. The service provider takes care of all maintenance, downtime, upgradation, and security. This is the web browsers distributing applications to end-users. It is installed by cloud clients, and they will allow them to run on the cloud platform. But this phase is not mandatory since software support and reduced maintenance are required (Figure 18.2).

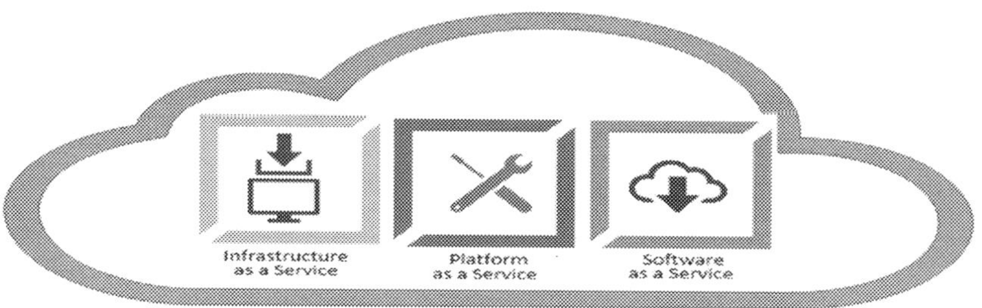

FIGURE 18.2 Cloud computing service model.

ii. *PaaS (platform as a service)*:

User management applications are provided with data in PaaS. Most of the time consumer needs to launch and manage their own cloud-based applications where PaaS comes into the frame. The service provider meets all of the hardware, networking, and OS requirements. The user can use any language of choice for the programming. The PaaS services are cheaper than the SaaS services. It is the part of platform for application creation and delivery as a service open to programmers or developers. We can use the basement to effectively develop, deploy, test, and manage SaaS applications. The key attributes of PaaS are point-and-snap equipment which enables the programmer to develop web-based applications.

ii. *IaaS (infrastructure as a service)*:

Virtualization and networking services are offered by the provider in IaaS-based service hardware, while the user takes care of OS, software, and data. Cloud computing provides both physical and virtual machines. The individual machines are accessed by hypervisors grouped into pools and operated by friendly operating networks. Cloud computing incorporates representations of the operating system on the individual computers and the programming of applications. Infrastructure as a service offers tools such as firewalls, IP addresses, monitoring systems, storage, bandwidth, and virtual machines—all of which are made available to customers on time-based charges.

18.1.1.2 Deployment Models

We've already heard about the Introduction to Cloud Computing and the Service Model in the section above; now we're going to hear about the three types of deployment models: public cloud, private cloud, and hybrid cloud [14] (Figure 18.3).

FIGURE 18.3 Deployment models of cloud computing.

i. *Public cloud*:
   - Internet provider makes available electronic services such as computing, storage, and software to the general public.
   - Any user can login to and use the services.
   - You bill for the amount of money you're using.
   - Users are less in charge of the results.

ii. *Private cloud*:
   - The provider offers hosting services to fewer firewall users.
   - Private cloud eliminates security risks.
   - It gives more control over the results.
   - Usually used by data security-focused organizations.

iii. *Hybrid cloud*:
   - Hybrid cloud computing involves utilizing a mix of private and public cloud resources, as the name suggests. Some services are hosted with private cloud, and others are hosted with the public cloud.
   - Hybrid cloud infrastructure allows companies to retain sensitive data in private space and other data in public space while combining the best in both worlds.

18.1.1.3 Cloud Computing Security Challenges

The major challenges in cloud computing are [11]:

i. *Security and privacy*:

   If we say security and privacy, we speak about client data that is stored in data centers of CSPs. A CSP would comply with guidelines not to exchange personal data or other data that matters to the users. The data centers must be safe, and a CSP should preserve the privacy of the data.

ii. *Availability and reliability*:

   The CSP's data and service will be available at all times regardless of the external situation, which is the optimal condition. The computing resource should be open to users and should be consistent in their operability. The complexities of cloud computing are generally on the CSP side, rather than on the consumer's side.

iii. *Portability*:

   This ensures that if client wants to migrate from one CSP to another, consumer data or services should not be locked in by the provider and the transfer should be quick. Data regulations are numerous in different countries.

iv. *Computing performance*:

   Cloud computing is a computing service on demand and facilitates multi-tenancy, so output will not be impacted by new consumer acquisition. The CSP would retain ample resources to support all users and any adhoc requests.

v. *Service quality*:

   The quality of service will be high and is a big end-user concern. Cloud computing's entire ecosystem is presented in virtual environments, and therefore, the CSP should deliver what's promised in terms of service, whether it's calculating resources or satisfying customers.

vi. *Interoperability*:

CSP's offerings should be versatile enough to fit with other networks and systems that other CSPs offer. The data pipeline should be easy to integrate and should drive performance improvement. In cloud computing, there are many problems, such as big data, long-haul storage, and storage data issues, but it is still the greatest computing platform available to date. Cloud computing is the greatest technology of our generation, with plenty of ups and downs. When more clients come on board to use cloud services, CSPs offer the services to all the clients without any resource limitations.

18.1.2 Cloud Computing Data Security and Threats Risks

The security of cloud computing refers to the security enforced on cloud computing. Cloud protection provides application, software, and processes with support and protection in a simplified way, and protects data from insecure attacks. Cloud protection came into being due to the massive network of cloud computing systems operating online and needing adequate routine maintenance. There is only one server where all user data are stored, and cloud storage protection manages these sensitive storage resources in order to utilize resources, and to prevent user data to get leaked or accessed by other users [15].

18.1.2.1 Security Risks

Although cloud computing can deliver substantial cost-saving benefits for small businesses, pay-as-you-go access to advanced software and powerful hardware service does come with some security risks. You will hold these top five security issues in mind when assessing potential providers of cloud-based services.

i. *Secure data transfer*:

All communication between your network and any service you connect in the cloud will go via the Internet. Make sure that your data is always moving on a safe channel; just link your browser to the provider through a URL beginning with "https." Also, your data should always be encrypted and authenticated using industry standard protocols, like IPSec (Internet Protocol Security), specifically established to protect Internet traffic.

ii. *Secure software interfaces*:

The Cloud Security Alliance (CSA) recommends that you be aware of the user interfaces or Application Programming Interface (API) [11] used for cloud services interaction. Reliance on a limited collection of interfaces and APIs exposes organizations to a range of confidentiality-, credibility-, availability-, and accountability-related security issues, the group says in its top threats to cloud computing. CSA suggests studying how any cloud provider that you are considering implements protection in its operation, from strategies for authentication and access control to policies for tracking activities.

iii. *Secure stored data*:

Your data will be encrypted safely when it is on servers of the provider and when it is being used by the cloud service. Forrester reports in Q&A: Demystifying Cloud Security that few cloud services have protection for data being used inside the application or for disposal of your data. Ask future cloud providers how they protect the data not only when it's in transit but also when it's on their servers and accessible by cloud-based applications. Find out too if, e.g., by removing the encryption key, the providers can safely dispose of your data.

iv. *User access control*:

Data stored on the server of a cloud provider may theoretically be accessed by that company's employee, so you do not have any of the normal administrative controls on these individuals. First, carefully consider the quality of the data that you send out into the cloud. Second, follow the advice from research firm Gartner to ask providers for information about the people who handle your data and the extent of access they have to it.

v. *Data separation*:

Every cloud-based service shares resources, namely, space on servers of the provider and other parts of infrastructure of the provider. Hypervisor software is used to build virtual containers for each of its clients on the provider's hardware. But CSA states that "attacks have arisen in recent years that target shared infrastructure inside cloud storage environments." The provider uses compartmentalization methods, like data encryption, to prevent other customers from accessing the virtual container.

18.1.2.2 Cloud Threats

A threat is an act of violence in which an attempt to elicit a negative response is suggested. It is an intention expressed to inflict damage or loss upon authorized resource [10].

Some common threats to cloud computing are as follows:

i. *Security*:

Businesses that have data offloaded to the cloud are exposed to potential security breaches like hacking. This occurs if the providers of cloud services fail to fix security problems in the service they provide. Those who work for a CSP are exposed to a company's sensitive details.

ii. *Outages*:

Though CSPs claim services without disruption, they are not entirely free of outages. Cloud computing delays in infrastructure can be disruptive for companies. An enterprise's operations could be completely halted if the cloud fails to provide the employees with the service. Businesses could lose possible customers of their online business due to outages. You can expect outages in cloud computing. Companies must ensure that the service provider has the appropriate measures in place to avoid disruptions to the service.

iii. *Malicious insiders*:

CSP workers may have differing access to the company information stored in the cloud. So a malicious insider could gain access to, and could damage, confidential information about a company. CSPs should be open on the recruiting process, providing access, and tracking of users.

iv. *Abuse*:

Cloud-based servers can be used for launching attacks and distributing malicious programs. CSPs have to take action to keep the registration processes exclusively for their customers. It could be critical in monitoring and stopping future attacks.

v. *Servers*:

In the cloud computing environment, data from multiple clients can reside in the same physical server. So a person who gets access to one client can also get access to all other clients.

vi. *Confusion with terminology*:

While there are plenty of people talking about cloud computing, many people are still finding it difficult to grasp what it really means. But not only embracing cloud computing without knowing it, companies will. It is vital they take the time before making any step into cloud computing to understand the concepts.

18.1.2.3 Privacy and Security

i. *Identity management*:

To control access to information and computing resources, each organization must have its own identity management system. Cloud providers either use federation or Single Sign On (SSO) [11] technologies to incorporate the customer's identity management system into their own network, or have their own identity management solution [11].

ii. *Physical and personnel security*:

Providers ensure that physical devices are sufficiently protected and do not only limit access to these devices as well as all related customer data, but also database access.

iii. *Availability*:

Cloud providers promise consumers that their data and software can be accessed routinely and predictably.

iv. *Application security*:

Software providers ensure that cloud-based applications are safe through the introduction of testing and approval procedures for outsourced or bundled computer code. This also includes protection measures for operation in the production environment.

v. *Privacy*:

Finally, companies ensure that all sensitive data (e.g., credit card numbers) are obscured and that only approved users have complete access to the data. In addition, digital identities and credentials must be protected in the cloud as should any data be collected or produced by the provider regarding customer activity.

vi. *Legal issues*:

Furthermore, providers and consumers must address legal issues, such as contracts and e-discovery, and relevant laws, which may differ by country [11].

18.1.2.4 Advantages of Cloud Computing Security

In recent technologies, cloud computing has generated a tremendous boom, which helps the business to operate online, save time, and be cost-effective. Therefore, in addition to having many benefits of cloud computing systems, cloud computing security also plays an important role in this technology, which means that consumers use it without any fear or stress [12].

i. *Protection against DDoS attack*:

Several organizations are facing the distributed denial of service (DDoS) assault that is a big challenge that hampers company data before hitting the customer they want. That's why cloud computing security plays a major role in data safety, because it cleans data in the cloud server before entering cloud applications by erasing the data hack threat.

ii. *Data security and data integrity*:

Cloud servers are real and easy targets to fall into data breach trap. Data will be compromised without due treatment, and hackers will get their hands on it. Cloud security servers provide the best-quality security protocols to help protect sensitive information and maintain the integrity of data.

iii. *Flexible feature*:

Cloud computing security gives the optimum flexibility when it comes to data traffic. Hence, the versatility in a happening server crash is obtained during high traffic usage. The consumer also gets scalability, which results in cost reduction when the high traffic flow stops.

iv. *Availability and 24/7 support system*:

Cloud servers are available and provide the best backup solution allowing users as well as clients to benefit from the constant 24/7 support system. Big organizations need to look for the best security solutions for cloud computing that will give them reliability, security, and availability. The cloud servers need to be combined with the latest information protection features to provide a robust cloud data solution.

18.2 ELECTRONIC HEALTH RECORDS IN CLOUD COMPUTING

The implementation of electronic health records (EHRs) in healthcare facilities has increased dramatically over the past decade. EHRs have many considerable advantages in comparison with conventional health records, such as economy, normative, efficiency, and accessibility. The electronic approach offers a convenient and omnipresent access to health data, which enhances healthcare science. EHRs are typically confidential and private because they provide patient identification and extremely sensitive details. However, users do not have a physical control over their data when using the cloud computing services. CSPs aren't completely trusted,

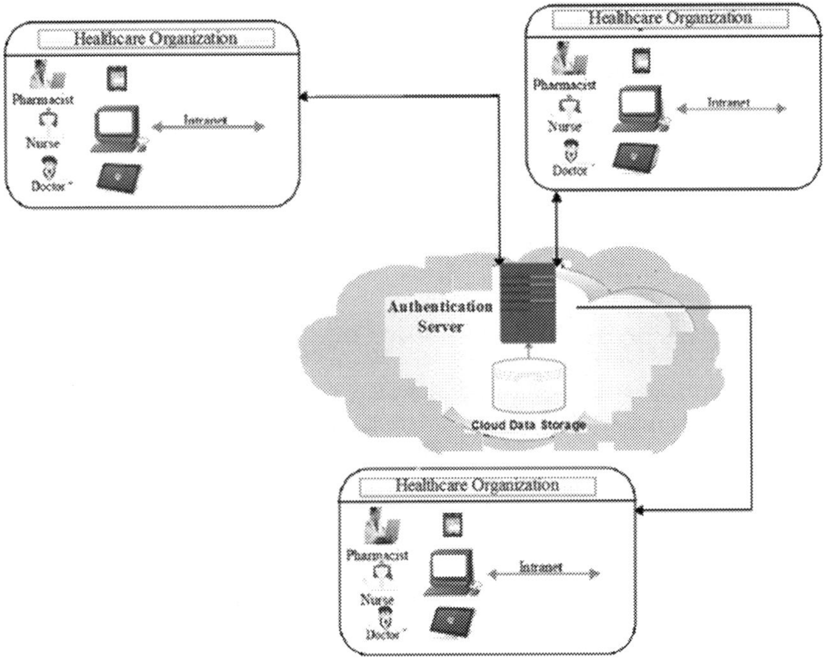

FIGURE 18.4 Cloud computing-based HER system.

although the cloud infrastructure is much more robust than personal computing devices (Figure 18.4).

18.2.1 Benefits and Risks of Cloud Computing in Healthcare

While it has been around for more than a decade, people still refer to cloud computing as the future for enterprise. Perhaps it's more appropriate to say cloud computing is quickly becoming a new norm for companies across the globe. Every day companies and industries are gradually migrating their data to a cloud or hybrid server. The healthcare sector is no exception. When healthcare professionals adapt to the ever-changing technology world, they have begun to incorporate cloud technologies into their work. For healthcare professionals, the advantages of cloud networks are abundant but so are the possible risks. Cloud infrastructure has the ability to radically change the way healthcare is provided and so does the way healthcare providers and cloud providers protect themselves against possible dangers. In that end, we've assembled a list of some of the advantages and threats cloud computing brings to the healthcare field. EHRs and the ability to electronically exchange health information can help you deliver higher-quality and safer patient care while creating tangible improvements for your organization. EHRs help providers manage patient care better, and provide better healthcare through [14]:

1. Providing accurate, up-to-date, and complete information at the point of care about the patients.
2. Facilitating fast access to patient information for more organized, more efficient treatment.
3. Sharing the details privately with patients and other professionals.
4. Helping providers treat patients more accurately, reduce medical errors, and provide more reliable care.
5. Improving contact and coordination between patient and provider, as well as comfort in healthcare.
6. Making prescribing healthier and more effective.
7. Helping to facilitate legible, accurate documentation, and precise, standardized coding and billing.
8. Improving data protection and patient health.
9. Helping providers are to improve productivity and work–life balance.
10. Enabling suppliers to increase performance and achieve their business objectives.
11. Reducing costs by reduced paperwork, enhancing protection, reducing test duplication, and improving health.

18.2.1.1 Benefits

i. *Data storage capacity*:

Although the healthcare needs are 24/7, sometimes the cold and flu season needs more attention from the healthcare provider. The cloud may scale up or down data storage and traffic, depending on the needs of the client. Healthcare providers are thus able to meet their network specifications to fit their service requirements.

ii. *Scalability of service*:

Clients who use the same cloud network can easily share data among themselves. This would be a major benefit in circumstances where healthcare providers need to exchange patient knowledge with each other. The data can be exchanged with those wanting to see it, thereby allowing greater communication to provide healthcare solutions.

iii. *Collaboration*:

Clients who use cloud systems can undoubtedly move information between others. In situations where well-being organizations need to impart clinical data to one another, this would be a significant advantage. Information can be shared to anybody, including those required to see it, permitting the association to rapidly give social insurance arrangements.

iv. *AI and machine learning*:

It takes a lot of time to handle the vast volume of data that healthcare providers deal with time that could be spent with patients. As more cloud providers integrate artificial intelligence (AI) and machine learning (ML) into their services, they can help alleviate some of the burdens. Healthcare providers may use these tools to evaluate and respond to the huge quantities of unstructured data that they are using.

18.2.1.2 Risk

i. *Implementation*:
   Switching from an on-site system to the cloud involves changing the entire way you perform tasks. Healthcare companies preparing to introduce a cloud solution will ensure everyone is up to speed on how to operate efficiently on the cloud.
ii. *Security dangers*:
   Cloud networks have security tools that search, alert, and fix suspicious behavior. These aren't fine, though.
iii. *HIPAA compliance*:
   The Health Insurance Portability and Accountability Act (HIPAA) will extend to all cloud-based health solutions. This covers security measures, which also apply to patient privacy policies, law enforcement, and procedures for violating the notice. To ensure compliance with HIPAA, HIPAA tenants need to be recognized by both the healthcare providers and cloud providers.
iv. *Availability and control*:
   Against all odds, cloud systems still go down. Healthcare companies need their data to be available at any stage, and any interruption on the side of the cloud infrastructure would affect profitability negatively. This is also true for business-owned physical installations, but businesses must rely not on themselves on the cloud provider to bring the service back online.

18.2.2 Challenges Faced by Blockchain Technology

i. *Scalability and storage capacity*:
   The blockchain data is available to everyone on the chain making the data transparent, which is not a desirable outcome for a decentralized network. The data stored on the blockchain would include patient medical history, documents, laboratory results, X-ray reports, MRI results, and many other reports; all of these voluminous data are to be stored on the blockchain, which would significantly affect blockchain's storage capacity [16].
ii. *Lack of social skills*:
   Very few people understand the way the blockchain technology operates. This technology is only in its early stages and continues to develop. In addition, the change from trustworthy EHR systems to blockchain technology would take time as hospitals, or any other health institutes would need to move their systems to blockchain entirely.
iii. *Absence of worldwide characterized norms*:
   There is a lack of clearly established standards. Since this technology is still in its early stages and is continually changing, there is no clear standard for it. Because of this, it would also take more time and effort to implement this technology in the healthcare sector. It would include the approved standards from foreign authorities, which neglect any technology's standardization process [17]. These universal standards will be useful in determining

the data size, data format, and type of data that would be stored on the blockchain. However, adaptation of this technology would be easier due to the established requirements, as they could be easily implemented in the organization.

18.2.2.1 Security and Privacy Requirements of E-Health Data in Cloud

In the new Enormous Information Age, data overload requires the cloud infrastructure to reappropriate healthcare information. Given the immense support the cloud offers, it also includes the possible risks to healthcare data security and privacy. Much of the possible attack involves data divulgence, denial of service (DoS) attack, cloud malware intrusion attack, middle cryptographic attack, spoofing, and collusion attack. The CSP and various administration organizations have introduced an array of safety measures and recommendations to improve patient to organizational assurance. There are three cloud server types, namely, trusted servers, semi-trusted servers, and untrusted servers. A secure server is one that can be fully trusted with no access to data, and this can be attributed to internal adversaries to the data stored in the healthcare. Semi-trusted servers are straightforward but inquisitive servers that obtain healthcare data by plotting with malicious clients, while untrusted servers are not trustworthy without components of privacy mechanisms and are safe from assaults [18].

18.3 ELECTRONIC HEALTH RECORD SECURITY USING BLOCKCHAIN

The greatest problem facing healthcare systems worldwide is how to share medical data with known and unknown stakeholders for different purposes while maintaining data confidentiality and patient privacy protection [19]. Blockchain has long been an important field of research, and the advantages it offers have been used by a variety of different industries. Similarly, because of security, privacy, confidentiality, and decentralization, the healthcare sector is bound to benefit immensely from the blockchain technology. The EHR systems, nonetheless, face data security, integrity, and management issues. Within this paper, we explore how to use the blockchain technology to transform the EHR systems and may be a solution to these problems.

Frameworks and applications for blockchain have already become the preferred addition to various clinical and pharmaceutical operations. Healthcare organizations use this to create end-to -end monitoring of the drug supply chain, verify the outcomes of clinical trials, and avoid fraud in health insurance. Similarly, implementation of blockchain in EHRs will revamp many conventional data management practices and include three new approaches for providing tech-driven treatment. Clinics, hospitals, laboratories, pharmacies, and even insurance companies do so manually. Besides being vulnerable to human error, the shortcomings in this approach are as follows:

- There is no one list where you can find the data of all the locations in the order in which it was originally entered.

- While details are available on the medication ever given to the patient, it may be unknown which medications the patient is taking now.
- While data protection standards are higher than ever before, it's hard to track who registered what, and when.

An EHR can be managed in a blockchain scenario, using distributed ledger, cryptographic identity management, and smart contracts. Smart contracts are self-executing codes programmed to execute a command/action when exposed to a trigger (which may be some input of data). The EHR based on a smart contract will submit feedback on medications, issues, allergy lists, etc. to a community-wide trusted open-source database, so medical record additions and subtractions are well known and auditable across organizations [18].

The EHR thus established could show data from any referenced database in the ledger, rather than merely showing data from a single database. The information provided by EHR on the distributed ledger of a blockchain authorization will be properly reconciled across the network, with the assured validity from the data generation point to the point of use, without manual human interference. Whenever a new record gets added by a healthcare service provider, a patient can check the proposed record before approving or refusing it and can allow sharing of records between providers. This system does not store blockchain medical data. Instead, the record signature is stored, which ensures an unaltered record copy is obtained. This also provides ownership of the knowledge to the patient. The blockchain solution for the healthcare industry significantly reduces the time taken to access information about patients, increases device interoperability, and enhances data quality [19].

18.3.1 Medicalchain

Medicalchain is an electronic health database blockchain. Medicalchain would require various healthcare agents such as doctors, hospitals, laboratories, pharmacists, and insurers to request permission to access medical records and communicate with them [19]. Every interaction is auditable, transparent, and efficient, and will be documented as a transaction on the distributed ledger of medicalchain. This is based on the permission-based architecture of Hyperledger Fabric, which allows varying levels of access; control of patients who can view their records, how much they see, and for what length of time. Currently, the blockchain is evolving [19].

- The medicalchain aims at offering patients and healthcare providers the following main benefits [19]:
- Only the patient's private key will access the data; even if the database is compromised, the patient's data would be unreadable (it's all encrypted).
- Patients have a full control over access to healthcare data; patients should verify who sees their data and what they see.
- Instant transfer of medical data where each member of blockchain's distributed healthcare network will have the same data for the patient: decreased risk of error and enhanced patient care.

18.3.1.1 Medicalchain Features

Medicalchain provides all the following functionalities [19]:

- Improved privacy and access control that enables users to set up access permissions and enables other providers to write data to their blockchain;
- Real-time and secure patient contact doctor where patients can share their information, seek second opinions, and communicate remotely via a safe channel with medical professionals;
- Robust Health Record Licensing by patients where patients can opt to license their EHRs to various parties, say, research pharmaceutical companies;
- Shared Medicalchain App Development Platform where third-party developers build decentralized applications on the Medicalchain platform. Possible health applications may include drug tests, interactivity with wearable data, fast tracking software, and diet and dietary guidance. Such applications can be catered for unique details about the device.

18.3.1.2 Process Flow

The basic process used in a medicalchain transaction and data entry is shown in Figure 18.5.

- Data is created where only the wearable system of the patient might produce data, where a doctor might write a note to the blockchain, or where a pharmacist might dispense medicine.
- The data is encrypted and given a patient's blockchain identity; this data is sent to cloud storage.
- The blockchain ID is used when requesting data to access the encrypted data.
- The data is eventually decrypted, and displayed on the computer or application concerned.

18.4 EXISTING HEALTHCARE DATA PREDICTIVE ANALYTICS USING MACHINE LEARNING TECHNIQUES

Data science plays an important part in healthcare data mining and data science. Although clinical data is imprecise and very information rich, this data can be useful, but also needless failure to obtain information. Rich abstraction of real data progressively replaced health data and result generated rendering the prediction disease. Information mining is used for research and develops clinical data which helps systematically determine the relationship. The first draft was prepared and predicted secure health system using conventional security systems. Take maximum storage and keep the data protection and privacy layer mix. Disguise encryption, security mechanisms, and granular access control; allow data encryption and the injection point for authentication, and various processes are integrated [20].

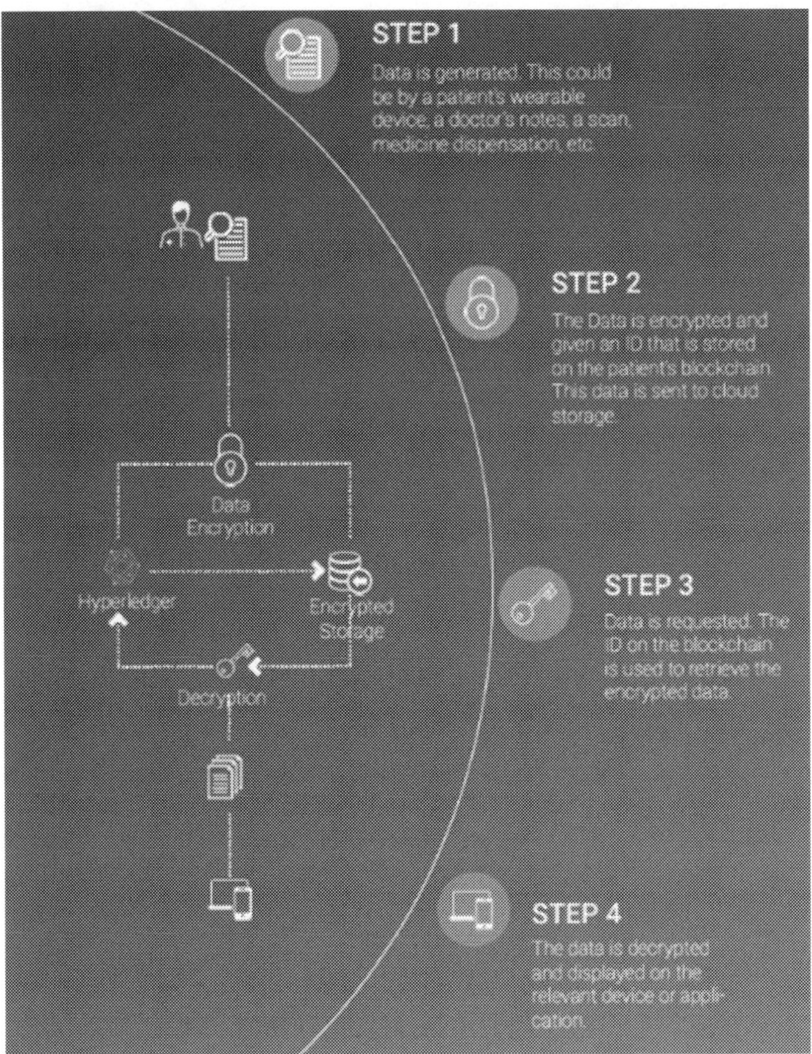

FIGURE 18.5 Process flow of medicalchain [19].

18.4.1 Analytical Relation among EHR, AI, ML, and NLP

Despite the variety of AI applications in clinical trials and healthcare facilities, they fall into two main categories: structured data analysis (including images, genes, and biomarkers) and unstructured data analysis (such as records, medical reports, or patient surveys to supplement structured data). The former method is driven by ML and deep learning algorithms, while the latter is based on the practice of advanced natural language processing (NLP) (Figure 18.6).

ML algorithms chiefly extract features from data, such as patient's "traits" and medical outcomes of interest.AI in healthcare has long been dominated by logistic

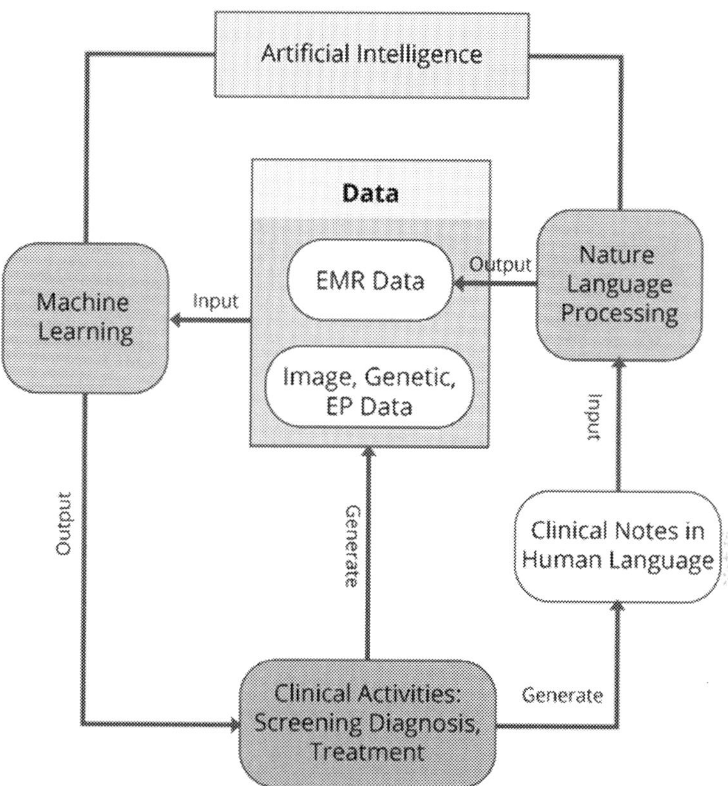

FIGURE 18.6 Analytical relations between EHR, AI, ML, and NLP [19].

regression, the simplest and most common algorithm when classification of things is required. It was user-friendly, simple to finish, and easy to translate. However, the situation has changed in recent years, with Support Vector Machine (SVM) [19] and neural networks taking the lead (Figure 18.7).

Until recently, healthcare AI technologies primarily discussed a few types of diseases: cancer, disease of the nervous system, and the largest cardiovascular disease. Current developments in AI and NLP, and in particular the development of deep learning algorithms, have transformed the healthcare industry into multi-sphere use of AI approaches, from data flow management to drug discovery [19].

18.4.2 Some Common Machine Learning EHR Algorithms

Because of developments in the design of neural network architecture, we can now incorporate the raw input into a vector space using various network architectures based on the characteristics of the given tasks. For example, we use Convolution neural network (CNN)-based models such as AlexNet, Inception, or ResNet to retain spatial information; recurrent neural network (RNN)-based models, etc. for computer vision problems [21].

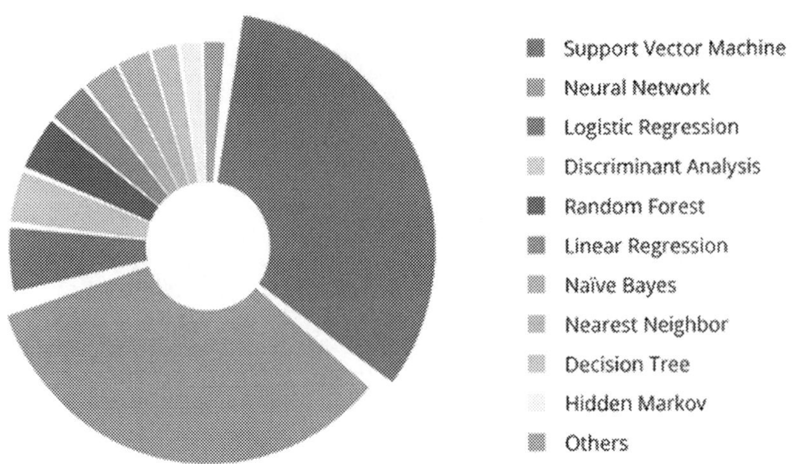

FIGURE 18.7 The most popular ML algorithms used in the medical literature [19].

i. *Sparse auto encoder*:

This is an unsupervised learning of representation mainly used for the engineering stage of the app. It is used for the reduction of nonlinear dimensionality and comes as a better alternative to other conventional techniques of reducing dimensionality, like the principle component analysis (PCA).

ii. *Convolution neural networks (CNNs)*:

CNNs are special algorithms that perform extremely well in problems about image classification. In the EHR sense, CNNs will provide good results in the analysis of medical images such as mammography, MRI images, and CT scans. They can be used from medical images to identify and distinguish malignant cancer cells with the benign cells.

iii. *Recurrent neural networks (RNNs)*:

For certain data types in EHR, like clinical reports, input data do not have the same duration for use with simple ANN. For instance, some medical applications may require processing vast amount of text (such as clinical notes, web-based medical queries platforms) to find keywords that are applicable to standard clinical entities such as the International Statistical Classification of Diseases (usually abbreviated as **ICD**) codes and Current Procedures Terminology (CPT) codes [21].

iv. *Deep transfer learning: solving the small labeled dataset issue*:

One of the main impediments to EHR data in ML is having sufficiently labeled data to practice. For example, if we examine Computed Tomography (CT) scans to identify a malignant tumor, we may not be able to find enough documented events that can be used to train a deep learning model. Transfer learning is a technique of deep learning, which takes intuition from human learning and uses information acquired from one problem to another. In a world of profound thinking, we can use the weights learned by modeling one problem to another.

18.4.3 Challenges for Machine Learning Approaches in EHR

Difficulties may arise in specific facilities [20]:

i. *Data retention*:

Health data must remain available for a minimum duration of five years, which means organizations need to take a long-term approach to data stewardship and keep track of where, by whom, and for what reason the data is accessed. Medical data management software enables users to create access rights and procedures, such as those that give members of various hospital departments temporary data viewing capabilities. These products can index data and notes, and monitor when the system has been entered. Organizations must set up processes to periodically sort the data so that it can be deleted or change and anonymize it so that it can be used in new ways.

ii. *Data management*:

Health services are facing bigdata problems that can impact patient safety. All data collected by health organizations need to be identified, organized, deduced, and accurately reviewed, and made available for different uses: medical, billing, administrative, and the amount and velocity of big data makes this job harder. Some hospitals now hire patient safety specialists, but in addition to providing medical experience, these positions require people to consider how data processing techniques can improve patient safety or impede it.

iii. *Data accessibility*:

All data management techniques are short provided they do not result in information that is available and in the correct reporting format. It is important to enable data analysts to access the data they need and share what the data shows.

18.4.4 Techniques for EHR Tasks

i. *Clinical adverse event detection*:

One of hospitals' primary tasks is to detect a clinical incident in real time. All causes of clinical events can be found embedded in historical data in the EHR, including treatment, diagnosis, and adverse drug effects. Important medical conditions may be interpreted as adverse changes in the medical condition of the patient.

ii. *Clinical adverse event prediction*:

Clinical adverse event prediction subtask via a learning algorithm is to predict the occurrence of diseases, a method that predicts the probability that patients may contract those diseases due to their current clinical condition. The basic aim is to predict future events from longitudinally diverse events (hospitalization, risk of suicide, risk of heart failure, etc.).In order to provide patient-centered support for each form of data element, intelligent support system would require a support system.

iii. *EHR-driven phenotyping*:

Clinical phenotyping is a method of assessing characteristics of the diseases. This process is carried out by opinions of experts and many years of research which has already established phenotypes of each disease. However, with the diversification and polymorphism of emerging diseases coupled with human genetic variants, there is a great need to find other methods for phenotyping the disease as well as the phenotype of individual patients using huge data stored in EHRs. Several researchers have employed approaches that provide a combination of perspectives from health professionals and automated processes.

iv. *Patient's features representation*:

A precise patient representation and stratification is very important for conducting an adverse clinical event prediction or any other EHR mission. Due to the high heterogeneity, it is highly erroneous to feed the EHR raw data directly into a deep network to perform clinical tasks such as forecasts, clinical trial recruitment, or disease detection. Therefore, a feature learning system that can reflect the features of the patient with less knowledge overlap has to be built from the vastly heterogeneous EHR data before performing these clinical support tasks with deep networks.

v. *Medication information extraction*:

The method of obtaining this knowledge from the EHR, however, can be a lengthy tiresome method as the data is buried deep in EHR's clinical accounts, and encounters with patients, ICU discharges, and charted incidents. The function of a computerized identification of adverse drug events includes three main tasks: Named Entity Identification (NER), a process of detecting key drug mentions, recognizing these named events as a process of identifying the meaning of these mentions, and finding the connection between them. The drug information extraction system aims to determine the names of the drugs and their signatures such as dosages, length, reasons for prescription, complications, doses, route of administration, and set unit.

vi. *Integrating EHR social network and web data*

Through social associates, it is more likely that a patient shares health experiences like adverse drug outcomes than with his doctor. Through the proliferation of social networks, there are massive, untapped medical insights that can be used by clinical support systems along with the hospital's EHR. Although the medical research community agrees that social networks should form part of the EHR, the ways in which this can be done remain a highly debated topic. The problems of this resistance, due to spelling errors, imprecise definitions, and vague or casual use of medical terminology, are high noise. Some clinical tasks can also rely on more structured EHR data than on social data.

The depiction of ML works done just as their methodologies is point-by-point in Table 18.1 [21].

TABLE 18.1
Studies that Covered ML Approaches to EHR [35]

Study	Year	Approach
Deep EHR: a survey of recent advances in deep learning techniques for HER analysis	2018	This study covered the aspects of EHR data, recent deep learning in healthcare projects. This paper describes some of deep learning algorithms used in health informatics and some of the clinical tasks that leverage deep learning methods. This paper does not provide a technical detailed implementation
Deep learning for health informatics [14]	2017	This study covered different deep learning architectures used in health informatics. This study also covered various software packages that are used to provide deep neural network implementations. Mainly, this study covered various clinical tasks that deep learning is used into like medical imaging pervasive sensing.
Opportunities and challenges in developing deep learning models using EHR data: a systematic review [22]	2018	A literature review covered clinical analytics tasks that leverage EHR data. It also covers various deep algorithms that are famously used in health informatics
Deep learning in pharmacogenomics: from gene regulation to patient stratification [23]	2018	This study also covered opportunities and challenges of deep learning in healthcare. This study focused mainly on issues of pharmacogenomics and drug targets
Current study: deep learning for EHR Analytics	2019	We describe the challenges of deep learning to EHR data. The focus of the paper is to provide technical intuition on the utilization of deep learning algorithms to each task of EHR clinical knowledge discovery. We try to cover each task that clinicians perform on daily basis. This will help EHR-based application developer to consider a task-oriented approach than a data-oriented approach in analytics development

18.5 LITERATURE REVIEW

Ashutosh Dhar et al. [24] proposed a novel system of changed blockchain models reasonable for Internet of Things (IoT) gadgets for clinical area. The extra protection and security properties in their model depended on the cutting-edge cryptographic natives. The arrangements given make IoT applications for clinical information and exchanges increasingly secure and unknown over a blockchain-based system. Coming up next are key difficulties that expository arrangements must deliver to give clinically significant experiences.

Yu Rang Park et al. [25] introduced a private blockchain arrangement utilizing Ethereum and led check utilizing the de-distinguished personal health records (PHRs). So as to check the viability of blockchain-based PHR management, PHRs at a time were stacked in an exchange between the clinic and patient hubs and proliferated to the entire system. They acquired and investigated the time required for information exchange and proliferation on the blockchain to organize and affirm that it is conceivable to make sure about trade of PHR information in a private blockchain arrangement.

Dinh C et al. [26] introduced a model usage utilizing Ethereum blockchain in genuine clinical information sharing on a versatile application with Amazon distributed computing. The outcome shows that their proposition gives a viable answer for solid clinical information trades on portable mists while saving touchy clinical data against potential dangers.

Xiaochen Zheng et al. [27] proposed a calculated structure for sharing individual unique well-being information utilizing blockchain innovation enhanced by distributed storage to share the well-being-related data in a protected and straightforward way. Additionally they presented an information quality review module that is dependent on AI systems to have a command over information quality.

Jingwei Liu et al. [26] proposed a blockchain-based security protecting information sharing for electronic medical records (EMRs), called blockchain-based privacy-preserving data sharing (BPDS). In BPDS, the first EMRs are put away safely in the cloud and the files are saved in a carefully designed consortium blockchain. Security investigation shows that BPDS is a protected and powerful approach to acknowledge information sharing for EMRs.

Harleen Kaur et al. [14] proposed a blockchain-based stage, which is proposed by creators that can be utilized to store electronic clinical records in cloud situations and the executives. In this investigation, they have proposed a model for the well-being informative blockchain-based structure for distributed computing conditions. Their commitments incorporate the proposed arrangement and the introduction of things to come course of clinical information at blockchain. This paper gives a review of the treatment of heterogeneous well-being information, and a portrayal of inside capacities and conventions.

Anastasia Theodouli et al. [22] introduced the security possibilities of blockchain innovation to encourage private medicinal services information imparting to social insurance information get to consent taking care of. They have indicated how the proposed framework has included qualities in patient data integrity, workflow robotization, and responsibility. Security of the proposed framework is likewise improved by empowering the mentioning information substances utilizing blockchain to check with the uprightness of the information that is gotten too.

Alex Roehrs et al. [23] introduced the usage and assessment of a PHR model that coordinates the appropriated well-being records utilizing blockchain innovation and the open EHR interoperability standard. Likewise, they follow the OmniPHR delineate engineering model, which gives the foundation and supports the usage of cross-worked PHR. Execution results showed that information dispersed by means of a blockchain could be recoupled with low normal reaction time and high accessibility.

Md. Abdur Rahman et al. [29] introduced a protected treatment structure that will permit a patient to claim and control his/her own information with no confided in outsider. With the help of blockchain, the system will be insusceptible to a solitary purpose of disappointment or unapproved get to. For security, the structure utilizes blockchain-based appropriated exchanges to protect the clinical information security, proprietorship, age, stockpiling, and sharing. Their test outcomes show that it can bolster an adequately huge number of clients without extensive increment in mean preparing time.

A. H. Mohsin et al. [30] proposed a framework to build up a common key that could be reproduced by the authentic gatherings before the procedure of finding and treatment starts. The information in the conclusion and treatment process is encoded and put away in a blockchain utilizing the common key. It utilizes the sibling intractable function families (SIFF) to set up a mutual key, and uses the Hyper record Fabric to store encoded information. Their proposition meets the trustworthiness, accessibility, and protection necessities of clinical information.

Rui Guo et al. [31] proposed an attribute-based signature scheme with multiple authorities for preserving patient privacy in an EHR system on blockchain. Moreover, there are different specialists without a confided in single or focal one to create and disperse open private keys of the patient, which maintains a strategic distance from the escrow issue and complies with the method of circulated information stockpiling in the blockchain. The examination exhibits the presentation and the expense of these convention increments straightly with the quantity of specialists and patient traits too.

Qi Xia et al. [32] proposed a blockchain-based information-sharing structure that adequately addresses the entrance control difficulties related to touchy information put away in the cloud utilizing permanence and inherent self-rule properties of the blockchain. A permissioned blockchain permits access to just check clients. Framework assessment shows that their plan is lightweight, adaptable, and effective for clinical information security.

Qi Xia, Emmanuel et al. [33] proposed MeDShare, a framework that tends to the issue of clinical information sharing among clinical enormous information in a trustless condition. The proposed framework is blockchain-based and gives information provenance, inspecting and control for shared clinical information in cloud archives among large information elements. The exhibition of MeDShare is practically identical to the current forefront answers for information sharing among cloud specialist co-ops.

Guy Zyskind et al. [18] proposed the decentralized individual information the executive's framework that guarantees client's own and control their clinical information. They executed a convention that utilizes blockchain into a robotized get to control administrator that doesn't require trust in an outsider. In their framework, they have used to convey directions, e.g., putting away, questioning, and sharing information. Furthermore, the blockchain perceives the clients as the proprietors of their own information.

Meng Shen et al. [34] proposed a clinically encoded picture recovery plot dependent on blockchain for security assurance. The emergency clinic separates the clinical picture highlights of various sorts and encodes. At the point when the outsider proposes a recovery demand, the administration starts the keen agreement to transmit the solicitation for the client, and feeds back the recovery result as indicated by the picture closeness.

Hirokazu Madokoroa et al. [35] presented a method using accelerated-KAZE, generation of codebook self-organizing maps (SOMs), and semantic recognition of multiple objects using category maps created using counter propagation networks (CPNs). The experimentally obtained results revealed that the mean recognition of medical image using region of interest (RoI) accuracy was 71.96% after optimizing key parameters.

Xiao Yue et al. [36] proposed an engineering which empowers control of information by using blockchain stage as capacity framework-driven access as one access-control model, and brought together and basic indicator-driven mapping as capacity model. They likewise bring up that secure MPC (multi-party computing) is one promising answer for empower un-confided in outsider to lead calculation over patient information without abusing protection.

Vishal Patel et al. [16] proposed a picture sharing utilizing blockchain with confirmation of-stake conspire for making sure about clinical picture calculation build up a system for cross-space picture sharing that utilizes a blockchain as a dispersed information store to set up a record of radiological examinations and patient-characterized get to authorizations. Additionally in the examination, numerous points of interest of utilizing blockchain in clinical area are talked about.

Kai Fan et al. [17] proposed information sharing and cooperation by means of blockchain strategy. Likewise, they have proposed a productive protection saving and sharing plan dependent on blockchain, which can ensure client's security contained in their clinical information by using the blend of access control convention and encryption innovation.

Hongyu Li et al. [37] proposed a novel blockchain-based data preservation system (DPS) for clinical information. With the proposed DPS, clients can save a significant information in interoperability, and the inventiveness of the information can be checked if altering is suspected. What's more, they use information stockpiling procedures and an assortment of cryptographic calculations to ensure client security. The presentation assessment results demonstrated that the current framework is likewise financially savvy with solid clinical information security insurance.

Gayathri Nagasubramanian et al. [38] introduced the keyless mark framework utilized in the proposed framework for guaranteeing the mystery of computerized marks likewise guaranteeing the parts of verification. Besides, information trustworthiness is overseen by the proposed blockchain innovation. The outcomes show that the reaction time of the proposed framework with the blockchain innovation is practically half shorter and the cost of capacity is about 20% not exactly the ordinary procedures.

HaoWang et al. [39] proposed a safe electronic well-being record (EHR) framework dependent on the quality-based cryptosystem and blockchain innovation. In their framework, they use characteristic-based encryption (ABE) and personality-based encryption (IBE) to encode clinical information, and use character-based mark (IBS) to actualize advanced marks. To achieve different functions of attribute-based encryption (ABE), identity-based encryption (IBE), and identity-based signature (IBS) in one cryptosystem, they introduce a new cryptographic primitive, called combined attribute-based/identity-based encryption and signature (C-AB/IB-ES).

Md. Mehedi Hassan Onik et al. [40] presented permission blockchain architecture designed to manage the EHRs. The architecture was proposed for both the Italian Regulation on Electronic Health Record and the recently introduced GDPR (General Data Protection Regulation). They have designed a blockchain-based architecture for the decentralized management of EHRs. Their proof-of-concept network represents the core architecture of a federated EHR management system.

Ching et al. [20] thoroughly discussed opportunities and challenges in using deep learning for biology and medicine; although this study was much exhaustive, it did not elaborate more on the technical side of the processes involved.

18.6 OBJECTIVES OF THE STUDY

The objectives of the study are:

- Achieving a secure EHR storage and retrieval process.
- Integration of ML with RoI logic to derive valuable knowledge from digital medical records.
- Suggesting a highly secure EHR storage network for third parties to provide safe storage and sharing of medical data.
- Testing our proposed algorithm by comparing the existing image encryption algorithms to standard parameters.

18.7 IMPLEMENTATION MODEL

Various standard journal research papers based on previous studies, their drawbacks, and weaknesses are reviewed. A new, efficient, and successful structure for EHR's data exchange method is proposed. This work focuses on an innovative architecture for security of medical data and privacy with cloud system performance.

18.7.1 Proposed Working Stages of Proposed Methodology

1. Collect medical services from EMR.
2. Classify the data by their forms, such as X-ray, MRI scans, and other medical data.
3. Apply ML algorithms on EHR to find the specific field of useful image information for different uses of EHR information.
4. Apply blockchain security mechanism to extracted EHR data before being stored in cloud storage (third-party storage).
5. Build cloud storage implementation model according to EHR data needs. It separates and stores different types of EHRs by their type of record.

18.8 CONCLUSION

Third-party storage system like cloud storage solves capacity problem for storing huge amounts of data like medical records. But some big issues are still in sight, such as data protection and data security. This research offers a viewpoint on the state of protection for blockchain-based health records. This methodology provides a groundbreaking new approach to security and privacy. The recent technical developments have resulted in a deluge of medical data from diverse fields. However, the reported data is poorly annotated, noisy, and unstructured from divergent sources.

The data is thus not completely leveraged to evaluate actionable information, which can be used in clinical applications. These data recorded in the EHRs of hospital consist of patient information, clinical notes, charted events, drugs, procedures, laboratory test results, diagnostic codes, and so forth. Traditional ML and statistical approaches have failed to provide information that can be used by clinicians to treat patients because they need to obtain features supported by an expert opinion before creating a model of benchmark tasks. With the emergence of methods of deep learning, one needs to consider how deep learning can save lives. This work provides the basic information, previous studies, and recent issues about the algorithms related to medical data security, blockchain, and ML.

REFERENCES

1. L. Brunese, F. Mercaldo, A. Reginelli, A. Santone, "A blockchain based proposal for protecting healthcare systems through formal methods", *Procedia Computer Science*, 159: 1787–1794 Elsevier, 2019.
2. L. Ismail, H. Materwala, "A review of blockchain architecture and consensus protocols: Use cases, challenges, and solutions", *MDPI/symmetry*, 11: 1198, 2019.
3. J. Mishra. "Modified Chua chaotic attractor with differential operators with non-singular kernels," *Chaos, Solitons & Fractals*, 125: 64–72, 2019.
4. J. Mishra. "Fractional hyper-chaotic model with no equilibrium", *Chaos, Solitons & Fractals*, 116: 43–53, 2018.
5. M. Holbl, M. Kompara, A. Kamišali and L.N. Zlatolas, "A systematic review of the use of blockchain in healthcare", *MDPI / Symmetry*, 10: 470, 2018.
6. J. Mishra. "Analysis of the fitzhugh nagumo model with a new numerical scheme", *Discrete & Continuous Dynamical Systems-S*, 13: 781, 2018.
7. P. Jangbari, D. Patel, "Review on region of interest coding techniques for medical image compression", *International Journal of Computer Applications*: 134, 2016.
8. D. Yee, S. Soltaninejad, D. Hazarika, "Medical image compression based on region of interest using better portable graphics (BPG)", *IEEE International Conference on Systems, Man, and Cybernetics (SMC)*, 2017.
9. P.M. Pradhana, C.H. Chengb, J.R. Mitchell, "A region of interest based approach for texture analysis of medical images in space–frequency domain", *Biomedical Signal Processing and Control*, 51: 222–234 Elsevier, 2019.
10. A. Mishra et al., "An authentication mechanism based on client-server architecture for accessing cloud computing", *International Journal of Emerging Technology and Advanced Engineering*, 2 (7): 95–99, July 2012, ISSN 2250-2459.
11. A. Mishra et al., "Secure cloud storage architecture for digital medical record in cloud environment using blockchain", *Elsevier SSRN International Conference on Intelligent Communication and Computation Research*, Available at http://dx.doi.org/10.2139/ssrn.3565922, April 1, 2020.
12. A. Mishra et al., "An enhanced DDoS TCP flood attack defence system in a cloud computing", *Elsevier SSRN International Conference on Intelligent Communication and Computation Research*, Available at http://dx.doi.org/10.2139/ssrn.3565916, April 1, 2020.
13. A. Mishra et al., "A review on DDOS attack, TCP flood attack in cloud environment", *Elsevier SSRN International Conference on Intelligent Communication and Computation Research*, Available at https://ssrn.com/abstract=3565043, March 31, 2020.
14. H. Kaur, M.A. Alam, R. Jameel, "A proposed solution and future direction for blockchain-based heterogeneous Medicare data in cloud environment" *Journal of Medical Systems*, Springer, 42: 156 Springer, 2018.

15. K. Patel, A. Alabisi, "Cloud computing security risks: Identification and assessment", *Journal of New Business Ideas & Trends*, 17: 11–19, 2 September 2019.
16. V. Patel, "A framework for secure and decentralized sharing of medical imaging data via Blockchain consensus", *Health Informatics Journals*, 25(4): 1398–1411, December 2018.
17. K. Fan, S. Wang, "MedBlock: Efficient and secure medical data sharing via blockchain", *Journal of Medical System*, 42: 136 Springer, 2018.
18. G. Zyskind, O. Nathan, A.S. Pentland, "Decentralizing privacy: Using blockchain to protect personal data", *IEEE CS Security and Privacy Workshops* 2015.
19. Crypt Bytes Tech. Medical Chain - A blockchain for electronic health records, 2017. https://medium.com/crypt-bytes-tech/medicalchain-a-blockchain-for-electronic-health-records-eef181ed14c2.
20. T. Ching et al., "Opportunities and obstacles for deep learning in biology and medicine", *Journal of the Royal Society Interface*, 15(141): 2170387, 2018.
21. G. Harerimana, J.W. Kim, H. Yoo, B. Jang, "Deep learning for electronic health records analytics", *IEEE Access*, 7, 2019, ISSN: 2169-3536.
22. A. Theodouli, S. Arakliotis, K. Moschou, "On the design of a blockchain-based system to facilitate healthcare data sharing", 2018.
23. A. Roehrs, C.A. da Costa, R. da Rosa Righi, "Analyzing the performance of a blockchain-based personal health record implementation", *Journal of Latex Class Files*, 0(0), October 2018.
24. A.D. Dwivedi, G. Srivastava, S. Dhar, "A decentralized privacy-preserving healthcare blockchain for IoT", *Sensors* 2019, www.mdpi.com/journal/sensors.
25. Y.R. Park, E. Lee, W. Na, "Is blockchain technology suitable for managing personal health records? Mixed-methods study to test feasibility", *Journal of Medical Internet Research*, 21(2):e12533, 2019.
26. D.C. Nguyen, P.N. Pathirana, "Blockchain for secure EHRs sharing of mobile cloud based E-health systems", *Special Section on Healthcare Information Technology for the Extreme and Remote Environments*, 2019, IEEE.
27. X. Zheng, R.R. Mukkamala, "Blockchain-based personal health data sharing system using cloud storage", *2018 IEEE 20th International Conference on e-Health Networking, Applications and Services (Healthcom)*-©2018, IEEE.
28. J. Liu, X. Li, L. Ye, H. Zhang, "BPDS: A blockchain based privacy-reserving data sharing for electronic medical records", arXiv:1811.03223v1, 8 November 2018.
29. Md. A. Rahman, M. Shamim Hossain, E. Hassanain, "Blockchain-based mobile edge computing framework for secure therapy applications", *IEEE Access*, 6: 2169–3536, 2018.
30. H. Tian, J. He, Y. Ding, "Medical data management on blockchain with privacy", *Journal of Medical Systems*, 43: 26 Springer, 2019.
31. R. Guo, H. Shi, Q. Zhao, and D. Zheng, "Secure attribute-based signature scheme with multiple authorities for blockchain in electronic health records systems", *IEEE Access, 6*: 2169–3536, 2018.
32. Q. Xia, E.B. Sifah, A. Smahi, S. Amofa and X. Zhang, "BBDS: Blockchain-based data sharing for electronic medical records in cloud environments", *MDPI/Information*, 8: 44, 2017.
33. Qi Xia, E.B. Sifah, K.O. Asamoah, J. Gao, "MeDShare: Trust-less medical data sharing among cloud service providers via blockchain", *IEEE Access*, 5: 2169–3536, 2017.
34. M. Shen, Y. Deng, L. Zhu, X. Du, N. Guizani, "Privacy-preserving image retrieval for medical IoT systems: A blockchain-based, approach", *IEEE Network*, 33: 27–33, 2019.
35. H. Madokoroa, A. Kainumaa, K. Satoa, "Non-rectangular RoI extraction and machine learning based multiple object recognition used for time-series areal images obtained using MAV", *Procedia Computer Science*, 126: 462–471, 2018.

36. X. Yue, H. Wang, "Healthcare data gateways: Found healthcare intelligence on blockchain with novel privacy risk control", *Journal of Medical Systems*, 40: 218 Springer, 2016.
37. H. Li, L. Zhu, M. Shen, F. Gao, X. Tao, S. Liu, "Blockchain-based data preservation system for medical data", *Journal of Medical System*, 42: 141 Springer, 2018.
38. G. Nagasubramanian, R.K. Sakthivel, R. Patan, A.H. Gandomi, M. Sankayya, B. Balusamy, "Securing e-health records using keyless signature infrastructure blockchain technology in the cloud", *Neural Computing and Applications*, 3/2020, Springer, 2018.
39. Md. M.H. Onik, S. Aich, J. Yang, C.-S. Kim, H.-C. Kim, "Privacy protection and management of medical records using blockchain technology", *Big Data Analytics for Intelligent Healthcare Management*, B978-0-12-818146-1.00008-8, Elsevier, 2019.
40. M. Ciampi, A. Esposito, F. Marangio, G. Schmid, M. Sicuranza, "A blockchain architecture for the Italian EHR system", IARIA, 2019.

19 Mathematical Model to Avoid Delay Wound Healing by Infinite Element Method

Manisha Jain
Vellore Institute of Technology, Bhopal (M.P.)

Pankaj Kumar Mishra
Amity University Madhya Pradesh

Ramakant Bhardwaj
Amity University Kolkata (W.B)

Jyoti Mishra
IPS College of Technology & Management

CONTENTS

The human body is one of the most complicated and complex biological structures on earth. Each organ consists of various suborgans and soft tissues, which are made by biological porous flexible materials. Not only tissues, but every organ of the human body also has its mechanism to work properly. The working of the organs depends on various nutrients, fluids, enzymes, and many more [1]. The latest computational algorithms and technology will make it possible to study the correlation between the cells on a computer [2]. Mathematical modeling of various pre- and post-disease problems can help the biologists to analyze the success rate of the biological phenomena. For the mathematical modeling and analysis of the clinical problems, precise and effective computational tools are required, which can be used in the study of surgical assessment [3]. Tissue engineering is very useful research in the new era. In this regard, mechanical interactions between the tissues and the fluid have been analyzed. Tissue engineering processes are integrally complicated in nature and consist of many physiochemical procedures, which proceed at different events oscillating from small part like molecule to cell body [4].

Healthy skin is the most important and essential need of the human being to maintain the body's thermal regulatory system, mechanical forces, infections, fluid balancing, and all other important functions. It is also responsible for the flexibility of joint function and maintaining the position of the palm or foot sole [5].

A wound is an injury accidental or surgical (Figures 19.1and 19.2), which can be represented as a fragmentation in the skin layers, i.e., the epidermis, dermis, subcutaneous tissues, fascia, muscles, or even bone. Surgical wound is an injury in which skin is wavering, cut, or penetrated [6].

Any wound heals naturally by the stimulation of dermal and epidermal tissues. Delayed wound healing requires the medical involvement. Diabetes mellitus or peripheral vascular disease may be the main reason for the delayed wound healing. Proper functioning of the skin organs will be affected due to extensive burn injuries. As a result, various health risk factors may take place. The phase of wound healing starts automatically and normally in a probable and well-timed manner. Any disturbance in the phases may cause either a chronic wound or a keloid scar [7] (Figure 19.3 and 19.4).

The functioning of enzymes plays an important role in the analysis of any disease. Enzymes are very important factors in the field of clinical chemistry. The role of enzymes

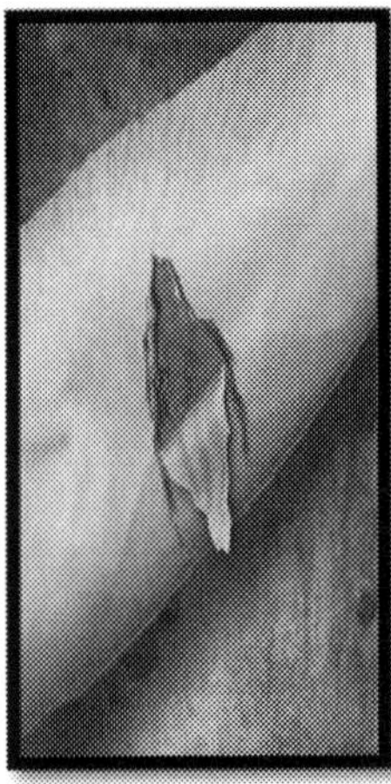

FIGURE 19.1 Accidental wound.

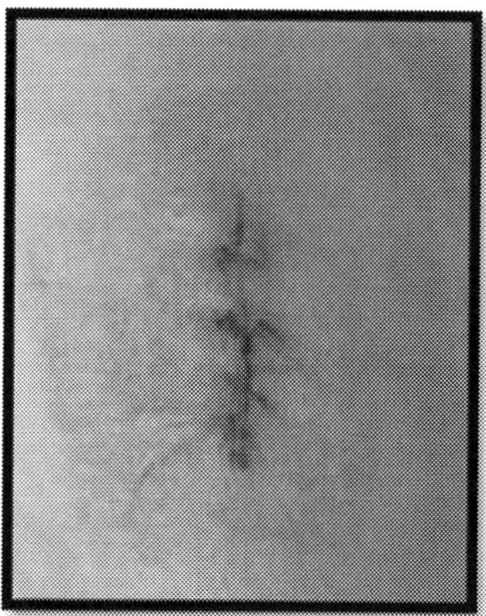

FIGURE 19.2 Surgical wound healing.

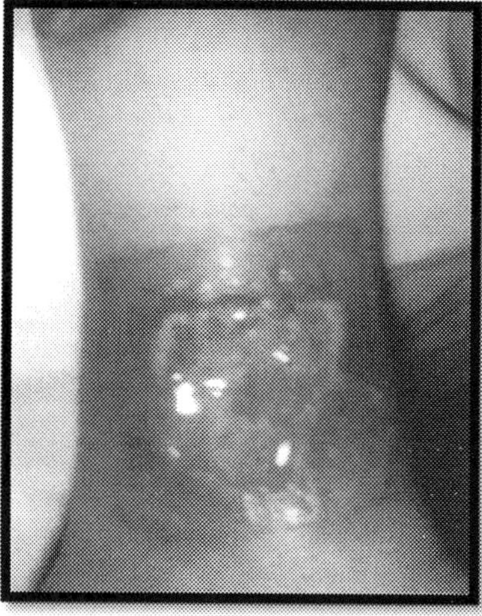

FIGURE 19.3 Chronic ulcer.

cannot be neglected in the case of the wound-healing process. Proper functioning of the enzymes depends upon two major factors, namely, pH value and temperature [8].

Experimental studies suggested that effective wound healing occurs at the normal body temperature (37°C). The temperature falls below the normal body core temperature or rise above 42°C will result in the delayed wound healing [9]. Infection, scarring, and poor wound healing may result due to the delayed wound healing.

In the wound-healing process, wound bed preparation is an important phase because it creates an optimal wound-healing environment. Timely healing of wound depends on the preparation of wound bed; the wound bed depends on body

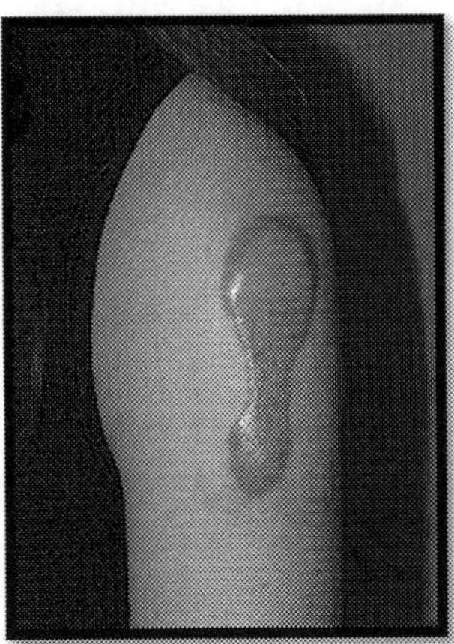

FIGURE 19.4 Keloid scar.

temperature and atmospheric temperature: For the temperature of outside temperature, the wound bed temperature can be controlled by proper cleansing and dressing of wounds from time to time.

The temperature of the wound bed can be controlled by proper dressing materials and the appropriate temperature of the cleansing solution [27]. Timely healing depends on ambient (atmospheric) temperature, the temperature of the dressing material, and the temperature of wound immediately after dressing removal.

Evaporation process plays an important role during the entire process. Frequently wound dressing stimulates fast wound healing. Maintaining the proper rate of evaporation helps to maintain moisture of the wound. Excess or less amounts of evaporation rate are not suitable for the acute wound. Excess evaporation causes wound dryness, and less evaporation causes wound wetness all the time [10–12].

All the above-mentioned facts suggest that temperature regulation has its own importance in the wound-healing process. Enzymes functioning, growth factors, and ultimately wound bed temperature depend on the body core temperature. Varying atmospheric temperature, rates of evaporation, and cleansing solution temperature affect the timely healing of wound [13–23].

19.1 MATHEMATICAL MODELING TO STUDY ISSUE TEMPERATURE DEVIATION DURING WOUND HEALING

19.1.1 Statement of the Problem

Heat and mass transfer equation in terms of time-dependent partial differential equation given by Perl [24] is very useful to analyze the tissue temperature of human body by applying the suitable boundary and initial conditions:

$$\text{Div}(K \text{ grad } T) + m_b c_b (T_b - T) + S = \rho c \frac{\partial T}{\partial t}. \tag{19.1}$$

19.2 BOUNDARY CONDITIONS AT THE OUTER SURFACE AND INNER CORE

Conduction, convection, radiation, and evaporation are the main reasons for heat loss. Maximum heat loss occurs from the surface of the skin, i.e., from the epidermis layer. Therefore, the net flux calculated at the normal to the skin surface [25] is

$$-K\frac{\partial T}{\partial n} = h(T - T_a) + LE \quad \text{for} \quad t > 0. \tag{19.2}$$

The human body maintains its body core temperature at 37°C; therefore, the boundary condition at the inner core is

$$T(X,t) = T_b \text{ for } t \geq 0;\ X = (x, y, z). \tag{19.3}$$

19.2.1 Initial Condition

The wound is covered before cleansing process; i.e., heat loss is zero at time $t = 0$ and hence, the initial condition is given by

$$T(X,0) = T_b;\ X = (x, y, z) \tag{19.4}$$

The human skin has subcutaneous, dermis, and epidermis layers. The physiological parameters such as thermal conductivity, metabolic rate, and blood mass flow have different values in different layers of the skin. Subcutaneous layer has constant values; epidermal, and the middle layer dermis layers have variable values, which can be calculated by Lagrange's interpolation polynomial. The epidermis layer has no blood vessels so its value is assumed as zero (Table 19.1).

Physiological properties in normal and wounded tissues have different values. Just after the surgery or wound, these values are negligible. These values are time dependent and will increase with time.

Mathematically physiological parameters are taken as follows:

$$K(X,t) = \varsigma(t)\sum_{d=0}^{1}\alpha_d X^d,\ M(X,t) = \psi(t)\sum_{d=0}^{1}\beta_d X^d,\ S(X,t) = \zeta(t)\sum_{d=0}^{1}\gamma_d^e X^d. \tag{19.5}$$

TABLE 19.1
List of Symbols in Thermal Systems [25]

H	Heat transfer coefficient	L	Latent heat
Ta	Atmospheric temperature	E	Rate of evaporation
$\partial T/\partial n$	Rate of change in temperature *T* along the normal to the skin surface	Tb	Body core temperature
S	Effect of metabolic heat generation	*mbcb (Tb-T)*	Blood mass flow
K	Thermal conductivity	ρ,	Density
C	Specific heat of tissue	*mb*	Blood mass flow rate
Cb	Specific heat of blood	*X (x,y,z)*	Coordinate

The values for α_d, β_d, and γ_d are calculated for layer-wise.

The value of the physiological function depends on time and space. For normal tissues, K, M, and S are considered as constant (Eq. 19.6), whereas for wounded region, they depend on increasing functions of time (Eq. 19.7),

$$\varsigma(t)=1, \psi(t)=1 \text{ and } \varsigma(t)=1 \tag{19.6}$$

$$\varsigma(t)=\left(v_0+v_1e^{-vt}\right), \quad \psi(t)=\left(\mu_0+\mu_1e^{-\mu t}\right), \quad \zeta(t)=\left(\theta_0+\theta_1e^{-\theta t}\right), \tag{19.7}$$

where ν, μ, and θ are the constants that are used to determine the change in the growth rates of K, M, and S. ν_o, ν_1, μ_o, μ_1, θ_0, and θ_1 are calculated by applying suitable conditions.

In short, to study the behavior of such complex organisms, a powerful mathematical tool is required. Mathematical modeling of biological system can be done using the finite element method (FEM) very effectively and accurately as a method because it allows us to divide a large domain into small elements. FEM deals with the laws of physics for space- and time-dependent problems. Partial differential equations deal with the time- and space-dependent problems and various shapes of the geometries. Therefore, FEM is a suitable tool to handle mathematical modeling as it deals with the PDE. Discretization processes estimated the PDE's (Partial Differential Equation) terms of numerical equations. The solutions of such PDE are real in nature. FEM deals with approximations. FEM is very useful for structural analysis, thermal system analysis, flow analysis, thermomechanical process analysis, and many more. In the process of discretization, it is very important to remove singularities. In the case of singularity, it is difficult to get the numerical solution of these problems accurately. The FEM allows local refinement to handle singularities of the system. Local refinements involve numbers of elements and large computer memory. Singularities can be removed by introducing hybrid elements and global–local elements. The infinite element method (IEM) is the best option to use forthe complex non linear geometry as it deals with various geometric singularities. Infinite elements are an extension of finite elements. A mesh or web can be refined easily with the IEM, but on the other hand, the stiffness matrix should be calculated efficiently to approximate the solution at the singular point [26–34].

19.3 USE OF THE FINITE ELEMENT METHOD AND THE INFINITE ELEMENT METHOD

By using Ritz variational method, Perl's bioheat Eq. (19.1) is discretized and then further the FEM can be applied with suitable boundary conditions. The variational form of the equation is obtained as follows:

$$I^e=\frac{1}{2}\iint_{\Delta^e}\left[K^e\left\{\left(\frac{\partial T^e}{\partial x}\right)^2+\left(\frac{\partial T^e}{\partial y}\right)^2\right\}+M^e(T_b-T^e)^2-2S^eT^e+\rho c\frac{\partial (T^e)^2}{\partial t}\right]dxdy$$

$$+\frac{1}{2}\int_{b1}\left[h(T^e-T_a)^2+2LET^e\right]dx+\int_{b2}\left[\bar{\varepsilon}^eT^e\right]dy+\int_{b3}\left[\bar{\theta}^eT^e\right]dy, \tag{19.8}$$

where Δ^e is the region contained in eth element, and $b1$, $b2$, and $b3$ are the boundaries of the region. For triangular and rectangular elements, the field variable T depends on T_i, T_j, T_k, and T_i, T_j, T_k, T_l respectively.

The equation (19.8) can be written as

$$I^e = I_K^e + I_M^e - I_S^e + I_\rho^e + I_\delta^e + I_\sigma^e + I_\Omega^e, \tag{19.9}$$

where

$$\left.\begin{aligned}
I_K^e &= \frac{1}{2}\iint_{\Delta^e}\left[K^e\left\{\left(\frac{\partial T^e}{\partial x}\right)^2+\left(\frac{\partial T^e}{\partial y}\right)^2\right\}\right]dxdy\\
I_M^e &= \frac{1}{2}\iint_{\Delta^e}\left[M^e(T_b-T^e)^2\right]dxdy,\ I_S^e = \iint_{\Delta^e}\left[S^eT^e\right]dxdy\\
I_\rho^e &= \frac{1}{2}\iint_{\Delta^e}\left[\rho c\frac{\partial (T^e)^2}{\partial t}\right]dxdy,\ I_\delta^e = \frac{1}{2}\int_{\Omega 1}\left[h(T^e-T_a)^2+2LET^e\right]dx\\
I_\sigma^e &= \int_{\Omega 2}\left[\bar{\varepsilon}^e T^e\right]dy,\ I_\Omega^e = \int_{\Omega 3}\left[\bar{\theta}^e T^e\right]dy,
\end{aligned}\right\}. \tag{19.10}$$

Width of the wound is considered as finite width, and length is extended to infinity to develop an infinite region. Further, the region can be divided into mixed elements, viz., triangular and rectangular elements of finite nodes (n_n) and finite numbers of elements (n_e) (Figure 19.5).

Following expression is used in MATLAB® to generate grid for automatic nodal connectivity for triangular and rectangular elements

```
xnode=41; ynode=11; y1=ynode-1; x3=xnode-2; x1=xnode-1;
x2=xnode+1; x4=(xnode-2)*2+1; xnel=(xnode-2)*2+1;
nel=xnel*(ynode-1);
foriiy=1:y1
foriix=1:x3
nodes1((iiy-1)*xnel+iix, 1)=(iiy-1)*xnode+iix;
nodes1((iiy-1)*xnel+iix, 2)=(iiy-1)*xnode+iix+1;
nodes1((iiy-1)*xnel+iix, 3)=(iiy)*xnode+iix+1;
end
foriix=x1
nodes2((iiy-1)*xnel+iix, 1)=(iiy-1)*xnode+iix;
nodes2((iiy-1)*xnel+iix, 2)=(iiy-1)*xnode+iix+1;
nodes2((iiy-1)*xnel+iix, 3)=(iiy)*xnode+iix+1;
nodes2((iiy-1)*xnel+iix, 4)=(iiy)*xnode+iix;
end
foriix=xnode:x4
nodes1((iiy-1)*xnel+iix, 1)=(iiy-1)*xnode+iix+2;
nodes1((iiy-1)*xnel+iix, 2)=(iiy-1)*xnode+iix+1;
nodes1((iiy-1)*xnel+iix, 3)=(iiy-1)*xnode+iix-(xnode-1);
end
```

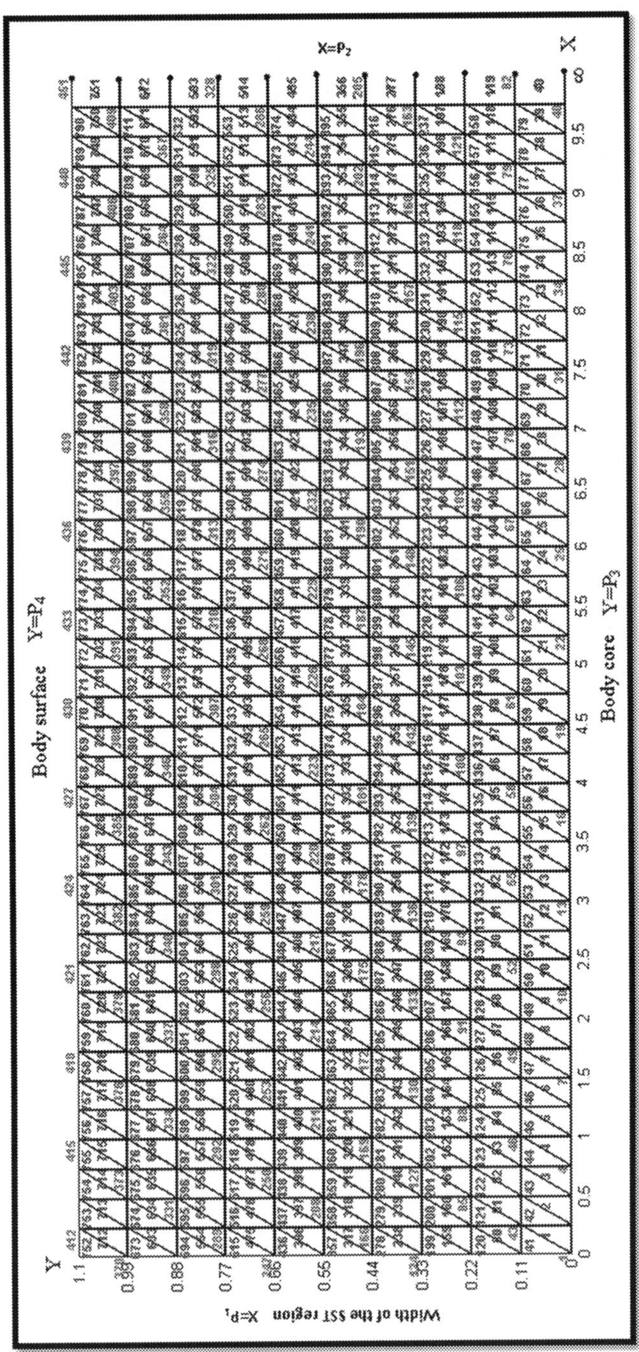

FIGURE 19.5 Discretization scheme of SST region where y-axis is finite and x-axis is treated as infinite domain.

The nodal coordinates are given as follows:

$x_1 = 0,\ x_{i+1} = x_i + 0.25$ for $i = 1(1)39,\ x_{41} = \infty$ (along x-axis)
$y_1 = 0,\ y_{i+1} = y_i + 0.11$ for $j = 1(41)412$ (along y-axis).

Accordingly, nodal coordinates for the remaining nodes can be obtained using the above expressions. The total number of nodes (n_n) is 451; out of the 10 nodes, 41(41)451 are at infinity.

19.4 SHAPE FUNCTIONS

The shape functions for each infinite rectangular element can be obtained through isoparametric mapping. Using isoparametric mapping, the rectangular element in the region is extended to infinity in global coordinates. For each infinite rectangular element, global coordinates are assumed as follows:

$$\left.\begin{aligned} x &= M_i(\xi,\eta)x_i + M_j(\xi,\eta)x_j + M_k(\xi,\eta)x_k + M_l(\xi,\eta)x_l \\ y &= \overline{M}_i(\xi,\eta)y_i + \overline{M}_j(\xi,\eta)y_j + \overline{M}_k(\xi,\eta)y_k + \overline{M}_l(\xi,\eta)y_l \end{aligned}\right\}, \tag{19.11}$$

where $-1 \le \xi \le 1$ and $-1 \le \eta \le 1$

where the mapping functions are given as follows:

$$\begin{aligned} M_i(\xi,\eta) &= \frac{(1-\eta)(-\xi)}{(1-\xi)}, \quad M_j(\xi,\eta) = \frac{(1+\eta)(-\xi)}{(1-\xi)}, \\ M_k(\xi,\eta) &= \frac{(1+\eta)(1+\xi)}{2(1-\xi)}, \quad M_l(\xi,\eta) = \frac{(1-\eta)(1+\xi)}{2(1-\xi)} \\ \overline{M}_i(\xi,\eta) &= \frac{(1-\eta)(1-\xi)}{4}, \quad \overline{M}_j(\xi,\eta) = \frac{(1+\eta)(1-\xi)}{4}, \\ \overline{M}_k(\xi,\eta) &= \frac{(1+\eta)(1+\xi)}{4}, \quad \overline{M}_l(\xi,\eta) = \frac{(1-\eta)(1+\xi)}{4}. \end{aligned} \tag{19.12}$$

The following linear piecewise approximate shape function for rectangular element in local coordinates ξ and η is assumed as (Figures 19.6 and 19.7) [15].

$$T^e(\xi,\eta) = N_iT_i + N_jT_j + N_kT_k + N_lT_l = \left[N(\xi,\eta)\right]\overline{T}^{\varepsilon}, \tag{19.13}$$

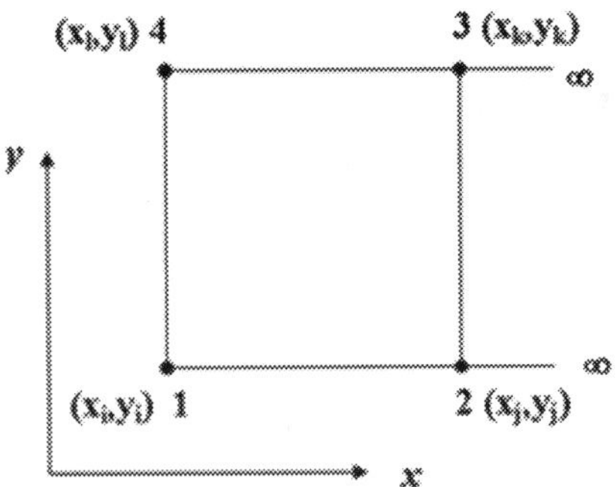

FIGURE 19.6 A four-node rectangular element in infinite region defined in global coordinate.

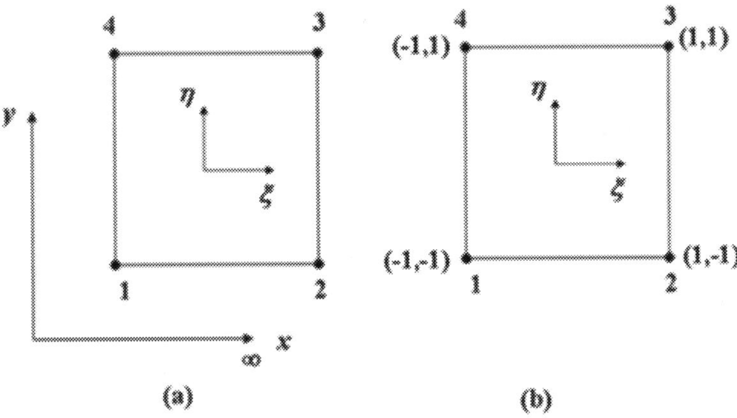

FIGURE 19.7 A four-node rectangular element showing (a) the translation to natural coordinate and (b) the natural coordinates of each node.

where

$$
\begin{aligned}
N_i(\xi,\eta) &= \frac{(1-\xi)(1-\eta)}{4}, \quad N_j(\xi,\eta) = \frac{(1+\xi)(1-\eta)}{4} \\
N_k(\xi,\eta) &= \frac{(1+\xi)(1+\eta)}{4}, \quad N_l(\xi,\eta) = \frac{(1-\xi)(1+\eta)}{4}
\end{aligned} \tag{19.14}
$$

$$N = \begin{bmatrix} N_i & N_j & N_k & N_l \end{bmatrix},$$

$$\overline{T}^e = \begin{bmatrix} T_i & T_j & T_k & T_l \end{bmatrix}'$$

19.5 MATRIX FORMATION

Differentiating I^e (Eq. 19.9) with respect to each nodal temperature of eth element, we get

$$\frac{dI^e}{d\overline{T}^e} = \sum_r \frac{dI_r^e}{d\overline{T}^e}; r = K, M, S, \rho, \delta. \tag{19.15}$$

The matrix form of Eq. (19.15) is as follows:

$$\frac{dI^e}{d\overline{T}^e} = \left[A_1^e\right]\left[\overline{T}^e\right] + \left[A_2^e\right]\left[\overline{T}^e\right] - \left[A_3^e\right] + [A_4^e]\left\{\frac{\partial \overline{T}^e}{\partial t}\right\} + [A_5^e]\left[\overline{T}^e\right] + [A_6^e], \tag{19.16}$$

where

$$\left.\begin{aligned}
\left[A_1^e\right]_{pxp} &= \iint_{\Delta^e} K^e\left[B^e\right]'\left[B^e\right]dxdy, \quad \left[A_2^e\right]_{pxp} = \iint_{\Delta^e} M^e\left[N^e\right]'\left[N^e\right]dxdy \\
\left[A_3^e\right]_{px1} &= \iint_{\Delta^e}\left(M^e T_b + S^e\right)[N^e]' dxdy, \quad \left[A_4^e\right]_{pxp} = \iint_{\Delta^e} \rho c[N^e]'\left[N^e\right]dxdy
\end{aligned}\right\} \tag{19.17}$$

$$\left.[A_5^e]_{pxp} = \int_e h[N^e]'[N^e]dx, \quad [A_6^e]_{px1} = \int_e (LE - hT_a)[N^e]' dx\right\}$$

$$N = \begin{cases} \begin{bmatrix} N_i & N_j & N_k \end{bmatrix} & ; p = 3 \\ \begin{bmatrix} N_i & N_j & N_k & N_l \end{bmatrix} & ; p = 4 \end{cases}$$

$$\overline{T} = \begin{cases} \begin{bmatrix} T_i & T_j & T_k \end{bmatrix}' & ; p = 3 \\ \begin{bmatrix} T_i & T_j & T_k & T_l \end{bmatrix}' & ; p = 4 \end{cases}$$

$$\left[B^e\right]_{2\times p} = \begin{cases} \begin{bmatrix} \dfrac{\partial N_i^e}{\partial r} & \dfrac{\partial N_j^e}{\partial r} & \dfrac{\partial N_k^e}{\partial r} & \dfrac{\partial N_l^e}{\partial r} \\ \dfrac{\partial N_i^e}{\partial s} & \dfrac{\partial N_j^e}{\partial s} & \dfrac{\partial N_k^e}{\partial s} & \dfrac{\partial N_l^e}{\partial s} \end{bmatrix}; p = 3 \\ \begin{bmatrix} \dfrac{\partial N_i^e}{\partial \xi} & \dfrac{\partial N_j^e}{\partial \xi} & \dfrac{\partial N_k^e}{\partial \xi} & \dfrac{\partial N_l^e}{\partial \xi} \\ \dfrac{\partial N_i^e}{\partial \eta} & \dfrac{\partial N_j^e}{\partial \eta} & \dfrac{\partial N_k^e}{\partial \eta} & \dfrac{\partial N_l^e}{\partial \eta} \end{bmatrix}; p = 4 \end{cases}$$

$$p = \begin{cases} 3, \text{ for triangular elements} \\ 4, \text{ for infinite rectangular elements} \end{cases}.$$

19.6 ASSEMBLY OF ELEMENTS

The region has been divided into n_e total number of elements using n_n total nodes. All the elements are assembled to get integral I as follows (see Figure 19.5):

$$I = \sum_{e=1}^{n_e} I^e; \quad n_e = 790 \tag{19.18}$$

Extremizing I, we get

$$\left[\frac{dI}{d\overline{\overline{T}}}\right]_{n_n x1} = \sum_{e=1}^{n_e} \left[D^e\right]'_{n_n \times p} \left[\frac{dI^e}{d\overline{T}^e}\right]_{p\times 1} = 0; \; n_n = 451 \tag{19.19}$$

$$\frac{dI}{d\overline{\overline{T}}} = \left[\frac{\partial I}{\partial T_1} \quad \frac{\partial I}{\partial T_2} \quad . \; . \; . \; . \; . \; . \; . \quad \frac{\partial I}{\partial T_{n_n}}\right]', \overline{\overline{T}} = \left[T_1 \ldots\ldots\ldots T_{n_n}\right]'.$$

Here, *[D^e]'* shows the transpose of matrix *[D^e]*,

where $p = 3$; for triangular elements; $p = 4$; for rectangular elements

$$D^e = \begin{cases} \begin{bmatrix} 0 & 0 & 0 & 0 & . & . & 1 & 0 & 0 & . & . & 0 & 0 & 0 & 0 \\ 0 & 0 & 0 & 0 & . & . & 0 & 1 & 0 & . & . & 0 & 0 & 0 & 0 \\ 0 & 0 & 0 & 0 & . & . & 0 & 0 & 1 & . & . & 0 & 0 & 0 & 0 \end{bmatrix}_{p\times n_n} \\ \begin{bmatrix} 0 & 0 & 0 & 0 & . & . & 1 & 0 & 0 & 0 & . & . & 0 & 0 & 0 \\ 0 & 0 & 0 & 0 & . & . & 0 & 1 & 0 & 0 & . & . & 0 & 0 & 0 \\ 0 & 0 & 0 & 0 & . & . & 0 & 0 & 1 & 0 & . & . & 0 & 0 & 0 \\ 0 & 0 & 0 & 0 & . & . & 0 & 0 & 0 & 1 & . & . & 0 & 0 & 0 \end{bmatrix}_{p\times n_n} \end{cases}.$$

19.7 FORMATION OF SIMULTANEOUS DIFFERENTIAL EQUATIONS IN TIME

Matrix form of Eq. (19.16) is derived for mixed elements with proper shape functions. Using (19.16–19.17), the system of simultaneous differential equation is obtained as follows:

$$[X]_{n_n\times n_n}\left[\frac{d\overline{\overline{T}}}{dt}\right]_{n_n\times 1} + [Y]_{n_n\times n_n}\left[\overline{\overline{T}}\right]_{n_n\times 1} = [Z]_{n_n\times 1}. \tag{19.20}$$

The Crank–Nicolson method is the suitable numerical method to solve transient heat equation using MATLAB®.

19.8 NUMERICAL RESULTS AND DISCUSSIONS

Analysis of temperature variation during the healing process is done at different atmospheric temperatures. The values of physical and physiological parameters are taken from Tables 19.2 and 19.3 [11].

The equation may solve for various atmospheric temperatures, rate of evaporation and corresponding blood mass flow, and rate of metabolic activities.

Some important results have been drawn after calculating temperature variation for wound healing:

1. Based on the experimental facts, the current model includes body core, body surface, and surrounding temperature for wound healing.
2. Temperature regulation is controlled by the thermal conductivity, blood mass flow rate, rate of metabolism, evaporation, latent heat, etc.
3. Experiment results [35, 13] show wound temperature drips expressively on dressing change for a few minutes and takes about 40 minutes for a freshly cleansed wound to return to the normal temperature. It takes about 3 hours for cell mitotic division to restart [35].
4. The graph shows a drop in tissue temperature, rapidly about 20 minutes for the abnormal region (wound area). More heat loss is observed in the wounded area. Wound temperature increases very slightly till about 3 hours and 20 minutes, and after, it becomes steady. The tissue temperature of the normal region becomes almost stable after 20 minutes.

TABLE 19.2
Values for Physical and Physiological Parameters [25]

Thermal Conductivity (cal/cm/min°C)	Heat Transfer Coefficient h (cal/cm²/min°C)	Specific Heat of Tissues c (cal/g°C)
$K_1 = 0.060$, $K_2 = 0.045$, $K_3 = 0.030$	0.009	0.830
Blood Density of Tissues ρ (g/cm³)	**Latent Heat L (cal/g)**	**Body Core Temperature T_b (°C)**
1.090	579.0	37

TABLE 19.3
Value of M, S, and E for Different Atmospheric Temperature [25]

Atmospheric Temperature T_a (°C)	Rate of Evaporation E (g/cm²/min)	Blood Mass Flow Rate M (cal/cm min°C)	Rate of Metabolism S (cal/cm³/min¹)
15	0	0.0030	0.0357
23	$0, 0.24 \times 10^{-3}, 0.48 \times 10^{-3}$	0.0180	0.0180
33	$0.24 \times 10^{-3}, 0.48 \times 10^{-3}$	0.0315	0.0180

5. More heat loss is observed at the skin surface (epidermis) due to conduction, convection, radiation, and evaporation.
6. For the same rates of evaporation, the decline in tissue temperature is more at lower atmospheric temperature.
7. The fall in tissue temperature is more for higher rates of evaporation at the same atmospheric temperature (Figures 19.8 and 19.9).

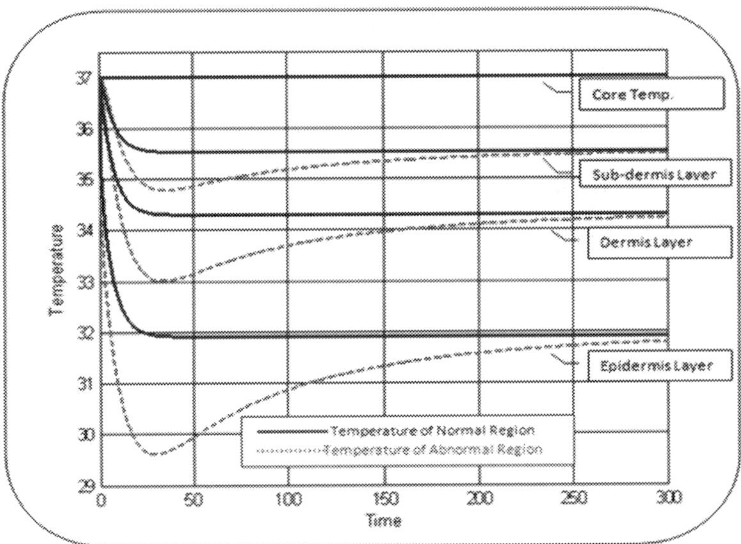

FIGURE 19.8

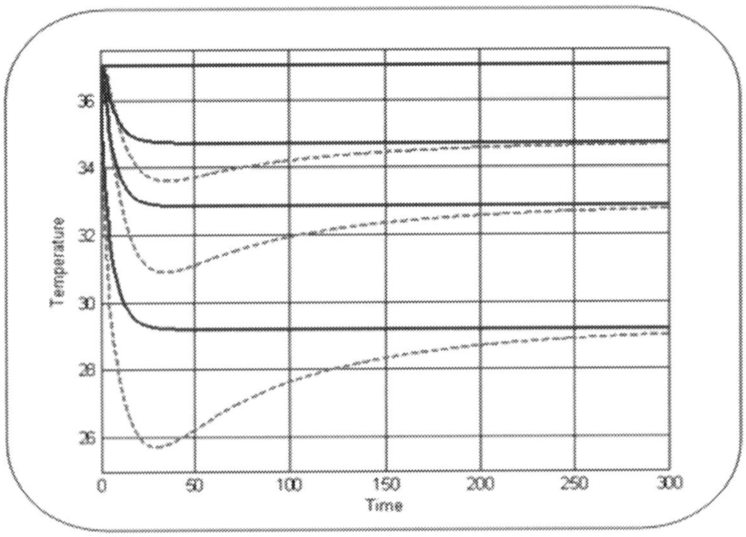

FIGURE 19.9 Graph temperature variation between operated tissues.

REFERENCES

1. A.C. Guyton, J.E. Hall, (2000), *Text Book of Medical Physiology*, 10th ed. W. B. Saunders Company.
2. P.J. Kolston, (January, 2000), Finite-element modelling: a new tool for the biologist, *Philosophical Transactions of the Royal Society of London. Series A: Mathematical, Physical and Engineering Sciences* 358 (1766): 611–631.
3. V. Vavourakis, B. Eiben et al., (2016), Multiscale mechano-biological finite element modelling of oncoplastic breast surgery—numerical study towards surgical planning and cosmetic outcome prediction, *PLoS One* 11 (7): e0159766.
4. J.A. Sanz-Herrera, E. Reina-Romo, (2011), Cell-biomaterial mechanical interaction in the framework of tissue engineering: Insights, computational modeling and perspectives, *International Journal of Molecular Sciences* 12(11): 8217–8244.
5. H. Sorg[1], D.J. Tilkorn, S. Hager, J. Hauser, U. Mirastschijski, (2017), Skin wound healing: An update on the current knowledge and concepts, *European Surgical Research* 58(1–2): 81–94.
6. K.S. Midwood, L.V. Williams, J.E. Schwarzbauer, (2004), Tissue repair and the dynamics of the extracellular matrix, *The International Journal of Biochemistry & Cell Biology* 36(6): 1031–1037.
7. A. Desmouliere, C. Chaponnier, G. Gabbiani, (2005), Tissue repair, contraction, and the myofibroblast, *Wound Repair and Regeneration* 3(1):7–12.
8. R. Fitridge, M. Thompson, eds. *Mechanisms of Vascular Disease: A Reference Book for Vascular Specialists.* University of Adelaide Press; 2011 ISBN-13: 978-0-9871718-2-5.
9. W. McGuinness, et al., (2004), Influence of dressing changes on wound temperature, *Journal of Wound Care* 13 (9): 383–385.
10. W. Cheng-chuan, P.U. Zhi-biao, LIU Hong-bin et al., (1998), Experimental study on maintaining physiological moist effect of moist exposed burn therapy/moist exposed burn ointment on treating burn wound, *The Chinese Journal of Burns Wounds & Surface Ulcers* 10(4): 18–20.
11. G. Kammerlander, A. Andriessen, P. Asmussen, (2005), Role of the wet-to-dry phase of cleansing in preparing the chronic wound bed for dressing application, *Journal of Wound Care* 14 (8): 349–352.
12. W. Cheng-chuan, P.U. Zhi-biao, L. Hong-bin et al., (1999), Experimental study on moist burn therapy/moist exposed burn ointment on burn wound water evaporation, *The Chinese Journal of Burns Wounds & Surface Ulcers* 11(1): 1–3.
13. R. Gannon, (2007), Wound cleansing: Sterile water or saline? *Nurse Times* 103(9): 44–46.
14. S. Jain, P.K. Mishra, V.V. Thakare, J. Mishra, (2019), Design of microstrip moisture sensor for determination of moisture content in rice with improved mean relative error, *Microwave and Optical Technology Letters* 61 (7): 1764–1768.
15. S. S. Rao, *The Finite Element Method in Engineering*, 4th Edition, Elsevier.
16. S. Jain, P.K. Mishra, V.V. Thakare, (2018), Rice moisture detection based on oven drying technique using microstrip ring sensor, *Engineering Vibration, Communication and Information Processing* (*Lecture Notes in Electrical Engineering*) 478: 99–109.
17. S. Jain, P.K. Mishra, V.V. Thakare, (2017), Design and development of microstrip sensor with triple frequency for determination of rice grains moisture content, *International Journal of Mechanical and Production Engineering Research and Development (IJMPERD)* 7 (5): 375–380.
18. P.K. Khare, G. Bangar, P. Mishra, J. Mishra, (2008), Exploration of space charge behavior of polyvinyl carbazole using measurement of thermally stimulated discharge current, *Indian Journal of Physics* 82 (10): 1301–1308.

19. P.K. Mishra, J. Mishra, P.K. Khare, (2009), Study of relaxation phenomena of polyvinyl carbazole (PVK) foil electret using TSDC technique, *Solid State Physics (India), Proceeding. of DAE Solid State Physics Symposium* 1349: 607–608.
20. P.K. Mishra, J. Mishra, P.K. Khare, (2010), Study of steady state conduction of pure and malachite green doped polyvinyl carbazole samples, *AIP Conference Proceedings* 1349: 7575–758.
21. J. Mishra, P.K. Mishra, P.K. Khare, (2010), Study of characterization of pure and malachite green samples using spectroscopic studies, *AIP Conference Proceedings* 1349: 561–562.
22. J. Mishra, P.K. Mishra, P.K. Khare, (2012), TSDC and dielectric relaxation study of pure and malachite green doped polyvinyl carbazole samples, *Invertis Journal of Renewable Energy* 2 (3), 171–175.
23. P.K. Mishra, J. Mishra, P.K. Khare, (2012, July–September), Study of electrical conduction and transient currents of pure and malachite green doped polyvinyl carbazole samples, *Invertis Journal of Renewable Energy* 5 (3): 138–142.
24. W. Perl, (1962), Heat and matter distribution in body tissues and the determination of tissue blood flow by local clearance method, *Journal of Theoretical Biology* 2: 201–235.
25. V.P. Saxena, K.R. Pardasani, R. Agarwal, (1988), Unsteady state heat flow in epidermis and dermis of human body, *Proceedings of the Indian Academy of Sciences – Mathematical Sciences* 98(1), 71–80.
26. P.K. Mishra, J. Mishra, P.K. Khare, (2012), Transient currents study of pure and malachite green doped PVK samples for the use in microelectronics applications, *FIER Special Issue*: 171–175.
27. P.K. Mishra, J. Mishra, P.K. Khare, R. Kathal, H. Pandey and P.K. Khare, (2013), "TSDC and X-ray diffraction analysis of pure and malachite green sensitized polyvinyl carbazole films, *AIP Conference Proceedings* 1512: 792–793.
28. P.K. Mishra, J. Mishra, P.K. Khare, (2013), Spectroscopic studies of pure and malachite green sensitized polyvinyl carbazole films, *AIP Conference Proceedings*, 1512: 610–611.
29. C. Bhat, R.B. Puriya, P. Mishra, S.K. Jain, A. Srivastava, (2014), Electronic and transport properties of BN sheet on adsorption of amine (NH2) group, *IEEE Conference Proceedings of Industrial and Information Systems (ICIIS)*, 1–4.
30. P.K. Mishra, R. Kathal, J. Mishra, (2015), TSDC analysis on pure and ZnO doped PMMA-PVDF films, *Advanced Science Letters* 21: 2930–2932.
31. J. Mishra, P.K. Mishra, (2015), Eletroactive properties of pure and malachite green doped PVK samples, *Advanced Science Letters* 21: 2933–2936.
32. R. Kathal, P.K. Mishra, J. Mishra, (2015), Preparation, characterization and cyclic voltammetric studies on PVDF-PMMA Polymer Nano-composite blends, *Advanced Science Letters* 21: 2943–2946.
33. P.K. Mishra, J. Mishra, (2016), Investigation of the molecular mechanisms, trapping and charge storage behavior of polymer, quantum matter, *American Scientific Publisher* 3: 335–338.
34. P.K. Mishra, R. Kathal, J. Mishra, (2016), Effect of ZnO nanoparticles on PMMA-PVDF polymer blend using scanning electron microscopy, Quantum Matter, *American Scientific Publisher* 3: 417–418.
35. E. Vella, D. Harrison, W. McGuiness, (2004), Influence of dressing changes on wound temperature, *Journal of Wound Care* 13: 383–385.

20 Data Classification Framework for Medical Data through Machine Learning Techniques in Cloud Computing

Saurabh Sharma and Harish K. Shakya
Amity School of Engineering & Technology

Ashish Mishra
Gyan Ganga Institute of Technology and Sciences

CONTENTS

20.1 INTRODUCTION

Cloud computing is a local server or personal computer [1], instead; it is used for saving remote servers using large numbers of hosts on the Internet, whose applications are to store and access data. Cloud computing provides information technology and access to resources. Cloud customers are the only services that we pay them to deliver and use. Some models and services make it easier for users for cloud computing (Figure 20.1).

20.1.1 Features of Cloud Computing

Here are some standard features of cloud computing [2]:

i. *High scalability*:
   This means that the system demanded massive resources without necessity for human interaction with every service provider.
ii. *High availability and reliability*:
   The availability of servers reduces the possibility of infrastructure failure and is more and more reliable.
iii. *Agility*:
   It distributes resources among users and works very quickly.

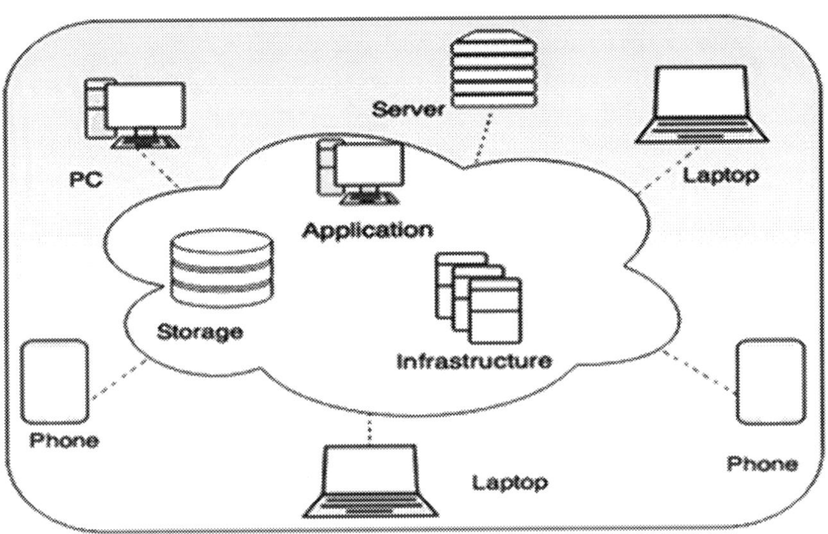

FIGURE 20.1 Cloud computing.

iv. *Multi-sharing*:

It reduces costs by utilizing distributed computing to impart basic framework to various clients and applications at work and all the more productively.

v. *Maintenance*:

There is an easy maintenance of cloud computing applications that do not need to be installed on each computer and can be accessed from different locations because they can ultimately reduce costs.

vi. *Low cost*:

This is cost-effective as there is no need for the company to set up its infrastructure. It pays according to the consumption of resources.

20.1.2 Advantages of Cloud Computing

i. *Fast implementation*:

If you can sometimes cut with, it takes months or years with the cloud to develop or implement an application, so application runs do things faster.

ii. *Instant scalability*:

With cloud resources or not, you can always reduce the scale. Resources and users depend upon their needs, but cloud capacity does not.

iii. *Access anywhere*:

Built-on cloud applications are designed to be accessed from anywhere; all you need is an Internet connection on a mobile device.

iv. *No upfront costs*:

Previously, deploying an application for architecture to buy software licenses, etc., buy the necessary hardware, but with cloud dramatically, all costs are reduced and, in some cases, have been eliminated.

v. *Maintenance free*:

Traditionally, you have to worry about needing to patch with the latest release of your software, the need to upgrade your hardware, and the hardware level to plug in your system, but you cloud with its own hardware; it will manage and maintain all its cloud provider.

20.1.3 Categories of Service Model

The models of cloud services are classified into three basic models [3] (Figure 20.2):

i. Software as a service (SaaS),
ii. Platform as a service (PaaS),
iii. Infrastructure as a service (IaaS).

i. *Software as a service (SaaS)*:

SaaS is known as "on-demand software." It is a software delivery model. In this model, applications are facilitated by the cloud specialist co-op and elevated to clients on the Internet. Mother, related information, and

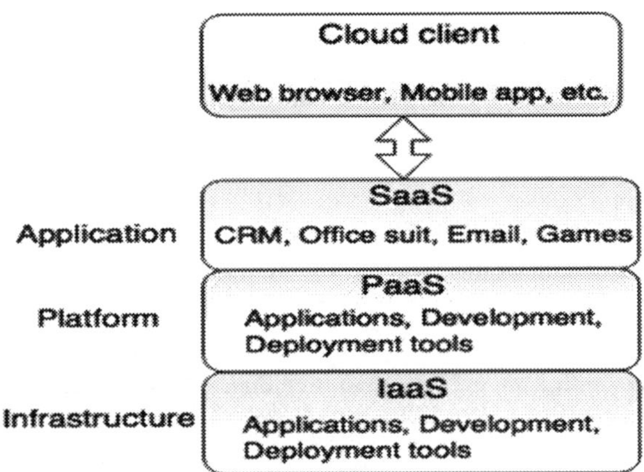

FIGURE 20.2 Categories of cloud computing.

programming are facilitated midway on a cloud server. Clients can get to SaaS utilizing a meager customer through an Internet browser. Office suite, email, games, and so on are programming applications that help through the Internet [3].

The estimation of the law depends on a month-to-month or yearly expense for simple mother purchasing and a lower cost for associations, which permit access to business usefulness on authorized applications. SaaS requires almost no equipment since programming is facilitated remotely, so associations don't need to put resources into extra equipment. There is no compelling reason to require low upkeep costs for the mother and explicit programming or equipment adaptations. SaaS applications are subject to web association. They are not usable without a web association. It is hard to switch between SaaS merchants.

ii. *Platform as a service (PaaS)*:

PaaS is anything but difficult to create. Engineers can concentrate on foundational improvement and development without stressing over it. In PaaS, designers need an Internet connection with manufacture of a PC and an application. PaaS is simpler to create. Developers can focus on the turn of events and development without stressing over the framework.

In PaaS, the engineer requires a PC and an Internet association to begin building applications. One engineer can compose the applications according to the stage given by the PaaS merchant. Henceforth, moving the form to another PaaS seller is an issue.

iii. *Infrastructure as a service (IaaS)*:

IaaS includes servers, storage, networks, and operating systems as a way to cloud computing infrastructure. Clients can get to these assets on-request administration on the Internet or distributed computing stage. In case, you buy total holdings as opposed to buying servers, programming, data center areas, or system gadgets.

In IaaS, clients can browse a CPU, memory stockpiling setup in the necessary robust structure. Without much of a stretch, clients can access the considerable processing power accessible on the IaaS cloud stage. The IaaS-distributed computing stage model is reliant on the accessibility of the Internet and virtualization administrations.

20.1.4 Types of Cloud

Types of cloud computing come under the deployment model. The cloud computing model and administration model sending are the working models. Distributed computing is open to end clients who can utilize sending models and administration models. The sorts of access to the cloud are characterized by organization models. The accompanying four techniques can be gotten to in the cloud:

i. Public cloud;
ii. Hybrid cloud;
iii. Private cloud;
iv. Community cloud.

i. *Public cloud*:

In the public cloud, the framework and administrations are accessible to the overall population. An open cloud is available to all. Thus, it might be less secure. This cloud is reasonable for data, which isn't sensitive [4]. The public cloud is a service of cloud computing that is open for everyone, and it can be used as general services with minimal authentication by any user.

The public cloud is more affordable than a private cloud or hybrid cloud since it has similar assets with different clients. Since it gives an adaptable method to clients, it isn't easy to consolidate with a private cover over an open cloud. It offers many assets in various areas, and on the off-chance that one asset falls flat, the other is utilized as it is dependable. The public cloud services are offered by big companies such as Google, Microsoft, IBM, and Amazon (Figure 20.3).

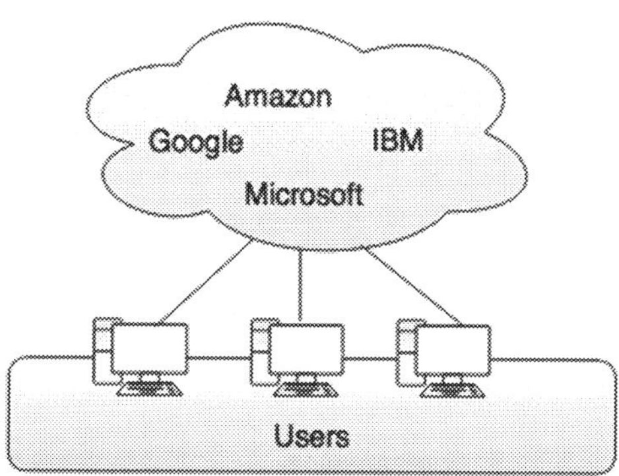

FIGURE 20.3 Public cloud computing model.

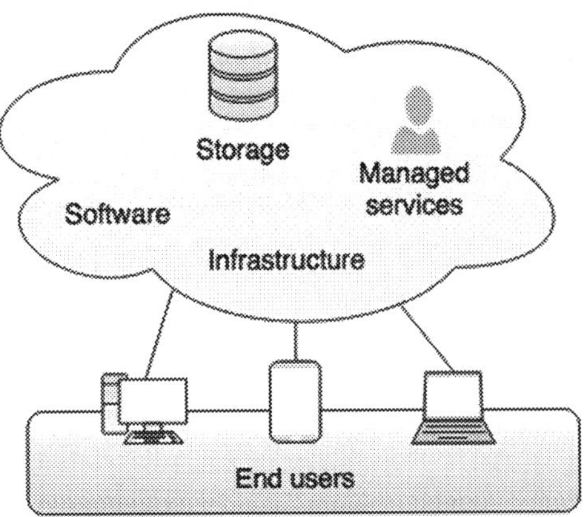

FIGURE 20.4 Private cloud computing model.

ii. *Private cloud*:

Private cloud is a framework of cloud computing where all the services are bounded for limited users only—available structures and administrations inside private cloud associations. Cloud works just in a specific association [5] (Figure 20.4).

A single organization shares private clouds since assets are profoundly secure. In the open cloud, you have more command over your assets and equipment in the private cloud since it just arrives at an association limit. Sending private cloud all-inclusive is troublesome, and it is available locally. Private cloud costs more than an open cloud.

iii. *Hybrid cloud*:

A hybrid cloud is a combination or fusion of public and private clouds. The hybrid cloud is a type of cloud deployment model used for processing important activities using private clouds and carrying noncore activities using public clouds [6] (Figure 20.5).

It is adaptable because it highlights both open and private clouds. Security assets are due to private cloud, and versatile assets are due to the public cloud. The expense of half-breed cloud is not a precisely private cloud. Half-and-half cloud organizing is progressively mind-boggling, as both closed and open veils of mist are accessible.

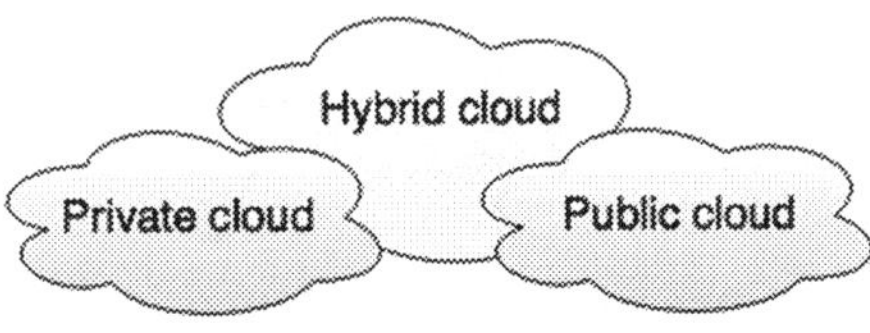

FIGURE 20.5 Hybrid cloud computing model.

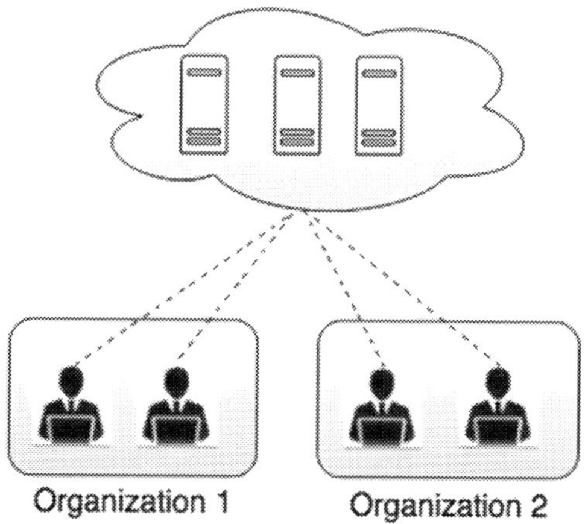

FIGURE 20.6 Community cloud computing model.

iv. *Community cloud*:

It enables community cloud frameworks and administrations that are available from bunch associations. It shares the foundation between a few associations from a particular network. It is overseen inside and is utilized or worked by numerous associations by outsiders or a blend of them [6] (Figure 20.6).

A community cloud is a framework for sharing cloud assets and capacities among numerous associations. Cloud is more secure than an open cloud; however, it is less secure than a private cloud.

20.2 DATA STORAGE IN CLOUD COMPUTING

Cloud storage administration empowers you to spare information on an offside stockpiling framework. An outsider finishes this information. This information can be gotten to by a web administration application programming interface (API) [7,8].

20.2.1 Storage Devices

The storage devices are categorized as follows:

i. *Block storage devices*:

Such devices provide raw storage customers. It contrasts to make crude stockpiling volume. Volume is a perceived unit of information storage.

ii. *File storage devices*:

File storage devices are given to the client as a document to keep up their document framework, namely system record framework (likewise Network File System (NFS)) [9].

20.2.2 Storage Classes of Cloud

Following are the classifications of storage capacity classes:

i. *Unmanaged cloud storage*:
   It is a predetermined storage client; this is known as unmanaged distributed storage. Can't design customers or set up their document framework or change drive properties?
ii. *Managed cloud storage*:
   Managed cloud storage provides on-demand online storage space. The user can identify the format system to the user as partitions and raw disks.

20.2.3 Creating Cloud Storage System

Cloud storage framework stores various duplicates of information on numerous servers in multiple areas. Data is put away in different areas so that if a frame comes up short, it can change the pointer area where the item is placed elsewhere. The cloud supplier utilizes virtualization programming to gather the capacity resources into the distributed storage framework. This framework is called storage GRID. Capacity GRID is a virtualization layer that produces stockpiling from various capacity gadgets in a single administration framework. It oversees information on Common Internet File System (CIFS) [10] and Internet NFS record frameworks.

20.2.4 Virtual Storage Containers

Virtual capacity compartments offer elite distributed storage frameworks. Virtual capacity holders are creating logical unit number (LUN) of gadgets, records, and articles that are produced.

20.3 VIRTUALIZATION IN CLOUD COMPUTING

- For consolidating neighborhood and system assets information stockpiling virtualization;
- For gathering physical capacity gadgets into the single unit;
- For arriving at the significant level of accessibility or improving accessibility utilizing virtualization;
- Improving execution utilizing virtualization;
- Using virtualization utilizing stripping and storing;
- Capacity improvement.

Virtualization is the foundation of distributed computing, and distributed computing makes virtualization progressively effective with the assistance of productive [11–13]; it gives answers for significant difficulties in the field of information security and security insurance. Virtualization is a duplicate of the equipment inside the product program. It permits the job of different PCs on one PC. In a document or web server, there is the double utilization of obtainment, support, deterioration, yet additionally vitality and floor, but our motivation is to make a virtual web or record server; by taking into account such most extreme equipment assets, adaptability, and security as an improvement, there is a decrease in cost. Effective utilization of assets, expanded security, versatility, issue-free testing, simple administration, expanded adaptability, advantages of shortcoming separation, and quick sending are the important characteristics of virtualization.

Many aspects of virtualization propose both advantages and disadvantages to security. Cloud in the cloud customer's reward has many of the critical challenges for investigating the use and computing of public cloud IT resources [14]. This segment covers the subsequent topics:

- Increased security flaws;
- Low operating governance control;
- Limits between portability of cloud providers;
- Multi-sectored regulatory and legal issues.

Protected images of shared virtual machine (VM) in cloud computing security requirements describe security issues in various layers in cloud computing. Virtualization image 13 is defined to be proposed as an outline (conceptual) design used for the cloud.

20.4 SECURITY ISSUE IN CLOUD COMPUTING

Figure 20.7 represents security is the biggest challenge in cloud computing because data and code of users reside in cloud provider premises (data center). Security is always a significant concern in the open-system architecture (Figure 20.7).

Figure 20.8 represents security risk has become a top major concern in cloud computing in 2019 by the latest survey of International Data Corporation with more than 1,100 analysts over worldwide given the study and report showing security is the biggest challenge in cloud computing [15,16] (Figure 20.8).

In the graph, it shows security is the biggest challenges in cloud computing that demotivate IT industries and cloud users towards using cloud services [17].

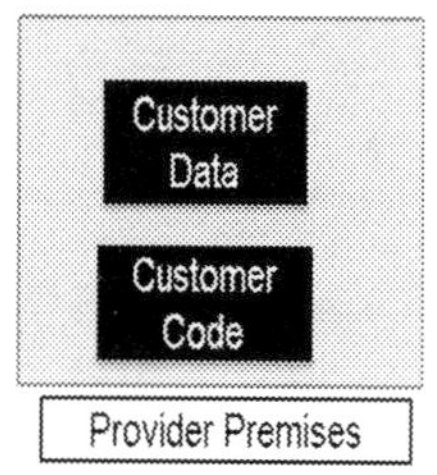

FIGURE 20.7 Provider premises: cloud service provider (data center), customer: cloud consumer.

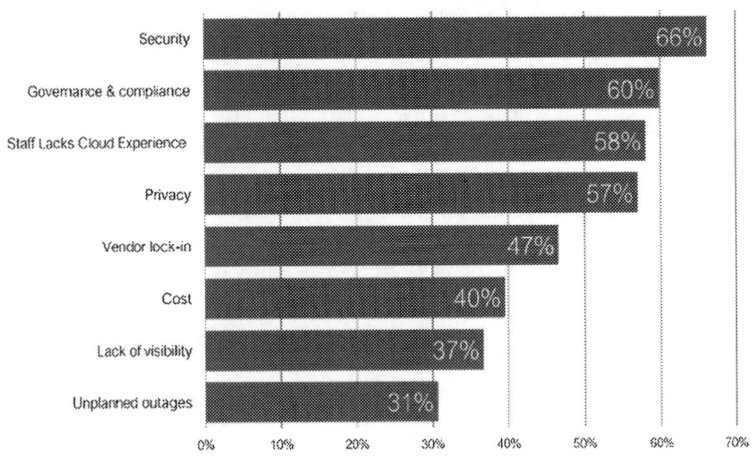

FIGURE 20.8 Security risk has become a top major concern in cloud computing in 2019. (https://www.idc.com.)

20.4.1 Common Security Requirement

Security has four major parameters:

i. Availability;
ii. Confidentiality;
iii. Integrity;
iv. Authentication.

i. *Authentication in cloud services*:
   It shows that only authorized users are allowed to access given services, intended for an authorized user. Unauthorized users are not allowing to access provided services [18].
ii. *Confidentiality in cloud services*:
   It shows data should not be disclosed to unauthorized users, and the privacy of the system should be maintained [19].
iii. *Integrity in cloud services*:
   It shows no modification and fabrication allowed in the data when cloud user and cloud service providers are communicating with each other [20].
iv. *Availability in cloud services*:
   It shows resources should not be made available from an authorized user by any malicious activity [21].

20.5 DATA CLASSIFICATION

Data classification, information security, sensitivity categorization, and arrangement of data are based on its level. Appropriate baseline security control data classification helps determine the security of the data. While data classification has been used for decades for organizations helping to protect security with appropriate levels, sensitive or critical data, lack of specificity in traditional rankings and the process of data classification allows organizations to categorize their stored data by sensitivity and business impact in order to determine the risks associated with the data [22,23]. Data classification is a starting point for maintaining confidentiality-specific settings [24].

Data classification is a fundamental step in cybersecurity risk management. It is an organization that involves the identification of proprietary or operated data that is processed in an information system and is stored. It also includes an assessment of data sensitivity and potential effects arising from compromise, damage, or misuse (Figure 20.9).

All institutional data should be classified into one of three sensitivity levels or classifications [14]:

i. *Restricted data*:

   It is confined as information when unapproved exposure must be arranged; the danger of making changes or information demolition of the university or its associates might be huge. Instances of information incorporate information and security understandings, ensuring information secured by state or government protection guidelines. The most significant levels of security controls ought to be applied to confined details.

ii. *Private data*:

   The information must be named private because of unapproved exposure, adjustment, or decimation of information that might be a danger to the university or the medium to its subsidiaries. Of course, all institutional knowledge that is not expressly confined or delegated open information ought to be treated as close to home information. A suitable degree of security control ought to be applied to individual information [14].

iii. *Public data*:

   The information should be delegated open when there are practically zero hazards from unapproved exposure, adjustment, or the college and its partners with the erased information. Free information models incorporate official statements, course data, and research distributions.

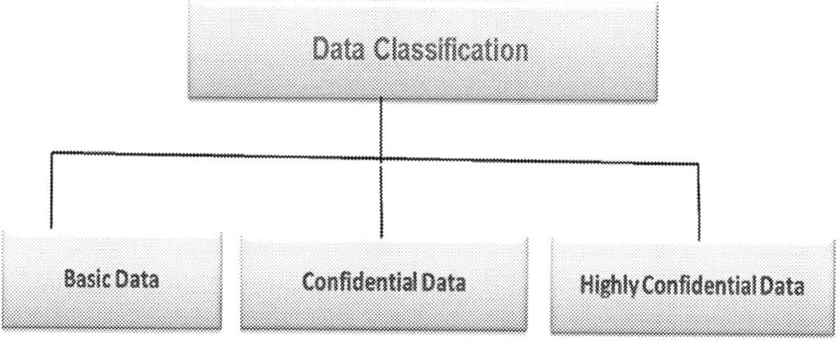

FIGURE 20.9 Data classification terminology.

> While no control is required to diminish accessible information or secure protection, some degree of power is needed to forestall unapproved alteration or annihilation of open details [25].

Full awareness of the association's prerequisites for fruitful information characterization in an association and the association's information resources requires an exhaustive comprehension of the home. Information is available in one of three essential advances: rest, procedure, and travel. The three states require specific answers for information order, yet the standards applying to information grouping must be the equivalent for each. The information, which is delegated secret, is required to stay private throughout the procedure. Data can likewise be organized or unstructured. Run of the mill grouping methodology for databases and organized information are found in spreadsheets that are less tedious than overseeing records, source code, and unstructured data, e.g., email and—ordinarily, more unstructured information than classified information in associations. Regardless of whether the information is organized or unorganized, this information affectability is essential to oversee associations. When appropriately executed, information grouping is done to guarantee that delicate or classified information to convey resources for people in general or with the expectation of complimentary aides they made do with more review than information resources.

20.5.1 Classification Applied to Information Types

Notice of security classification can be joined with two clients' three data and framework data, and it may be applied very well to data in electronic or nonelectronic. It can likewise be utilized as information given the suitable security classification of the data framework. While a characterized record or asset that groups information generally has various levels, the present order ought to build up the most elevated level by and large arrangement. For instance, the document must be named confined to touchy information (Table 20.1).

i. *Confidential (Restricted)*:

Likewise included is data named private or limited, which can be unfortunate for at least one person and/or associations whenever traded off or lost information is incorporated. Such data is frequently given on a "need-to-know" premise and may include the following:

TABLE 20.1
Data Classification Types

Sensitivity	Terminology Model 1	Terminology Model 2
High	Previously used : Confidential	Presently used : Restricted
Medium	Previously used : For internal use only	Presently used : Sensitive
Low	Previously used : Public	Presently used : Unrestricted

- Personal information includes individual distinguishing data, such as standardized savings or nationally recognizable proof numbers, identification numbers, Visa numbers, driver's permit numbers, clinical records, and medical coverage ID numbers;
- Financial records, budgetary record number, e.g., check or venture account number;

  Business content, e.g., records or information, which is one of a kind or novel licensed innovation;
- Legal information, including potential lawyer unique material;
- Includes validation information, e.g., individual biometric vital documents, a private cryptographic key, username or secret key, or another distinguishing proof succession;

  Information that is regularly named secret is administrative and consistent prerequisites for taking care of that information;

  Sensitive data, e.g., documents and information which won't seriously affect an individual and/or association on the off-chance that they are lost or wrecked, and are to be delegated medium affectability. This may incorporate such data as follows:
- Most of the personalities can be erased or conveyed to people (without an emergency mail or post box order of messages), aside from email.
- Documents and documents that don't contain classified information.

By and large, it is remembered for this order, which is not classified. This classification may incorporate most business information; in light of the fact that most records are overseen by day or by analysis, they can be delegated touchy. All inside the business association, information can be appointed touchy naturally, with the general population or except for information looking after secrecy.

ii. *Public (unrestricted)*:

Publicly characterized data information and records are not basic to business prerequisites or tasks. This characterization may remember information for the type of promoting materials or press declarations that are deliberately given for their utilization. What's more, an email administration arrangement may incorporate information put away by spam, like email messages.

Table 20.2 shows potential ramifications of the definition for every security target privacy, recommendations, trustworthiness, and accessibility [26].

20.5.2 Medical Data set Classification

Human health, to a great extent, utilized clinical information in determination. In this way, it has assumed a significant job in clinical design, just like doctors. As needs are to improve the inquiry system in medical offices to more readily anticipating the nature of the illness or related conclusion. Grouping has a significant capacity for AI (artificial intelligence) and information mining. Learn arrangement demonstrating training information, which makes a few blunders as possible, is as

TABLE 20.2

Potential Impact Definitions for Security Objectives [26]

Security Objective	Low	Moderate	High
Confidentiality Utilization of data, including gadgets to secure individual protection and exclusive data, and to ensure lawful limitations on exposure. [27]	The unauthorized revelations of data might be relied upon to have a substantial, unfavorable impact on hierarchical tasks, authoritative resources, or people—unapproved data exposure.	There might be a critical genuine unfriendly impact on hierarchical activities, authoritative resources, or people.	The unauthorized revelation of data might be relied upon to cause natural or unfriendly offensive consequences for scholarly activities, hierarchical resources, or people. [28]
Integrity Inappropriate data alteration or annihilation, and keep away from the consideration of data to make a counter and genuineness. [26]	Unapproved adjustment or annihilation can be relied upon to have a noteworthy restricted antagonistic impact on data, hierarchical resources, or people to complete authoritative tasks. [29]	Unapproved adjustment. Data annihilation might probably have a remarkable genuine antagonistic impact on hierarchical activities, authoritative resources, or people. [27]	To direct hierarchical tasks of unapproved change or annihilation when positive or unfriendly offensive impacts might be typical on data, authoritative property, or people. [30]
Availability Access and dependable access to data about the time of utilization. [31]	Information or data frameworks or use, hierarchical operational interruption of access to authoritative resources, or might be relied upon to have a noteworthy restricted unfriendly impact on people. [32]	There might be a noteworthy genuine unfriendly impact on hierarchical tasks, authoritative resources, or people using data or data frameworks. [33]	A genuine or disastrous obstruction in the utilization or utilization of data or data frameworks can be normal—unfavorable impacts on scholarly activities, hierarchical resources, or people. [34]

of now being applied to disregard the report. There are many arrangement calculations developed and utilized in clinical applications. An electronic well-being record (EHR (electronic health record)) is put away in advanced electronic structure version of clinical information, where patients are resolved with test outcomes. There are chronicled records of prescriptions, just as treatment history. These records are considered to have especially delicate individual information of patients [33,35]; accordingly, they ought to be very much secured EHR continuous patient center records which have approved clients to give moment data in school. This clinical information order paper expects to exhibit cloud conditions with the goal that information privacy is ensured.

The classification approach involves blockchain security algorithms that include privacy to protect the data generated using privacy and security factors. However, many researchers are working on either dimensional space or unbalanced data. Because of this, sometimes, it cannot be correctly estimated or the classification of incurable diseases depends on both factors that are equally important. So even now, there are more factors to solve these biomedical challenges, but two factors require an improvement.

20.5.3 Challenges of Medical Data Classification

Existing systems encountered have challenges related to data classification and security, which are as follows:

- A significant test of the strategies is used to secure protection in the cloud, and the number of highlights you have to think about secrets of the proposition.
- A classifier second test in the cloud condition is implemented because the classifier must guarantee unsafe security data.
- Cloud registering relies upon the hazard related to information arrangement, which is significant and tested this revelation or user's data, e.g., client IDs and utilization designs.
- Another test is identified with the advances that give security; this nature has included the trait of cloud information. The strategy ought to be successful in managing those qualities instead of numerical characteristics.
- A noteworthy test identifies with building mistake oncology. Subsequently, it has been prescribed to incorporate the mistake indicators and improvements to make viable ontologies.

20.5.4 EHR Information Extraction through Machine Learning Approaches

A few late examinations have concentrated on separating clinically essential data from clinical notes utilizing a learning machine. The info learning structure that speaks to a wide range of tools is a crucial issue. The strategy of info attributes for the device for every datum point, known as highlights, is utilized as a learning set. In customary AI, these administrations are structured by hand dependent on area information. Basics arranged element extraction from computerized information escalated learning, as it proceeded to talk about in the accompanying segment. The principle subtasks include the following:

i. *Single concept extraction*:
   The clinical incorporates free exercises with unique work-organized clinical ideas: e.g., the removal of ailments, medications, or methods. Numerous past examinations have applied traditional regular language preparing Natural language processing (NLP) systems to make different levels of progress; however, there is a considerable space for additional improvement given the multifaceted nature of clinical notes.
ii. *Temporal event extraction*:
   On this sub-tab of each concentration, the idea manages progressively complex issues, e.g., predetermined for the perspectives on the most recent couple of months.
iii. *Relation extraction*:
   Relation Extraction is a crucial cog in the field of Natural Language Processing (NLP) and linguistics. It's widely used for tasks such as Question Answering Systems, Machine Translation, Entity Extraction, Event Extraction, Named Entity Linking, Coreference Resolution, Relation Extraction, etc.
iv. *Abbreviation expansion*:
   A few new restorative briefs in clinical content that organized ideas for extraction have been found to require elaboration before mapping. There can be several potential clarifications for every truncation; contraction extension can be an overwhelming errand.

20.5.5 Security and Privacy of Classified Data

Human components can be separated dangers increasingly related internal or outer, purposeful, or inadvertent. In this manner, working with experts is done. In light of medicinal services, unique risks in IT frameworks must be borne when creating strategies to ensure patients' well-being data. Potential perils of social insurance records can be characterized in danger associated with the accompanying:

i. A human factor—representatives or programmers;
ii. Natural disasters—earthquakes, fires, etc;
iii. Specialized failures—framework crashes.

This is a significant assignment to ensure touchy information, precisely the situations when simple access to PCs and systems went a huge increment in instances of psychological oppression and digital fear-mongering. Another serious issue of data security is one-time approval. Typically, work begins with the gadget or IT framework. In this way, the client may represent a genuine risk to the authentication specifically, open sort of field, e.g., clinical offices, with the goal that the framework and EHR information be helpless against a programmer assault. To guarantee a significant security level, mainly clinical data, data frameworks must be predictable to determine the status. Such an observing structure is finished by interruption location intrusion detection system (IDS), which continually tries to screen the client's activity and afterward, to confirm the individual's identity. The methodology concerns EHR protection issues; client recognition depends on PC client confirmation and interruption profiling. Data about the client reliance profile is between keystrokes. A host-based intrusion detection system (HIDS) is an intrusion detection system that is capable of monitoring and analyzing the internals of a computing system as well as the network packets on its network interfaces, similar to the way a network-based intrusion detection system (NIDS) operates.

In the wake of characterizing the information, the application is to find secret information security strategies and becomes an indispensable piece of any information security arrangement procedure. It requires extra thoughtfulness regarding security for private data and how information is put away and transmitted in the cloud with conventional design. This segment gives fundamental data about a portion of the advancements that can robotize authorization endeavors to help secure information named secret. As appeared in the accompanying figure, these advances are conveyed on-premises or cloud-based arrangements, or can be sent in a half-and-half manner, some of which are on-premises and cloud. The utilization of specific advancements, e.g., in encryption is as follows:

i. *Encryption gateway*:

The encryption passages re-embed all entrances to make cloud-based information work in layers giving encryption administrations. This methodology should not be mistaken for a virtual private system (VPN); the encryption doors are intended to provide a straightforward layer to cloud-based arrangements. Overseeing encryption can give entryway information, and the remainder of the encryption arrangements to give being delegated secret. The encryption passage is put in the report to provide stream encryption/decoding administrations between the client's gadget and application server farms. VPNs are fundamentally on-premises arrangements, e.g., encryption arrangements. They are intended to furnish an outsider with power over which encryption keys help diminish the danger of being a supplier with the two sources of information and critical administration. Such arrangements are planned like encryption, to work all the more straightforwardly between the first clients and the administration.

ii. *Data loss prevention*:

Data loss (sometimes referred to as data leakage) is an important consideration, and the prevention of malicious and accidental external data loss through insiders is paramount for many organizations. This sort of information misfortune avoidance (DLP (data loss prevention)) innovations arrangement can help guarantee that email administrations don't get information that is named private. To enable the association with DLP, DLP offers in the existing items. Utilization of such DLP highlights approaches that can be made without any preparation or by programming suppliers effectively utilizing the provided layouts.

DLP technologies can catchphrase through inside and out substance examination coordinating, and word reference coordinating, other substance tests to decide if the substance is a regular articulation assessment, and an infringement of authoritative DLP approaches. For instance, DLP can help forestall this information misfortune as follows:

- National identification number;
- Social security;
- Information related to banking;
- Credit card number;
- IP address.

Some DLP advances additionally give the capacity to revoke DLP arrangements (for instance, an association is required to transmit data to the government disability number of a finance processor). What's more, it is conceivable to arrange DLP ought not to get data before endeavoring to send those touchy data.

A portion of the advantages of design-based information security of the board includes the following:

Secure sensitive information:

Clients can straightforwardly ensure their information utilizing rights of the executives empowered applications. No further advances are required.

Travels with data protection:

Live client control is access to your information, regardless of whether it very well may be in the cloud, foundation, or client's work area. Associations can decide to encode their information and alternatives to restrict utilization according to business necessity.

Default information security policies:

Administrators can use standard policies for many common business scenarios, such as "company no further" and "read-only confidential." Use that as a support for a comprehensive set of rights to allow flexibility in defining read rights custom usage, copy, print, save, edit, and so forth.

By and large, it doesn't produce as much enthusiasm as different information grouping subjects, and all the more energizing innovation themes. Be that as it may, you can get unique advantages from information order, e.g., consistence abilities, better approaches to deal with the association's assets, and encourage relocation to the cloud. Even though information arrangement can make an unpredictable endeavor and practical usage requires a chance appraisal, fast and straightforward efforts can likewise profit. In any information order exertion, information must attempt to comprehend the association's necessities and determine the information that is put away the handling abilities, and the whole association to communicate.

Supporting information characterization is much more significant for the executives than for IT to attempt. An idea can fundamentally be delegated to review work. Yet, there are a few accessible arrangements that can decrease the exertion measure required to execute an information characterization model effectively.

It might likewise be significant that information groupings are related to information maintenance, ought to be tended to when cloud and cloud arrangements help diminish chance. A few information security advancements, like encryption, are moved to the cloud. In rights, the executives and information misfortune counteraction arrangements can help decrease cloud computing risks. In any case, this paper was explicitly talked about on the cross-breed condition. A blend of on-premises and viable cloud-based information order advancements can help diminish the chance for associations of any size to search for a choice to keep up put-away information and clients. Highly sensitive information should be stored on-premises cloud and under a different set of controls. It can be a hybrid environment, a way of future possibility, and rely on effective data management, which is the key to useful data classification.

20.6 LITERATURE REVIEW

Zhang et al. [5] tell us concerning the cloud-layer engineering. It comprises four layers, namely, the equipment layer, foundation layer, stage layer, and application layer. What're more, there are different difficulties of distributed computing innovation.

Subashini and Kavitha [6] dealt with the examination of issues identified with the cloud administration conveyance model—SaaS, PasS, and IaaS and its security. The model was associated with significant problems identified with the mother's safety, such as arranging security, personality the board and sign-on procedure, validation and approval, information region, and information confinement. On account of PaaS, while creating and conveying applications, we can evaluate chance security utilizing unique measurements. On account of IaaS, the creator suggests actualizing shields by both supplier and buyer. They have a thought for the eventual fate of research in the field of an incorporated security model for SaaS and IaaS to discover.

Phaphoom et al. [36] have realized that distributed computing is a mix of virtualization, as predefined ideas of circulation frameworks and web administrations and arrangements. A significant region of worry is for cloud engineering, virtualization, information the board and security issues, and methods. There is an approach for issue identified with virtualization and proposed information to the executives.

Gebauer et al. [17] went through the analysis of the implementation of various risk factors, the impact of cloud environmental responsibility with its characteristics, and the use of technology. However, a detailed description of the reliable system used is not mentioned in its impact paper.

Morsy et al. [20] explained the architectural design issues by clarifying cloud security, properties, service models, and union case. They do this for the problem of virtualization and service-oriented technology company with isolation and multiuser-related security feature. They are suggesting safety models based on the structural changes, but with detailed safety reports and associated risks.

Takabi et al. [21] explained, due to the outsourcing of data and applications, virtualization service-level agreements (SLAs) and regulation, a severe security and privacy issue came across cloud service providers. Unified organizational security management techniques come as a solution to this problem to the proposed secured and trusted services.

Vaquero et al. [24] stated issues in IaaS layer of cloud computing. There are three essential areas of IaaS, namely, virtualization, network, and physical space. They enlightened the issues and solutions related to the IaaS layer. For ensuring the security of infrastructure components, they define access control and encryption techniques.

Dabur et al. [14] mentioned that the cloud environment has its own set of risks, threats, and vulnerabilities. Working with Cloud technologies adds more dangers and threats to it. But it does not help us in providing an exact amount of understanding of related issues.

Zissis and Leakkas [25] used a trusted third-party (TTP) security algorithm to rectify the threat related to the user-specific security. And they try to solve this security issue.

Pearson [37] discussed the challenges related to safety, trust, and confidentiality of cloud with traditional mechanisms. Less flexibility and dynamism traditionally raised as an issue.

Fernandes et al. [38] provided the real-time examples and briefly discussed the related security issues. Safety issues are suggested in a real-time and practical implementation. They analyzed various studies in the industry that were published between 2008 and 2012.

Aguilar et al. [39] discussed the security associated with virtualization and data about different issues. These issues are settled, e.g., remote stockpiling confirmation; guaranteeing respectability; addressing responsibility; and saving security, devotion, and proposals.

Ramachandra et al. [40] featured the intricate design, problematic nature, and benefited assets that cloud all posture extraordinary dangers to the cloud on-screen character. They focused on that every single applicable on-screen character has joint duty regarding security. They center on the current security issues and difficulties, and control measures.

Guy Zyskind et al. [41] Decentralization of which one doesn't propose individual information the board framework for clients to make themselves and assume responsibility for their clinical knowledge. They said that a conventional blockchain actualized that guarantees a computerized get to control chief doesn't require trust in an outsider. In their framework, they utilized the directions given, e.g., shared capacity, question, and information. Furthermore, clients' blockchain perceive as proprietors of their knowledge.

Meng Shen et al. [42] proposed a scrambled clinical picture recovery conspires depending upon blockchain to ensure protection. The emergency clinic brings various sorts and qualities of Ankript's clinical picture. When an outsider proposes a recovery demand, the administration dispatches a lot to make a solicitation to the client and sends back the recovery results that are dependent on picture similarity.

Hirokazu Madokoroa et al. [43] presented a method of introducing instantaneous accelerated-KAZE, self-organized generation of map (SOM) codebook, and over-object recognition using maps created using chain counter-propagation networks (CPN). After optimizing experimentally, the major parameters showed that the results obtained identifying medical field utilization of interest (ROI (region of interest)) showed 71.96% accuracy.

Md. Mehedi Hassan Onik et al. [27] presented permission blockchain architecture that is designed for managing EHRs. Architectural Record Electronic Health recently started proposed for Italian regulation (General Data Security Regulation) on GDPR. They showed a blockchain-based architecture built for decentralized management of EHR. The proof-of-concept network represents the core architecture of the federal EHR management system.

20.7 OBJECTIVES OF THE STUDY

The objectives of the study are as follows:

- Determining the sensitivity of the different types of data;
- Automatically extracting and classifying the information from data through various machine learning approaches;
- Reviewing the cloud computing technology and data storage process across cloud computing;
- Storing data in cloud environment according to the sensitivity of the data using various security mechanisms, like blockchain;
- Analyzing the proposed algorithm with the standard parameters by comparison with the existing techniques of data classification;
- Evaluating the security requirements for the data storage across the cloud.

20.8 IMPLEMENTATION MODEL

Cloud services serve every type of organization. Privacy and, therefore, various companies and government raise the issue of sensitive knowledge integrity to prefer to shop in the cloud. The use of compassionate knowledge for malicious users is the knowledge that is easy to delete, and changes occur during sharing, so in return, there is a danger of privacy. Besides, the way data has been secured; it is also necessary to classify it. The sensitivity level has been proposed to organize data in a cloud environment as encryption technology is used based on the data classification model.

20.8.1 Proposed Model

Cloud is used to analyze various security environment issues that are categorized into three aspects, namely, privacy, integrity, and availability. Data classification algorithms in machine learning such as NLP, Support Vector Machine (SVM), k-nearest neighbors (KNN), Naïve Bayes, and Random Forest are involved in this research. In this research, the model using the machine learning approach is for monitoring and offering secure data classification. This approach is used for the level of sensitivity to classify data.

The proposed architecture uses data as input, which will be processed by learning a machine algorithm—machine-learning algorithms to match user data with the existing data sets, which are stored in the data dictionary. Based on the data set, the user data is classified as the original, possible, and highly classified based on their sensitivity. These confidential data will be stored in a cloud environment. In cloud storage, we can offer a cloud memory bucket system. Different data from the cloud storage repository are classified in a different place in the order.

Step 1: Importing the data set:

In the given article *.txt file, the most optimal data set is used, which is taken from Kaggle or Google data set. Only the useful data is extracted, including private data, and the title is added together and saved in the file.

Step 2: Preliminary text exploration:

Important common and uncommon words are used; the pre-processing text should be noted well.

Step 3: Most common and uncommon words:

Most common terms mean that the words are repeated and not in terms of everyone's ability to keep complete information or data. The custom stopping word list, general rules, and the default English stopping word must be reported and investigated.

Step 4: Text pre-processing:

Sparsity: Whenever we are looking for any keyword based on the number of times, the giant matrix gets zero values in many places of the word.

Python NLTK library comprises the default list of stop words.

Previously, we have seen the concept of "most common words," which is much helpful to add on more; let us see how to make a list of stop words and add some custom stop words.

Step 5: Data exploration:

To gain information on repeated words, we will assume the text corpus prepared by pre-processing.

20.9 CONCLUSION

Data classification has been used for many decades to help organizations safeguard sensitive or critical information with suitable protection. Regardless of whether data is processed or stored in conventional on-premises systems or the cloud, data classification is an initial starting point for taking the confidentiality (and potentially the availability and integrity) of data based on risk to the organization. This research provides insight into data classification using different machine learning approaches and categorizes data for public and private organizations to consider when moving data to the cloud storage. It outlines a process through which customers can build data classification categories with secure data storage using security techniques. This study gives the basic approaches for developing a system for efficient data classification with data security aspects that provides critical ideas for researchers.

REFERENCES

1. Statista, size of the cloud computing and hosting market worldwide from 2010 to 2020, 2017, URL https://www.statista.com/statistics/500541/worldwide-hosting-and-cloud-computing-market/, [Accessed 07-Jul-2018].
2. J. Willis, Who coined the phrase cloud computing? 2008, URL http://www.johnmwillis.com/cloud-computing/who-coined-the-phrase-cloud-computing, [Accessed on 26-Jan-2018].

3. A. Regalado, Who coined 'cloud computing'? 2011, URL https://www.technologyreview.com/s/425970/who-coined-cloud-computing, [Accessed on 26-Jan-2018].
4. Gartner, Gartner forecasts worldwide public cloud services Revenue, 2017, URL https://www.gartner.com/newsroom/id/3815165, [Accessed on 08-Apr-2018].
5. Q. Zhang, L. Cheng, R. Boutaba, Cloud computing: State-of-the-art and research challenges, *J. Internet Serv. Appl.* (ISSN: 1869-0238) 1 (1), 7–18, 2010 http://dx.doi.org/10.1007/s13174-010-0007-6.
6. S. Subashini, V. Kavitha, A survey on security issues in service delivery models of cloud computing, *J. Netw. Comput. Appl.* (ISSN: 1084-8045) 34 (1), 1–11, 2011 http://dx.doi.org/10.1016/j.jnca.2010.07.006.
7. M. Avram, Advantages and challenges of adopting cloud computing from an enterprise perspective, *Procedia Technol.* (ISSN: 2212-0173) 12, 529–534, 2014 http://dx.doi.org/10.1016/j.protcy.2013.12.525.
8. Forbes Technology Council, 13 biggest challenges when moving your business to the cloud, URL https://www.forbes.com/sites/forbestechcouncil/2017/06/05/13-biggest-challenges-when-moving-your-business-to-the-cloud/#3a8e0e219b0e, 2017.
9. N. Phaphoom, X. Wang, S. Samuel, S. Helmer, P. Abrahamsson, A survey study on major technical barriers affecting the decision to adopt cloud services, *J. Syst. Softw.* (ISSN: 0164-1212) 103, 167–181, 2015 http://dx.doi.org/10.1016/j.jss.2015.02.002.
10. C.-L. Hsu, J.C.-C. Lin, Factors affecting the adoption of cloud services in enterprises, *Inf. Syst. E-bus. Manag.* (ISSN: 1617-9846) 14 (4), 791–822, 2016 http://dx.doi.org/10.1007/s10257-015-0300-9.
11. J. Mishra. Modified Chua chaotic attractor with differential operators with non-singular kernels *Chaos, Solitons & Fract.* 125, 64–72, 2019.
12. Z. Xiao, Y. Xiao, Security and privacy in cloud computing, *IEEE Commun. Surv. Tutor.* (ISSN: 1553-877X) 15 (2), 843–859, 2013. http://dx.doi.org/10.1109/SURV.2012.060912.00182.
13. C. Modi, D. Patel, B. Borisaniya, A. Patel, M. Rajarajan, A survey on security issues and solutions at different layers of cloud computing, *J. Super Comput.* (ISSN: 1573-0484) 63 (2), 561–592, 2013 http://dx.doi.org/10.1007/s11227-012-0831-5.
14. K. Dahbur, B. Mohammad, A.B. Tarakji, A survey of risks, threats and vulnerabilities in cloud computing, *Proceedings of the 2011 International Conference on Intelligent Semantic Web-Services and Applications*, ISWSA '11, ACM, New York, NY, USA, ISBN: 978-1-4503-0474-0, 2011, 1–6, http://dx.doi.org/10.1145/1980822.1980834.
15. S.A. Alatresh, A.B. Ali, M.A. Ambarek, Cloud computing adoption by business organization: A systematic review, *Aust. J. Basic Appl. Sci.* 11 (13), 17–28, 2017 http://dx.doi.org/10.22587/ajbas.2017.11.13.3.
16. RightScale, RightScale 2018 state of the cloud report, URL https://www.rightscale.com/lp/state-of-the-cloud?campaign=7010g0000016JiU, [Accessed on 08-Apr-2018], 2018.
17. B. Grobauer, T. Walloschek, E. Stocker, Understanding cloud computing vulnerabilities, *IEEE Secur. Priv.* (ISSN: 1540-7993) 9 (2), 50–57, 2011 http://dx.doi.org/10.1109/MSP.2010.115.
18. F. Liu, J. Tong, J. Mao, R. Bohn, J. Messina, L. Badger, D. Leaf, *NIST Cloud Computing Reference Architecture (SP 500-292)*, National Institute of Standards & Technology, Gaithersburg, MD 20899-8930, USA, 2011, URL http://ws680.nist.gov/publication/get_pdf.cfm?pub_id=909505, [Accessed on 10-Feb-2018].
19. R. Mogull, J. Arlen, A. Lane, G. Peterson, M. Rothman, D. Mortman, Security guidance for critical areas of focus in cloud computing, *Cloud Security Alliance*, 2017, URL https://downloads.cloudsecurityalliance.org/assets/research/security-guidance/security-guidance-v4-FINAL.pdf, [Accessed on10-Feb-2018].

20. M.A. Morsy, J. Grundy, I. Müller, An analysis of the cloud computing security problem, *Proceedings of APSEC 2010 Cloud Workshop*, 1–6, 2010, URL https://www.cs.auckland.ac.nz/~john-g/papers/cloud2010_1.pdf, [Accessed on 11-Feb-2018].
21. H. Takabi, J.B.D. Joshi, G.-J. Ahn, Security and privacy challenges in cloud computing environments, *IEEE Secur. Priv.* (ISSN: 1540-7993) 8 (6), 24–31, 2010 http://dx.doi.org/10.1109/MSP.2010.186.
22. I.M. Khalil, A. Khreishah, M. Azeem, Cloud computing security: A survey, *Computers* (ISSN: 2073-431X) 3 (1), 1–35, 2014 http://dx.doi.org/10.3390/computers3010001.
23. C.A. Ardagna, R. Asal, E. Damiani, Q.H. Vu, From security to assurance in the cloud: A survey, *ACM Comput. Surv.* (ISSN: 0360-0300) 48 (1), 1–50, 2015 http://dx.doi.org/10.1145/2767005.
24. L.M. Vaquero, L. Rodero-Merino, D. Morán, Locking the sky: A survey on IaaS cloud security, *Computing* (ISSN: 1436-5057) 91 (1), 93–118, 2011 http://dx.doi.org/10.1007/s00607-010-0140-x.
25. D. Zissis, D. Lekkas, Addressing cloud computing security issues, *Future Gener. Comput. Syst.* (ISSN: 0167-739X) 28 (3), 583–592, 2012 http://dx.doi.org/10.1016/j.future.2010.12.006.
26. Computer Security Division Information Technology Laboratory National Institute of Standards and Technology, Standards for Security Categorization of Federal Information and Information Systems, FIPS Publication 2004.
27. M. Ciampi, A. Esposito, F. Marangio, G. Schmid, M. Sicuranza, A blockchain architecture for the Italian EHR system, IARIA, *HEALTHINFO 2019: The Fourth International Conference on Informatics and Assistive Technologies for Health-Care, Medical Support and Wellbeing*, 2019, ISBN: 978-1-61208-759-7.
28. A. Mishra et al., An authentication mechanism based on client-server architecture for accessing cloud computing, *Int. J. Emerging Technol. Adv. Eng*, ISSN 2250-2459, 2 (7), 95–99, July 2012.
29. K. Hashizume, D.G. Rosado, E. Fernández-Medina, E.B. Fernandez, An analysis of security issues for cloud computing, *J. Internet Serv. Appl.* (ISSN: 1869-0238) 4 (1), 1–13, 2013 http://dx.doi.org/10.1186/1869-0238-4-5.
30. J. Mishra. Fractional hyper-chaotic model with no equilibrium, *Chaos, Solitons & Fractals* 116, 43–53, 2018.
31. A. Mishra et al., Secure cloud storage architecture for digital medical record in cloud environment using blockchain, *Elsevier SSRN International Conference on Intelligent Communication and Computation Research*, Available at http://dx.doi.org/10.2139/ssrn.3565922, April 1, 2020.
32. A. Mishra et al., An enhanced DDoS TCP flood attack defence system in a cloud computing, *Elsevier SSRN International Conference on Intelligent Communication and computation Research*, Available at http://dx.doi.org/10.2139/ssrn.3565916, April 1, 2020.
33. J. Mishra. Analysis of the Fitzhugh Nagumo model with a new numerical scheme, *Discrete & Continuous Dynamical Systems-S*, 781, 2018.
34. A. Mishra et al., A review on DDOS attack, TCP flood attack in cloud environment, *Elsevier SSRN International Conference on Intelligent Communication and computation Research*, Available at https://ssrn.com/abstract=3565043, March 31, 2020.
35. W. Huang, A. Ganjali, B.H. Kim, S. Oh, D. Lie, The state of public infrastructure-as-a-service cloud security, *ACM Comput. Surv.* (ISSN:0360-0300) 47 (4), 1–31, 2015 http://dx.doi.org/10.1145/2767181.
36. N. Phaphoom, X. Wang, P. Abrahamsson, Foundations and technological landscape of cloud computing, *ISRN Softw. Eng.* 2013, 2013 http://dx.doi.org/10.1155/2013/782174, [Accessed on 10-Feb-2018].

37. S. Pearson, Privacy, security and trust in cloud computing, in: S. Pearson, G. Yee (Eds.), *Privacy and Security for Cloud Computing*, Springer, London, ISBN: 978-1-4471-4189-1, 2013, 3–42, http://dx.doi.org/10.1007/978-1-4471-4189-1_1.
38. D.A.B. Fernandes, L.F.B. Soares, J.V. Gomes, M.M. Freire, P.R.M. Inácio, Security issues in cloud environments: A survey, *Int. J. Inf. Secur.* (ISSN: 1615-5270) 13 (2), 113–170, 2014 http://dx.doi.org/10.1007/s10207-013-0208-7.
39. E. Aguiar, Y. Zhang, M. Blanton, An overview of issues and recent developments in cloud computing and storage security, in: K.J. Han, B.-Y. Choi, S. Song (Eds.), *High Performance Cloud Auditing and Applications*, Springer, New York, NY, USA, ISBN: 978-1-4614-3296-8, 2014, 3–33, http://dx.doi.org/10.1007/978-1-4614-3296-8_1.
40. G. Ramachandra, M. Iftikhar, F.A. Khan, A comprehensive survey on security in cloud computing, *Procedia Comput. Sci.* (ISSN: 1877-0509) 110, 465–472, 2017.
41. G. Zyskind, O. Nathan, A.S. Pentland, Decentralizing privacy: Using blockchain to protect personal data, *CS Security and Privacy Workshops*, IEEE, 2015.
42. M. Shen, Y. Deng, L. Zhu, X. Du, and N. Guizani, "Privacy-preserving image retrieval for medical IoT systems: A blockchain-based, approach", *IEEE Network*, 33, 27–33, 2019.
43. H. Madokoroa, A. Kainumaa, K. Satoa, Non-rectangular RoI extraction and machine learning based multiple object recognition used for time-series areal images obtained using MAV, *Procedia Computer Science*, 126, 462–471, Elsevier 2018.

Index

Printed in the United States
By Bookmasters